Removal of Pollutants from Saline Water

Removal of Pollutants from Saline Water
Treatment Technologies

Edited by

Shaik Feroz and Detlef W. Bahnemann

CRC Press
Taylor & Francis Group
Boca Raton London New York

CRC Press is an imprint of the
Taylor & Francis Group, an **informa** business

First edition published 2022
by CRC Press
6000 Broken Sound Parkway NW, Suite 300, Boca Raton, FL 33487-2742

and by CRC Press
2 Park Square, Milton Park, Abingdon, Oxon, OX14 4RN

© 2022 Taylor & Francis Group, LLC

CRC Press is an imprint of Taylor & Francis Group, LLC

Library of Congress Cataloging-in-Publication Data
A catalog record for this title has been requested

ISBN: 978-1-032-02835-4 (hbk)
ISBN: 978-1-032-02836-1 (pbk)
ISBN: 978-1-003-18543-7 (ebk)

DOI: 10.1201/9781003185437

Typeset in Times
by codeMantra

Contents

SECTION I   Seawater

SECTION II   Brackish Water

SECTION III   Industrial Saline Wastewater

Contents

Foreword #1

Salts and minerals are essential for life. We find them everywhere in nature and industrial processes. Even if most of the worldwide water is not useful to quench our thirst because of salinity and it is often seen as an impurity salts are increasingly recovered from water and wastewater as valuable materials. Besides the improving effort to reach Sustainable Development Goals (SDGs) and to improve desalination processes, the reduction of pollutants from saline water is becoming of higher importance.

The first time when I observed the challenges in the removal of pollutants in saline water and wastewater was during my doctoral studies at the Institute for Solar Energy Research Hameln/Hannover and later at the Institute of Technical Chemistry at the University of Hannover, where also the path of the authors Feroz Shaik and Detlef W. Bahnemann and myself came across. During the scientific experiments with photochemical reactors, we often had to recognize the effect of salinity on the photochemical degradation of impurities. A book with an overview of removing pollutants from saline water would have helped me a lot that time.

Therefore, I am very happy to find with this book about the removal of pollutants from saline water an overview dealing with scientific background and findings as well as technical processes and solutions for typical applications. Seawater, brackish water, oil produced, and other industrial saline wastewater are often treated with classical membrane, ion exchange, evaporation, or electrodialysis technologies, while further advanced technologies are arising. The engagement to treat industrial wastewater is increasing worldwide. Typical water and wastewaters with the need for desalination or with high contamination of salts are found in seawater desalination, in well water of the oil and gas industry, in food processing like conserving food, in leather tanneries landfill leachates, and others.

To find more economical and useful solutions to treat saline water and to recover salts for further usage, we need advanced technologies to remove also pollutants from saline water, especially for Zero Liquid Discharge processes. This book summarizes the actual knowledge of scientific insights and applied technologies.

Dr.-Ing. Gerd Sagawe
June 24, 2021
EnviroChemie, Member of the Board (1996-2021)
German Water Partnership, Member of the Board
VDMA, Member of the Board of directors of Water
and Wastewater Technology Department

Foreword #2

Water is a precious and life-saving component needed for the survival of life on the earth. Of late, as its existence is in question due to overutilization and pollution, there is an extensive need for recovery process and saline water naturally becomes an obvious attractive abundant source to be dealt with for treatment.

This book is a collection of various technologies that are implemented to treat the pollutants present in saline water. The authors and editors put up tremendous efforts in bringing up this unique manuscript in front of academicians and practicing professionals. Most of the text written is based on the real-time experiences of the contributors. I am delighted to present this book to the global audience. My heartiest congratulations to the editors for coming with this brilliant idea and best wishes for this book and their future endeavors. I am proud to see the success of my beloved student Dr. Shaik Feroz (one of the editors of this book) in presenting his fourth book.

Prof. V.S.R.K. Prasad
Director
Indian Institute of Petroleum and Energy
India.
June 10, 2021

Preface

The challenge of providing and ensure clean, fresh water is rapidly growing worldwide due to increase in population growth. Urbanization and subsequent growth in industrial activities have created tremendous stress on water resources. Water, which is the vital commodity for the very existence of life on earth and a necessity for economic, social development, and environmental sustainability, is becoming scarce. As per published reports, the Earth's hydrosphere represents 1.38 billion km^3 of water. About 97% of Earth's hydrosphere is seawater, with oceans covering approximately 71% of Earth's surface. Fresh water covers 3% of the hydrosphere, with 69.5% in the form of ice or permanent snow. The other 30.1% is groundwater estimated at 10 million km^3. Lakes and rivers represent 0.4% of freshwater resources, i.e., around one 1/150000th of water on Earth.

Water availability includes the issues of both water quantity and quality. Desalination of saline water, which includes seawater, brackish water, and industrial saline water, offers the potential to add significantly to freshwater availability in the regions of the world where water availability is scarce. Desalination technologies are expensive and once considered as a last resort solution are becoming increasingly affordable. Saline water normally consists of a variety of chemical components at different concentration levels depending on the source and geographical locations. The salinity and chemical composition widely vary among seawater, brackish water, and industrial saline water. The quality of raw feed water plays a vital role in desalination process and the technology used. The minimum the pollutants present, especially the scaling and fouling potential compounds, the better the efficiency of desalination process. Therefore, it is bound to remove or treat the pollutants present in raw saline water before being fed to desalination process. The disposal of industrial saline wastewater is another major challenging aspect. Saline wastewater and concentrated brine solutions from various industrial sources have become a serious cause of pollution and a threat to coastal and surface water ecosystems. These saline industrial wastewaters are posing serious disposal problems with growing human health and environmental concerns. The disposed saline wastewaters can harm the marine environment. The pollutants present in the industrial saline wastewater need to be treated before disposal.

Keeping in view the importance of the removal of pollutants present in saline water other than salts, the idea of this textbook has come into force. Most of the books available are on desalting of saline water and not much reported on the treatment of pollutants present in various saline waters. This book is broadly classified into three sections. Chapters 2–9 in the first section are entirely focused on seawater. Characteristics of seawater, nature of pollutants, assessment of scaling, and fouling potential of pollutants present in seawater are discussed. Treatment technologies including novel green technologies for the removal of pollutants from seawater. Application of Artificial Intelligence (AI) and Response Surface Methodology (RSM) to optimize the process parameters are presented. Section II involving Chapters 10 and 11 emphasizes solely on brackish water. Characteristics and technologies employed for the removal of pollutants from brackish waters are presented. Chapters 12–20 of Section III covers the characteristics of various industrial saline wastewaters. Treatment technologies employed for the removal of pollutants from different industrial saline wastewaters are discussed.

In many ways, this text is collective of the latest and advanced information on various technologies/techniques/methods that are adopted for the removal of pollutants from saline waters (seawater, brackish water, and industrial saline wastewaters) including some aspects of desalting. The text also involves some case studies on industrial assessment techniques for scaling and fouling. AI and RSM techniques to optimize the process parameters are presented. Application of polymer composites, natural and low-cost adsorbents, and natural plants for the removal of pollutants present in saline waters are discussed.

MATLAB® is a registered trademark of The MathWorks, Inc. For product information,
please contact:
The MathWorks, Inc.
3 Apple Hill Drive
Natick, MA 01760-2098 USA
Tel: 508-647-7000
Fax: 508-647-7001
E-mail: info@mathworks.com
Web: www.mathworks.com

Acknowledgments

This book could not have come into reality without the contribution of a number of persons directly and indirectly. All their works and support are duly acknowledged here.

The editors express sincere thanks to Dr. Gerd Sagawe and Dr. V.S.R.K. Prasad for providing the forewords to this book.

Thanks to family members and friends for their kind support in all means.

List of Abbreviations

CHAPTER 1

OPW   Oil-produced water

CHAPTER 2

ACC	Acetyl-CoA carboxylase
AEDS	Atopic eczema/dermatitis syndrome
AMPK	5-Adenosine monophosphate-activated protein kinase
BOD	Biochemical oxygen demand
CIP7A1 gene	Hepatic cholesterol 7a-hydroxylase
COD	Chemical oxygen demand
DD	Dichloro-diphenyl-trichloroethane
DO	Dissolved oxygen
EC	Electrical conductivity
HDL	High-density lipoprotein
HMGCR	3-Hydroxy-3-methyl-glutaryl-CoA reductase
IgE	Immunoglobulin E
IGF-1R	Insulin-like growth factor 1 receptor
IL-1β	Interleukin 1 beta
IL-6	Interleukin 6
LDL	Low-density lipoprotein
MDA	Malondialdehyde
RBP4	Retinol binding protein 4
TC	Total cholesterol
TDI	Tolerable daily intake
TDS	Total dissolved solids
TEAC	Trolox equivalent antioxidant capacity
TEWL	Transepidermal water loss
TG	Triglycerides (TG)
TNF-α	Tumor necrosis factor α
TSS	Total suspended solids
WHO	World Health Organization

CHAPTER 3

AOC	Assimilable organic carbon
ASTM	American Society for Testing and Materials
ATP	Adenosine triphosphate
BGP	Bacterial growth potential
BW	Backwash
CIP	Cleaning-in-place

DAF	Dissolved air flotation
DOC	Dissolved organic carbon
EEM	Excitation-emission matrix
FCM	Flow cytometry
FEEM	Fluorescence excitation-emission matrices
FI	Fluorescence index
FI	Fouling Index
FTIR	Fourier transform infrared spectroscopy
IAP	Ion activity product
IP	Ion product
LC-OCD	Liquid chromatography-organic carbon detection
LSI	Langelier saturation index
mBFR	Membrane biofilm formation rate
MFI	Modified fouling index
MFS	Membrane fouling simulator
MTC	Mass transfer coefficient
MWCO	Molecular weight cutoff
NDP	Normalized permeate flow
NOM	Natural organic matter
NTU	Nephelometric Turbidity unit
PAN	Poly-acrylonitrile
PES	Polyether sulfone
PF	Plugging factor
POC	Particulate organic carbon
RC	Regenerated cellulose
RO	Reverse osmosis
S&DSI	Stiff-Davis stability index
SDI	Silt density index
SI	Saturation index
Sr	Supersaturation ratio
SUVA	Specific UV absorbance
SWRO	Seawater reverse osmosis
TEP	Transparent exopolymer particles
TOC	Total dissolved organic carbon
UF	Ultrafiltration
UV	Ultraviolet
XL	Extra large
XRD	X-ray diffraction

CHAPTER 4

AQP	Aquaporin
CAOW	Closed air, open water cycle
CDI	Capacitive deionization
CNM	Carbon-based nanomaterial
CNT	Carbon nanotube

CWOA	Closed water, open air cycle
DO	Dissolved oxygen
EC	Electrical conductivity
ECt	Electrical conductivity/salinity threshold
ED	Electrodialysis
EDR	Electrodialysis reversal
EPS	Extracellular polymeric substances
ET	Evapotranspiration
FO	Forward osmosis
GDP	Gross domestic product
HDH	Humidification-dehumidification
IX	Ion exchange
MBBR	Moving bed biofilm reactor–membrane bioreactors
MBR	Membrane bioreactor
MED	Multiple effect distillation process
MENA	Mediterranean, Middle East, and North African
MF	Microfiltration
MFC	Microbial fuel cells
MSF	Multistage flash
NF	Nanofiltration
NORM	Naturally occurring radiative material
NP	Nanoparticle
OWOA	Open water, open air cycle
PET	Polyethylene terephthalate
RO	Reverse osmosis
SODIS	Solar disinfection of drinking water
SWE	Solar-driven water evaporation
TDS	Total dissolved solids
tRNA	Transfer ribonucleic acid
UF	Ultrafiltration
UV	Ultraviolet
VCD	Vapour compression distillation
WHO	World Health Organization

CHAPTER 5

AOPs	Advanced oxidation processes
CCD	Central composite design
COD	Chemical oxygen demand
DoE	Design of experiments
ECD	Electrochemical degradation
PEC	Photoelectrochemical
RSM	Response surface methodology
TDS	Total dissolved solids
TOC	Total organic carbon
UV	Ultraviolet

CHAPTER 6

AOP	Advanced oxidation processes
CCD	Central composite design
COD	Chemical oxygen demand
DoE	Design of experiments
GA	Genetic algorithm
GWO	Grey Wolf Optimizer
NSGA	Non-dominated sorting genetic algorithm
RMSE	Root mean squared error
RO	Reverse osmosis
TOC	Total organic carbon
UV	Ultraviolet

CHAPTER 7

BOD	Biological oxygen demand
CNT	Carbon nanotube
COD	Chemical oxygen demand
DCMD	Direct contact membrane distillation
ENM	Electrospinning nanofiber membrane
FD	Freeze distillation
FO	Forward osmosis
GO	Graphene oxide
MD	Membrane distillation
MED	Multiple effect distillation
MF	Microfiltration
MOF	Metal-organic framework
MSF	Multistage flash
NF	Nanofiltration
PP	Polypropylene
PTFE	Polytetrafluoroethylene
PVC	Polyvinylchloride
PVDF	Polytetrafluoroethylene
PVDF	Polyvinylidene fluoride
RO	Reverse osmosis
SMCL	Secondary maximum contaminant level
SWM	Spiral-wound membrane
TDS	Total dissolved solids
TFC	Thin-film composite
TOC	Total organic carbon
TP	Temperature polarisation
TSS	Total suspended solids
UF	Ultrafiltration
USEPA	U.S. Environmental Protection Agency
ZLD	Zero liquid discharge

CHAPTER 8

2D	Two-dimensional
AGMD	Air gap membrane distillation
CNT	Carbon nanotube
COD	Chemical oxygen demand
DCMD	Direct contact membrane distillation
FO	Forward osmosis
GO	Graphene oxide
h-BN	Hexagonal boron nitride nanotubes
ICP	Internal concentration polarization
LbL	Layer-by-layer
MD	Membrane distillation
NF	Nanofiltration
PA	Polyamide
PAA	Poly(acrylic acid)
PAN	Polyacrylonitrile
pH	Potential of hydrogen/ power of hydrogen
pHEMA	Polyhydroxyethyl methacrylate
PVDF	Poly(vinylidenefluoride)
RO	Reverse osmosis
SGMD	Sweeping gas membrane distillation
TOC	Total organic carbon
UF	Ultrafiltration
VMD	Vacuum membrane distillation

CHAPTER 9

DM	Dynamic membrane
EC	Electrocoagulation
EDS	Energy-dispersive X-ray spectroscopy
FTIR	Fourier Transform Infrared
GC-MS	Gas chromatography–mass spectrometry
MALDI-TOF-MS	Matrix-assisted laser desorption ionization-time of flight mass spectrometry
MBR	Membrane bioreactors
MNPs	Micro-nanoplastics
MPs	Microplastics
MPSS	Munich plastic sediment separator
MS	Mass spectroscopy
NPs	Nanoplastics
PA	Polyamide
PAM	Polyacrylamide
PE	Polyethylene
PET	Polyethylene terephthalate

PP	Polypropylene
PS	Polystyrene
PTFE	Polytetrafluoroethylene
PVC	Polyvinyl chloride
PVFD	Polyvinylidene fluoride
SEM	Scanning electron microscopy
TD-PTR-MS	Thermal desorption-Proton transfer reaction-mass spectroscopy
TMP	Transmembrane pressure
WWTPs	Wastewater treatment plants

CHAPTER 10

CDI	Capacitive deionization technology
DCMD	Direct contact membrane distillation
EC	Electrical conductivity
ED	Electrodialysis
FO	Forward osmosis
HDH	Humidification and dehumidification
MBR	Membrane bioreactor
MED	Multiple effect distillation
MED-MVC	Multiple effect distillation with mechanical vapour compression
MED-TVC	Multiple effect distillation with thermal vapor compression
MPN	Most probable number
mSAC	Maximum salt adsorption capacity
MSFD	Multistage flash distillation
MVC	Mechanical vapor compression
NF	Nanofiltration
OMBR	Osmotic membrane bioreactor
PV	Pervaporative distillation
RO	Reverse osmosis
TDS	Total dissolved solids
US EPA	United States Environmental Protection Agency
USGS	United States Geological Survey
VMD	Vacuum membrane distillation
WHO	World Health Organization

CHAPTER 11

AOPs	Advanced oxidation technologies
BOD	Biochemical oxygen demand
CDI	Capacitive deionization
COD	Chemical oxygen demand
EC	Electrocoagulation

EDR	Electro dialysis reversal
HA	Humic acid
IEM	Ion exchange membranes
MCDI	Membrane capacitive de-ionization
MF	Microfiltration
NF	Nanofiltration
NOM	Natural organic matter
RO	Reverse osmosis
TDS	Total dissolved solids
TOC	Total organic carbon
UF	Ultrafiltration

CHAPTER 12

COD	Chemical oxygen demand
DO	Dissolved oxygen
FTIR	Fourier Transform Infrared
MF	Microfiltration
MPPE	Macro-porous polymer extraction
NF	Nanofiltration
OPW	Oil-produced water
PU	Polyurethane
RO	Reverse osmosis
RPM	Revolutions per minute
SEM	Scanning electron microscope
TDS	Total dissolved solids
TGA	Thermal gravimetric analysis
UF	Ultrafiltration

CHAPTER 13

BOD	Biological oxygen demand
COD	Chemical oxygen demand
DCMD	Direct contact membrane distillation
ED	Electrodialysis
EDR	Electrodialysis reverse flow
FO	Forward osmosis
HPM	High-pressure membrane
LPM	Low-pressure membrane
L-S	Loebe Sourirajan
MBR	Membrane bioreactor
MBR-RO	Membrane bioreactor–Reverse osmosis
MBR-UF	Membrane bioreactor–Ultrafiltration
MD	Membrane distillation
MF	Microfiltration

NF	Nanofiltration
PC	Polarized concentration
RO	Reverse osmosis
STP	Sewage treatment plant
TFC	Thin-film composition
TFC-FO	Thin-film composite–Forward osmosis
TMP	Transmembrane pressure
TOC	Total organic carbon
UF	Ultrafiltration
WWTP	Wastewater treatment plant

CHAPTER 14

ASW	Artificial sea water
BOD	Biological oxygen demand
CAT	Catalase
COD	Chemical oxygen demand
EPS	Exopolysaccharide
ETP	Effluent treatment plant
GSH	Glutathione
HFT	Hydraulic fracturing technology
MF	Microfiltration
NF	Nanofiltration
PE	Polyethylene
PPCP	Pharmaceutical and personal care product
PSU	Practical salinity unit
PTFE	Polytetrafluoroethylene
RO	Reverse osmosis
SOD	Superoxide dismutase
TDS	Total dissolved solids
UF	Ultrafiltration
WWCSH	Wastewater from the chicken slaughterhouse

CHAPTER 15

BOD	Biological oxygen demand
DO	Dissolved oxygen
EPDM	Ethylene propylene diene monomer
FAO	Food and Agriculture Organization
GAC	Granular dynamic carbon channels
PPCP	Pharmaceuticals and personal care product
PSU	Practical salinity unit
RSF	Rapid sand channels
SSF	Slow sand channels
VOC	Volatile organic carbon

CHAPTER 16

BOD	Biological oxygen demand
COD	Chemical oxygen demand
TDS`	Total dissolved solids

CHAPTER 17

COL	Color
FTIR	Fourier transform infrared
LSAE	Land-based saline aquaculture effluent
PBC	*Parkia biglobosa* seeds coagulant
SAL	Salinity
SEM	Scanning electron microscopy
TDSP	Total dissolved and suspended particles
TUR	Turbidity

CHAPTER 18

COD	Chemical oxygen demand
CSA	Coconut shell activated carbon
DO	Dissolved oxygen
DSA	Data seed activated carbon
EOR	Enhanced oil recovery
FTIR	Fourier transform infrared
GC-MS	Gas Chromatography-Mass spectroscopy
LMW	Light molecular weight
NORM	Naturally occurring radioactive materials
OPW	Oil-produced water
RO	Reverse osmosis
SAR	Sodium adsorption ratio
SEM	Scanning electron microscopy
TDS	Total dissolved solids
TOC	Total organic carbon
XRD	X-ray diffraction

CHAPTER 19

AOX	Adsorbable organic halogen
BPS	BisphenolS
CAP	Chloramphenicol
CNT	Carbon nanotube
MB	Methylene blue
NOM	Natural organic matter
NOR	Norfloxacin
PMS	Peroxymonosulfate

PS	Persulfate
RO	Reverse osmosis
SAL	Salbutamol
SCuCoO	S-doped Cu-Co bimetallic oxides
SR-AOP	Sulfate radical advanced oxidation processes
TBL	Terbutaline
TDS	Total dissolved solids
TOC	Total organic content
US/SR-AOP	Ultrasonic/sulphate radical advanced oxidation processes
UV	Ultraviolet
UV/PMS	Ultraviolet/peroxymonosulfate
UV/PS	Ultraviolet/persulfate

CHAPTER 20

2,4-DCP	2,4-Dichlorophenol
ABR	Anaerobic baffled reactor
AOPs	Advanced oxidation processes
COD	Chemical oxygen demand
CPC	Compound parabolic collectors
DNAN	Dinitroanisole
DOC	Dissolved organic carbon
DW	Demineralized water
EBCT	Empty bed contact time
ED	Electrode distance
ED	Electrodialysis
EF	Electro-Fenton
EFP	Electro Fenton process
FC	Freeze crystallization
FO	Forward osmosis
HF	Heterogeneous Fenton
HFP	Hybrid Fenton Method
HPLC-DAD	High-performance liquid chromatography with diode-array detection
HRT	Hydraulic retention time
MBBR	Moving-bed biofilm reactor
MCr	Membrane crystallization
MD	Membrane distillation
MPAC	Magnetic powdered activated carbon
MTBE	Methyl t-butyl ether
NHE	Normal hydrogen electrode
PCW	Petrochemical wastewater
PFP	Photo-Fenton process
PYR	Pyrimethanil
S/N	Signal-to-noise
SFP	Sono-Fenton process

TDS	Total dissolved solids
TEP	Triethyl phosphate
TOC	Total organic carbon
TP	Transition products
US	Ultrasound
UV	Ultraviolet
ZVI	Zero-valent iron

Editors

Prof. Dr. Shaik Feroz works at Prince Mohammad Bin Fahd University, Kingdom of Saudi Arabia. Dr. Feroz obtained his doctorate in the field of Chemical Engineering from Andhra University, India in 2004 and Post-Doc Research Fellow from Leibniz University, Germany in 2015; M. Tech in Chemical Engineering from Osmania University, India, in 1998; and B. Tech in Chemical Engineering from S. V. University, India, in 1992. Post-Graduate Diploma in Environmental Studies from Andhra University in 2003.

Dr. Feroz has expertise in process engineering; plant design and troubleshooting; quality control using advance analytical equipment; wastewater treatment; solar energy systems (PV & CSP) for desalination; hot water systems and water treatment; synthesis of nano photocatalysts; simultaneous treatment of wastewater and production of hydrogen; environmental impact assessment; design, delivery, and management chemical and process engineering programs; and tailor-made industrial-based training programs related to chemical and process engineering.

Dr. Feroz has more than 190 publications to his credit in journals and conferences of international repute and supervised 4 PhD research works with another 4 ongoing. He is associated as the Principal Investigator/Co-investigator to a number of research projects and also involved as "Technology transfer agent" by Innovation Research Center, Sultanate of Oman. He has a significant contribution as editorial member and peer reviewer for various reputed international journals and conferences. He was awarded "Best Researcher" by the Caledonian College of Engineering for the year 2014. Dr. Feroz has a total experience of 30 years in teaching, research, and industry.

Prof. Dr. Detlef W. Bahnemann is Head of the Photocatalysis and Nanotechnology Research Unit at the Institute of Technical Chemistry, Leibniz University of Hannover (Germany), Director of the Megagrant Laboratory "Photoactive Nanocomposite Materials" at St. Petersburg State University (Russia), Supernumerary Professor in the Faculty of Natural Sciences at the Leibniz University of Hannover (Germany), and Distinguished Professor at Shaanxi University of Science & Technology, Xi'an (Peoples Republic of China). He is the author of more than 500 publications in refereed journals and in books and more than 100 other publications with a current Hirsch-Factor (h-Index) of 85 (papers have been cited more than 47,000 times altogether) according to Scopus and Web of Science. According to Google Scholar Citations, Prof. Bahnemann is a Highly Cited Author with his current Hirsch-Factor (h-Index) being 100, whose papers have been cited more than 63,000 times altogether. He has presented more than 400 lectures at scientific meetings, in various universities, research institutes and at industrial research and development centers (approx. 50% invited keynote and plenary lectures). Prof. Bahnemann is editor of five scientific books.

Amongst other honors, he received an Honorary Professorship at the Robert Gordon University in Aberdeen/Scotland (United Kingdom), an Erudite Professorship,

Mahatma Gandhi University, Kottayam (India), a Guest Professorship at the
Tianjin University in Tianjin (China), and an Honorary Professorship at Xinjiang
Technical Institute of Physics & Chemistry, Chinese Academy of Science (China).
Prof. Bahnemann is DeTao Master of Photocatalysis, Nanomaterials and Energy
Applications (China), Visiting Professor under Academic Icon Programme to the
Department of Civil Engineering, Faculty of Engineering, University Malaya in Kuala
Lumpur (Malaysia), Visiting Research Professor in the School of Chemistry and
Chemical Engineering, Queens University Belfast (United Kingdom), and Visiting
Professor in the Chemistry Department, College of Science, University of Dammam
(Saudi Arabia). In 2017, Dr. Bahnemann received the Medal "To the Memory of
Academician N. M. Emanuel" presented by the Russian Academy of Sciences and
M. V. Lomonosov Moscow State University (Russian Federation). Currently, he is
Visiting Scholar in the School of Chemistry and Chemical Engineering, Queens
University Belfast (United Kingdom), Fellow of the European Academy of Sciences
(EURASC), and Member of the Academia Europaea (AE).

Prof. Dr. Detlef W. Bahnemann is the Associate Editor of the journal *Catalysis
Letters,* Associate Editor of the journal *Photochem,* Editor in Chief for the
"Photocatalysis" section of the journal *Catalysts,* Executive Board Member of
Journal of Physics: Energy, and an editorial board member of several other journals.

Contributors

Almotasembellah Abushaban
The International Water Research
    Institute
Mohammed VI Polytechnic University
Ben Guerir, Morocco

Shahin Ahmadi
Department of Environmental Health
Zabol University of Medical Sciences
Zabol, Iran

Faizan Ahmed
Department of Mechanical Engineering
College of Engineering
Prince Mohammad Bin Fahd University
Al Khobar, Kingdom of Saudi Arabia

Ahmed Yousuf Khalfan Al-Busaidi
Department of Mechanical & Industrial
    Engineering
College of Engineering
National University of Science and
    Technology
Muscat, Sultanate of Oman

Syed Murtuza Ali
Department of Mechanical & Industrial
    Engineering
College of Engineeirng
National University of Science and
    Technology
Muscat, Sultanate of Oman

S. Ameen
Institute of Soil and Environmental
    Sciences
University of Agriculture
Faisalabad, Pakistan

Detlef W. Bahnemann
Photocatalysis and Nanotechnology
Gottfried Wilhelm Leibniz University
Hannover, Germany

B. Bharathiraja
Department of Chemical Engineering
Vel Tech High Tech Engineering
    College
Chennai, Tamil Nadu, India

Murthy Chavali
NTRC-MCETRC and Aarshanano
    Composites Technologies Pvt. Ltd.
Guntur, Andhra Pradesh, India

M. Chithra
Department of Chemical
    Engineering
Kongu Engineering College
Erode, Tamil Nadu, India

P. Divya Sruthi
Andhra Pradesh State Warehousing
    Corporation
Vijayawada, Andhra Pradesh, India

Susmita Dutta
Department of Chemical
    Engineering
National Institute of Technology
    Durgapur
Durgapur, India

Manoj Kumar Enamala
Bioserve Biotechnologies (India) Private
    Limited
Hyderabad, Telangana, India

Izharul Haq Farooqi
Department of Civil Engineering
Zakir Hussain College of Engineering
    and Technology, Aligarh Muslim
    University
Aligarh, Uttar Pradesh, India

Z.U.R. Farooqi
Institute of Soil and Environmental
    Sciences
University of Agriculture
Faisalabad, Pakistan

Shaik Feroz
Department of Mechanical Engineering
College of Engineering
Prince Mohammad Bin Fahd University
Al Khobar, Kingdom of Saudi Arabia

Joshua O. Ighalo
Department of Chemical Engineering
Nnamdi Azikiwe University
Awka, Nigeria
and
Department of Chemical Engineering
University of Ilorin
Ilorin, Nigeria

Chinenye Adaobi Igwegbe
Department of Chemical Engineering
Nnamdi Azikiwe University
Awka, Nigeria

P. Ilić
Management
PSRI Institute for Protection and
    Ecology of the Republic of Srpska
Banja Luka, Bosnia and Herzegovina

K. Jeevith
Department of Chemical
    Engineering
Kongu Engineering College
Erode, Tamil Nadu, India

Varghese Manappallil Joy
Department of Chemical
    Engineering
National Institute of Technology
    Durgapur
Durgapur, India
and
Department of Mechanical and
    Industrial Engineering
College of Engineering
National University of Science and
    Technology
Muscat, Sultanate of Oman

V.P. Kamalakannan
Department of Chemical
    Engineering
A. C. Tech Campus,
    Anna University
Chennai, Tamil Nadu, India

Afzal Husain Khan
Department of Civil Engineering
Jazan University
Jazan, Saudi Arabia

Nadeem Ahmad Khan
Department of Civil Engineering
Jamia Millia Islamia Central
    University
New Delhi, India

Saif Ullah Khan
Department of Civil Engineering
Zakir Hussain College of
    Engineering and Technology,
    Aligarh Muslim University
Aligarh, Uttar Pradesh, India

Chandrasekhar Kuppam
School of Civil and Environmental
    Engineering
Yonsei University
Seoul, Republic of Korea

Chandrasekar Kuppan
Division of Chemistry, Department of
  Sciences & Humanities
Vignan's Foundation for Science,
  Technology and Research
Guntur, Andhra Pradesh, India

**Saisanthosh Vamshi Harsha
Madiraju**
Department of Civil Engineering
The University of Toledo
Toledo, Ohio

Mohd Salim Mahtab
Department of Civil Engineering
Zakir Hussain College of Engineering
  and Technology, Aligarh Muslim
  University
Aligarh, Uttar Pradesh, India

Mohammad Saood Manzar
Department of Environmental
  Engineering
Imam Abdurrahman University
Dammam, Saudi Arabia

H. Munawar
Indonesian Research Centre for
  Veterinary Science
IAARD-Ministry of Agriculture
Jakarta, Indonesia

K. Nagamalleswara Rao
Centre for Disaster Mitigation and
  Management
Vellore Institute of Technology
Vellore, Tamil Nadu, India

M. Nasir Mangal
Water Supply, Sanitation and
  Environmental Engineering
  Department
IHE Delft Institute for Water Education
Delft, Netherlands

M. Naveen Kumar
Department of Chemical Engineering
Kongu Engineering College
Erode, Tamil Nadu, India

Mohammed Nayeemuddin
School of Civil Engineering
Universiti Sains Malaysia
Gelugor, Penang, Malaysia

M. Neeraja
Department of Chemical Engineering
Kongu Engineering College
Erode, Tamil Nadu, India

D. Nešković Markić
Faculty of Health Science
Pan-European University
  "APEIRON"
Banja Luka, Bosnia and Herzegovina

Okechukwu Dominic Onukwuli
Department of Chemical
  Engineering
Nnamdi Azikiwe University
Awka, Nigeria

Abhiram Siva Prasad Pamula
School of Civil and Environmental
  Engineering
Oklahoma State University
Stillwater, Oklahoma

Palaniandy Puganeshwary
School of Civil Engineering
Universiti Sains Malaysia
Gelugor, Penang, Malaysia

Bilal Rajput
Department of Civil Engineering
Zakir Hussain College of Engineering
  and Technology, Aligarh Muslim
  University
Aligarh, Uttar Pradesh, India

Lakkimsetty Nageswara Rao
Department of Mechanical and
    Industrial Engineering
College of Engineering
National University of Science and
    Technology
Muscat, Sultanate of Oman

Parul Sahu
Salt and Marine Chemicals
    Division
CSIR – Central Salt and Marine
    Chemicals Research Institute
Bhavnagar, Gujarat, India

A. Sai Kumar
Bioserve Biotechnologies (India) Private
    Limited
Hyderabad, Telangana, India

Sergio G. Salinas-Rodriguez
Water Supply, Sanitation and
    Environmental Engineering
    Department
IHE Delft Institute for
    Water Education
Delft, Netherlands

S. Samraj
Department of Chemical
    Engineering
MVJ Engineering College
Bengaluru, Karnataka, India

Poosalaya Sangadi
Division of Chemistry, Department of
    Sciences & Humanities
Vignan's Foundation for Science,
    Technology and Research
Guntur, Andhra Pradesh, India

K. Senthil Kumar
Department of Chemical Engineering
Kongu Engineering College
Erode, Tamil Nadu, India

Meenakshi Singh
Department of Botany
The M.S. University of Baroda
Vadodara, Gujarat, India

L.J. Stojanović Bjelić
Faculty of Health Science
Pan-European University "APEIRON"
Banja Luka, Bosnia and Herzegovina

B. Sudeep
Department of Chemical Engineering
Kongu Engineering College
Erode, Tamil Nadu, India

Shadab Usmani
Department of Civil Engineering
Zakir Hussain College of Engineering
    and Technology, Aligarh Muslim
    University
Aligarh, Uttar Pradesh, India

M. Venkata Ratnam
Department of Chemical Engineering
Mettu University
Metu, Ethiopia

Loreen O. Villacorte
Denmark, Global Technology and
    Development
Grundfos Holding A/S
Bjerringbro, Denmark

Mo Washeem
Department of Civil Engineering
Zakir Hussain College of Engineering
    and Technology, Aligarh Muslim
    University
Aligarh, Uttar Pradesh, India

A. Zafar
Institute of Soil and Environmental
    Sciences
University of Agriculture
Faisalabad, Pakistan

Section I

Seawater

1 Introduction

Shaik Feroz
Prince Mohammad Bin Fahd University

Detlef W. Bahnemann
Gottfried Wilhelm Leibniz University

CONTENT

Fresh water is essential for the survival of all living creatures across the globe. The oceans hold about 97% of the world's overall water in the form of saline water, with fresh water accounting for just 3%. Nearly 69% of fresh water is in the form of glaciers and can be found in the polar regions and higher mountain ranges. Underground water accounts for about 30% of fresh water, while surface water contained in reservoirs and rivers accounts for just 1%. There are a lot of sectors, which involve the utilization of fresh water such as industrial, agricultural, oil and natural gas sectors. However, the availability of fresh water from natural water bodies is reduced due to population growth, water pollution and improper management of water resources. In addition to these factors, there are also other factors such as geographical location and climate changes due to which many regions in the world are facing water scarcity issues [1]. This water crisis can be overcome firstly by proper conservation and management of the naturally available fresh water and then secondly by the generation of fresh water to meet the demand of all sectors. The generated fresh water must be free from all kinds of pollutants and should be of low cost.

Saline water is becoming one of the most important resources, which is supported for the production of potable water through desalination. Researchers have started developing treatment technologies since the early 17th century, but due to prospective freshwater shortage, improving the efficacy of saline water treatment gained momentum in the past few decades. The quality of saline water influences the performance of desalination plants. The resources for saline water vary from surface (seawater, brackish) and ground (brackish and oil produced water) with variable compositions depending upon the geological factors. The technologies used to desalinate the saline water are costly and also energy intensive. The energy requirement for the production of potable water depends on the quality of feed water, levels of water treatment and the technology involved in the plant.

Seawater is becoming an increasingly significant water source globally due to high stress on freshwater availability and contamination issues. The desalination of seawater would require the highest amount of energy. But since the quantity of available surface water and groundwater is not sufficient to meet the existing demand for

DOI: 10.1201/9781003185437-2

fresh water, desalination of seawater is practiced globally irrespective of the energy consumed during this process [2]. A lot of efforts are being carried out to desalinate the vast resources of saline water in different parts of the world, especially in water stress arid regions.

Due to sustained human and industrial activities on the coastal line, the quality of seawater is getting deteriorated. The quality of the raw seawater strongly influences the ultimate quality of the discharge water to be disposed of [3]. Desalination processes are very sensitive toward the composition of feed water, especially on the scale and fouling forming substances. The main challenges still existing in the desalination technologies are scaling of pipelines, equipment and fouling of membranes, due to the presence of various inorganic and organic pollutants. Fouling is the deposition of organics, inorganic or biological material on the membrane, and scaling is the precipitation of inorganic material onto the membrane surface [4].

Fouling and scaling can be marked as the major problem in reverse osmosis desalination. It affects the overall performance of the plant. Fouling can be due to the microorganisms present in the seawater, which consists of bacteria that can settle on the membrane and grow, besides that the plugging of the suspended particles, organics and colloid. Scaling can also result to membrane due to the inorganic compounds present in the seawater, such as calcium carbonate, calcium fluoride, barium sulfate and strontium sulfate [5]. Dissolved and dispersed hydrocarbons, heavy metals, radionuclides, dissolved formation minerals, salts, waxes, microorganisms present in seawater are gradually becoming one of the major threats not only for desalination plants but also for aquatic as well as human life.

Fouling due to the presence of organics detrimentally affects the desalination systems by decreasing permeate flux, reducing energy efficiency due to high-pressure drop and causing serious damage to the desalination systems. High-molecular humic substances (organic contaminants) present in the saline water are the main contributors of fouling to the membrane systems. If the total organic carbon of saline water is less than 0.5 mg/L, the presence of microbial foultants is minimal, and the feed saline water has very low biofouling potential. If the total organic carbon values are in the range of 0.5 and 2 mg/L, the saline water has moderate fouling potential. If the values exceed 2 mg/L, the biofouling potential will be very high [6]. The environmental dangers due to chemical pollution in seawater have become increasingly serious and directly related to the dumping of industrial wastewaters. Marine toxins produced by natural phytoplankton are another challenging pollution aspect in the seawater. There has been an increased interest in finding suitable techniques to remove these toxins from the seawater [7].

Brackish water is another source of saline water that can be treated and produce fresh water. It is more saline than freshwater and less salty than the sea and ocean water. The average salinity of the brackish water ranges between 0.5 and 30 ppt. Along with salts, brackish water may also contain organics, inorganic and biological pollutants. The sources of these pollutants are due to human and industrial activities. The disposal of domestic waste, untreated effluents, runoff from agricultural fields, leachates, etc., which contains a variety of pollutants eventually land up in brackish waters. The treatment of brackish water is challenging to maintain sufficient water quantity for human needs, especially in arid regions. With effective treatment technology, brackish water can be used for ecosystem services and human consumption, including irrigation, industry and land management practices.

The salinity of oil-produced water (OPW) that accompanies the crude oil poses a greater challenge to the oil industry. OPW contains a lot of hazardous inorganic and organic contaminants which cannot be discharged directly into the environment. Pollutants include dissolved and dispersed oil components, grease, heavy metals, radionuclides, hydrofluoric chemicals, dissolved formation minerals, salts, dissolved gases, scale products, waxes, microorganisms and dissolved oxygen. The composition will vary widely as a function of geologic formation, the lifetime of the reservoir and the type of hydrocarbon produced. In the oil fields, part of the produced water is reinjected for reservoir pressure maintenance. However, large amounts remain and cause a disposal problem. If the produced water is disposed of in shallow aquifers, a problem of contaminating potable water reservoirs is likely to happen. Therefore, the reuse/reclaim/recycling/recharging of the treated OPW in arid areas could be a promising development. Membrane filtration, thermal treatment, gas floatation, hydro cyclones, biological aerated filters, adsorption, ion exchange, chemical oxidation, electro dialysis, etc. are some of the technologies used for OPW treatment.

Agro-food, pharmaceuticals, meat, fish processing and leather industries also generate highly saline wastewater. Saline wastewater is generated in the industries due to the usage of brine solution in the finished product. Mining, steel and electroplating industries generate saline wastewater with a lot of heavy metals and salt ions [8]. Untreated saline effluents mostly consist of salt, in addition to other chemicals used in pretreatment methods such as chlorine compounds [9]. The wastewater with high salinity and organic pollutant discharge without prior treatment causes adverse effects to marine life, water quality and agriculture. Deep-well injection, irrigation of plants tolerant to high salinity (Reed bed technology), discharge into the sea, concentration into salts through pond evaporation are some of the conventional disposal methods for saline effluents [10].

Due to adverse impact created by industrial saline wastewater effluents on the environment, stringent regulations are now made compulsory by the government agencies for the treatment of industrial saline wastewater before being disposed to environment. Saline water was conventionally treated through physicochemical and biological methods such coagulation, adsorption, sand filters, reverse osmosis, nanofiltration, membrane bioreactors, etc. Salt concentrations and turbidity present in saline water are a major concern for the efficiency of these treatment methods [11].

The removal of pollutants from saline water with conventional treatment techniques is sometimes ineffective in removing the trace pollutants. Efficient removal of recalcitrant organic pollutants present in low concentration in saline water requires the development of innovative technologies. Green technologies (solar-based advanced oxidation processes, phytoremediation, eco-technology, green remediation) for saline water treatment are practical methodologies based on nontoxic chemical processes. Green technologies usually include the use of renewable energy, plants, including trees and grasses, to remove contaminants from saline water. These can be used to clean up all the salts and accompanied toxic metals, pesticides, oil, polyaromatic hydrocarbons and landfill leachates. The term "Green technology" is relatively new and has been used in recent decades.

This book provides a comprehensive material on the latest and advanced information of various treatment technologies employed for the removal of pollutants from saline water. It includes some case studies on the utilization of novel low-cost adsorbents, polymers, plants and artificial intelligence concepts in the treatment processes.

REFERENCES

1. Shannon, M.A., Bohn, P.W., Elimelech, M., Georgiadis, J.G., Mariñas, B.J, Mayes, A.M., Science and technology for water purification in the coming decades, *Nature.* 452(2), 187–200 (2008).
2. Shahzad, M.W., Burhan, M., Ang, L., Ng, K.C., Energy-water-environment nexus underpinning future desalination sustainability, *Desalination.* 413, 52–64 (2017).
3. Missimera, T.M., Malivab, R.G., Environmental issues in seawater reverse osmosis desalination: Intakes and outfalls, *Desalination.* 434, 198–215 (2018).
4. Goh, P.S., Lau, W.J., Othman, M.H.D.D., Ismail, A.F., Membrane fouling in desalination and its mitigation strategies, *Desalination.* 425, 130–155 (2018).
5. Feroz, S., Al Harthy, W., Baawain, B., Al Saadi, S., Varghese, M.J., Rao, L.N., Experimental studies for treatment of seawater in a recirculation batch reactor using TiO_2 P25 and polyamide, *International Journal of Applied Engineering Research.* 10(10), 26259–26266 (2015).
6. Voutchkov, N., *Pretreatment for Reverse Osmosis Desalination*, Elsevier, Amsterdam (2017).
7. Camacho-Muñoz, D., Lawton, L.A., Edwards, C., Degradation of okadaic acid in seawater by UV/TiO_2 photocatalysis – Proof of concept, *Science of the Total Environment.* 733, 139346 (2020).
8. Moosavirad, S.M., Sarikhani, R., Shahsavani, E., Mohammadi, S.Z, Removal of some heavy metals from inorganic industrial wastewaters by ion exchange method, *Journal of Water Chemistry and Technology.* 37, 191–199 (2015).
9. Alberti, F., Mosto, N., Sommariva, C, Salt production from brine of desalination plant discharge, *Desalination Water Treatment.* 10, 128–133 (2009).
10. Afanga, H., Zazou, H., Titchou, F.E., Rakhila, Y., Akbour, R.A., Elmchaouri, A., Ghanbaja, J., Hamdani, M., Integrated electrochemical processes for textile industry wastewater treatment: system performances and sludge settling characteristics, *Sustainable Environment Research.* 30, 2 (2020).
11. Lefebvre, O., Moletta, R, Treatment of organic pollution in industrial saline wastewater: A literature review, *Water Research.* 40(20), 3671–3682 (2006).

2 Characteristics of Saline Water and Health Impact

L.J. Stojanović Bjelić and D. Nešković Markić
Pan-European University "APEIRON"

P. Ilić
PSRI Institute for Protection and Ecology
of the Republic of Srpska

Z.U.R. Farooqi
University of Agriculture

H. Munawar
IAARD-Ministry of Agriculture

CONTENTS

DOI: 10.1201/9781003185437-3

2.1 INTRODUCTION: BACKGROUND OF PROBLEMS

Water is the most present substance in every living organism, makes up about 70% of the body weight of most organisms on average, is the most widespread natural resource and is essential for all living organisms. Undoubtedly, life originated in water and the further course of evolution originated from this solvent (Rajković and Lačnjevac 2010, Dalmacija 1999). Climate change, environmental pollution, fast economic development, population growth, and the resulting rapid rise of levels of water use creating high water contamination have raised concerns about the unsustainability of present water use (Ngo et al., 2016). The quality of coastal waters, from a scientific research point of view, is very important. Chemical, physical and biological processes that take place in this environment can have a diverse impact on the marine ecosystem as well as on the health and quality of life of people in direct contact with these environments.

The first recorded study in the Mediterranean on the relationship between the quality of recreational water and health effects was conducted in Alexandria, Egypt, in 1976. It was a retrospective study of the Alexandria Hospital for Infectious Diseases and was based on comparing the risk of typhoid fever in swimmers on polluted and unpolluted beaches. The results showed that there was a significant risk of contracting typhus from bathing in polluted water and that those most affected were in the younger age group of 10–19 years. Large amounts of wastewater are discharged directly into the sea. Saline (>2% w/v NaCl) wastewaters are discharged from many oil and gas producing industries, seafood processing, textile dyeing, tanneries, drinking water treatment plants from reverse-osmosis processes or ion exchange and oil-gas production and seafood processing (Xiao and Roberts, 2010).

The idea of management of sustainable water usually involves the progress of the health of groundwater and surface water, identification of alternative resources of water (e.g. stormwater, rainwater, recycled and desalinated water), maintenance of freshwater supplies and implementation of environmental educational campaigns and related policies. Faced with many major challenges, the water industry encourages the use of green technologies that encourage efficient forms of water recycling and reuse. This includes various innovations in water collection, treatment, distribution and drainage systems, as well as improved approaches to management, assessment strategies, policies and regulations.

There is an emergency need to use and develop suitable eco(green) technologies with promotion of design, system of production and supply because the main cause of water shortage is the unsustainable and disorganized model of production and consumption. The deficiency of freshwater use from natural sources such as surface water and groundwater water hase forced the industry to develop other sources of water such as desalinated water, stormwater, rainwater and recycled water, which in most cases need large treatment before use (Ngo et al., 2016).

A large number of biological and physical-chemical technologies can be used for saline wastewater treatment. Biological processes are usually the most cost-effective and sustainable. Anaerobic processes can be better than aerobic processes because they can be used to treat contaminants, which cannot be removed in aerobic conditions. They consume less energy and can produce sources of new energy in the form of fuel alcohol or methane. The presence of large amounts of salt in these wastewaters has been observed as inhibitory for conventional anaerobic procedures. Many of the salt-tolerant mechanisms available for aerobic organisms are energetically expensive and are not practical in marine ecosystems.

Biological purification processes are very well applicable for the treatment of wastewater, with low cost, no or with little need for chemicals and null production of secondary contaminants compared with second physical and chemical purification processes. Anaerobic purification processes are better than aerobic purification processes, with low energy use, low sludge output and high tolerance to the different of influent water quality. Value is also the production of energy, as methane gas. For saline wastewater, salt was considered to be an important inhibitor of anaerobic purification treatment. Research by Xiao and Roberts (2010) shows that the classic bacterial species can tolerate low-to-medium concentration of salt in wastewater from tannery and textile dyeing, fish processing and oil-produced water (Xiao and Roberts, 2010).

2.2 MOVEMENT OF WATER IN NATURE: HYDROLOGICAL CYCLE

Water is defined as a liquid that takes the shape of the vessel in which it is located, and as such can have any shape and different color variations. People need water for many functions, such as regulating body temperature, improving the body's metabolism and providing minerals that the body needs. There are many sources of water, such as surface water, spring water and seawater. Meanwhile, seawater can also be a good source of water. It is useful because it can provide minerals that are essential for health. Seawater is usually associated with the following characteristics: low temperature, high purity and richness of nutrients, i.e. useful elements. Its location leads to the fact that it has a high activity of bacteria and phytoplankton in charge of sterilization or no bacterial activity. Less photosynthesis of plant plankton, consumption of nutrients and a lot of organic decomposition cause an abundance of nutrients to remain there.

The abundance of inorganic material becomes greater as the depth of seawater increases. These characteristics have attracted the attention of many researchers, especially for many useful minerals, which include magnesium (Mg), calcium (Ca), potassium (K), chromium (Cr), selenium (Se), zinc (Zn) and vanadium (V). Seawater is richer in minerals compared to other water sources. People usually consume drinking water that is in the form of bottled drinking water (such as mineral water) or tap water. Drinking water sold by suppliers is expected to contain a good content of nutrients and to be safe for consumption because suppliers have a production permit from the authorities. Surprisingly, it has been noted that some drinking water available on the market has a low mineral content. This is probably due to the usual exposure of drinking water to a large number of purification methods, such as reverse osmosis and filtration, which removes the mineral content in it.

Mineral water that does not undergo the extensive required process is taken entirely from groundwater and receives mineral ions from its sources such as rocks. It has also been found to be low in minerals (Hwang et al., 2009). However, the mineral content of the water may vary depending on the geographical locations and the treatment processes that have taken place. It is promising that seawater can offer a lot of minerals for the production of drinking water and other byproducts of seawater. The production of refined seawater usually involves a desalination process, followed by a mineralization process. Seawater with high concentration of mineral salts usually treated by means such as reverse osmosis, electrodialysis or low vacuum temperature, to produce a safe concentration of water for consumption (Katsuda et al., 2008, Aris et al., 2013).

The global ocean covers 71% of the Earth's surface and contains 97% of all surface water on Earth (Costello et al., 2010). Freshwater flows into the ocean include direct runoff from continental rivers and lakes, seepage from groundwater, runoff, melting of icebergs from polar ice sheets, melting of sea ice and direct precipitation that is mostly rain but also includes snowfall. Evaporation removes fresh water from the ocean. Of these processes, evaporation, precipitation and runoff are the most significant at present. Using current best estimates, 85% of surface evaporation and 77% of surface rainfall occur across the ocean (Trenberth et al., 2007). Accordingly, the ocean dominates the global hydrological cycle. The water that leaves the ocean by evaporation condenses in the atmosphere and falls as precipitation, ending the cycle.

Hydrological processes can also vary over time, and these time variations can manifest as changes in global sea levels if the net freshwater content of the ocean changes. Precipitation is caused by the condensation of atmospheric water vapor and is the largest single source of fresh water entering the ocean ($\sim$530,000 km^3/year). The source of water vapor is surface evaporation, which has a maximum over the subtropical oceans in the trade wind regions (Yu, 2007). Equatorial trade winds carry water vapor evaporated in the subtropical zone to the Intertropical Convergence Zone (ITCZ) near the equator, where intense surface warming from the sun causes the growth of warm moist air, producing frequent convective thunderstorms and heavy rain (Xie and Arkin, 1997). Heavy rains and high temperatures support and affect life in tropical rainforests (Malhi and Wright, 2011). Evaporation is intensified as the global mean temperature rises (Yu, 2007). The water retention capacity of the atmosphere increases by 7% for each degree Celsius of heating, according to the Clausius-Clapeyron ratio. Increased moisture content in the atmosphere leads to changes in the intensity, frequency and duration of precipitation (Trenberth et al., 2007) and leads to an increase in global precipitation by 2%–3% for each Celsius degree of warming (Held and Soden, 2006).

Figure 2.1 shows the water cycle in nature, which begins with evaporation from water surfaces and from the earth and on average about 516,000 km^3 of water evaporates annually, mostly from sea surfaces. This water condenses in the atmosphere and returns in the form of precipitation (rain, snow, hail). Condensed water from the land partly evaporates and returns to the atmosphere (evaporation), it is partly absorbed into the surface parts of the earth, then it binds in plants and evaporates by the process of transpiration, and partly it flows into surface waters (rivers, lakes) and partly it penetrates into deeper aquifers layers forming groundwater. From the underground layers, the water partially reaches the surface again (spring, source), which closes the circular flow of water in nature.

As seen in Table 2.1, of the total amount of water circulating in nature, the biggest part is in oceans, about 97.6%, in the form of ice about 2%, and almost negligible part belongs to surface water and groundwater (Jovičić and Jovanović, 2017).

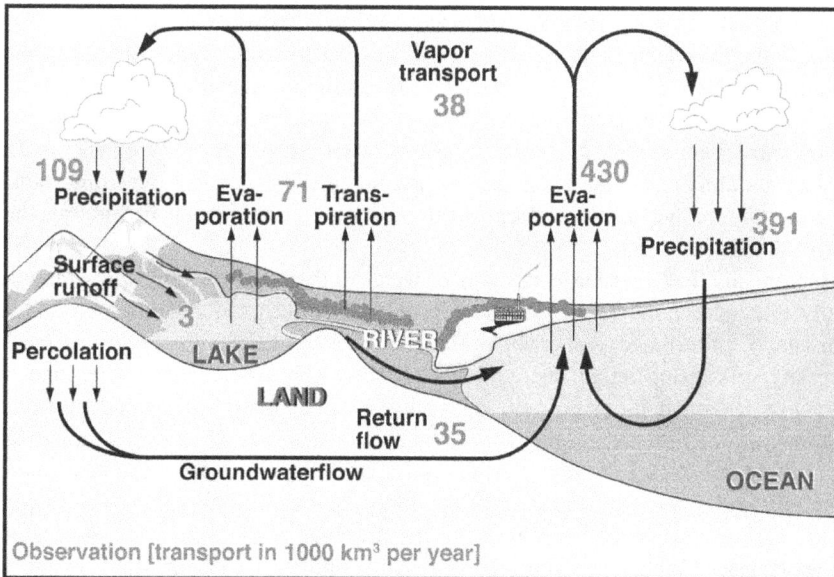

FIGURE 2.1   Water cycle (see Bengtsson, 2010).

TABLE 2.1
Distribution of Water in the Land in
Percents (Jovičić and Jovanović, 2017)

Type of Water	%
Ice (polar glaciers)	78.10
Groundwater	21.03
Soil moisture	0.45
Lakes (freshwater)	0.37
Rivers	0.005
Atmosphere	0.039

Large-scale anthropogenic pollution has not left large water basins such as the seas and oceans without worrying consequences. Not so long ago, it was still believed that the capacity of the world's seas and oceans to the pollution was so great that there was no danger that their pollution could occur. However, the events in the last decades of the 20th century completely refuted that. The pollution of the sea and ocean is done in two ways: directly and indirectly (Cochran et al., 2019). Direct pollution is done through the discharge of wastewater from cities and industries directly into seas and oceans, accidents of tankers and other ships, by depositing various forms of waste in the water of the seas and oceans or on their bottom. These are often quantities of waste substances that are truly enormous, but the world sea is not seen as anyone's property. Control on the high seas, in international waters, is difficult to enforce.

Indirect pollution of the seas and oceans is done in two basic ways:(i) through the inflow of polluted water from various watercourses into the seas and oceans, and (ii) through precipitation. In this case, we will neglect the importance of transmitting pollution to the sea from the air. Pesticides reach the marine environment by leaching from agricultural land. These substances are foreign material in seawater and pose a danger to wildlife because they are poorly degradable and often accumulate in seawater and bioaccumulate in marine organisms. For pesticides, it is important to note that their concentration increases through trophic chains, which increases their harmful effect.

Organisms that filter large amounts of water in their diet or respiration are especially endangered by pesticides. The group of chlorinated hydrocarbons which consists of chlorinated pesticides and polychlorinated biphenyls stands out by the intensity of causing lethal and sublethal effects. These pesticides are among the strongest chemical mutagens. In sea pollution in general, chlorinated pesticides occupy a significant place. They are very stable in seawater, and the most important representatives of this group of pesticides are DDT and its metabolites, hexachlorobenzene, lindane, cyclodiene, aldrin, etc. (Stemmler and Lammel, 2009). As early as the early 1960s, scientists warned of the presence of chlorinated pesticides and polychlorinated biphenyls in marine organisms and in the marine environment in general (Monton et al., 2005). Thus, these compounds are present in the air, ice and snow of the Arctic and Antarctic, as well as in the planktonic and benthic organisms of the Arctic. The highest concentrations of these substances in marine organisms are found in the adipose tissue of Arctic fish and mammals.

Exploitation of oil from offshore wells, as well as the transport of large quantities of oil by sea, is inevitably associated with the danger of platform accidents, as well as oil tankers transporting the oil, and with unforeseeable and unpredictable consequences of such accidents but also some procedures in oil transport not related to tanker accidents (Qu and Deng, 2001). These phenomena are such in their manifestations that whenever they happened, they attracted the attention of public media around the world.

In addition to marine pollution by wastewater, there is an increasing number of introductions of various plant species by humans into the sea (Klein and Verlaque, 2008). Out of a total of 85 plant species introduced into Mediterranean waters by humans, about 9 species have become pests, including the aggressive algae species – green tropical algae (Caulerpa taxifolia) (Boudouresque and Verlaque, 2002). With its spread, it pushed other sea grasses from those habitats, and fish do not eat it due to the strong poison it contains. In the natural environment, green tropical algae rarely exceed a height of 25 cm and wither if the sea temperature drops below 20°C. This type of algae in the Mediterranean Sea grows two to three times higher and survives at a temperature of 10°C.

2.3   GENERAL CHARACTERISTICS OF SEAWATER

2.3.1   PHYSICAL COMPOSITION OF SEAWATER

Seawater or salt water is water from the sea or ocean. On average, seawater in the world's oceans has a salinity of about 3.5% (35 g/L, 599 mM). This means that each kilogram of seawater has about 35 g of dissolved salts (mostly sodium ions (Na^+) and chloride ions (Cl^-)). The average surface density is 1.025 kg/L. Seawater is thicker than fresh and pure water (density 1.0 kg/L at 4°C) because dissolved salts increase the mass more in relation to the volume. The freezing point of seawater decreases as the salt concentration increases. At typical salinity, it freezes at −2°C. The coldest seawater ever recorded (in liquid form) was in 2010 in a stream below the Antarctic glacier and measured −2.6°C. The pH of seawater is usually limited to the range between 7.5 and 8.4, as seen in Figure 2.2.

However, there is no universally accepted reference pH scale for seawater and the difference between measurements based on different reference scales can be up to 0.14 (Malhi and Wright, 2011).

2.3.2   CHEMICAL COMPOSITION OF SEA WATER

Seawater contains more dissolved ions than all types of fresh water (Trenberth et al., 2007). However, the ratios of solutes differ dramatically. For example, although seawater contains about 2.8 times more bicarbonate than river water, the percentage of baking soda in seawater as a ratio of all dissolved ions is far lower than in river water. Bicarbonate ions make up 48% of the solution in river water, but only 0.14% in seawater. These differences are due to the difference in retention times of certain

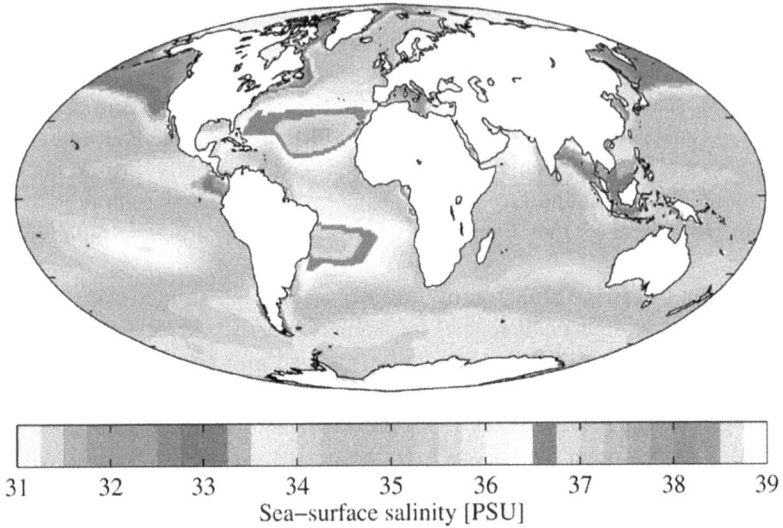

31   32   33   34   35   36   37   38   39
Sea–surface salinity [PSU]

FIGURE 2.2   Annual mean values of surface salinity of the world's seas (see Levitus et al., 2010).

substances in sea and river water, sodium and chloride have a very long residence time, while calcium (important for carbonate formation) tends to settle much faster (Held and Soden, 2006). The most numerous dissolved ions in seawater are sodium, chloride, magnesium, sulfate and calcium (Jovičić and Jovanović, 2017). Small amounts of other substances have been found in seawater, including amino acids at concentrations of up to 2 µg of nitrogen atoms per liter that are thought to have played a key role in the origin of life (UNICEF, 1999).

2.3.3   MICROBIOLOGICAL COMPONENTS OF SEAWATER

The first study by the Institute of Oceanography in 1957 sampled water at various locations in the Pacific Ocean. Direct microscopic examinations using bacterial and culture counts were used, with direct numbers showing in some cases up to 10,000 times the value of seeded cultures. These differences are attributed to the appearance of bacteria in the aggregates, the selective action of culture media and the presence of inactive cells. A large number of spiral shapes were observed under the microscope, but they were not processed. The discrepancy in the numbers obtained using these two methods is well known in this and other studies (Batut, 2012). During the 1990s, improved microbial detection and identification techniques by studying only small DNA fragments allowed researchers participating in the census of marine life to identify thousands of previously unknown microbes that are usually present in only a small number. This has revealed a much greater diversity than previously thought, so a liter of seawater can hold more than 20,000 species. Laboratories estimate that the number of different species of bacteria in oceans could exceed 5–10 million (Roy and Tim, 2012).

Bacteria are found at all depths in the water column, as well as in sediments, some of which are aerobic, others anaerobic. Most move freely, primarily by swimming, but some exist as symbionts within other organisms – examples are bioluminescent bacteria. Cyanobacteria played an important role in the evolution of ocean processes, enabling the development of stromatolites and oxygen in the atmosphere. Some bacteria interact with diatoms and form a critical bond in the movement of silicon in the ocean. One anaerobic species, *Thiomargarita namibiensis*, plays an important role in the decay of hydrogen sulfide eruptions from diatomaceous sediments on the Namibian coast, generated by high phytoplankton growth rates in the current zone of the Benguela current, eventually falling to the seabed.

Bacterial archaea have surprised marine microbiologists with their survival and success in extreme environments, such as hydrothermal vents on the ocean floor. Alkalotolerant marine bacteria such as *Pseudomonas* and *Vibrio* spp. survive in the pH range of 7.3–10.6, while some species will only grow at pH 10–10.6 (Pilson, 2012). Archaea also exist in pelagic waters and can make up as much as half of ocean biomass, apparently playing an important role in ocean processes (Pinet, 1996). In 2000 sediments from the ocean floor, the Archaea species was discovered, which decomposes methane, an important greenhouse gas, which has a great contribution to atmospheric warming. Some bacteria break down the rocks of the seabed, affecting the chemistry of seawater. Oil spills and water spills containing human sewage and chemical pollutants have a pronounced impact on the life of microbes in the vicinity, as well as on the collection of pathogens and toxins that affect all forms of marine life (Kato, 1966).

Pandoravirus salinus, a type of very large virus, a genome much larger than any other type of virus, was discovered in 2013. Like other very large Mimivirus and Megalovirus viruses, Pandoravirus infects amoebae, but its genome, which contains 1.9–2.5 megabases of DNA, is twice as large as that of Megalovirus and differs greatly from other large viruses in appearance and genome structure.

The ocean has a long history of disposing of human waste under the assumption that its enormous size makes it capable of absorbing and diluting all harmful material (Tada et al., 1998). While this may be true on a small scale, large amounts of wastewater that are routinely disposed of have damaged many coastal ecosystems and endangered them. Pathogenic viruses and bacteria such as *Escherichia coli*, *Vibrio cholerae* causing cholera, hepatitis A, hepatitis E and poliovirus occur in such waters, along with protozoa such as *Giardia lamblia* and cryptosporidium. These pathogens are routinely found in large water vessels and easily spread when wastewater is discharged (Jannasch and Jones, 1959, Marine Biological Laboratory, 2006).

2.4 PHYSIOLOGICAL AND PATHOPHYSIOLOGICAL EVENTS CAUSED BY THE INTRODUCTION OF SEAWATER INTO THE BODY

Accidental consumption of small amounts of clean seawater is not harmful, especially if seawater is taken together with a larger amount of fresh water. However, drinking seawater to maintain hydration is counterproductive. The renal system actively regulates sodium chloride in the blood in a very narrow range of about 9 g/L. In most open waters, concentrations vary slightly around typical values of about 3.5%, far higher than the body can handle, and much beyond what the kidney can process. A point that is often overlooked in claims that the kidney can excrete NaCl in concentrations of 2% is that the intestines cannot absorb water in such concentrations, so there is no benefit to drinking such water. Drinking seawater temporarily increases the concentration of NaCl in the blood. This signals the kidney to excrete sodium, but the concentration of sodium in seawater is above the maximum concentration of the kidneys. Eventually, the concentration of sodium in the blood rises to the level of toxicity, removing water from the cells and interfering with nerve conduction, which results in the development of status epilepticus or fatal cardiac arrhythmias (Farrell and Bower, 2003).

2.5 THE MOST COMMON ELEMENTS AS POLLUTANTS IN SEAWATER AND THEIR IMPACT ON HEALTH

2.5.1 ALUMINUM

Aluminum salts are used as flocculants in water treatment. The amount of aluminum is directly proportional to the amount used in water treatment, as well as the efficiency of the processing process. The maximum permissible concentration of aluminum in drinking water is 0.2 mg/L. Recent studies have shown that an increased amount of aluminum in the body can lead to Alzheimer's disease.

2.5.2 ANTIMONY

Antimony is introduced into the organism by inhalation, much less often by ingestion through water or food. Trivalent antimony compounds are more toxic than pentavalent ones. Exposure of the organism to antimony dust in a longer period of time leads to bronchitis and emphysema of the lungs with the formation of pleural adhesions, and a certain number of subjects also experience skin changes known as antimony dermatitis. The maximum permissible concentration in drinking water is 0.003 mg/L.

2.5.3 ARSENIC

Arsenic is found in groundwater and surface water in the form of its compounds. Inorganic compounds lead to neoplastic transformation of cells of living organisms, while organic compounds do not have such an effect. Arsenic most often enters the water through erosion processes; from mines; and as waste from the pharmaceutical industry, the paint industry, cars and pesticides.

2.5.4 COPPER

Copper, like the previous element, enters the water through mines, once the presence of copper in drinking water was connected with old copper water pipes. Depending on the hardness of water, pH, temperature and concentration of anions, water from copper pipes can contain even up to several mg/L of copper, and if it has stood in copper pipes for 12 or more hours, the concentration can be more than 22 mg/L. The maximum permissible concentration of copper in drinking water is 2 mg/L. Copper in larger quantities changes the taste of water; however, water can be acceptable for drinking and will not negatively affect health at higher concentrations than recommended.

2.5.5 CYANIDES

Cyanides usually reach the water by polluting it with industrial wastewater. Cyanides, although a highly toxic compound, are converted to much less toxic thiocyanates in the gastrointestinal tract. A very small number of studies have dealt with the maximum concentration in drinking water, and it is 0.05 mg/L. Cyanides in high concentrations lead to vitamin B12 deficiency and the consequent occurrence of megaloblastic anemia, while in high concentrations they lead to disorders of the thyroid gland and central nervous system.

2.5.6 FLUORIDE

Fluorides are effective in the prevention of caries, when ingested in large quantities, toxic effects appear in the form of fluorosis of teeth and skeletal bones. It is taken into the body in the form of fluoride or fluoride compounds, and the main way of intake is water and dental preparations.

2.5.7 CHROMIUM

Chromium is found in water in the form of its salts. They could be hexavalent and trivalent chromium salts. Hexavalent chromium salts have a malignant potential, as shown by experimental studies conducted on animals.

2.5.8 MANGANESE

Manganese is one of the essential trace elements; it has a role in the metabolism of many minerals, mainlycalcium and phosphorus; and it participates in maintaining reproductive functions. It is taken into the body by inhalation, food and rarely water. The limit value of manganese in water is 0.05 mg/L. It is ingested due to occupational exposure, although there are no comprehensive studies to confirm its toxicity.

2.5.9 LEAD

Lead in water can be found originating from lead pipes, pipes that contain only a lead component or from a tap or connection apparatus. It can also enter the water from soldered or corroded pipes. By increasing the pH value of the water distribution system, the introduction of lead into the human body can be prevented.

2.5.10   ORGANIC SUBSTANCES IN WATER AND THEIR IMPACT ON HEALTH

Organic substances that can be found in water are divided into following groups: chlorinated alkanes, chlorinated ethenes, chlorinated benzenes, aromatic hydrocarbons and mixed substances. According to their properties, carbon tetrachloride and 1.2-dichloroethane should be singled out from the group of chlorinated alkanes.

2.5.11   CARBON TETRACHLORIDE

Carbon tetrachloride is most often found in water when chlorine is used to disinfect water. Animal studies have shown that it has the potential to lead to malignant cell transformation causing malignant diseases; on the other hand, some studies show that it is not genotoxic and its allowable value of 1 mg/kg and tolerable daily intake (TDI) of 0.71 µg/kg of body weight, of which 10% belongs to drinking water.

2.5.12   VINYL CHLORIDE

Vinyl chloride in water is measured in µg/L. Many studies show the ability of vinyl chloride to lead to malignant cell transformation, and it is believed that it is associated with the development of lung adenocarcinoma, hepatocellular carcinoma and neoplasms of the central nervous system.

2.5.13   BENZENE

The concentration of benzene in drinking water does not exceed 5 µg/L. People are most often exposed to these compounds through food and air, less often with water. Its carcinogenicity has been experimentally proven in the human population, most often in acute cases of exposure leading to malignancy of the central nervous system, while chronic exposure leads to hematological disorders in the form of pancytopenia, aplastic anemia and acute leukemia. Exposure to benzene is mainly through food and air, and much less through drinking water.

2.5.14   POLYCYCLIC AROMATIC HYDROCARBONS

This large group of organic compounds reaches surface waters through the atmosphere and pollutes them. Benzo(a)pyrene should be singled out from this group of polycyclic aromatic hydrocarbons. Benzo(a)pyrene represents a small fraction of the total polycyclic aromatic hydrocarbons.

2.5.15   NITRATES

Nitrates represent the end product of metabolic processes involving organic substances that contain nitrogen, most often amino acids. Detection of nitrate in water, either for drinking or for other uses, means the so-called "old" water pollution with organic substances of animal or human origin.

2.5.16 NITRITES

Nitrites are the product of ammonia oxidation and nitrate reduction. They are a sign of recent water pollution. They convert hemoglobin to methemoglobin which can lead to methemoglobinemia. They are formed by oxidation of ammonia or reduction of nitrates and are always a sign of "fresh" fecal water pollution.

2.5.17 AMMONIA

Ammonia is a sign of recent, acutely occurring water pollution, but it is also a major indicator of fecal water pollution. It is one of the main indicators of the health aspect of fecal water pollution.

2.5.18 HYDROGEN SULFUR

This compound in water is a sign of organic pollution, but it can also occur due to the reduction of inorganic compounds and can be found both in surface and deep waters. Water contaminated with this compound has specific organoleptic properties in terms of the smell of rotten eggs (Knežević and Tanasković, 1997).

2.6 PHYSICOCHEMICAL PROPERTIES OF SEAWATER

2.6.1 pH

pH is a measure of the concentration of hydrogen ions in water or a measure of how acidic or basic the water is. Values are displayed on a scale from 0 to 14, where 7 is the neutral value (Talling, 2010). Fresh natural waters have a pH in the range between 6.5 and 8.5. Water pH is important because it affects the solubility and availability of nutrients and how aquatic organisms can use them (Stone et al., 2013). The pH of surface water can be relatively higher in waters with small discharges since water is rich in dissolved substances characteristic of groundwater. The pH value can indicate with great sensitivity the variations in water quality and how it is affected by dissolved substances (WHO, 2006).

2.6.2 TURBIDITY AND TOTAL SUSPENDED SOLIDS

Water turbidity and transparency are determined by the concentration and nature of total suspended solids (TSS). TSS contains soluble organic compounds, as well as fine particles of organic and inorganic matter (Matta, 2014). TSS and turbidity differ over time based on biological activities in the water system and the types of sediments transmitted by surface runoff. Turbidity is high under the influence of precipitation. Turbidity can be associated with TSS, and thus turbidity is used as an indirect measurement for TSS. Turbidity refers to the amount of suspended material, which interferes with the penetration of light into the water column. Greater turbidity can cause stratification of temperature and dissolved oxygen (DO) in water bodies. High TSS surface water levels absorb heat from sunlight, which increases the water temperature and reduces dissolved oxygen levels. This results in the loss of the water body's ability to support aquatic life (Cavanagh and Harding, 2014).

Electrical conductivity (EC) can be used to assess total dissolved substances (TDS) in water. EC is determined by the amount of dissolved solids in water. EC is related to the content of salts and minerals, i.e. the higher the salt content, the higher the EC will be (Anhwange et al., 2012). An increase in water conductivity indicates the addition of mineral salts to the water. The value of TDS in mg/L is about half of the electrical conductivity (mS/cm). TDS is formed due to the ability of water to dissolve salts and minerals, and these minerals produce an unwanted taste in water. The presence of high levels of TDS is not desirable because it causes stones on water pipes, heaters, boilers and household appliances (Mohsin et al., 2013).

2.6.3 SALTS OF CALCIUM (CA) AND MAGNESIUM (MG)

Calcium (Ca) and magnesium (Mg) salts are largely responsible for the total hardness (TH) of water. The total hardness is the concentration of multivalent metal cations in solution. Bicarbonates and carbonates of Ca and Mg give temporary hardness, while sulfates, chlorides and other anions produce permanent hardness. Sources of Ca and Mg in natural water are various types of rocks, industrial waste and sewage. Ca compounds become stable when CO_2 is present in the water, but calcium concentrations decrease when $CaCO_3$ precipitates due to the rise in water temperature. Mg salts are formed by the weathering of rocks containing Mg minerals and from some rocks where CO_3 is found (Uchchariya and Saksena, 2012).

2.6.4 SODIUM (NA)

Sodium (Na) is one of the most abundant elements on earth and is highly soluble in water. An increase in Na in surface waters can arise from sewage and industrial wastewater. Na ions mainly reach water bodies from sodium salts present in rocks and occasionally as a result of industrial and domestic activities. The WHO guidelines do not specify limits for Na in surface waters. But for drinking water, the average taste threshold for Na is about 200 mg/L. Potassium (K) levels in water bodies are generally very low compared to Na because potassium salts are rarely found in rocky deposits. High amounts of K in drinking water cause laxative effects. Potassium salts are widely used in industry and agriculture. K deposits enter fresh water through industrial discharges and runoff from cultivated fields (Mustapha and Usman, 2014).

2.6.5 CHLORIDES (CL)

Chlorides occur in most fresh waters, such as sodium or calcium salts. The high chloride content in the water sample may be due to pollution by sewage and municipal wastewater rich in chloride. Excess chloride gives a salty taste to water and drinks. Chloride concentration can be used as an important parameter to detect sewage contamination, prior to another test such as biochemical oxygen demand (BOD) and chemical oxygen demand (COD). A maximum of 250 mg/L of Cl is allowed in surface waters according to EPA 2001 (Verma et al., 2013).

2.7 HEALTH ASPECTS OF THE SEAWATER

Many researchers have examined the effects of seawater on human health. Studies have been conducted on both the human population and experimental animals, in which the impact of different modes of exposure to seawater and many substances present in it has been assessed.

2.7.1 INFLUENCE OF SEAWATER ON CHOLESTEROL PROFILE

The most promising benefits that can be achieved by the intake of seawater are that it is able to improve the cholesterol profiles in the serum, i.e. the liver. Exposure to seawater leads to decreased levels of triglycerides (TG), levels of lipoprotein cholesterol without high density (non-HDL-Chol) and total cholesterol (TC) in serum as per the studies performed on animal models. Drinking water produced from seawater containing Mg of 600 and 1,000 ppm can reduce cholesterol levels by 18% and 15%, respectively (Kimura et al., 2004). Interestingly, a study of seawater consumption in hypercholesterolemic individuals demonstrated that it can reduce TC and low-density lipoprotein (LDL) and reduce lipid peroxidation in these subjects. Mechanisms for improving the cholesterol profile are related to the regulation of the hepatic receptor of low-density lipoprotein and cholesterol-7α-hydroxylase, as well as the expression of the CIP7A1 gene, which are involved in cholesterol catabolism. Seawater intake resulted in higher excretion of cholesterol and bile acids in the stool, thus reducing TC levels (Hsu et al., 2011). Seawater reduces the lipid content in hepatocytes by activating AMP-activated protein kinase, inhibiting the synthesis of cholesterol and fatty acids.

2.7.2 INFLUENCE OF SEAWATER ON THE CARDIOVASCULAR SYSTEM

Seawater provides protection against cardiovascular disease by reducing levels of TC, TG, atherogenic index and malondialdehyde (MDA), while increasing the antioxidant capacity of serum trolox (TEAC). The molecular mechanism of its cardiovascular protection is through the regulation of hepatic low-density lipoprotein receptors (LDL receptors) and CIP7A1 gene expression (Hsu et al., 2011). Its cardioprotective effects have been further demonstrated when its use can reduce abnormal cardiac architecture and apoptosis and improve the signaling of cardiac survival similar to insulin-like growth factor-1 receptor (IGF-1R) (Shen et al., 2012). Seawater may also improve cardiovascular hemodynamics in a study conducted by Katsuda et al.

The antiatherogenic effects of seawater are associated with 5-adenosine monophosphate-activated protein kinase (AMPK), stimulation and consequent inhibition of phosphorylation of acetyl-CoA carboxylase (ACC) (Shen et al. 2012). AMPK plays an important role in lipid metabolism by inhibiting 3-hydroxy-3-methyl-glutaryl-CoA reductase (HMGCR) and ACC, and then inhibiting cholesterol production.

2.7.3 Prevention of Atherogenesis

Prevention of atherogenesis can avoid serious health problems, including coronary heart disease and stroke. Seawater has antiatherogenic properties due to the existence of many useful minerals and ions like Mg and Ca. Hence, this could be an explanation of the widely promoted theory that it improves cardiovascular protection. Atherogenesis is the formation of plaque in the inner mucosa of the artery, in which mast cells are deposited, which phagocytose fat globules, cholesterol, cellular waste products, calcium and other substances. Studies in which subjects were treated with seawater in addition to drug therapy have shown that this combined approach leads to the prevention of the process of atherogenesis (Shen et al., 2012, Miyamura et al., 2004). The application of seawater with hardness of 300, 900 and 1,500 ppm significantly reduces the atherogenic index [(TC-HDL-C)/HDL-C], improves cardiovascular hemodynamics and reduces blood pressure (Chang et al. 2011). Hypertensive rats treated with seawater for 8 weeks had slightly lower mean blood pressure values compared to the control group. Decreased fats and lipids in the blood may be associated with decreased blood pressure. Although the seawater used in the study contained quite a lot of salt, the blood pressure did not increase.

In another study, the use of seawater did not affect blood pressure. Moreover, seawater can also prevent thrombotic disorder by suppressing the release of inhibitors of plasminogen activator type 1 (Ueshima et al., 2003). Many combinations of minerals in seawater, such as Mg, Ca and Na, are associated with a decrease in blood pressure. Increased sodium levels can induce hypertension, although the addition of Mg can lower blood pressure by suppressing adrenergic activity and possibly natriuresis (Itoh et al., 1997). Interestingly, high Mg content can lower blood pressure.

Seawater has properties in the prevention of obesity and has been proven to reduce fat and body weight (Yuan et al., 2016, Ha et al., 2013). It has been recognized in nature as a possible therapeutic agent against obesity (Yun, 2010). The study reported that seawater was able to significantly reduce lipid accumulation in in vitro and in vivo models. A study with obese mice found that seawater with a hardness of 1,000 ppm managed to reduce body weight by 7%. There was also an increase in plasma adiponectin levels and a decrease in the levels of residues, RBP4 and plasma fatty-acid binding proteins (Ha et al., 2013). The results suggest that anti-obesity activities are mediated by modulation of the expression of obesity-specific molecules. The magnificent effects of seawater on obesity were further proven when it was found to stimulate mitochondrial biogenesis, a component that controls the release of energy associated with lipid metabolism. Thus, this suggested that Mg and Ca are not the main factors for fat reduction, because the roles of many elements in seawater have yet to be clarified. However, the available findings of the clinical study showed that there was no significant difference in TG levels and body weight between treated subjects and controls.

2.7.4 Influence of Seawater on Skin Changes

Seawater also shows a great capacity for treating skin problems. In a study involving patients with atopic eczema/dermatitis syndrome (AEDS) treated with seawater, improvement in skin symptoms such as inflammation, lichenification and cracking of the skin was observed (Hataguchi et al., 2005). Patients with AEDS usually have an imbalance of different essential minerals in their hair, and some also have toxic minerals. From that study, seawater intake restored essential minerals such as selenium and reduced levels of toxic minerals such as mercury and lead in treated patients. In another study, seawater intake reduced allergic skin reactions and levels of total IgE in serum, Japanese cedar pollen-specific IgE, interleukin-4 (IL-4), IL-6, IL-13 and IL18 in patients with allergic rhinitis, compared to the intake of distilled water in which no such effects are observed (Kimata et al., 2001).

An in vivo study found that seawater can recover an atopic skin lesion by improving symptoms such as edema, erythema, dryness, itching, transepidermal water loss (TEWL), reduced epidermal thickness and inflammatory cell infiltration. Its use can reduce allergic reactions when a decrease in total levels of IgE and released histamine has been observed. It also leads to inhibitory regulation of IgE, histamine and proinflammatory cytokines (tumor necrosis factor α (TNF-α), IL-1β and IL-6) in serum.

2.7.5 EPIDEMIOLOGICAL SIGNIFICANCE OF WATER

Water is essential for life. An adequate, secure and accessible supply must be available to all. Improving access to safe drinking water can result in significant health benefits. Every effort should be made to make drinking water quality as safe as possible (Fenwick, 2006).

Many people struggle to gain access to safe water. Supplying clean and purified water to every home may be the norm in Europe and North America, but in developing countries, access to both clean water and sanitation is not the rule, and water-borne infections are common. Two and a half billion people do not have access to improved sanitation, and more than 1.5 million children die from diarrhea every year (George et al., 2001). According to the WHO, the death rate from water-related diseases exceeds 5 million people a year. Of that, more than 50% are microbiological intestinal infections, and cholera stands out in the first place.

Generally speaking, the greatest microbial risks are associated with the ingestion of water contaminated with human or animal feces. Discharges of wastewater into freshwater and coastal seawater are major sources of fecal microorganisms, including pathogens (Grabow, 1996, Seas et al., 2000).

Acute microbial diarrhea is a major public health problem in developing countries. People suffering from diarrhea are the ones with the least financial resources and the poorest hygienic facilities. Children under the age of five, primarily in Asian and African countries, are most affected by water-borne microbial diseases (Medema et al., 2003).

Water-borne microbial diseases also affect developed countries. In the United States, it is estimated that 560,000 people suffer from severe water-borne diseases every year, and 7.1 million from mild to moderate infections, resulting in 12,000 deaths annually (Olsson et al., 1998).

As seen in Table 2.2, the most common pathogenic microorganisms that contaminate water are bacteria: *Salmonella, Shigella, Escherichia coli, Vibrio*

TABLE 2.2
Major Intestinal Diseases Whose Pathogens Are Transmitted by Water (Cabral, 2010)

Illness	Bacterial Agent
Cholera	*Vibrio cholerae*, serovariants O1 and O139
Typhoid fever	*Salmonella enterica* (ssp. *typhi, typhimurium, paratyphy*)
Vibrio gastroenteritis	*Vibrio parahaemolyticus*
Bacillary dysentery	*Shigella dysenteriae*
	Shigella flexneri
	Shigella boydii
	Shigella sonnei
Acute diarrhea and gastroenteritis	*Escherichia coli*, serotypes: O148, O157 and O124

cholere, Yersinia enterocolitica and *Compylobacter jejuni*; viruses: *Adenoviruses, Enteroviruses, Hepatitis A, Rotaviruses*; protozoa: *Giardia, Cryptosporidium* spp., *Entamoeba histolytica*; helminth: *Dracunculus medinensis*; and others. Cyanobacteria (blue-green algae) flowers, which produce two types of toxins, can also be found in the water:

- **Hepatotoxin**, which is produced by members of the genus *Microcystis, Oscillatoria, Anabaena* and *Nodularia*.
- **Neurotoxin**, produced by representatives of the genus *Anabaena, Oscillatoria, Nostoc, Cylindrospermum* and *Aphanizomenon*.

2.8 CONCLUDING REMARKS

The world's population is expanding very fast. This affects the pollution of the environment, both inland and seawater. Although the global amount of water is considered sufficient for the current population from the perspective of the overall hydrological cycle, the world's water resources are concentrated in certain areas, and in other places, there is a serious shortage of water. In addition to water shortages, improperly managed freshwater resource systems have caused significant water pollution. The problems are even worse in developing countries because they release large amounts of wastewater directly into inland waters or the marine ecosystem.

The two main uses of the coastal marine environment by man are for recreation and as a source of seafood. Pollution of seawater by sewage and (in some cases) industrial wastewater has raised concerns by public health authorities about the risk to human health associated with the use of seawater. In the case of coastal recreational waters, this concern has been significantly emphasized in recent decades, especially in the case of tropical and subtropical geographical regions, due to a combination of socioeconomic development factors. Increasing emphasis is placed on the water ecosystem, especially seawater, as the main recreational program in certain areas. The rapid and increasingly uncontrolled development of coastal areas due to mass tourism has resulted in a deterioration in the quality of neighboring water bodies resulting in an increase in municipal wastewater, causing moderate to severe pollution. The mixed nature of the tourist population also contributes to a larger number of pathogenic microorganisms that are found in the sewage system that originates from coastal areas and summer resorts and discharges into the sea. The increased incidence of numerous diseases, both among the local population and tourists, is attributed to this, which, rightly or wrongly, is variously attributed to bathing in polluted waters. Other activities such as diving, spearfishing and the like, which are now expanding far, have increased the need in terms of area and depth and added new dimensions to risk assessment procedures to deal with different patterns of exposure among various recreational groups.

In the future, in order to determine the quality of seawater, it is necessary to conduct detailed, comprehensive research that will include physicochemical and microbiological parameters of water quality by periodic testing of water samples over a longer period of time. Maintaining high-quality seawater is essential for maintaining the normal functioning of ecosystems and health safety.

REFERENCES

Anhwange, B. A., E. B. Agbaji and E. C. Gimba. 2012. Impact assessment of human activities and seasonal variation on River Benue, within Makurdi Metropolis. *International Journal of Science and Technology* 2(5):248–254.

Aris, A. Z., R. C. Y. Kam, A. P. Lim, and S. M. Praveena. 2013. Concentration of ions in selected bottled water samples sold in Malaysia. *Applied Water Science* 3(1):67–75.

Batut, M. J. 2012. Health safety of drinking water from central water supply systems in the Republic of Serbia in 2012. http://www.batut.org.rs/download/izvestaji/higijena/Zdravstvena%20ispravnost%20vode%20za%20pice%202012.pdf (accessed March 15, 2021).

Bengtsson, L. 2010. The global atmospheric water cycle. *Environmental Research Letters* 5(2):025202.

Boudouresque, C. F. and Verlaque, M. 2002. Biological pollution in the Mediterranean Sea: Invasive versus introduced macrophytes. *Marine Pollution Bulletin* 44(1):32–38.

Cabral, J. P. 2010. Water microbiology. Bacterial pathogens and water. *Internal Journal of Environmental Research in Public Health* 7(10):3657–3703.

Cavanagh, J. E. and J. S. Harding. 2014. Effects of suspended sediment on freshwater fish. New Zealand Ltd and West Coast Regiona, Landcare Research. https://ir.canterbury.ac.nz/bitstream/handle/10092/15020/1445-WCRC129%20Effects%20of%20suspended%20sediment%20on%20freshwater%20fish.pdf?sequence=2 (accessed March 15, 2021).

Chang, M.-H., B.-S. Tzang, T.-Y. Yang et al. 2011. Effects of deep-seawater on blood lipids and pressure in high-cholesterol dietary mice. *Journal of Food Biochemistry* 35(1):241–259.

Cochran, J. K., H. J. Bokuniewicz, and P. L. Yager. 2019. *Encyclopedia of Ocean Sciences*. Academic Press, Boston, MA.

Costello, M. J., A. Cheung, and N. De Hauwere. 2010. Topography statistics for the surface and seabed area, volume, depth and slope, of the world's seas, oceans and countries. *Environmental Science and Technology* 44(23):8821–8828.

Dalmacija, B. 1999. *Natural Water Quality Problems. Drinking Water Quality, Problem and Solutions*. Monograph. Institute of Chemistry, Novi Sad.

Farrell, D. J. and L. Bower. 2003. Fatal water intoxication. *Journal of Clinical Pathology* 56(10):803–804.

Fenwick, A. 2006. Waterborne diseases—Could they be consigned to history? *Science* 313:1077–1081.

George, I. and P. Crop Servais. 2001. Use of β-D-galactosidase and β-D-glucuronidase activities for quantitative detection of total and faecal coliforms in wastewater. *Canadian Journal of Microbiology* 47:670–675.

Grabow, W. O. K. 1996. Waterborne diseases: Update on water quality assessment and control. *Water SA* 22:193–202.

Ha, B. G., E. J. Shin, J. E. Park, and Y. H. Shon. 2013. Anti-diabetic effect of balanced deep-sea water and its mode of action in highfat diet induced diabetic mice. *Marine Drugs* 11:4193–4212.

Hataguchi, Y., H. Tai, H. Nakajima, and H. Kimata. 2005. Drinking deep-sea water restores mineral imbalance in atopic eczema/dermatitis syndrome. *European Journal of Clinical Nutrition* 59(9):1093–1096.

Held, I. M. and Soden, B. J. 2006. Robust responses of the hydrological cycle to global warming. *Journal of Climate* 19(21):5686–5699.

Hsu, C.-L., Y.-Y. Chang, C.-H. Chiu et al. 2011. Cardiovascular protection of deep-seawater drinking water in high-fat/cholesterol fed hamsters. *Food Chemistry* 127(3):1146–1152.

Hwang, H. S., H. A. Kim, S. H. Lee, and J. W. Yun. 2009. Antiobesity and antidiabetic effects of deep-sea water on ob/ob mice. *Marine Biotechnology* 11(4):531–539.

Hwang, H. S., S. H. Kim, Y. G. Yoo et al. 2009. Inhibitory effect of deep-sea water on differentiation of 3T3-L1 adipocytes, *Marine Biotechnology* 11(2): 161–168.

Itoh, K., T. Kawasaki, and M. Nakamura. 1997. The effects of high oral magnesium supplementation on blood pressure, serum lipids and related variables in apparently healthy Japanese subjects. *British Journal of Nutrition* 78(5):737–750.

Jannasch, H. W. and Jones, G. E. 1959. Bacterial populations in sea water as determined by different methods of enumeration 1. *Limnology and Oceanography* 4(2):128–139.

Jovičić, M. and Jovanović, D. J. 2017. Kvalitet površinskih voda u Srbiji: Medjunarodna konferencija "Kvalitet voda", Zbornik radova, sažetak, str: 89–99.

Kato, K. 1966. Studies on calcium content in sea water: I. Chelatometric determination of calcium in sea water. *Boletim do Instituto Oceanográfico* 15(1):25–28.

Katsuda, S.-I., T. Yasukawa, K. Nakagawa et al. 2008. Deep-sea water improves cardiovascular hemodynamics in Kurosawa and Kusanagi-hypercholesterolemic (KHC) rabbits. *Biological and Pharmaceutical Bulletin* 31(1):38–44.

Kimata, H., H. Tai, and H. Nakajima. 2001. Reduction of allergic skin responses and serum allergen-specific IgE and IgE-inducing cytokines by drinking deep-sea water in patients with allergic rhinitis. *Oto-Rhino-Laryngologia Nova* 11(6):302–303.

Kimura, M., H. Tai, K. Nakagawa et al. 2004. Effect of drinking water without salt made from deep sea water in lipid metabolism of rats, in *Proceedings of the MTS/IEEE TechnoOcean '04: Bridges across the Oceans (Ocean '04)*, pp. 320–321, Kobe, Japan.

Klein, J. and M. Verlaque. 2008. The Caulerpa racemosa invasion: a critical review. *Marine Pollution Bulletin* 56(2):205–225.

Knežević, T. and M. Tanasković. 1997. *Health Aspect of Drinking Water, Preparation of Drinking Water in the World of New Standards and Norms*, Monograph. Institute of Chemistry, Novi Sad, pp. 40–54.

Levitus, S., R. A. Locarnini, and T. P. Boyer et al. 2010. *World Ocean Atlas 2009*.

Malhi, Y. and J. Wright. 2011. Spatial patterns and recent trends in the climate of tropical rainforest regions. *Philosophical Transactions of the Royal Society, London B* 359(1443):311–329.

Marine Biological Laboratory. 2006. Ocean microbe census discovers diverse world of rare bacteria. *ScienceDaily*. https://www.sciencedaily.com/releases/2006/08/060829081744.htm. (accessed March 15, 2021).

Matta, G. (2014). A study on physico-chemical characteristics to assess the pollution status of river Ganga in Uttarakhand. *Journal of Chemical and Pharmaceutical Sciences* 7(3):210–217.

Medema, G. J., P. Payment, A. Dufour et al. 2003. *Assessing Microbial Safety of Drinking Water Improving Approaches and Method. Safe Drinking Water: An Ongoing Challenge.* WHO & OECD, IWA Publishing, London, pp. 11–45.

Miyamura, M., S. Yoshioka, A. Hamada et al. 2004. Difference between deep seawater and surface seawater in the preventive effect of atherosclerosis. *Biological and Pharmaceutical Bulletin* 27(11):1784–1787.

Mohsin, M., S., Safdar, F., Asghar, and F. Jamal 2013. Assessment of drinking water quality and its impact on resident's health in Bahawalpur City. *International Journal of Humanities and Social Science* 3(15):114–128.

Montone, R. C., S. Taniguchi, C. Boian, and R. R. Weber. 2005. PCBs and chlorinated pesticides (DDTs, HCHs and HCB) in the atmosphere of the southwest Atlantic and Antarctic oceans. *Marine Pollution Bulletin* 50(7):778–782.

Mustapha, A. and B. Usman. 2014. Sources and pathway of environmental pollutants into surface water resources: A review, *Journal of Environments, Asian Online Journal Publishing Group* 1(2), 54–59

Ngo, H. H., W. Guo, R. Y. Surampalli, and T. C. Zhang (Eds.). (2016). *Green Technologies for Sustainable Water Management.* American Society of Civil Engineers, Reston, VA.

Olsson, P.-E., P. Kling, and C. Hogstrand. 1998. Mechanisms of heavy metal accumulation and toxicity in fish. In: Langston, W. J. and M. J. Bebianno (Eds.) *Metal Metabolism in Aquatic Environments.* Thomson Science, London, pp. 321–350.

Pilson, M. E. 2012. *An Introduction to the Chemistry of the Sea.* Cambridge University Press, Cambridge.

Pinet, P. R. 1996. *Invitation to Oceanography.* West Publishing Company, St. Paul, MN, pp. 126, 134–135.

Qu, W. Z., and S. G. Deng. 2001. Disastrous ocean pollution of petroleum. *Journal of Natural Disasters* 10(1):69–74.

Rajković, M. B. and Č. Lačnjevac. 2010. *General and Inorganic Chemistry.* Prosveta, Beograd, pp. 75–84.

Roy, C. and J. Tim. 2012. *Marine Geochemistry.* Blackwell Publishing, Hoboken, NJ.

Seas, C., M. Alarcon, J. C. Aragon et al. 2000. Surveillance of bacterial pathogens associated with acute diarrhea in Lima, Peru. *International Journal of Infectious Diseases* 4:96–99.

Shen, J.-L., T.-C. Hsu, Y.-C. Chen et al. 2012. Effects of deep-sea water on cardiac abnormality in high-cholesterol dietary mice. *Journal of Food Biochemistry* 36(1):1–11.

Stemmler, I. and G. Lammel. 2009. Cycling of DDT in the global environment 1950–2002: World Ocean returns the pollutant. *Geophysical Research Letters* 36(24):L24602.

Stone, N., J. L., Shelton, B. E., Haggard, and H. K. Thomforde. 2013. *Interpretation of water Analysis Reports for Fish Culture.* SRAC Publication No. 4606, SRAC, Stoneville, MS, pp. 1–12.

Tada, K., M., Tada, and Y. Maita. 1998. Dissolved free amino acids in coastal seawater using a modified fluorometric method. *Journal of Oceanography* 54(4):313–321.

Talling, J. F. 2010. pH, the CO_2 system and freshwater science. *Freshwater Reviews* 3:133–146.

Trenberth, K. E., L. Smith, T. Qian, A. Dai, and J. Fasullo. 2007. Estimates of the global water budget and its annual cycle using observational and model data. *Journal of Hydrometeorology* 8:758–769.

Uchchariya, D. K. and D. N. Saksena 2012. Study of nutrients and trophic status of Tighra Reservoir, Gwalior (Madhya Pradesh), India. *Journal of Natural Sciences Research* 2(8):97–111.

Ueshima, S., H. Fukao, K. Okada and O. Matsuo. 2003. Suppression of the release of type-1 plasminogen activator inhibitor from human vascular endothelial cells by Hawaii deep sea water. *Pathophysiology* 9(2):103–109.

UNICEF. 1999. *A Water Handbook*, Technical Guidelines Series No. 2. UNICEF, New York.

Verma, P. U., D. K. Chandawat, and H. A. Solanki. 2013. Pollution status of Nikol Lake located in Eastern Ahmedabad. *International Journal of Innovative Research in Science, Engineering and Technology* 2(8):3603–3609.

WHO. 2006. Guidelines for drinking-water quality. http://www.who.int/water_sanitation_health/dwq/gdwqvol32ed.pdf (accessed March 15, 2021).

Xiao, Y. and D. J. Roberts. 2010. A review of anaerobic treatment of saline wastewater. *Environmental Technology* 31(8–9):1025–1043.

Xie, P. and P. A. Arkin. 1997. Global precipitation: A 17-year monthly analysis based on gauge observations, satellite estimates, and numerical model outputs. *Bulletin of the American Meteorological Society* 78(11):2539–2558.

Yu, L. 2007. Global variations in oceanic evaporation (1958–2005): The role of the changing wind speed. *Journal of Climate* 20(21):5376–5390.

Yuan, H., S. Chung, Q. Ma, L. Ye, and G. Piao. 2016. Combination of deep-sea water and *Sesamum indicum* leaf extract prevents high-fat diet-induced obesity through AMPK activation in visceral adipose tissue. *Experimental and Therapeutic Medicine* 11(1):338–344.

Yun J. W. 2010. Possible anti-obesity therapeutics from nature—A review. *Phytochemistry* 71(14–15):1625–1641.

3 Methods for Assessing Fouling and Scaling of Saline Water in Membrane-Based Desalination

Sergio G. Salinas-Rodriguez and M. Nasir Mangal
IHE Delft Institute for Water Education

Loreen O. Villacorte
Grundfos Holding A/S

Almotasembellah Abushaban
Mohammed VI Polytechnic University

CONTENTS

DOI: 10.1201/9781003185437-4

3.1   INTRODUCTION

Nonconventional water resources such as seawater are increasingly relied upon for producing drinking water, for irrigation and for industrial purposes. The world's desalination capacity in 2020 was about 101 million m^3/d of which about 60% corresponded to seawater desalination. About 57% of seawater desalination relied on reverse osmosis (RO) as a desalinating technology and the rest by thermal processes (Global Water Intelligence, 2020). Many RO plants run smoothly. But many others, both old and new, suffer from membrane fouling. Membrane fouling is still the main "Achilles heel" for an overall application of this technology (Schippers et al., 2021).

To prevent and control the occurrence of membrane fouling, pretreatment in RO plants is essential. Additionally, methods and tools can help significantly by monitoring the performance of the pretreatment with regard to fouling control and process optimization. Pretreatment can take place in the form of media filters with or without coagulation, membrane filtration with or without inline coagulation (e.g., ultrafiltration) and dissolved air flotation in combination with the previously mentioned two options. Along with the increase in the number of desalination plants witnessed (reaching a cumulative total of about 21,000 plants in 2020), the capacity of the plants has increased significantly over time. We have seen a growing preference for extra-large (XL) plants (capacity > 50,000 m^3/d). More XL seawater RO (SWRO) plants are expected in the future. This means reliable pretreatment systems and monitoring tools will be essential for these XL plants, as cleaning-in-place (CIP) of membrane modules more than once per year is difficult (Salinas-Rodriguez, 2021).

In the next sections, examples of tools, tests, methods, equipment and approaches for water quality assessment that have been developed and applied in full-scale RO plants are described.

Fouling may result in a variety of problems, such as the need for (frequent) membrane cleaning, the reduction of production capacity and/or plant availability, a higher energy consumption during treatment, a decrease in produced water quality, making RO installations less reliable and finally a frequent replacement of the RO membranes.

The causes of fouling in RO membranes can be classified into five categories (Schippers et al., 2021):

1. Particulate fouling due to suspended and colloidal matter.
2. Inorganic fouling due to iron and manganese.
3. Biofouling due to growth of bacteria.
4. Organic fouling due to organic compounds, e.g., polymers.
5. Scaling due to deposition of sparingly soluble compounds.

In a RO membrane system, fouling and scaling may manifest in three ways:

1. increasing the differential pressure across the feed spacer in spiral wound elements due to a mechanism named "clogging," possibly resulting in membrane telescoping damage;
2. increasing the membrane resistance (decreasing normalized permeability (K_w) or mass transfer coefficient (MTC)) due to deposition and/or adsorption of material on the membrane surface, resulting in higher required feed pressure to maintain capacity; and
3. increasing the normalized salt passage due to concentration polarization in the foul layer, resulting in higher salinity in product water.

3.2   PARTICULATE FOULING

This fouling is caused due to the presence of particles in the RO feed water. Parameters like suspended matter (mg/L), turbidity and particle counting are useful tools in characterizing the fouling potential of water. In particular, turbidity is a useful tool parameter in monitoring suspended matter concentration; unfortunately, low turbidity did not and will not guarantee a low fouling potential (Salinas-Rodriguez et al., 2021).

3.2.1   Silt Density Index (SDI)

The SDI was introduced by DuPont (Permasep Products) at the request of the U.S. Bureau of Reclamation. Initially, the test was called the Fouling Index (FI). It was intended to characterize the fouling potential of feed water to DuPont's hollow fine fiber RO permeators (membrane elements). The target contaminants were suspended and colloidal matter. Later on, manufacturers of spiral wound elements and different hollow fiber elements recommended this test as well and formulated maximum levels for SDI to minimize suspended and colloidal fouling to enable good long-term performance (Schippers et al., 2014). Currently, SDI < 5 has been set as a recommendation for the performance of pretreatment systems for RO and NF, and preferably SDI < 3.

The SDI testing procedure is described in the American Society for Testing and Materials (ASTM). The latest version for SDI testing is from 2014 (code 4189-14). The method describes that the SDI test can be used as an indication of the quantity of *particulate* matter (size greater than 0.45 μm) in water, and it should be used for relatively low (<1.0 NTU) turbidity waters such as well water, filtered water or clarified effluent samples. As the nature of particulate matter in water may vary, the ASTM method indicates that the test is not an absolute measurement of the quantity of particulate matter (ASTM D4189-14, 2014). Furthermore, it is clearly mentioned that the test is not applicable to permeate from RO and UF systems. This recommendation is not always followed in practice as pretreatment systems using membrane filtration are often assessed via the SDI test (Salinas-Rodriguez et al., 2021, Schippers et al., 2014). In some cases, high SDI values were obtained after UF pretreatment that could not be attributed to the "lack of integrity" of the system. In practice, high SDI indicates that fouling might occur, and low SDI does not guarantee that fouling will not occur.

SDI is measured by filtering water through a 47 mm diameter hydrophilic membrane (mixed cellulose nitrate or cellulose acetate) with 0.45 μm pores, in "dead-end" filtration mode, at a constant pressure of 210 kPa (30 psi, 2 bar). A typical scheme for performing an SDI test is illustrated in Figure 3.1.

The ASTM standard provides some guidelines regarding the recommended membranes for the SDI test. The method describes that, for a range of pressures (91. 4–94.7 kPa), the water flow should be around 25–50 seconds/500 mL. Based on this information, the recommended permeability of the filters at 20°C was calculated to be 21,911 L/m^2/h/bar to 45,405 L/m^2/h/bar (Salinas-Rodriguez et al., 2019).

The SDI$_T$ is calculated from the following equation:

$$\text{SDI}_T = \frac{\%\text{PF}}{T} = \frac{\left(1 - \dfrac{t_1}{t_2}\right) \times 100}{T} \tag{3.1}$$

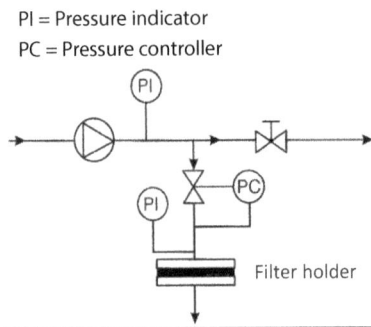

PI = Pressure indicator
PC = Pressure controller

Filter holder

FIGURE 3.1   Scheme of an SDI apparatus (left) and photo of automatic SDI/MFI equipment (right). (Adapted from Salinas-Rodriguez, S.G., *Particulate and Organic Matter Fouling of SWRO Systems: Characterization, Modelling and Applications*, CRC Press/Balkema, Delft, 2011. With permission.)

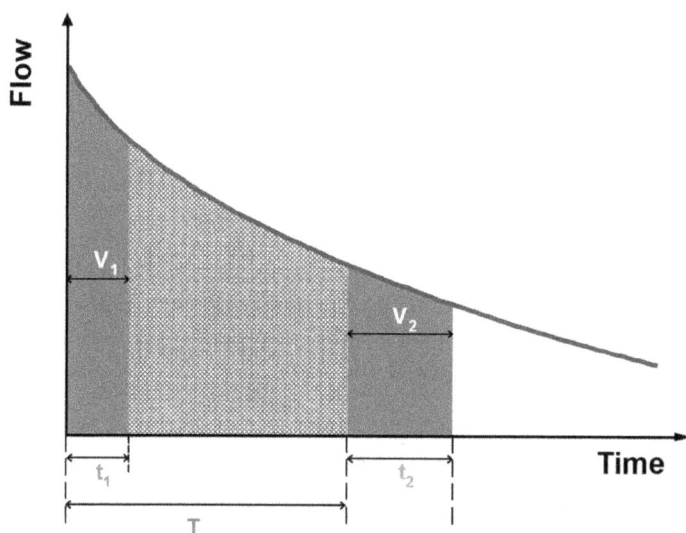

FIGURE 3.2 Schematic representation of a filtration SDI test. (Adapted from Alhadidi, A. et al., *J. Membr. Sci.*, 381, 142–151, 2011a. With permission.)

where t_1 is the filtration time of initial filtered volume (min), t_2 is the filtration time of second filtered volume (min), T is the total filtration time (min) and %PF is the percentage of plugging factor. SDI measures the decline in filtration rate expressed in percentage per minute although it is usually reported without units. The filtration flow over time is illustrated in Figure 3.2.

The limitations of the SDI test are well documented (Salinas-Rodriguez et al., 2019, Rachman et al., 2013, Alhadidi et al., 2011a, b, Nahrstedt and Camargo Schmale, 2008, Schippers and Verdouw, 1980) and include:

- no correction for test water temperature;
- the result is heavily dependent on the test membrane permeability;
- not applicable for testing high fouling feed water, e.g., raw water – ASTM recommends that turbidity should be <1 NTU;
- not applicable for testing UF permeate, which is increasingly being used in desalination pretreatment;
- no linear relation with colloidal/suspended matter;
- fouling potential of particles smaller than 0.45 μm are not measured;
- it is not based on any filtration mechanism.

3.2.2 Modified Fouling Index (MFI)

MFI has been developed to overcome the main deficiencies of the SDI test (Schippers, 1989). The $MFI_{0.45}$ test uses the same equipment and the same 47 mm diameter hydrophilic 0.45 μm membrane (white, mixed cellulose nitrate or cellulose acetate) as the SDI test. It considers that pore blocking occurs initially, followed by cake/gel filtration, and finally, cake/gel blocking and/or enhanced compression occurs. The MFI was adopted by the ASTM as a standard method in 2015 (designation D8002-15) and contrary to the SDI can be applied to measure the particulate fouling potential of ultrafiltration permeate.

The MFI makes use of the general equation describing cake filtration.

$$\frac{t}{V} = \frac{\eta \cdot R_m}{\Delta P \cdot A} + \frac{\eta \cdot I}{2 \cdot \Delta P \cdot A_m^2} \cdot V \tag{3.2}$$

Equation 3.2 gives a straight line when t/V is plotted against V and has been widely applied since suggested by Underwood in 1926 (Underwood, 1926) to test for cake filtration and to obtain information on the permeability of the cake deposited. Carman (1938) defined the gradient of line (b) as:

$$b = \frac{\eta \cdot I}{2 \cdot \Delta P \cdot A_m^2} \tag{3.3}$$

The gradient of the line was adopted by Schippers et al. (1981) to define the modified fouling index (MFI) as an index of the fouling potential of feedwater containing particles, when fixed reference values are used for pressure (ΔP, 2 bar), water viscosity ($\eta_{20^\circ C}$) and (effective) membrane area A_m ($13.8 \times 10^{-4} \, m^2$).

In the MFI (Equation 3.4), the fouling index I is the product of the specific resistance of the cake (α) and the concentration of particles (C_b) in the feedwater and is assumed to be independent of pressure. An advantage of using I is that in most cases it is impossible to determine C_b and α accurately. The fouling index I is a function of the dimension and nature of the particles present in feedwater and directly correlated to their concentration (Schippers and Verdouw, 1980).

$$MFI = \frac{\eta \cdot I}{2 \cdot \Delta P \cdot A_m^2} \tag{3.4}$$

The I value is determined from the stage of cake/gel filtration and is defined as the minimum slope (tan α) in the curve t/V versus V, where t = total filtration time and V = total filtered volume (see Figure 3.3). The MFI can be calculated by replacing the I value in Equation 3.4 and applying the reference values of pressure, membrane area and temperature.

3.2.3 MFI-UF Constant Flux

Considering that particles much smaller than 0.45 µm were responsible for the fouling rate observed in practice, the MFI-UF at constant flux was developed (Salinas-Rodriguez et al., 2011, 2015, Boerlage, 2001). This was supported by the measurement of MFI with membranes of different pore sizes varying from 0.8 µm down to 0.05 µm for RO feedwater which resulted in respective MFI values increasing from 4 to 4,500 s/L² (Schippers et al., 1981). Consequently, the MFI-UF test was developed using a hollow fiber poly-acrylonitrile ultrafiltration membrane with a

t/V (s/L)

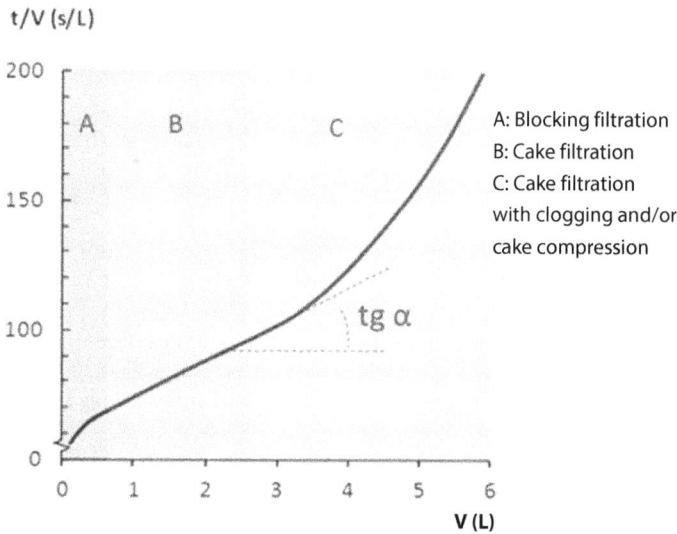

FIGURE 3.3   Filtration curve *t/V* versus *V*. (Adapted from Schippers, J.C., Salinas-Rodriguez, S.G., Boerlage, S.F.E. & Kennedy, M.D., *Why MFI Is Edging SDI as a Fouling Index. Desalination & Water Reuse*. Faversham House with the Cooperation of the International Desalination Association, York, 2014, Schippers, J.C., *Vervuiling van hyperfiltratiemembranen en verstopping van infiltratieputten*, Keuringinstituut voor waterleidingartikelen KIWA N.V, Rijswijk, 1989. With permission.)

13,000 Da molecular weight cut off (PAN 13 kDa) to capture these smaller particles (Boerlage et al., 2000). The pores of the PAN 13 kDa membrane are *circa* 1,000 times smaller than the pores of the existing $MFI_{0.45}$ and can therefore retain smaller particles (Boerlage, 2007).

MFI-UF measured at constant flux has been developed because the index measured at constant pressure is not correct and cannot be used in prediction calculations. The MFI-UF constant flux test was proposed initially using a hollow fiber PAN 13 kDa membrane as the reference membrane (Boerlage et al., 2004) and further developed by Salinas-Rodriguez (2011) using flat 25 mm diameter PES UF membranes with 5, 10, 30 and 100 kDa molecular weight cut off (MWCO) (see Figure 3.4).

The MFI-UF is the most sensitive index to measure the particulate fouling potential of RO feedwater and can be used to predict the rate of fouling of RO membrane systems. 10 kDa UF membranes made of polyether sulfone have been successfully tested and applied in full-scale desalination plants for measuring the MFI-UF of RO feed water. The flux rate at which the test is performed is very relevant as the higher the flux rate, the higher the MFI-UF value. For predicting the rate of RO systems, a flux rate similar to the average flux rate of the RO system is ideal. In addition, the quantity of particles accumulating on the surface of the membrane needs to be quantified by measuring the deposition factor as described in literature (Salinas-Rodriguez et al., 2011, 2015, Schippers, 1989).

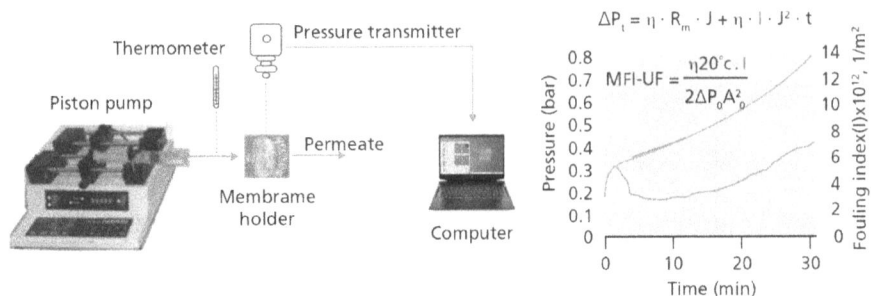

FIGURE 3.4 Filtration set-up to measure MFI-UF at constant flux. Filtration flux can vary between 10 and 300 L/m²/h. (Adapted from Salinas-Rodriguez, S.G., *Particulate and Organic Matter Fouling of SWRO Systems: Characterization, Modelling and Applications*, CRC Press/Balkema, Delft, 2011. With permission.)

3.3   ORGANIC FOULING

3.3.1   MEMBRANE FOULING BY ORGANIC MATTER

Organic matter is one of the major causes of fouling in reverse osmosis (RO) membranes. Accumulation of organics on membrane surface can cause substantial membrane permeability decline. It can also facilitate biofilm development by serving as a conditioning or growth platform for biofilm-forming bacteria (Villacorte et al., 2017a).

In seawater, organic foulants typically originate from three major sources, namely: (i) natural organic matter, (ii) organics introduced by human activities and (iii) organics from chemicals introduced in the treatment processes. Natural organic matter, or NOM, is a collective name for a variety of complex organic molecules in soil and water, produced from the decomposition of plants, animals and microbial materials. It can be found in all natural water sources where animal and plant materials break down. NOM is often the dominant type of organic contaminant in surface water sources.

Human activities can introduce organic foulants into water sources such as uncontrolled or accidental discharge of municipal sewage or wastewater from industries or from vessels navigating the body of water. The most common and perhaps most detrimental of these organic contaminants are oily compounds which can impact both the operation and integrity of the membrane units. Typical RO membranes cannot tolerate more than 0.1 mg/L of oil and grease compounds (DuPont, 2021). Organic foulants can also originate from the treatment processes within the RO plant. These organic sources are typically chemicals introduced to optimize the performance of the pretreatment process such as organic coagulant aids (e.g. cationic polymers) to enhance sedimentation, flotation or filtration processes. Anti-scalants applied to the feedwater to minimize scale formation in the RO may contain organic polymers which can directly, or if combined with other applied chemicals in the process, cause organic fouling issues.

Since organic contaminants can originate from different sources, it is important to identify, monitor and understand their occurrence in the source water and through the pretreatment processes of a seawater RO plant. Measurement and characterization of organic matter are therefore the main focus in organic fouling investigations.

3.3.2 TYPES OF ORGANIC MATTER

Various literature has defined organic matter in a large variety of sizes, functional groups and origins. Therefore, analytical methods can be quite complex which sometimes leads to irreproducible results (Schäfer, 2001). In terms of size, organic matter has been divided into particulate and dissolved (solutes) components. No natural cut-off exists between these two fractions and the distinction is arbitrary, based on the filtration of the sample through 0.45 μm filter. Generally, the dissolved fraction is of greater abundance than the particulate fraction. Overlapping the dissolved and particulate fractions is the colloidal or macromolecular fraction, which consists of suspended solids that are operationally considered solutes. Colloidal organic matter in natural waters typically comprises living and senescent organisms, cellular exudates and partially-to-extensively degraded detrital material (Aiken, 2002). An overview of the size distribution of several key biological and organic components is shown in Figure 3.5.

Wiesner et al. (1992) identified four functional groups of organic matter which can cause fouling in membrane systems: proteins, amino sugars, polysaccharides and humic substances. Polysaccharides are among the most detrimental foulants because of cake and gel layer formation on the membrane. Humic substances may also play a major role in the fouling process: firstly due to gel formation when the solubility limit is exceeded in the concentration polarization layer, and secondly due to adsorption (Matthiasson, 1983).

Currently, organic matter analyses and characterization may involve quantitative and qualitative measurements ranging from simple spectrophotometric to more advanced chromatographic techniques. An overview of the various organic

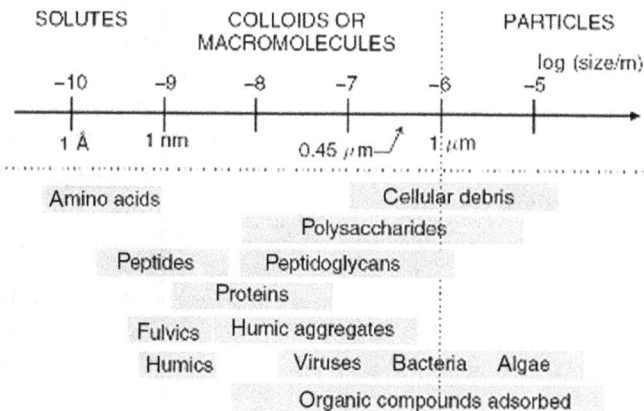

FIGURE 3.5 Size range of various types of environmental colloids and particles (see Balnois et al., 1999).

Total Organic Carbon (TOC)

Dissolved Organic   Particulate Organic
Carbon, DOC (<0.45µm)   Carbon, POC (>0.45µm)

Low molecular weight (LMW) organics   Biopolymers (>> 1 kDa)

TOC/POC

FEEM   Building blocks   Humic substances   Proteins (PT)   Polysaccharides (PS)
              (0.3 - 0.5 kDa)   (0.5 - 1.2 kDa)
LC-OCD-UVD-OND

UV$_{254}$ LMW acids   LMW neutrals   Other   Glyco-   Acidic   Other
                    (<0.35 kDa)   (<0.35 kDa)   PT   proteins   PS   PS
TEP$_{0.4µm}$/TEP$_{10kDa}$

TEP + precursors

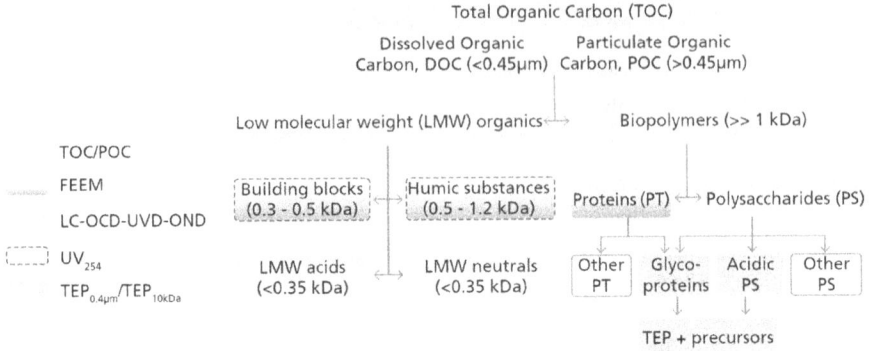

FIGURE 3.6 Overview of organic matter fractions and corresponding analytical techniques for identification and quantification. Legend: LC-OCD-UVD-OND = liquid chromatography with inline detectors for organic carbon, UV absorbance at 254 nm and organic nitrogen; FEEM = fluorescence excitation-emission matrices; TEP = transparent exopolymer particles. (Adapted from Villacorte, L.O., Boerlage, S.F.E. & Dixon, M.B., *Seawater Reverse Osmosis Desalination: Assessment & Pre-treatment of Fouling and Scaling*, IWA Publishing, London, 2021. With permission.)

matter fractions and the corresponding analytical methods applied in practice and membrane fouling research is illustrated in Figure 3.6 and further discussed in the following sections.

3.3.3 Conventional Monitoring Methods

The conventional methods of measuring bulk organic matter in the water are in terms of organic carbon concentration while measurement of a specific fraction is based on UV absorbance.

3.3.3.1 Organic Carbon

Organic carbon is the most common measure to monitor organic matter in the water. Carbon analyzers typically measure both organic and inorganic carbon (e.g. carbon within carbonates) in an aqueous solution. Measuring organic carbon, therefore, requires separating organics from the inorganic fraction. This is typically done in three stages, namely: (i) acidification, (ii) oxidation and (iii) detection and quantification. Current commercial TOC analyzers are already equipped with functionalities that automate all three of these stages. Therefore, analyzing for TOC is as simple as loading the samples and reading the results from the equipment. Moreover, online TOC analyzers where samples can be measured close to real time are already available. All TOC analyzers require proper regular maintenance to maintain precision. Moreover, analysis can be problematic at low concentrations (<100 µg/L), where contamination due to wash water, chemicals, atmosphere and sample vials can be higher than the value of interest.

The commonly used parameters for monitoring organic carbon are total organic carbon (TOC) and dissolved organic carbon (DOC). TOC refers to the organic carbon concentration of the water sample analyzed while DOC is the organic carbon

TABLE 3.1
Typical Concentrations of Organic Carbon in mg/L-C of Various Parts of the Aquatic Environment (Kördel et al., 1997)

Organic Carbon	River	Estuary	Coastal Sea	Open Sea Surface	Open Sea Deep	Sewage
DOC	10–20	1–5	1–5	1–1.5	0.5–0.8	100
POC	5–10	0.5–5	0.01–1.0	0.01–1.0	0.003–0.01	200
TOC	15–30	1–10	1–2.5	1–2.5	0.5–0.8	300

concentration after the sample is prefiltered through a 0.45 μm pore size membrane filter. In some cases, particulate organic carbon (POC) is also determined based on measured TOC and DOC concentrations (see Equation 3.5). Typical ranges of organic carbon concentration of the different fractions in various water sources are shown in Table 3.1.

$$POC = TOC - DOC \qquad (3.5)$$

In practice, TOC/DOC is routinely monitored inline or through grab samples in various locations of an RO plant. It gives a quick overview of the organic load in the water and its fate through the treatment processes without knowing its specific components. Therefore, a high TOC/DOC in the feedwater may not necessarily indicate high organic fouling potential on the RO membrane. Analytical errors are also generally high at low levels so a more advanced or specific technique is needed when higher accuracy is required for the investigation.

3.3.3.2 Ultraviolet (UV) Absorbance
TOC/DOC generally does not give any information about the concentration of humic substances in the water where it is typically the most abundant component. For many years, spectrophotometric measurement of ultraviolet (UV) absorbance through water samples has been used as an indicator of the abundance of humic substances in water treatment applications, including RO desalination. UV absorbance at 254 nm wavelength (UV_{254}) is a useful surrogate measure for DOC particularly in waters where humic substances are the dominant component. A derivative parameter known as the specific UV absorbance (SUVA) is often used to determine the relative abundance of the humic fraction or to determine the aromaticity of organic matter in the sample (Weishaar et al., 2003). This is calculated based on the ratio between UV_{254} absorbance and DOC concentration (Equation 3.6).

$$SUVA = \frac{UV_{254}}{DOC} \qquad (3.6)$$

UV absorbance is a fast and simple method that can be monitored routinely. However, it measures the aromatic compound preferentially and may give inaccurate results if the aromaticity is altered as in the case of most treatments. Non-humic organic components such as protein and polysaccharides are not detected by UV, so a separate method is required for other foulants in the water. Aromatic organic compounds are also considered as less problematic foulants in seawater compared to non-humic materials (e.g. polysaccharides), so monitoring UV_{254} or SUVA may not provide substantial insight into the organic fouling potential of seawater in an RO plant (Amy et al., 2011).

3.3.4 CHARACTERIZATION METHODS

Over the years, numerous analytical techniques have been introduced to characterize organic matter in aquatic systems, including saline water. Assessing the presence of protein-like and humic-like materials based on fluorescence excitation-emission matrices (FEEM) has been employed in various applications (Henderson et al., 2008, Lee et al., 2006). Semi-quantitative techniques such as liquid chromatography – organic carbon detection (LC-OCD) can be used to fractionate organic materials based on size and composition (Huber et al., 2011, Salinas-Rodriguez et al., 2009).

Alcian blue, a dye known to specifically bind with acidic polysaccharides and glycoprotein has been widely used to visualize and measure transparent exopolymer particles (TEPs) in seawater and lake water (Passow, 2002). Some studies used staining with fluorochrome labeled lectins to visualize and identify the carbohydrate components of extracellular polymeric substances from microbes (Villacorte et al., 2015b, Neu, 2000). Lectins are proteins extracted from various organisms which can bind specifically to one or more carbohydrate functional groups. Some functional groups that make up the organic matter can also be identified using Fourier transform infrared (FTIR) spectroscopy (Villacorte et al., 2015b, Lee et al., 2006).

The stickiness of macromolecular organic matter (e.g. biopolymers) can be studied by measuring interaction forces between organic material and other surfaces using an atomic force microscope. This technique had been applied to measure the adhesive strength of bacterial polysaccharides on polymeric membranes (Villacorte et al., 2015b, Yamamura et al., 2008).

The above mentioned characterization methods are typically costly and require laborious steps and a specialized laboratory to perform the analyses. Applications are so far widely reported in academic research and occasionally in practice when the desalination plant operator wanted to more deeply investigate the cause of fouling (e.g. membrane autopsy). The following sections discuss further some of the common characterization techniques.

3.3.4.1 Fluorescence Spectra

Organic matter can be characterized by differences in fluorescence spectra which is associated with different sources of organic compounds present in the water. Basically, a prefiltered (typically through 0.45 µm pore size membrane) water sample is excited by a light source to a specific wavelength at which fluorophores in organic compounds absorb light and subsequently emit the light at a longer wavelength. This is performed using a spectrofluorometer across a spectrum of light wavelengths. The corresponding emission intensities are measured and then plotted in a 3-D excitation-emission matrix (EEM). The signal corresponds to differences in structural characteristics of the dissolved organic matter. The location of EEM peaks provides a qualitative indication of types of organic molecules present in water samples (Westerhoff et al., 2001). For example, humic-like fluorescence peak could be clearly discriminated from protein-like peak in the EEM spectra. Fluorescence peak positioning and shift, peak intensity and peak broadening could be correlated to such structural signatures as molecular sizes or polycondensation and contents of aromatics, phenolics, carboxyl and hydroxyl functional groups (Chen et al., 2003). Typical examples of EEM spectra are illustrated in Figure 3.7 with peak assignments described in Table 3.2.

McKnight et al. (2001) introduced a parameter called the fluorescence index (FI) which is calculated based on the ratio of the fluorescence intensities (see Equation 3.7). An FI of between 1.7 and 2.0 indicates that the fluorescent organic materials are autochthonous (microbial origin) while an FI between 1.3 and 1.4 indicates that they are allochthonous (terrestrial origin).

$$FI = \frac{\left(Em\ 500\ nm,\ Ex\ 450\ nm\right)}{\left(Em\ 500\ nm,\ Ex\ 370\ nm\right)} \tag{3.7}$$

In general, fluorescence spectroscopy has been used as a rapid and sensitive method to characterize dissolved organic matter, but analysis has been limited to qualitative or semiquantitative spectral interpretation due to the broad nature of the fluorescence spectra without fine structure (Ohno and Bro, 2006). The analysis is limited to an organic compound that contains fluorophores. Other organic matter components such as polysaccharides do not fluoresce and could not be analyzed using the EEM spectra. Therefore, FEEM is typically not a standalone method for organic matter characterization.

3.3.4.2 Liquid Chromatography

Chromatography is a physical method of separation in which the components to be separated are distributed between two phases: one of which is stationary (stationary phase) while the other (the mobile phase) moves in a definite direction. Liquid chromatography is a common technique used in the organic matter characterization where the stationary phase is a column filled with adsorbent material while the mobile phase is the water sample. The most recent development involves the application of high sensitivity liquid chromatography – organic carbon detection (LC-OCD) method.

LC-OCD allows the fractionation of various important chemical functionalities of organic matter (Huber et al., 2011). This includes biopolymers, humics, building blocks, low molecular weight organic acids and low molecular weight neutrals (see Table 3.3). A typical LC-OCD chromatogram presents a graphical representation of the major classes of NOM highlighting corresponding signals at different retention times (see Figure 3.8). The LC-OCD technique is quite sensitive and can detect organic carbon concentrations down to ppb level. This also means that the preparation of sample containers and water sampling should be carefully implemented to avoid contamination.

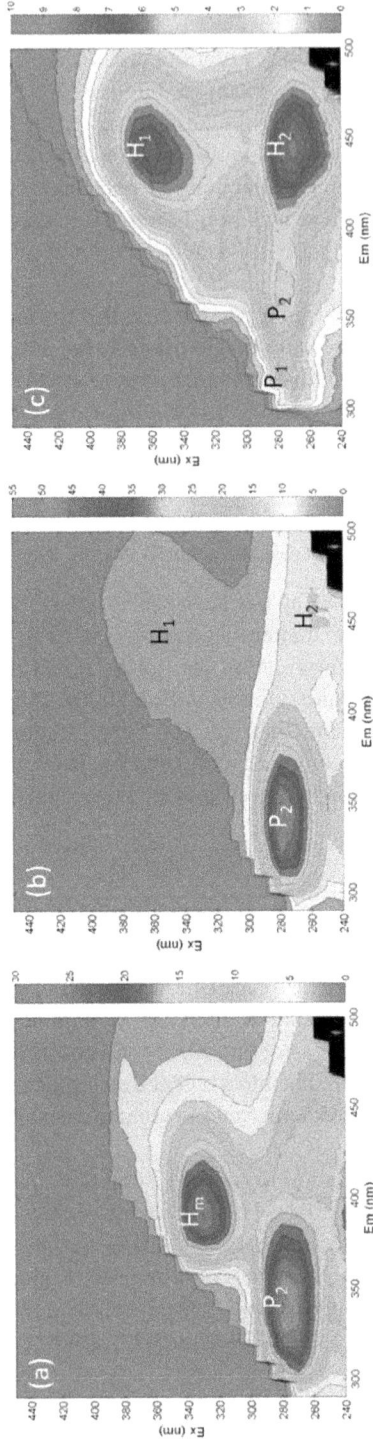

FIGURE 3.7 Typical EEM spectra of algal-derived organic matter from (a) *Alexandrium tamarense*, (b) *Chaetoceros affinis* and (c) *Microcystis* sp. (Adapted from Villacorte, L.O. et al., *Water Res.*, 73, 216–230, 2015b. With permission.)

TABLE 3.2
Typical EEM Peak Locations of Natural Organic Matter (Salinas-Rodriguez et al., 2009, Leenheer and Croué, 2003, Coble, 1996)

Code	Description	Fluorescence Range (nm)	
		Excitation	Emission
H_1	Humic-like primary peak	330–350	420–480
H_2	Humic-like secondary peak	250–260	380–480
H_m	Marine humic-like peak	300–330	380–420
P_1	Protein-like (tyrosine) peak	270–280	300–320
P_2	Protein-like (tryptophan) peak, phenol-like	270–280	320–350

TABLE 3.3
Descriptions of Organic Matter Fractions Measured by LC-OCD (Huber et al., 2011)

Organic Fraction	Typical Size (Da)	Typical Composition
Biopolymers	>20,000	Polysaccharides, proteins, amino sugars, polypeptides
Humic subst.	~1,000	Humic and fulvic acids
Building blocks	300–500	Weathering and oxidation products of humics
LMW neutrals	<350	Mono-oligosaccharides, alcohols, aldehydes, ketones, amino acids
LMW acids	<350	All monoprotic organic acids

This LC-OCD technique has been extensively applied in numerous studies for characterizing natural organic matter in surface water, its fate through the water treatment processes and its role on membrane fouling (Villacorte et al., 2009, 2015c, Salinas-Rodriguez et al., 2009). Further improvement of the technique includes the addition of UV (254 nm) and organic nitrogen detectors (Huber et al., 2011).

In seawater application, the high ionic strength of the water can modify the interaction of the OM with the mobile phase and size exclusion resin. Some of the observed effects include shifting on the elution time of the organic matter fractions (longer elution time) and a negative depression on the baseline at elution times corresponding to acids. Further modifications for the method were introduced specifically for saline water application, which allows an acceptable separation of the different organic fractions with a reported variation coefficient of <12% (Amy et al., 2011).

3.3.4.3 Transparent Exopolymer Particles (TEP)

Aquatic microbes such as algae and bacteria can produce colloidal organic matter known as extracellular polymeric substances (EPS). During algal blooms,

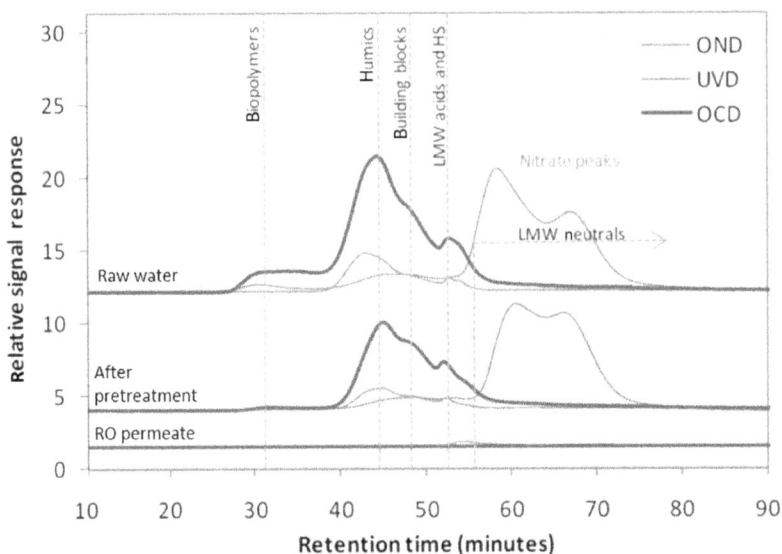

FIGURE 3.8 Typical example of LC-OCD chromatograms of a surface water source and through the treatment processes of an RO plant. (Modified from Villacorte, L.O., *Algal Blooms and Membrane Based Desalination Technology*, CRC Press/Balkema, Leiden, 2014. With permission.)

polysaccharides can comprise more than 80% of EPS production (Myklestad, 1995). These algal-derived polysaccharides have been identified to be hydrophilic, anionic muco-polysaccharides and glycoproteins, collectively known as transparent exopolymer particles (TEP).

Various methods have been introduced to quantify TEP and their precursors. The first-ever TEP method is a direct method based on Alcian blue staining and optical microscopic enumeration (Alldredge et al., 1993). This method provides useful information on the size-frequency distribution of TEP in seawater but is not feasible for quantifying TEPs $< 2\,\mu m$ and their precursors. The succeeding methods based on semiquantitative spectrophotometric techniques were able to address these issues. The method by Passow and Alldredge (1995) also known as $TEP_{0.4\,\mu m}$ is the most widely used. With additional sample preparation techniques (e.g. bubble adsorption, laminar shear), TEP precursors can be measured using such a method (Zhou et al., 1998). Two alternative methods were introduced by Arruda Fatibello et al. (2004) and Thornton et al. (2007), which are capable of measuring both TEP and their precursors without laborious sample pretreatment. However, the former is only applicable in freshwater samples while the latter requires a dialysis step for saline samples. Further modification of the method, known as TEP_{10kDa}, was later introduced to address various practical limitations of these methods through introduction of a concentration step by filtration through 10 kDa membrane (Villacorte et al., 2015a, 2017b). In principle, this method enables size fractionation of TEPs in seawater using a series of membranes with different pore sizes during the extraction step.

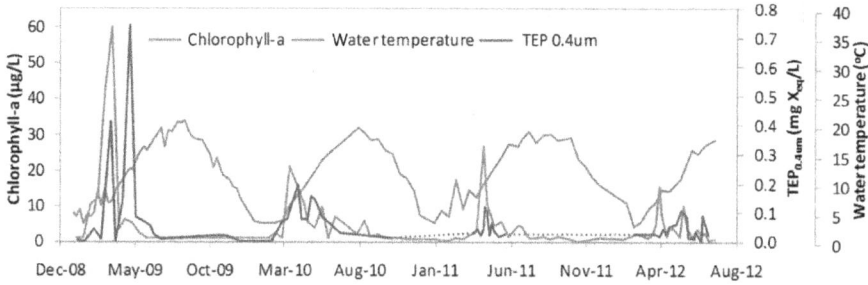

FIGURE 3.9 Variations of $TEP_{0.4\mu m}$, chlorophyll-a and water temperature at the intake water of seawater RO plant. (Adapted from Villacorte, L.O., *Algal Blooms and Membrane Based Desalination Technology*, CRC Press/Balkema, Leiden, 2014. With permission.)

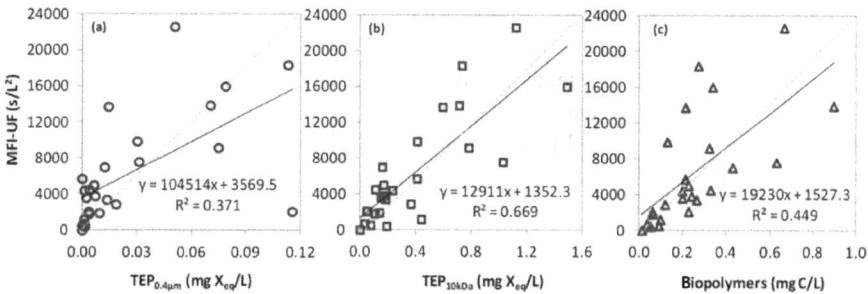

FIGURE 3.10 Correlation between membrane fouling potential base on MFI-UF in relation to (a) $TEP_{0.4\mu m}$, (b) TEP_{10kDa} and (C) biopolymers. (Adopted from Villacorte, L.O., *Algal Blooms and Membrane Based Desalination Technology*, CRC Press/Balkema, Leiden, 2014. With permission.)

Monitoring $TEP_{0.4\mu m}$ in a seawater desalination plant showed that the occurrence of TEP generally coincides with the seasonal algal bloom based on chlorophyll-a concentration Figure 3.9; Villacorte (2014). It also demonstrates that chlorophyll-a concentration is not a reliable indicator of the abundance of TEP, likely due to the fact that some bloom-forming algal species produce more TEP than others (Villacorte et al., 2015b).

TEP can be used as an indicator of the fouling potential of seawater and through the treatment processes. A significant correlation was reported between TEP_{10kDa} and MFI-UF, which is better than the observed correlation between biopolymers (measured by LC-OCD) and MFI-UF Figure 3.10 (Villacorte, 2014). $TEP_{0.4\mu m}$ showed a lower correlation with MFI-UF likely because a significant amount of colloidal TEPs ($<0.4\,\mu m$) were not measured.

More recent developments have shown that there is a potential to monitor TEP online using an auto-imaging technique (Thuy et al., 2017) or a cross-flow filtration unit with an integrated spectrophotometer (Sim et al., 2019). Although still not demonstrated in RO plants, online measurement techniques like these would be the next logical step toward routine TEP monitoring in seawater RO plants, particularly during algal bloom seasons.

3.4 BIOLOGICAL FOULING

3.4.1 INTRODUCTION

In seawater desalination, biofouling is defined as the biofilm accumulation on a membrane surface to such extent that exceeds the threshold of interference and leads to operational problems (Flemming, 1997). Biofilm formation usually starts with microorganisms adhering to surfaces and being ready to utilize any biodegradable dissolved matter from the water phase and converting them into metabolic products and biomass. Microorganisms can multiply to form a thick layer of slime called a biofilm (Flemming, 2011). Biofouling is considered as the "Achilles heel" of membrane processes as all other types of fouling can be removed by pretreatment while microorganisms can multiply even if 99.99% of removal is achieved through pretreatment (Flemming et al., 1997).

In general, biofouling starts in the first RO element in the feed-concentrate channel (Vrouwenvelder et al., 2008, 2011). Therefore, to monitor biofouling in full-scale RO plants, the operating performance data (such as the development of pressure drop) of the first element is monitored. However, this is very unspecific as pressure drop can be caused by other fouling mechanisms as well. To date, there is no standard method to monitor biological fouling in RO membrane systems (Abushaban, 2020, Dhakal et al., 2020). However, several approaches are used to monitor biofouling potential in RO feedwater and biofilm development on RO membrane surfaces.

3.4.2 METHODS FOR MEASURING BIOFOULING POTENTIAL IN SEAWATER

Measuring biofouling potential in the pretreatment and in the SWRO feedwater is an attractive approach because it can be used as an early biofouling detection method when biofouling potential in the SWRO feedwater exceeds a threshold value. Thus, it allows better control of biofouling by implementing corrective actions in the SWRO pre-treatment in time. Several methods were developed to measure biofouling potential in seawater such as assimilable organic carbon (AOC) and bacterial growth potential (BGP). However, these methods are still used as a research tool and have not been automated (Abushaban, 2020).

3.4.2.1 Assimilable Organic Carbon

AOC is a small fraction (0.1%–10%) of dissolved organic carbon, which is utilized by heterotrophic microorganisms for their growth (Wang et al., 2014). AOC comprises various low molecular weight organic molecules, namely sugar, organic acids and amino acids (Van der Kooij, 1992). The AOC concept was initially introduced by Van der Kooij et al. (1982) in drinking water.

Basically, AOC is measured by pasteurizing the sample at 70°C for 30 minutes, inoculating it with *Pseudomonas fluorescens* P17 bacteria, incubating it for 2 weeks and measuring bacterial growth using plate counting. Many studies have been carried out to optimize the incubation period of water samples, broaden the substrate utilization range of the inoculum and accurately monitor bacterial growth.

TABLE 3.4

Assimilable Organic Carbon Methods in Seawater (Weinrich et al., 2011, Jeong et al., 2013, Abushaban et al., 2019a, Quek et al., 2015)

References	Bacterial Inactivation	Culture	Detection Principle	Incubation Temperature
Weinrich et al. (2011)	Pasteurization (70°C for 30 minutes)	*Vibrio fischeri*	Bioluminescence	30°C
Jeong et al. (2013)	Pasteurization (70°C for 30 minutes)	*Vibrio harveyi*	Bioluminescence	25°C
Quek et al. (2015)	–	Indigenous microorganisms	Microbial electrolysis cell biosensor	20°C

In seawater, three methods have been developed to measure AOC in the SWRO pretreatment and feedwater (Table 3.4). Weinrich et al. (2011) and Jeong et al. (2013) developed fast (<day) methods using a single strain of bacteria (*Vibrio fischeri* and *Vibrio harveyi*, respectively) measured by bioluminescence. The advantage of using a single strain is that it allows normalization of the bacterial yield based on a carbon source. However, these two methods may underestimate AOC concentration in seawater as carbon utilization of a single strain is not reflecting the utilization of indigenous microorganisms in seawater. Therefore, Quek et al. (2015) used an indigenous inoculum and measured the bacterial growth of the indigenous microorganisms by microbial electrolysis cell biosensor.

Only a few micrograms of AOC concentration can lead to significant bacterial growth which may result in biofouling problems. The relationship between AOC concentration in the SWRO feed water and biofouling of SWRO membrane systems has not yet been defined. However, some trials have been made. Weinrich et al. (2016) monitored the operating parameters and biofilm formation on a bench-scale SWRO membrane and reported higher fouling on the SWRO membrane surface with 1,000 µg-C/L (as acetate) than with 30 µg-C/L in RO feed water. Moreover, Weinrich et al. (2015) observed an increase in the pressure drop (0.3–0.6 bar) within 4 months of operation of a pilot SWRO plant fed with 50 µg-C/L AOC concentration.

3.4.2.2 Bacterial Growth Potential

Bacterial growth potential (BGP) is a new version of AOC and follows the same concept and protocol of AOC; however, the reported result in BGP is not expressed as an AOC concentration but rather as equivalent carbon (glucose or acetate). This is to avoid the incorrect expression of AOC results as AOC is a mixture of several carbon types, while the calibration line is always established using one type of carbon (acetate or glucose) (Abushaban et al., 2019a).

BGP in seawater can be measured in four steps (Figure 3.11): bacterial inactivation at 70°C for 30 minutes, bacterial inoculation with 10,000 cells/mL (intact cell concentration measured by flow cytometry) using an indigenous microbial consortium from the raw seawater, incubation at 30°C and bacterial growth detection (Abushaban et al., 2021). Bacterial inactivation and inoculation are needed as the

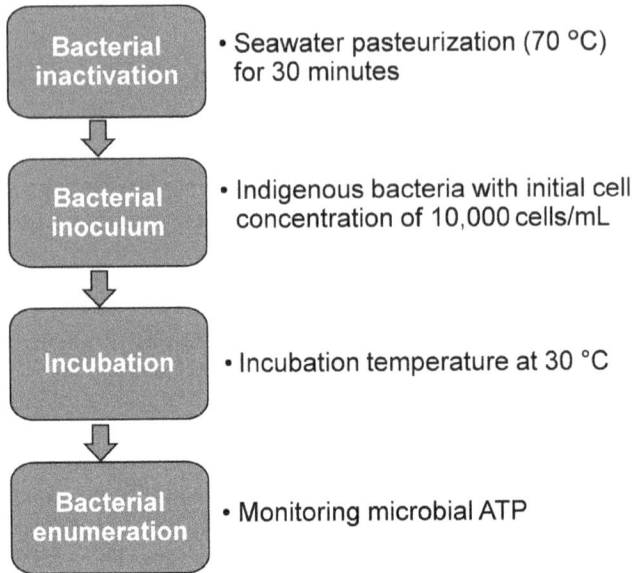

FIGURE 3.11   The procedures for measuring bacterial growth potential in seawater (see Abushaban et al., 2019a).

TABLE 3.5

Growth Potential Methods in Seawater (Dixon et al., 2012, Farhat et al., 2018, Abushaban et al., 2018, 2019b, Dhakal, 2017)

References	Bacterial Inactivation	Culture	Detection Principle	Incubation Temperature	Expressed Results
Dixon et al. (2012)	Filtration (0.2 μm)	Indigenous microorganisms	Turbidity	Not available	μg-C as acetate equivalent
Abushaban et al. (2017)	Pasteurization (70°C for 30 minutes)	Indigenous microorganisms	Microbial ATP	30°C	μg-C as glucose equivalent
Dhakal (2017)	Filtration (0.22 μm)	Indigenous microorganisms	Intact cell counts by FCM	30°C	μg-C as glucose equivalent
Farhat et al. (2018)	Filtration (0.2 μm)	Indigenous microorganisms	Total ATP and total cell count by FCM	30°C	μg-C as acetate equivalent

microbial population during SWRO pretreatment is not constant in terms of number and composition. Thus, microbial inactivation allows the standardization of the initial microbial population by adding a constant inoculum concentration.

Four BGP methods were developed recently using different detection principles including turbidity, microbial adenosine triphosphate (ATP) and cell count by flow cytometry (FCM) (Table 3.5). Abushaban et al. (2019a) reported that BGP in the

North Sea varied between 45 µg-C/L as glucose in the winter and 385 µg-C/L as glucose in the spring. The four mentioned BGP methods in Table 3.5 have been employed to assess the pretreatment and SWRO performance of different full-scale desalination plants. The removal of BGP through the pretreatment ranged between 40% and 80% in six SWRO plants located in the Middle East and Australia. BGP in the SWRO feedwater ranged between 70 and 200 µg-C/L (as glucose equivalent) (Abushaban et al., 2019a, b). Abushaban et al. (2020) tried to investigate the correlation between BGP in the SWRO feedwater and RO membrane operating parameters such as pressure differential and membrane permeability and reported that BGP measured in the SWRO feedwater (from 100 to 950 µg-C/L) led to an increase in the normalized pressure drop within 3 months.

The relationship between BGP in the SWRO feedwater and the cleaning in place (CIP) frequency of SWRO membrane systems was preliminarily explored based on the results of four desalination plants. It was estimated that BGP of 100 µg-C/L in the SWRO feedwater requires one per year CIP frequency. Accordingly, a safe level of BGP (<70 µg-C/L as glucose) was tentatively proposed (Abushaban, 2020). However, many more SWRO plants still need to be monitored at different locations to establish a robust correlation and threshold level.

3.4.2.3   Orthophosphate

Phosphate is one of the main nutrients used by microorganisms for their growth. It is considered to be a limiting nutrient and therefore eliminating phosphate concentration in the SWRO pretreatment could be directly linked to minor SWRO biofouling. There are several types of phosphate. However, orthophosphates (such as H_3PO_4, $H_2PO_4^-$, HPO_4^{2-} and PO_4^{3-}) are the most utilized fraction by microorganisms because they are biodegradable fraction (Maher and Woo, 1998). Phosphate, in general, and orthophosphate are not frequently measured through SWRO pretreatment processes due to the lack of methods that have a low limit of detection as orthophosphate is present in very low concentration in seawater. It has been reported that the maximum phosphate concentration measured in the Atlantic and Pacific oceans are 240 and 310 µg P/L, respectively (Rimmelin and Moutin, 2005). However, typical concentrations in seawater are usually below 33 µg P/L (Jacobson et al., 2009).

Orthophosphate was recently measured in seawater at the facility of Rijkswaterstaat (Lelystad, Netherlands) with a limit of detection of 0.3 µg/L (Abushaban et al., 2020). In short, molybdate reagent and ascorbic acid were added to seawater at a temperature of 37°C. The added molybdate and the orthophosphate present in seawater formed a phosphor-molybdate complex in the acidic environment after reduction with ascorbic acid and in the presence of antimony. This gave a blue-colored complex, which was measured at 880 nm using a 50 mm cuvette and a spectrophotometer. Abushaban et al. (2020) measured orthophosphate concentration along the pretreatment of a full-scale SWRO desalination plant and reported a concentration of 11 µg P/L in the seawater intake. High removal of orthophosphate (70%) was noticed in the dissolved air flotation (DAF) combined with 1–5 mg-Fe^{3+}/L coagulation. Orthophosphate concentration was lowered to 0.6 µg P/L through the pretreatment. However, antiscalant dosing increased orthophosphate concentration by 150%.

3.4.3 Tools for Measuring the Development of Biofilm

3.4.3.1 Membrane Fouling Simulator

The membrane fouling simulator (MFS) is a tool (Figure 3.12) that uses the same materials as spiral wound RO and NF membranes to validate membrane fouling. It has been demonstrated that biofouling development in both MFS units and spiral wound membrane modules is similar in terms of pressure drop and biofilm accumulation (Vrouwenvelder et al., 2009). The MFS unit has been used for characterizing the fouling potential of feedwater, comparing different pretreatment, evaluating fouling potential of different chemicals and testing newly developed membranes.

One of the advantages of MFS is that fouling can be monitored in the following three different ways: (i) the development of operational parameters such as pressure drop, (ii) nondestructive (visual and microscopic) observations using the glass window and (iii) analysis of coupons sampled from the membrane sheet in the MFS (Vrouwenvelder et al., 2006).

Several types of MFS units have been developed for different purposes. However, permeate production was not possible in the cell due to the fact that the maximum pressure on the MFS is 5 bar. To overcome this limitation, a new long channel membrane test cell was developed recently with 1 m long and five sections over the length. The new unit enables measuring permeate flux and salt passage over the test cell length (Kim et al., 2018, Siebdrath et al., 2017). However, it has been tested up to 15 bar. It should be noted that MFS units still cannot be used to simulate SWRO as high pressure is required and there is, therefore, a need for the development of a high pressure (up to 80 bar) simulator to represent membrane modules used for seawater desalination.

3.4.3.2 Membrane Biofilm Formation Rate

The mBFR is a tool used to evaluate the biofilm formation rate on an RO membrane surface including attachment of microorganisms and microbial growth (Kurihara et al., 2001). Seawater continuously flows into columns equipped with O-rings covered with the RO membrane, as shown in Figure 3.13. Over time, biofilm is collected

FIGURE 3.12 The membrane fouling simulator (MFS) with external dimensions of $0.07 \times 0.30 \times 0.04$ m and a transparent window (see Vrouwenvelder et al., 2006).

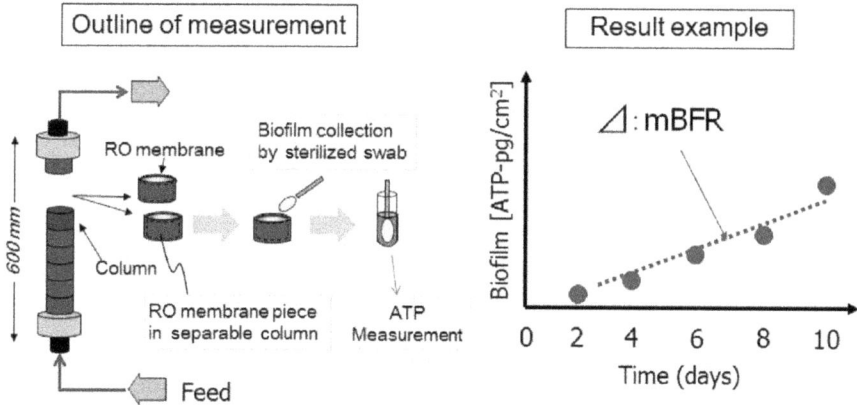

FIGURE 3.13 The schematic and procedure for measuring membrane biofilm formation rate (mBFR) (see Kurihara, 2020).

by swabbing and is suspended in 1 mL of distilled water (Ito et al., 2013). ATP concentration is measured to quantify the microbial content, and mBFR values are calculated based on the slope of the linear relationships between biomass (ATP-pg/cm^2) and time.

A correlation study between mBFR and the pressure differential was performed at a pilot-scale plant with and without chlorination (Kurihara, 2020), in which similar trends were reported. When chlorination/dichlorination was not applied, the pressure differential was more or less constant while mBFR was very low. However, when chlorination/dichlorination was applied, mBFR increased more than ten times and the pressure differential increased significantly. In addition, the relationship between mBFR and the chemical cleaning interval was investigated by Kurihara and Ito in six SWRO plants and reported that once or twice per year chemical cleaning is needed when the mBFR value is less than 10 pg/cm^2/d.

3.4.4 FUTURE OF BIOFOULING MONITORING

To monitor biofouling in full-scale SWRO desalination plants, measuring pressure drop and permeability of the first stage is a commonly applied practice. However, it cannot be used to mitigate biofouling but rather to monitor the development of biofouling. The methods that are discussed earlier in this chapter are still lab-based methods and cannot be used to monitor biofouling online.

Having an online device to monitor biofouling would be beneficial to control biofouling in an SWRO membrane system. Recently, online sensors were developed to monitor biofilm formation on RO membranes including electrical potential measurements (Sung et al., 2003), biosensors (Lee and Kim, 2011), ultrasonic time-domain reflectometry (Kujundzic et al., 2007) and online fluorimeter (Ho et al., 2004). However, the major challenges of these sensors are the fouling on the sensors over a long period of use, the need for frequent calibration of the sensors, the high pressure applied in an SWRO system and the use of spiral wound modules (Nguyen et al., 2012).

Therefore, to avoid these limitations, measuring biofouling potential methods (such as AOC and BGP) through SWRO pretreatment and in SWRO feedwater is a very attractive option. Moreover, it can be used as an early warning system, allowing adjustment of the operational conditions of the pretreatment processes to meet the required quality in RO feedwater and consequently resulting in better control of biofouling in RO systems.

3.5 SCALING POTENTIAL

3.5.1 MEMBRANE SCALING

Scaling is the precipitation (deposition) of the sparingly soluble salts on the membrane surface, resulting in (Kucera, 2010):

- Lowering the permeate production, due to a decrease in membrane permeability
- Increasing operational costs, due to higher operating pressure, cleaning costs, etc.
- Deteriorating the permeate water quality, due to increasing salt passage

Membrane scaling is a challenging problem both in seawater and brackish water RO applications. However, in treating brackish water, scaling is the key barrier for operating the RO at high recovery rates, which leads to:

- High-specific (electrical) energy consumption (kWh/m^3)
- More water abstraction and production of concentrate (waste)

Various types of scaling species such as calcium carbonate, calcium sulfate, barium sulfate, calcium phosphate, calcium fluoride, strontium sulfate and silica may precipitate in RO processes depending on the feedwater composition (Antony et al., 2011). As scaling is a concentration phenomenon, it starts in the last stage (tail elements) where the concentration of sparingly soluble salts is the highest. The high concentration of sparingly soluble salts exceeds their solubility limits, which as a result triggers crystallization.

3.5.1.1 Solubility, Solubility Product and Super Saturation

Solubility is defined as the maximum amount of a substance that dissolves in a given volume of solvent (e.g. water) under certain conditions (e.g. temperature) (Mittal, 2017). It is a chemical property that is influenced by various parameters such as temperature, pressure, pH, ion strength, etc. For instance, with an increase in temperature, solubility of some salts (e.g., $BaSO_4$) increases, while of some others (e.g. $CaCO_3$, $CaSO_4$) decreases. Solubility of salts in water is commonly expressed as mol of salt per liter of water (mol/L), gram of salt per liter or mL of water (g/L or g/mL) and gram of salt per gram of water (g/g), etc.

Solubility product (K_{SP}) is another term which is used to get information about the solubility of sparingly soluble salts. K_{SP} is the equilibrium constant of salts which represents the level of a salt that may dissociate from its ionic species. Salts with a low water solubility have low K_{SP} values, and vice versa. It is calculated as the mathematical multiplication of the molar concentrations of the dissociated ions raised to the power of their stoichiometric coefficients as described in Equation 3.9 for a dissolution reaction of Equation 3.8.

$$A_m B_{n(S)} \leftrightarrow mA_{(aq)} + nB_{(aq)} \tag{3.8}$$

$$K_{SP} = [A]^m [B]^n \tag{3.9}$$

Due to the fact that K_{SP} is temperature dependent, the value should always be accompanied by the temperature at which it was determined. In Table 3.6, K_{SP} values of some common scaling species at 25°C are presented.

Supersaturation is known as the driving force for the initiation of crystallization. Supersaturation develops when the concentrations of inorganic ions for a given scaling species exceed the equilibrium concentration or the solubility limit. In other words, a solution is referred to as supersaturated with respect to a given salt when the ion product (*IP*) of the salt exceeds the K_{SP}.

Based on the concentrations of scaling salt present in water, a water solution can be categorized as:

- **Saturated:** water is in equilibrium with a salt; no more can dissolve.
- **Under saturated:** water can dissolve more salt than present in the water.
- **Super saturated:** water contains more salts than can dissolve; precipitation may occur.

Theoretically, compounds will precipitate when the solubility is exceeded. However, it should be remembered that the crystallization process not only involves the supersaturated conditions but also depends on the precipitation kinetics (nucleation and crystal growth) (Koutsoukos, 2010).

TABLE 3.6
K_{SP} Values of Some Common Scaling Species at 25°C

Compound	Formula	K_{SP}
Calcium carbonate	$CaCO_3$	3.36×10^{-9}
Calcium sulfate	$CaSO_4$	9.1×10^{-6}
Barium sulfate	$BaSO_4$	1.1×10^{-10}
Calcium phosphate	$Ca_3(PO_4)_2$	2.0×10^{-29}
Calcium fluoride	CaF_2	5.3×10^{-9}
Strontium sulfate	$SrSO_4$	3.2×10^{-7}

3.5.2   METHODS TO ASSESS THE OCCURRENCE OF SCALING

3.5.2.1   Scaling Indices

There are a number of indices available to measure the scaling tendency of the sparingly soluble salts in a water solution. The most commonly used in RO applications
are:

- Saturation index (SI)
- Supersaturation ratio (Sr)

For $CaCO_3$ scaling, the following indices are also used:

- Langelier saturation index (LSI)
- Stiff-Davis stability index (S&DSI)

3.5.2.1.1   Saturation Index (SI)

SI is the logarithmic ratio between the ion activity product (IAP) and the thermodynamic solubility product (K_{sp}) of a sparingly soluble salt in water. For instance, when
$CaCO_3$ is the scaling species, SI can be calculated according to Equation 3.10.

$$\text{SI} = \log \frac{\text{IAP}}{K_{sp}} = \log \frac{\gamma_{Ca}\left[Ca^{2+}\right]\gamma_{CO_3}\left[CO_3^{2-}\right]}{K_{sp} \text{ of } CaCO_3} \tag{3.10}$$

For a water solution, when:

- SI = 0 the solution is just saturated or is in equilibrium.
- SI > 0 the solution is supersaturated, precipitation may occur.
- SI < 0 the solution is under saturated, more salt can be dissolved.

In Equation 3.10, γ represents the activity coefficient which is used to determine the
effective concentration of ions in a solution. The activity coefficient is dependent on
the ionic strength, as it decreases with an increase in the ionic strength. The γ can be
calculated by Equation 3.11.

$$\log \gamma = -\frac{0.5 \times Z_i^2 \times \sqrt{I}}{1 + \sqrt{I}} \tag{3.11}$$

where Z_i = charge (oxidation number) of ion i, I = ionic strength.
    The term "ionic strength" is defined as the total concentration of ions in a solution
and is calculated by Equation 3.12.

$$I = \frac{1}{2}\sum_i Z_i^2 \times C_i \tag{3.12}$$

where C_i = molar concentration of ion i.
    An empirical formula (Equation 3.13) is also sometimes used to roughly calculate
the ionic strength from the total dissolved solids (TDS) concentration.

$$I\left(\frac{\text{mol}}{\text{L}}\right) \approx 2.5 \times 10^{-5} \times \text{TDS}\left(\frac{\text{mg}}{\text{L}}\right) \tag{3.13}$$

3.5.2.1.2   Super Saturation Ratio (S_r)

S_r is the square root of the ratio between the ion activity product (IAP) and the thermodynamic solubility product (K_{sp}) of a sparingly soluble salt in water. For instance, when $CaCO_3$ is the scaling species, S_r can be calculated according to Equation 3.14.

$$S_r = \sqrt{\frac{\gamma_{Ca}\left[Ca^{2+}\right]\gamma_{CO_3}\left[CO_3^{2-}\right]}{K_{sp} \text{ of } CaCO_3}} \tag{3.14}$$

For a water solution, when:

- $S_r = 0$ the solution is just saturated or is in equilibrium.
- $S_r > 1$ the solution is supersaturated, precipitation may occur.
- $S_r < 1$ the solution is under saturated, more salt can be dissolved.

3.5.2.1.3   Langelier Saturation Index (LSI)

LSI is the most common method used for assessing the feedwater potential for calcium carbonate scaling in RO applications, and it is derived from a theoretical concept of saturation (Sheikholeslami, 2005). According to the ASTM method, LSI is applicable for water with total dissolved concentration (TDS) up to 10,000 mg/L (Singh, 2014). It is calculated by Equation 3.15.

$$\text{LSI} = \text{pH} - \text{pH}_S \tag{3.15}$$

where pH is the measured water pH and pH_s is the pH at saturation in calcium carbonate and is calculated by Equation 3.16.

$$\text{pH}_S = (9.3 + A + B) - (C + D) \tag{3.16}$$

where

$$A = \frac{(\log_{10}\text{TDS} - 1)}{10}$$

$$B = -13.12 \times \log_{10}(°C + 273) + 34.55$$

$$C = \log_{10}(Ca^{2+} \text{ as } CaCO_3) - 0.4$$

$$D = \log_{10}(\text{alkalinity as } CaCO_3)$$

For a water solution, when:

- LSI $= 0$ the solution is just saturated with $CaCO_3$.
- LSI > 0 the solution is supersaturated, $CaCO_3$ precipitation may occur.
- LSI < 0 the solution is under saturated, more $CaCO_3$ salt can be dissolved.

3.5.2.1.4   Stiff-Davis Stability Index (S&DSI)

The S&DSI is used to assess the scaling potential of calcium carbonate scaling for high saline water (TDS > 10,000 mg/L) (Singh, 2014). This method, similar to LSI, is based on the actual pH of the water solution and pH of the water solution saturated with respect to $CaCO_3$.

For a water solution, when:

- S&DSI = 0 the solution is just saturated with $CaCO_3$.
- S&DSI > 0 the solution is supersaturated, $CaCO_3$ precipitation may occur.
- S&DSI < 0 the solution is under saturated, more $CaCO_3$ salt can be dissolved.

3.5.2.2   Commercial Software Programs

A number of commercial programs are available which can be used to predict the scaling potential in RO. Most of these programs are developed by antiscalant suppliers and membrane manufacturers. The programs are:

- Genesys Membrane Master (MM5) – Genesys International
- Sokalan RO-Xpert-BASF
- Hyd-RO-dose – French Creek Software
- Flodose – BWA Water Additives
- Argo Analyzer (Winflow) – Suez
- Avista Advisor – Avista Technologies Inc.
- WAVE – DOW membrane projection software
- IMSDesign (Integrated Membrane Solutions Design) – Hydranautics membrane projection software

3.5.3   Methods to Monitor Scaling in RO Processes

3.5.3.1   Parameters for Monitoring Scaling

In practice, some indicators such as permeate flow, pressure drop, salt passage, etc. are used to monitor scaling (fouling) in RO processes. It is essential to normalize these parameters to a reference condition since they are also affected by other factors such as temperature, flow, etc. Normalizing the RO data helps operators to decide whether the observed variations in flow, pressure drop and salt rejection are due to scaling (fouling) or are due to different operating conditions. The equations to normalize RO data are given by the membrane manufacturers. The normalization equations provided in this chapter are based on the membrane manufacturer Hydranautics. It is worth mentioning that for the normalization, commercial programs are also available from the membrane manufacturers (e.g., RO DataXL provided by Hydranautics).

Normalized permeate flow is a useful parameter for monitoring scaling (fouling) in the RO unit. A decrease in the normalized permeate flow of the last element/stage can be an indication of scaling. As the permeate flow is related to the NDP and temperature, any increase/decrease in the permeate flow due to variations in NDP and temperature should be factored out. The normalized permeate flow can be calculated using Equation 3.17.

$$Q_N = Q_i \times \frac{\text{NDP}_r}{\text{NDP}_t} \times \frac{\text{TCF}_r}{\text{TCF}_t} \qquad (3.17)$$

where

Q_N = Normalized permeate flow at time t (m³/h);
Q_i = Actual permeate flow at time t (m³/h);
NDP_r = NDP at reference point (bar);
NDP_t = NDP at time t (bar);
TCF_r = Temperature correction factor at reference conditions;
TCF_t = Temperature correction factor at time t.

NDP is the actual net driving pressure in producing permeate water by passing the saline/concentrate water through the RO membrane. It is calculated as the average of the feed and concentrate pressure minus the osmotic pressure and the permeate pressure as shown in Equation 3.18.

$$\text{NDP} = P_f - \frac{\Delta P_{fc}}{2} - P_p - \pi_{fc} + \pi_p \qquad (3.18)$$

where

P_f = feed pressure (bar);
ΔP_{fc} = pressure drop (bar);
P_p = permeate pressure (bar);
π_{fc} = feed-concentrate osmotic pressure (bar);
π_p = permeate osmotic pressure (bar).

The feed-concentrate and permeate osmotic pressure can be calculated using Equations 3.19 and 3.20.

$$\pi_{fc} = 0.002654 \times C_{fc} \times \frac{(T+273)}{\left(1{,}000 - \dfrac{C_{fc}}{1{,}000}\right)} \qquad (3.19)$$

$$\pi_p = 0.002654 \times C_p \times \frac{(T+273)}{\left(1{,}000 - \dfrac{C_p}{1{,}000}\right)} \qquad (3.20)$$

$$C_{fc} = \frac{C_f + C_c}{2} \qquad (3.21)$$

where

C_f = TDS of the feed (mg/L);
C_c = TDS of the concentrate (mg/L);
C_p = TDS of the permeate (mg/L);
T = Temperature (°C)

TCF is calculated in Equation 3.22.

$$\text{TCF} = e^{2,700 \times \left(\frac{1}{298} - \frac{1}{273 + T}\right)} \tag{3.22}$$

Normalized pressure drop (ΔP) is monitored in the last element and/or last stage which is the difference between the feed pressure entering the last stage/last element and the concentrate pressure leaving the last element. When the water temperature and flows are constant, ΔP should be constant unless something deposits on the membrane surface/feed spacer and hence causes blockage in the passage of the flow. Therefore, any increase in the ΔP can be attributed to the occurrence of scaling. It is worth mentioning that ΔP may not be much help in case of an amorphous scale with a very thin layer on the membrane surface since the layer may not cause significant blockage for the passage of the flow. In this case, normalized permeate flow may decline considerably, but the increase in ΔP may not be noticeable.

$$\Delta P = \Delta P_t \times \left(\frac{Q_{fc,r}}{Q_{fc,t}}\right)^{1.4} \times \left(1 + \frac{T - 25}{100}\right) \tag{3.23}$$

where

ΔP_t = actual pressure drop at time t (bar)
$Q_{fc,r}$ = average reference feed/concentrate flow (m³/h)
$Q_{fc,t}$ = average feed/concentrate flow at time t (m³/h)

Salt passage is also used as a parameter to monitor the occurrence of scaling. At fixed recovery, when the feed conductivity and temperature are constant, an increase in salt passage could be attributed to scaling due to a concentration polarization effect. The salt passage should also be normalized, for instance, to factor out any increase in salt passage due to increase in feed temperature. Based on hydranautics, salt passage normalization can be done using Equation 3.24.

$$\% \, SP_N = \% \, SP_t \times \frac{Q_{p,t}}{Q_{p,r}} \times \frac{\text{TCF}_r}{\text{TCF}_t} \tag{3.24}$$

where

$\% \, SP_N$ = Normalized salt passage in percentage to standard conditions;
$\% \, SP_t$ = Actual salt passage at time t in percentage;
$Q_{p,r}$ = Permeate flow of the element at standard conditions;
$Q_{p,t}$ = Permeate flow of the element at actual conditions at time t.

3.5.3.2 Monitoring Systems

There are a number of monitoring devices available which can be used to continuously monitor scaling in RO applications. These monitoring systems are installed on the concentrate stream of the last stage in RO applications that provide an additional recovery to the overall recovery of the RO facility. Due to the provision of additional recovery, it is expected that scaling would generally occur first within the monitoring device before the actual membranes of the RO. These monitoring devices are:

- External scale guard (scale monitor)
- Internal scale guard (scale monitor)

External scale guard as shown in Figure 3.14 can be fed with the concentrate of the last stage of pilot or full-scale RO application. Scaling guard can be operated at additional recovery in the range of 1%–4%. Scaling is monitored by observed decrease in the normalized permeate flow (or membrane permeability) of the scale guard.

Internal scale guard is the last element in the last stage of RO for which the permeate flow and the ΔP across the element are measured. Permeate of the internal scale guard is collected separately from the permeate outlet (located on the concentrate side of the pressure vessel) and permeate of the other membrane elements in the pressure vessel is collected from the permeate outlet which is located on the feed side as illustrated in Figure 3.15. Internal scale guards are installed in several full-scale plants in the Netherlands. The increase in differential pressure and decrease in permeate flow is attributed to scaling. In case, the internal scale monitor suffers of

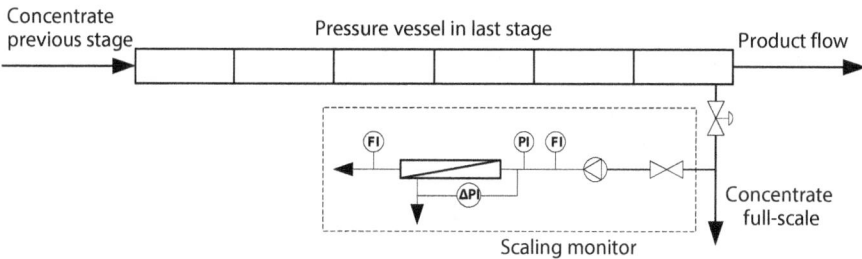

FIGURE 3.14 Schematic of the external scale guard. (Adopted from van de Lisdonk, C.A.C., van Paassen, J.A.M. & Schippers, J.C., *Desalination*, 132, 101–108, 2000. With permission.)

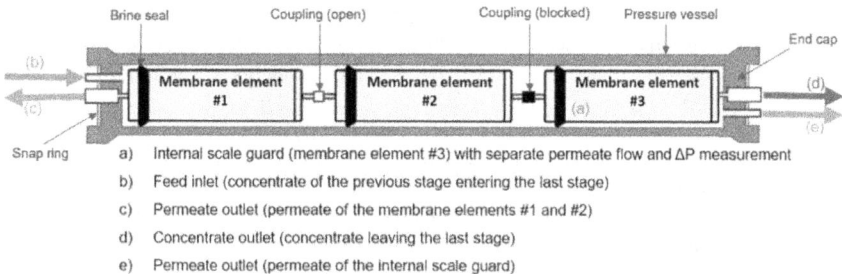

a) Internal scale guard (membrane element #3) with separate permeate flow and ΔP measurement
b) Feed inlet (concentrate of the previous stage entering the last stage)
c) Permeate outlet (permeate of the membrane elements #1 and #2)
d) Concentrate outlet (concentrate leaving the last stage)
e) Permeate outlet (permeate of the internal scale guard)

FIGURE 3.15 Schematic of the internal scale monitor in the last element of the last stage in RO systems. (Adapted from Mangal, N. et al., *Seawater Reverse Osmosis Desalination: Assessment & Pre-treatment of Fouling and Scaling*, IWA Publishing, London, (2021. With permission.)

scaling, it can affect the operation of the RO unit which is considered as the main drawback of this type of scale guard.

3.6 SUMMARY

The tools, methods and tests presented in this chapter can enable engineers, plant operators and scientists not only to design better plants but also to improve operation and monitoring of biological and particulate fouling in SWRO systems.

As we anticipate progress with technical challenges, the prospects are that seawater, brackish water, and waste water effluent desalination can continue to expand its contribution to meeting the global demand for safe and clean water.

REFERENCES

Abushaban, A. (2020). *Assessing Bacterial Growth Potential in Seawater Reverse Osmosis Pretreatment: Method Development and Applications.* CRC Press, Boca Raton, FL.

Abushaban, A., Mangal, M. N., Salinas-Rodriguez, S. G., Nnebuo, C., Mondal, S., Goueli, S. A., Schippers, J. C. & Kennedy, M. D. (2018). Direct measurement of ATP in seawater and application of ATP to monitor bacterial growth potential in SWRO pre-treatment systems. *Desalination and Water Treatment, 99,* 91–101.

Abushaban, A., Salinas-Rodriguez, S. G., Dhakal, N., Schippers, J. C. & Kennedy, M. D. (2019a). Assessing pretreatment and seawater reverse osmosis performance using an ATP-based bacterial growth potential method. *Desalination, 467,* 210–218.

Abushaban, A., Salinas-Rodriguez, S. G., Kapala, M., Pastorelli, D., Schippers, J. C., Mondal, S., Goueli, S. & Kennedy, M. D. (2020). Monitoring biofouling potential using ATP-Based bacterial growth potential in SWRO pre-treatment of a full-scale plant. *Membranes,* 10, 360.

Abushaban, A., Salinas-Rodriguez, S. G., Mangal, M. N., Mondal, S., Goueli, S. A., Knezev, A., Vrouwenvelder, J. S., Schippers, J. C. & Kennedy, M. D. (2019b). ATP measurement in seawater reverse osmosis systems: Eliminating seawater matrix effects using a filtration-based method. *Desalination, 453,* 1–9.

Abushaban, A., Salinas-Rodriguez, S. G., Pastorelli, D., Schippers, J. C., Mondal, S., Goueli, S. & Kennedy, M. D. (2021). Assessing pretreatment effectiveness for particulate, organic and biological fouling in a full-scale SWRO desalination plant. *Membranes,* 11, 167.

Almotasembellah Abushaban, M. Nasir Mangal, Sergio G. Salinas-Rodriguez, Chidiebere Nnebuo, Subhanjan Mondal, Said A. Goueli, Jan C. Schippers, Maria D. Kennedy (2017). Direct measurement of ATP in seawater and application of ATP to monitor bacterial growth potential in SWRO pre-treatment systems. *Desalination and Water Treatment,* 99, 91.

Aiken, G. (2002). Organic matter in ground water. In: USGS (ed.) *US Geological Survey Artificial Recharge Workshop,* pp. 21–23, Sacramento, CA.

Alhadidi, A., Blankert, B., Kemperman, A. J. B., Schippers, J. C., Wessling, M. & van der Meer, W. G. J. (2011a). Effect of testing conditions and filtration mechanisms on SDI. *Journal of Membrane Science, 381,* 142–151.

Alhadidi, A., Kemperman, A. J. B., Schippers, J. C., Wessling, M. & van der Meer, W. G. J. (2011b). The influence of membrane properties on the silt density index. *Journal of Membrane Science,* 384, 205–218.

Alldredge, A. L., Passow, U. & Logan, B. E. (1993). The abundance and significance of a class of large, transparent organic particles in the ocean. *Deep Sea Research (Part I, Oceanographic Research Papers),* 40, 1131–1140.

Amy, G. L., Salinas-Rodriguez, S. G., Kennedy, M. D., Schippers, J. C., Rapenne, S., Remize, P.-J., Barbe, C., Manes, C.-L. D. O., West, N. J., Lebaron, P., Kooij, D. V. D., Veenendaal, H., Schaule, G., Petrowski, K., Huber, S., Sim, L. N., Ye, Y., Chen, V. & Fane, A. G. (2011). Water quality assessment tools. In: Drioli, E., Criscuoli, A. & Macedonio, F. (eds.) *Membrane-Based Desalination – An Integrated Approach (MEDINA)* (pp. 3–32). IWA, New York.

Antony, A., Low, J. H., Gray, S., Childress, A. E., Le-Clech, P. & Leslie, G. (2011). Scale formation and control in high pressure membrane water treatment systems: A review. *Journal of Membrane Science,* 383, 1–16.

Arruda Fatibello, S. H., Henriques Vieira, A. A. & Fatibello-Filho, O. (2004). A rapid spectrophotometric method for the determination of transparent exopolymer particles (TEP) in freshwater. *Talanta,* 62, 81–85.

ASTM D4189-14 (2014). *Standard Test Method for Silt Density Index (SDI) of Water.* ASTM International, West Conshohocken, PA.

Balnois, E., Wilkinson, K. J., Lead, J. R. & Buffle, J. (1999). Atomic force microscopy of humic substances: Effects of pH and ionic strength. *Environmental Science & Technology,* 33, 3911–3917.

Boerlage, S. F. E. (2001). *Scaling and Particulate Fouling in Membrane Filtration Systems.* Swets & Zeitlinger Publishers, Lisse.

Boerlage, S. F. E. (2007). Understanding the SDI and modified fouling indices (MFI0.45 and MFI-UF). *IDA World Congress on Desalination and Water Reuse 2007 – Desalination: Quenching a Thirst Maspalomas,* Gran Canaria, Spain.

Boerlage, S. F. E., Kennedy, M. D., Aniye, M. P., Abogrean, E. M., El-Hodali, D. E. Y., Tarawneh, Z. S. & Schippers, J. C. (2000). Modified fouling index ultrafiltration to compare pretreatment processes of reverse osmosis feedwater. *Desalination,* 131, 201–214.

Boerlage, S. F. E., Kennedy, M., Tarawneh, Z., Faber, R. D. & Schippers, J. C. (2004). Development of the MFI-UF in constant flux filtration. *Desalination,* 161, 103–113.

Carman, P. C. (1938). Fundamental principles of industrial filtration (a critical review of present knowledge). *Transactions of the Institution of Chemical Engineers,* 16, 168–188.

Chen, J., LeBoeuf, E. J., Dai, S. & Gu, B. (2003). Fluorescence spectroscopic studies of natural organic matter fractions. *Chemosphere,* 50, 639–47.

Coble, P. G. (1996). Characterization of marine and terrestrial DOM in seawater using excitation-emission matrix spectroscopy. *Marine Chemistry,* 51, 325–346.

Dhakal, N. (2017). *Controlling biofouling in seawater reverse osmosis membrane systems.* PhD, Delft University of Technology.

Dhakal, N., Abushaban, A., Mangal, N., Abunada, M., Schippers, J. C. & Kennedy, M. D. (2020). *Membrane Fouling and Scaling in Reverse Osmosis. Membrane Desalination.* CRC Press, Boca Raton, FL.

Dixon, M. B., Qiu, T., Blaikie, M. & Pelekani, C. (2012). The application of the bacterial regrowth potential method and flow cytometry for biofouling detection at the Penneshaw desalination plant in South Australia. *Desalination,* 284, 245–252.

DuPont (2021). FILMTEC™ reverse osmosis membranes technical manual. In: *Water Solutions.* Form No. 45-D01504-en, Rev. 7 August 2021, www.dupont.com/water.

Farhat, N., Hammes, F., Prest, E. & Vrouwenvelder, J. (2018). A uniform bacterial growth potential assay for different water types. *Water Research,* 142, 227–235.

Flemming, H. C. (1997). Reverse osmosis membrane biofouling. *Experimental Thermal and Fluid Science,* 14, 382–391.

Flemming, H.-C. (2011). Microbial biofouling: Unsolved problems, insufficient approaches, and possible solutions. In: Flemming, H.-C., Wingender, J. & Szewzyk, U. (eds.) *Biofilm Highlights* (pp. 81–110). Springer, Berlin, Heidelberg.

Flemming, H. C., Schaule, G., Griebe, T., Schmitt, J. & Tamachkiarowa, A. (1997). Biofouling – The Achilles heel of membrane processes. *Desalination,* 113, 215–225.

Global Water Intelligence (2020). *32nd Worldwide Desalting Plant Inventory.* Media Analytics Ltd, Oxford.

Henderson, R. K., Baker, A., Parsons, S. A. & Jefferson, B. (2008). Characterisation of algogenic organic matter extracted from cyanobacteria, green algae and diatoms. *Water Research,* 42, 3435–3445.

Ho, B. P., Wu, M. W., Zeiher, E. H. & Chattoraj, M. (2004). *Method of monitoring biofouling in membrane separation systems.* 6,699,684. *U.S. Patent.* 2004 Mar 2.

Huber, S. A., Balz, A., Abert, M. & Pronk, W. (2011). Characterisation of aquatic humic and non-humic matter with size-exclusion chromatography – Organic carbon detection – Organic nitrogen detection (LC-OCD-OND). *Water Research,* 45, 879–885.

Ito, Y., Takahashi, Y., Hanada, S., Chiura, H. X., Ijichi, M., Iwasaki, W., Machiyama, A., Kitade, T., Tanaka, Y. & Kogure, M. K. (2013). *Impact of Chemical Addition on the Establishment of mega-ton per day Sized SWRO Desalination Plant in YT.* Kogure Kazuhiro, Kurihara, Masaru.

Jacobson, J. D., Kennedy, M. D., Amy, G. & Schippers, J. C. (2009). Phosphate limitation in reverse osmosis: An option to control biofouling? *Desalination and Water Treatment,* 5, 198–206.

Jeong, S., Naidu, G., Vigneswaran, S., Ma, C. H. & Rice, S. A. (2013). A rapid bioluminescence-based test of assimilable organic carbon for seawater. *Desalination,* 317, 160–165.

Kim LH, Nava-Ocampo, M, van Loosdrecht MCM, Kruithof JC, Vrouwenvelder JS (2018) The membrane fouling simulator: development, application, and early-warning of biofouling in RO treatment. *Desalination and Water Treatment* 126, 1–23. Available: http://dx.doi.org/10.5004/dwt.2018.23081.

Kördel, W., Dassenakis, M., Lintelmann, J. & Padberg, S. (1997). The importance of natural organic material for environmental processes in waters and soils (Technical Report). *Pure and Applied Chemistry,* 69, 1571–1600.

Koutsoukos, P. (2010). *Calcium Carbonate Scale Control in Industrial Water Systems. The Science and Technology of Industrial Water Treatment.* CRC Press, Boca Raton, FL.

Kucera, J. (2010). *Reverse Osmosis Membrane Fouling Control. The Science and Technology of Industrial Water Treatment.* CRC Press, Boca Raton, FL.

Kujundzic, E., Fonseca, A. C., Evans, E. A., Peterson, M., Greenberg, A. R. & Hernandez, M. (2007). Ultrasonic monitoring of early stage biofilm growth on polymeric surfaces. *Journal of Microbiological Methods,* 68, 458–467.

Kurihara, M. (2020). Sustainable seawater reverse osmosis desalination as green desalination in the 21st century. *Journal of Membrane Science and Research,* 6, 20–29.

Kurihara, M., Ito, Y. & Nakaoki, Y. (2001). *Innovative Biofouling Prevention on Seawater Desalination Reverse Osmosis Membrane.* International Desalination Association, Oxford.

Lee, J. & Kim, I. S. (2011). Microbial community in seawater reverse osmosis and rapid diagnosis of membrane biofouling. *Desalination,* 273, 118–126.

Lee, N., Amy, G. & Croué, J.-P. (2006). Low-pressure membrane (MF/UF) fouling associated with allochthonous versus autochthonous natural organic matter. *Water Research,* 40, 2357–2368.

Leenheer, J. A. & Croué, J.-P. (2003). Characterizing dissolved aquatic organic matter. *Environmental Science & Technology,* 37, 18A–26A.

Maher, W. & Woo, L. (1998). Procedures for the storage and digestion of natural waters for the determination of filterable reactive phosphorus, total filterable phosphorus and total phosphorus. *Analytica Chimica Acta,* 375, 5–47.

Mangal, N., Salinas-Rodriguez, S. G., Yangali Quintanilla, V., Schippers, J. C. & Kennedy, M. D. (2021). Ch 8 – Scaling. In: Salinas-Rodriguez, S. G., Schippers, J. C., Amy, G. L., Kim, I. S. & Kennedy, M. D. (eds.) *Seawater Reverse Osmosis Desalination: Assessment & Pre-Treatment of Fouling and Scaling* (pp. 207–239). IWA Publishing, London.

Matthiasson, E. (1983). The role of macromolecular adsorption in fouling of ultrafiltration membranes. *Journal of Membrane Science,* 16, 23–36.

McKnight, D. M., Boyer, E. W., Westerhoff, P. K., Doran, P. T., Kulbe, T. & Andersen, D. T. (2001). Spectrofluorometric characterization of dissolved organic matter for indication of precursor organic material and aromaticity. *Limnology and Oceanography,* 46, 38–48.

Mittal, B. (2017). Chapter 2 – Pharmacokinetics and preformulation. In: Mittal, B. (ed.) *How to Develop Robust Solid Oral Dosage Forms from Conception to Post-Approval* (pp. 17–37). Academic Press, Cambridge, MA.

Myklestad, S. M. (1995). Release of extracellular products by phytoplankton with special emphasis on polysaccharides. *Science of the Total Environment,* 165, 155–164.

Nahrstedt, A. & Camargo Schmale, J. (2008). New insights into SDI and MFI measurements. *Water Science and Technology: Water Supply,* 8, 401–412.

Neu, T. R. (2000). In situ cell and glycoconjugate distribution in river snow studied by confocal laser scanning microscopy. *Aquatic Microbial Ecology,* 21, 85–95.

Nguyen, T., Roddick, F. A. & Fan, L. (2012). Biofouling of water treatment membranes: A review of the underlying causes, monitoring techniques and control measures. *Membranes (Basel),* 2, 804–840.

Ohno, T. & Bro, R. (2006). Dissolved organic matter characterization using multiway spectral decomposition of fluorescence landscapes. *Soil Science Society of America Journal,* 70, 2028–2037.

Passow, U. & Alldredge, A. L. (1995). A dye-binding assay for the spectrophotometric measurement of transparent exopolymer particles (TEP). *Limnology and Oceanography,* 40, 1326–1335.

Passow, U. (2002). Transparent exopolymer particles (TEP) in aquatic environments. *Progress in Oceanography,* 55, 287–333.

Quek, S.-B., Cheng, L. & Cord-Ruwisch, R. (2015). Detection of low concentration of assimilable organic carbon in seawater prior to reverse osmosis membrane using microbial electrolysis cell biosensor. *Desalination and Water Treatment,* 55, 2885–2890.

Rachman, R. M., Ghaffour, N., Waly, F. & Amy, G. L. (2013). Assessment of silt density index (SDI) as fouling propensity parameter in reverse osmosis (RO) desalination systems. *Desalination and Water Treatment,* 51, 1091–1103.

Rimmelin, P. & Moutin, T. (2005). Re-examination of the MAGIC method to determine low orthophosphate concentration in seawater. *Analytica Chimica Acta,* 548, 174–182.

Salinas-Rodriguez, S. G. (2011). *Particulate and Organic Matter Fouling of SWRO Systems: Characterization, Modelling and Applications.* CRC Press/Balkema, Delft.

Salinas-Rodriguez, S. G. (2021). *Fouling and Scaling in Seawater Reverse Osmosis Desalination. The Source.* International Water Association (IWA), London.

Salinas-Rodriguez, S. G., Amy, G. L., Schippers, J. C. & Kennedy, M. D. (2015). The modified fouling index ultrafiltration constant flux for assessing particulate/colloidal fouling of RO systems. *Desalination,* 365, 79–91.

Salinas-Rodriguez, S. G., Boerlage, S. F. E. & Schippers, J. C. (2021). Ch 4 – Particulate fouling. In: Salinas-Rodriguez, S. G., Schippers, J. C., Amy, G. L., Kim, I. S. & Kennedy, M. D. (eds.) *Seawater Reverse Osmosis Desalination: Assessment & Pre-treatment of Fouling and Scaling* (pp. 85–122). IWA Publishing, London.

Salinas-Rodriguez, S. G., Kennedy, M. D., Schippers, J. C. & Amy, G. L. (2009). Organic foulants in estuarine and bay sources for seawater reverse osmosis – Comparing pre-treatment processes with respect to foulant reductions. *Desalination and Water Treatment,* 9, 155–164.

Salinas-Rodriguez, S. G., Sithole, N., Dhakal, N., Olive, M., Schippers, J. C. & Kennedy, M. D. (2019). Monitoring particulate fouling of North Sea water with SDI and new ASTM MFI0.45 test. *Desalination,* 454, 10–19.

Schäfer, A. I. (2001). *Natural Organics Removal Using Membranes: Principles, Performance, and Cost*. CRC Press, Boca Raton, FL.

Schippers, J. C. & Verdouw, J. (1980). The modified fouling index, a method of determining the fouling characteristics of water. *Desalination*, 32, 137–148.

Schippers, J. C. (1989). *Vervuiling van hyperfiltratiemembranen en verstopping van infiltratieputten*. Keuringinstituut voor waterleidingartikelen KIWA N.V., Rijswijk.

Schippers, J. C., Hanemaayer, J. H., Smolders, C. A. & Kostense, A. (1981). Predicting flux decline or reverse osmosis membranes. *Desalination*, 38, 339–348.

Schippers, J. C., Salinas-Rodriguez, S. G., Boerlage, S. F. E. & Kennedy, M. D. (2014). *Why MFI Is Edging SDI as a Fouling Index. Desalination & Water Reuse*. Faversham House with the Cooperation of the International Desalination Association, York.

Schippers, J. C., Salinas-Rodriguez, S. G. & Kennedy, M. D. (2021). Ch 3 – Fouling and pretreatment. In: Salinas-Rodriguez, S. G., Schippers, J. C., Amy, G. L., Kim, I. S. & Kennedy, M. D. (eds.) *Seawater Reverse Osmosis Desalination: Assessment & Pretreatment of Fouling and Scaling* (pp. 59–83). IWA Publishing, London.

Sheikholeslami, R. (2005). Scaling potential index (SPI) for $CaCO_3$ based on Gibbs free energies. *AIChE Journal*, 51, 1782–1789.

Siebdrath, N., Ding, W., Pietsch, E., Kruithof, J., Uhl, W. & Vrouwenvelder, J. S. (2017). Construction and validation of a long-channel membrane test cell for representative monitoring of performance and characterization of fouling over the length of spiral-wound membrane modules. *Desalination and Water Treatment*, 89, 1–16.

Sim, L. N., Suwarno, S. R., Lee, D. Y. S., Cornelissen, E. R., Fane, A. G. & Chong, T. H. (2019). Online monitoring of transparent exopolymer particles (TEP) by a novel membrane-based spectrophotometric method. *Chemosphere*, 220, 107–115.

Singh, R. (2014). *Membrane Technology and Engineering for Water Purification: Application, Systems Design and Operation*. Elsevier Science & Technology, Amsterdam.

Sung, J. H., Chun, M.-S. & Choi, H. J. (2003). On the behavior of electrokinetic streaming potential during protein filtration with fully and partially retentive nanopores. *Journal of Colloid and Interface Science*, 264, 195–202.

Thornton, D. C. O., Fejes, E. M., DiMarco, S. F. & Clancy, K. M. (2007). Measurement of acid polysaccharides in marine and freshwater samples using alcian blue. *Limnology and Oceanography: Methods*, 5, 73–87.

Thuy, N. T., Huang, C. P. & Lin, J. L. (2017). Visualization and quantification of transparent exopolymer particles (TEP) in freshwater using an auto-imaging approach. *Environmental Science and Pollution Research International*, 24, 17358–17372.

Underwood, A. J. V. (1926). A critical review of published experiments on filtration. *Transactions of the Institution of Chemical Engineers*, 4, 19.

van de Lisdonk, C. A. C., van Paassen, J. A. M. & Schippers, J. C. (2000). Monitoring scaling in nanofiltration and reverse osmosis membrane systems. *Desalination*, 132, 101–108.

Van der Kooij, D. (1992). Assimilable organic carbon as an indicator of bacterial regrowth. *American Water Works Association*, 84, 57–65.

Van der Kooij, D., Visser, A. & Hijnen, W. A. M. (1982). Determining the concentration of easily assimilable organic carbon in drinking water. *American Water Works Association*, 74, 540–545.

Villacorte, L. O. (2014). *Algal Blooms and Membrane Based Desalination Technology*. CRC Press/Balkema, Leiden.

Villacorte, L. O., Boerlage, S. F. E. & Dixon, M. B. (2021). Ch 6 – Algal blooms and RO desalination. In: Salinas-Rodriguez, S. G., Schippers, J. C., Amy, G. L., Kim, I. S. & Kennedy, M. D. (eds.) *Seawater Reverse Osmosis Desalination: Assessment & Pre-Treatment of Fouling and Scaling* (pp. 145–176). IWA Publishing, London.

Villacorte, L. O., Ekowati, Y., Calix-Ponce, H. N., Kisielius, V., Kleijn, J. M., Vrouwenvelder, J. S., Schippers, J. C. & Kennedy, M. D. (2017a). Biofouling in capillary and spiral wound membranes facilitated by marine algal bloom. *Desalination, 424,* 74–84.

Villacorte, L. O., Ekowati, Y., Calix-Ponce, H. N., Schippers, J. C., Amy, G. L. & Kennedy, M. D. (2015a). Improved method for measuring transparent exopolymer particles (TEP) and their precursors in fresh and saline water. *Water Research, 70,* 300–312.

Villacorte, L. O., Ekowati, Y., Neu, T. R., Kleijn, J. M., Winters, H., Amy, G., Schippers, J. C. & Kennedy, M. D. (2015b). Characterisation of algal organic matter produced by bloom-forming marine and freshwater algae. *Water Research, 73,* 216–230.

Villacorte, L. O., Ekowati, Y., Winters, H., Amy, G., Schippers, J. C. & Kennedy, M. D. (2015c). MF/UF rejection and fouling potential of algal organic matter from bloom-forming marine and freshwater algae. *Desalination, 367,* 1–10.

Villacorte, L. O., Kennedy, M. D., Amy, G. L. & Schippers, J. C. (2009). The fate of Transparent exopolymer particles (TEP) in integrated membrane systems: Removal through pretreatment processes and deposition on reverse osmosis membranes. *Water Research, 43,* 5039–5052.

Villacorte, L. O., Schippers, J. C. & Kennedy, M. D. (2017b). Appendix 3. Methods for measuring transparent exopolymer particles and their precursors in seawater. In: Anderson, D. M., Boerlage, S. F. & Dixon, M. B. (eds.) *Harmful Algal Blooms (HABs) and Desalination: A Guide to Impacts, Monitoring and Management. IOC Manuals and Guides No. 78* (pp. 17–52). Intergovernmental Oceanographic Commission of UNESCO, Paris.

Vrouwenvelder, J. S., Manolarakis, S. A., Van der Hoek, J. P., Van Paassen, J. A. M., Van der Meer, W. G. J., Van Agtmaal, J. M. C., Prummel, H. D. M., Kruithof, J. C. & Van Loosdrecht, M. C. M. (2008). Quantitative biofouling diagnosis in full scale nanofiltration and reverse osmosis installations. *Water Research, 42,* 4856–4868.

Vrouwenvelder, J. S., Van Loosdrecht, M. C. M. & Kruithof, J. C. (2011). Early warning of biofouling in spiral wound nanofiltration and reverse osmosis membranes. *Desalination, 265,* 206–212.

Vrouwenvelder, J. S., van Paassen, J. A. M., Kruithof, J. C. & van Loosdrecht, M. C. M. (2009). Sensitive pressure drop measurements of individual lead membrane elements for accurate early biofouling detection. *Journal of Membrane Science, 338,* 92–99.

Vrouwenvelder, J. S., Van Paassen, J. A. M., Wessels, L. P., Van Dam, A. F. & Bakker, S. M. (2006). The membrane fouling simulator: A practical tool for fouling prediction and control. *Journal of Membrane Science, 281,* 316–324.

Wang, Q., Tao, T., Xin, K., Li, S. & Zhang, W. (2014). A review research of assimilable organic carbon bioassay. *Desalination and Water Treatment, 52,* 2734–2740.

Weinrich, L. A., LeChevallier, M. & Haas, C. (2015). *Application of the Bioluminescent Saltwater Assimilable Organic Carbon Test as a Tool for Identifying and Reducing Reverse Osmosis Membrane Fouling in Desalination.* Water Reuse Research Foundation, Alexandria, VA.

Weinrich, L. A., LeChevallier, M. & Haas, C. N. (2016). Contribution of assimilable organic carbon to biological fouling in seawater reverse osmosis membrane treatment. *Water Research, 101,* 203–213.

Weinrich, L. A., Schneider, O. D. & LeChevallier, M. W. (2011). Bioluminescence-based method for measuring assimilable organic carbon in pretreatment water for reverse osmosis membrane desalination. *Applied and Environmental Microbiology, 77,* 1148–1150.

Weishaar, J. L., Aiken, G. R., Bergamaschi, B. A., Fram, M. S., Fujii, R. & Mopper, K. (2003). Evaluation of specific ultraviolet absorbance as an indicator of the chemical composition and reactivity of dissolved organic carbon. *Environmental Science & Technology, 37,* 4702–4708.

Westerhoff, P., Chen, W. & Esparza, M. (2001). Fluorescence analysis of a standard fulvic acid and tertiary treated wastewater. *Journal of Environmental Quality,* 30, 2037–2046.

Wiesner, M. R., Clark, M. M., Jacangelo, J. G., Lykins, B. W. & Marinas, B. J. (1992). Committee report: Membrane processes in potable water treatment. *Journal AWWA,* 84, 59–67.

Yamamura, H., Kimura, K., Okajima, T., Tokumoto, H. & Watanabe, Y. (2008). Affinity of functional groups for membrane surfaces: Implications for physically irreversible fouling. *Environmental Science & Technology,* 42, 5310–5315.

Zhou, J., Mopper, K. & Passow, U. (1998). The role of surface-active carbohydrates in the formation of transparent exopolymer particles by bubble adsorption of seawater. *Limnology and Oceanography,* 43, 1860–1871.

4 Green Technologies for Saline Water Treatment

Z.U.R. Farooqi, A. Zafar, and S. Ameen
University of Agriculture

P. Ilić
PSRI Institute for Protection and Ecology
of the Republic of Srpska

L.J. Stojanović Bjelić and D. Nešković Markić
Pan-European University "APEIRON"

CONTENTS

DOI: 10.1201/9781003185437-5

4.1   INTRODUCTION

Water is the most valuable fuel for sustaining life on Earth and driving human society's economic growth. Underground water accounts for 27% of fresh water, while surface water contained in reservoirs and rivers accounts for just 1%. Rapid developments in human lifestyles, along with urbanization and industrialization, have put a strain on the world's dwindling fresh water supplies. Furthermore, in several areas of the planet, the imminent climate shift has favoured salinization of both soil and water (Reca et al., 2018; Shahid et al., 2018).

Salinity is impacting crop development all over the world, with 20% of planted land and 33% of irrigated land damaged and depleted by salinity. Climate change, unsustainable use of groundwater (especially near the sea), increased use of low-quality water in agriculture, and widespread irrigation associated with intensive farming will all exacerbate this method. Excessive soil salinity decreases the production of many agricultural crops, including most vegetables, which are particularly vulnerable during their life cycle. The majority of vegetable crops have a low salinity threshold (ECt) (ranging from 1 to 2.5 dS/m in saturated soil extracts), and their salt resistance reduces as salty water is used for irrigation (Machado & Serralheiro, 2017). The salinity of land and water supply is one of the most environmentally costly obstacles for achieving global sustainability.

Salinity, a significant environmental concern in both arid and semi-arid areas is very stressful for plants and compounds along with other pressures such as water depletion, nutrient shortages, and soil alkalinity. Remediation is a technique for removing toxins from the root zone of plants in order to minimize tension and increase production. This method involves biological soil and water treatment, which often leads to increased soil penetration and salt leaching from the root zone. Any soil and water remediation solutions have been suggested, and they can be categorized into two categories: engineering-based remediation and green remediation. Green remediation is the use of plants to remove or contain chemical contaminants such as heavy metals, radioactive elements, organic compounds, and radioactive materials from soil or water. The use of halophytes for green remediation has recently gained prominence, especially in developing countries (Nouri et al., 2017).

4.2   WATER SALINITY IN THE ENVIRONMENT: SOURCES AND EFFECTS

Agriculture is the world's biggest water consumer. Irrigation of farm fields consumes more than two-thirds of overall water consumption. The inadequate supply of water supplies threatens the long-term viability of many irrigation areas in semi-arid regions. Aside from the shortage, there are significant questions about irrigation water's safety. Water salinization is a worldwide problem, but it is exacerbated in water-scarce places including arid and semi-arid regions, where groundwater is the primary source of water (Salomaa, 2004). Increased salinization of freshwater supplies due to overexploitation of aquifers and irrigation water returns is affecting many regions. If adequate steps are not taken immediately, the situation will worsen in the distant future because of global climate change.

Desalination of both brackish and seawater is becoming more popular in many irrigation districts as the cost of desalinated water has dropped significantly, making desalinated water more commercially viable. One of the most significant and productive agricultural systems is greenhouse vegetable processing. As a result, the total area of greenhouses has steadily increased in recent decades, especially in dry areas with extreme water scarcity, such as the Mediterranean basin. However, since vegetable crops are particularly vulnerable to water and salinity stresses, this practice is highly reliant on water irrigation supply, both in terms of quantity and efficiency. Furthermore, in recent decades, the proliferation of greenhouses has resulted in an increased demand for irrigation water in order to sustain high crop output standards, resulting in significant environmental and water supply issues (Reca et al., 2018).

4.2.1 SOURCES OF SALTS IN WATER

4.2.1.1 Natural

When we think of saline water, we naturally think of the oceans. Residents in states such as Colorado and Arizona, which are hundreds of miles from the Pacific Ocean, will enjoy a day at the beach by merely walking outside their front door, as they might be right next to a salty shore. There is a significant amount of highly salty water in the atmosphere of the western United States. Nearly 75% of New Mexico's groundwater is too saline for certain uses if it is not handled (Reynolds, 1962). Water in this region may have been left over from ancient days when salt seas dominated the western United States, and when rain infiltrates further into the earth, it may come into contact with rocks containing highly soluble salts, turning the water saline. For thousands of years, groundwater may live and travel, being as salty as ocean water.

Rainfall, rock weathering, seawater infiltration, and aerosol deposits all play a role in determining the ion content of inland surface waters (Williams, 2002; Millán et al., 2011). This condition is known as primary salinization if certain natural mechanisms are the cause of salinization. Salts are formed by the chemical weathering of minerals in soils and rocks. A few of the more important chemical weathering processes include dissolution (contact with water), hydrolysis (reaction with water), carbonation (reaction with dissolved carbon dioxide), acidification (reaction with proton), and oxidation-reduction (transfer of electrons). All of these reactions increase the amount of dissolved minerals in the soil solution and the bath water (Council, 1993).

Salts are formed in the ocean because of chemical weathering and the movement of water by rivers. However, rivers carry a negligible volume of salt relative to ocean waters, but there may be another factor at play. Weathering and seismic activities are the sources of salt. The ocean originated very early in Earth history, and as soon as water meets rock, weathering processes begin to leach and dissolve the soluble elements out of the rock (sodium, calcium, magnesium, potassium, etc.). While there is no chlorine or sulphur in minerals, there is enough in volcanic gases, which readily dissolves in water in the atmosphere to produce chloride and sulphate, and then rains into the ocean. We get a salty ocean as these cycles continue for billions of years. Surface seawater can get saltier if evaporation outweighs precipitation. Local seawater salinity is often increased if there aren't many waterways nearby and/or if the basin is constrained, since high-salinity seawater doesn't blend well with normal-salinity seawater throughout this situation. The Red Sea, for example, has a salinity of 40% on average. High-salinity water from confined basins eventually leaks out and enters global circulation. Owing to sluggish ocean mixing, higher and lower salinity waters may be detected a long distance from their source.

4.2.1.2 Anthropogenic

In terms of human impact, agricultural-induced changes in soil salinity play a major role in the salinity in rivers and aquifers since salt is mobilized to the water sources. Salinization of aquifers is caused by heavy pumping rather than by climate variability. Decreased groundwater levels, for example, can contribute to saltwater infiltration from the seas in coastal areas; industrial groundwater drainage can also raise salinity as salt concentrates. Anthropogenic practices are often the source of secondary salinization. Some of the more common causes include the use of salt water for irrigation of fields and insufficient drainage (Munns, 2009), seepage from canals, evaporation of salty water from the surface of soil leaves salts behind on the surface (Siyal et al., 2002), excessive slope, over irrigation, and intense rice cultivation in low water table areas (Siyal et al., 2002; Khan & Abdullah, 2003).

The irrigation water has 0.5 t/1,000 m^3 of salts with a salt concentration of 500 mg/L (i.e. 500 mg/kg). Meanwhile, crops need 6,000–10,000 m^3 of water per hectare annually, and each hectare of land can receive 3–5 tonnes of salt (Munns, 2009). Pakistan's agriculture sector is heavily dependent on the Indus Basin Irrigation System for its GDP contribution. The inefficient drainage scheme of the Indus Basin Irrigation System is one of the primary causes of salinity (Zaman & Ahmad, 2009). Owing to the reduced storage potential of lakes, water supply per acre is declining. Farmers are erecting a huge number of tube wells to meet the requirement for irrigating crops. Salinity is indeed a product of this. In most parts of Punjab and Sindh, salinity is at an all-time high (Ali, 2010). In Pakistan, the number of tube wells has surpassed 900,000. In Pakistan, it is projected that 61 billion cubic metres of groundwater are pumped per year. According to estimates, more than 70% of brackish groundwater in Pakistan's irrigated irrigation is drained by tube wells, causing secondary salinization issues (Farid et al., 2018). Due to a lack of canal water sources, farmers must use marginal to brackish in quality groundwater, which may result in sodification (due to sodium salts) and/or secondary salinization (due to soluble salts) (Zaman & Ahmad, 2009).

In this situation, climate variability influences the need for water; during a drought, more water may be pumped for agriculture or water supply. Extreme conditions such as droughts are becoming even more common and long-lasting because of climate change, as we have recently shown. Sea level rise and dams are two other factors that can affect salinity of rivers, all of which are influenced by climate change. Rising salinity is caused by several factors, some of which are interconnected. The removal of native, deep-rooted plants from catchments and its substitution with shallow-rooted agricultural plant species, along with the discharge of salty agricultural wastewater, causes the salinity of many freshwater streams, reservoirs, and rivers to increase, as well as the expansion of natural salt lakes. The salinization of certain fresh waters is caused by rising salty groundwaters. When water is redirected from inflows for agriculture and other uses, the salinity of several major natural salt lakes in drylands is rising. The building of impoundments can result in an increase in river salinity. Also, in temperate zones, such as Germany's River Werra, brine discharges from mining operations will salinate rivers (Schmitz, 1956; Williams, 2001).

4.2.2 Effects of Saline Water

4.2.2.1 Plants

Under salt tension, agricultural crops react in a variety of ways. Salinity affects not just the agricultural productivity of most crops, but also the physicochemical properties of the soil and the area's ecological equilibrium. Low crop production, low economic returns, and land degradation are all effects of salinity (Hu & Schmidhalter, 2002). Crop germination, plant development, and water and nutrient intake are all affected by salinity, which is the product of dynamic interactions between morphological, physiological, and biochemical processes (Akbarimoghaddam et al., 2011; Singh & Chatrath, 2001). Plant production is influenced by salinity in almost every way, including germination, vegetative growth, and reproductive development. Plants are subjected to ion toxicity, osmotic stress, nutrient (N, Ca, K, P, Fe, Zn) depletion, and oxidative stress as a result of soil salinity, which reduces water uptake.

Since phosphate ions precipitate with calcium ions in salinity, plant phosphorus (P) uptake is drastically reduced (Bano & Fatima, 2009). Sodium, arsenic, and boron, for example, have particularly harmful effects on plants. Excessive sodium concentration in cell walls will quickly cause osmotic stress and cell death (Munns, 2002). If the soil contains enough toxic material, plants vulnerable to these elements can be harmed even at low salt concentrations. Since certain salts are also plant nutrients, high salt levels in the soil can disrupt the plant's nutrient balance or prevent any nutrients from being absorbed (Blaylock, 1994). Salinity also has an effect on photosynthesis, primarily by reducing leaf region, chlorophyll material, and stomatal conductance, and to a lesser degree by lowering photosystem-II performance (Netondo et al., 2004). Salinity has a negative impact on reproductive growth because it inhibits microsporogenesis and stamen filament elongation, increases programed cell death in certain tissue forms, causes ovule abortion, and causes fertilized embryos to age prematurely.

Because of the poor osmotic ability of the soil solution (osmotic stress), particular ion effects (salt stress), nutritional imbalances, or a mixture of these reasons, the saline growth medium has many negative effects on plant growth (Ashraf, 2004). A reduction in fresh or dry weight occurs as salinity rises. Salt stress had the greatest impact on seedling development, shoot height, and root length (Saddiq et al., 2019). Salt marsh and sand-dune plants suffer from stunted growth due to reduced leaf scale, internode, and duration as a result of salinity tension (Cheng et al., 2020). Both of these causes have negative physiological and biochemical effects on plant growth and production, as well as molecular effects (Munns & James, 2003; Tester & Davenport, 2003). For plants developing in a saline medium, osmotic equilibrium is important. When this equilibrium is disrupted, cells lose their turgidity, become dehydrated, and eventually die. On the other side, salinity's negative impact on plant growth may be due to a reduction in the availability of photosynthetic assimilates or hormones to the developing tissues (Ashraf, 2004). Ion toxicity is caused by the substitution of Na^+ for K^+ in biochemical processes, as well as conformational modifications in proteins caused by Na^+ and Cl. The K^+ is a cofactor for many enzymes and cannot be replaced by Na^+. High K^+ levels are also necessary for tRNA binding to ribosomes and, as a result, protein synthesis (Zhu, 2002). Metabolic imbalance is caused by ion toxicity and osmotic stress, which contributes to oxidative stress (Chinnusamy et al., 2006).

4.2.2.2 Soil

The salinity of soil water is affected by soil composition, temperature, water usage, and irrigation practices. Farm sufficient water is at its best, and surface water salinity is at its lowest just after the soil is irrigated, for example. However, as plants use more soil water, the residual water becomes more tightly bound to the soil, making it more difficult for plants to access. Since salts become more concentrated in the residual surface water when water is soaked up by plants by transpiration or lost to the environment by evaporation, soil water salinity rises. As a result, evapotranspiration (ET) between irrigation cycles will raise salinity much further (Pearson, 2003).

Soil water salinity can affect soil physical properties by causing fine particles to accumulate together. The technique is known as flocculation, and it aids soil aeration, root penetration, and root growth. Although growing soil solution salinity increases soil accumulation and stabilization, at high amounts, it can be harmful and potentially lethal to plants. As a result, it's difficult to increase salinity to maintain soil structure without also solving potential plant health issues. Sodium has the same effect on soils as salinity. The key physical processes related to elevated sodium concentrations are soil dispersion and clay platelet and aggregate swelling. The forces that hold clay particles together are broken because there are too many heavy sodium ions within. Clay particles scatter as they separate, causing swelling and soil dispersion. As a result of soil dispersion, clay particles obstruct soil openings, decreasing soil permeability. The soil reforms and solidifies into an almost cement-like soil of little or no structure as clay dispersion occurs as a consequence of constant wetting and drying. The three main problems caused by sodium-induced dispersion are decreased penetration, reduced hydraulic conductivity, and surface crusting (Pearson, 2003).

Since salts that lead to salinity, such as calcium and magnesium, are smaller and cluster closer to clay particles, they do not have this impact. Since calcium and magnesium fight for the same spaces as sodium to adhere to clay particles, they seem to hold soil flocculated. Calcium and magnesium supplementation may help to mitigate sodium-induced dispersion. Saline water often has a negative impact on soil chemical properties, such as reducing water and mineral supply, changing the pH of the soil, and rendering it unsuitable for flora and fauna.

4.2.2.3 Soil Flora and Fauna

Soil microorganisms make up less than 0.5% of soil mass (w/w), but they are critical to soil resources and processes. Soil organisms include bacteria, archaea, fungi, protozoa, and viruses (Tate, 2000). Microorganisms participate in oxidation, nitrification, ammonification, nitrogen fixation, and other processes that result in soil organic matter decomposition and nutrient transformation (Amato & Ladd, 1994). They may also store carbon and nutrients in their biomass, which are mineralized after cell death by living microbes (Anderson & Domsch, 1980). The microbial population is negatively affected by salinity. High concentrations of soluble salts have two major impacts on microbes: osmotic influence and specific ion effects. Andronov et al. (2012), Batra and Manna (1997), Pathak & Rao (1998), Rousk et al. (2011), and Setia et al. (2011) have shown that salinity reduces microbial activity and biomass, and affects the composition of microbial communities. Since osmotic stress induces cell drying and lysis, salinity reduces microbial biomass (Batra & Manna, 1997; Laura, 1974; Pathak & Rao, 1998; Rietz & Haynes, 2003; Sarig et al., 1996; Sarig & Steinberger, 1994). Soil respiration declines when soil EC increases, according to some studies (Adviento-Borbe et al., 2006; Wong et al., 2009; Yuan et al., 2007). Setia et al. (2010) discovered that soil respiration was reduced by more than 50% at $EC_{1:55}$ dS/m. However, soil respiration was not significantly related to EC, according to Rietz and Haynes (2003), but the metabolic quotient (respiration per unit biomass) changed as EC increased.

4.3 REMEDIATION STRATEGIES/TECHNOLOGIES FOR SALINE WATER TREATMENT

Although irrigated agriculture has improved crop production significantly, inadequate and ineffective irrigation has wasted water, poisoned surface and groundwater, harmed productivity, and altered the ecosystem of vast swaths of property. Irrigation-related contamination of water sources is causing health hazards in many areas and driving up the expense of treating water for domestic and industrial use. Salts, fertilizers, herbicides, and contaminants are contaminating surface and groundwaters in many places. Surface waters' recreational and aesthetic appeal are also harmed by these toxins. Simultaneously, agriculture is being subjected to expensive restrictions in order to minimize emissions or manage wastes prior to discharge. Finding a suitable, safe location for such discharge is becoming increasingly difficult in certain cases, especially in developing countries. The potential usability of the total water source is reduced as salty and fresh waters are mixed. The use of contaminated water for irrigation reduces crop output efficiency while still presenting certain health risks to food users.

To address the above issues, innovative strategies to minimize unnecessary water usage and protect scarce water sources must be formulated and applied, as well as improved approaches to apply current practices more efficiently. Irrigation efficiency must be improved by the implementation of effective management policies, processes, and procedures, as well as through education and training. To maintain irrigated farmland and avoid contamination of related water supplies, effective salinity management policies must be enforced. Such interventions must be selected with an appreciation of how they impact the nature of land and water supplies, not just crop development, and with a knowledge of the natural processes at work in irrigated, geohydrologic systems, not just those on-farm. Such practices may be used to regulate salinity in the field rootzone, and others can be used to control salinity in broader management units including drainage schemes, river basins, and so on. In certain cases, there is no single method to ensure that salt water is used safely in irrigation. Many various methods and practices may be combined to create effective saline water irrigation systems; the best combination is determined by fiscal, climatic, social, edaphic, and hydrogeologic factors. The below are some of the techniques for treating salt water:

4.3.1 CONVENTIONAL METHODS

Some of the conventional effluent discharge methods are discussed briefly in the following sections.

4.3.1.1 Deep-Well Injection

This process is often used to dispose of extracted water from oil and gas projects, as well as radioactive and dangerous wastes (such as NORM contaminant salts) (Reeder 1981). The polluted waste is pumped into geologically segregated horizons (wells) that are geologically separated from possible underground drinking water supplies. Given environmental safety and water quality concerns, relying on these injection wells for long-term and environmentally sustainable brine disposal is getting more difficult and expensive (Brandt & Tait, 1997). For salt-water drainage, Weller (1993) created an underground injection cost model. This model calculates capital and maintenance costs, and then optimizes them depending on hydraulic fracturing geometry and reservoir conditions. Surface stresses, injection speeds, and disposal costs for deep-well injection are predicted using primary inputs. This model's cost estimates can be related to the corresponding costs of other options such as reverse osmosis (RO), thermal evaporation, and so on. These approaches become uneconomical and inappropriate for high-volume salt sources. Injection technologies will be rendered more possible with promising auxiliary technologies such as pre-treatment, pre-concentration, efficient partial demineralization, and improved treatment systems (i.e., de-oiling, dissolved organic removal).

4.3.1.2 Evaporation Ponds

Evaporation ponds are man-made ponds with a large surface area that utilize solar rays to evaporate water (Velmurugan & Srithar, 2008). Solar-powered evaporation ponds may be a viable choice for effluent disposal, particularly in arid and semi-arid areas where evaporation rates are high and land costs are low. Evaporation ponds have many advantages, including simplicity of construction and maintenance, reduced maintenance, and low cost. During the building of solar ponds, a suitable impervious substrate (clay, synthetic polymer) must be used to line the pond to prevent leakage. Surface salinity should not rise because of proper closure of adjacent parts and joints, and underlying water runoff from pond seepage is avoided (Shammas et al., 2010). Evaporation ponds/basins have also been seen to be useful in desalination plants for concentrating rejected brine to decrease volume and deliver salts as part of the treatment method (Abdulsalam et al., 2017). To mitigate the environmental hazards associated with saline disposal basins, it is critical to follow agreed engineering requirements when constructing these ponds. Gilron et al. (2003) devised a wind-assisted accelerated evaporation system to solve the problem of evaporation ponds' wide area requirements. Vertically installed, permanently wetted evaporation surfaces with large packing densities were engineered to greatly minimize the amount of land required for brine disposal. The evaporative potential per area footprint of a large-scale wind-aided accelerated evaporation test device built in this study was increased by a factor of >10. Evaporation ponds, particularly in inland areas, may be a viable means of saline effluent disposal if properly built and maintained.

4.3.1.3   Discharge into Surface Water/Municipal Sewer

Discharging effluent to a surface water source (river, bay, ocean, etc.) is found to be the most often used management technique for saltwater drainage owing to the lowest related expense as compared to other traditional approaches (Afrasiabi & Shahbazali, 2011). Such discharges into surface water sources, on the other hand, are acceptable as long as the amount of reject water is not too high in comparison to the volume of surface water. Prior to discharge, the compatibility of the concentrate to be discharged with the receiving water must be determined. Effluents with salinities beyond the surface water disposal levels can be diluted with low salinity water from wastewater treatment facilities, cooling towers, and other sources. Until discharge, salt water's dissolved oxygen (DO) levels should be increased to reduce harmful effects on obtaining water bodies (Mickley, 2001).

For coastal desalination plants, natural dispersion of brine by long sea outfalls is used as a cost-effective disposal strategy. To prevent adverse effects on aquatic flora and fauna, hydrodynamic analyses of brine effluents must be used to select the necessary structure and position of the water outfall device (Nikiforakis & Stamou, 2015). Shao et al. (2008) created a statistical model that simulates the accumulated effect of brine drainage into deeper coastal waters with a flat seabed. The location of the outfall farther offshore was observed to reduce the effect of shoreline salinity in a coastal area in their research. Fernández-Torquemada et al. (2009) studied the hypersaline plume produced by brine discharge in the shoreline or near beaches and the region of impact. Field campaigns and data collection were used to monitor the effluent dispersion from three seawater RO desalination plants. The plume behaviour was found to be dependent on discharge concentration, output level, and season.

It was discovered that increasing the dilution using seawater as a diluent decreased the discharge's region of control. To optimize brine dilution and, as a result, reduce the region of control, a diffuser was proposed to improve discharge velocity and mix in the close sector. González et al. (2011) looked at modern discharge technologies in order to reduce the environmental effects. The research looked at combining brine with a thermal effluent in a pipeline to achieve greater dilution and reduce the impact on the environment. The impact of increased salinity on seagrass development and survival was also considered. The findings of these tests, which included proper discharge device architecture and pipe design for homogeneous mixing, were considered to be promising and had less negative environmental consequences. Some small plants use another means of drainage, which is to refuse the effluent in urban sewage networks. However, such discharges are only possible if the saline effluents are non-toxic and do not interfere with the settling activity of the clarifier. The use of saline effluent in the wastewater system has the benefit of lowering the sewage's biochemical oxygen requirement (Ahmed et al., 2001). Combining high-salt brines with sewage effluents, on the other hand, may allow sewage pollutants and other particulates to accumulate into particles of varying sizes. This, in particular, has an effect on sedimentation rates and can conflict with light transmission in the water body, lowering phytoplankton productivity (Hashim & Hajjaj, 2005).

Since the expenditure estimates are based on desalinated brine (Panagopoulos et al., 2019), the relative expense effects of saline wastewater treatment can be extrapolated. Direct drainage of saline effluents through surface water or sewers is clearly less expensive than deep-well injection or solar evaporation; nevertheless, both activities have negative consequences for the marine ecosystem. Excessive salt discharge will impair the efficiency of the sewage system's biological treatment. As a result, before dumping saline effluents directly into surface water, sufficient pre-treatment should be performed, and the receiving water body's self-purification capability should be considered and not exceeded. However, in environmentally vulnerable environments, such discharges can be stopped at all costs.

4.3.2 ADVANCED TECHNIQUES

Technology advancements will broaden the range of solutions open to decision-makers at all levels. As a result, solutions to the major desalination technologies are also being researched in order to improve or substitute established desalination processes or fill niche requirements where conventional technologies are ineffective. Several alternate technologies have been suggested, with the primary goal of lowering desalination's energy requirements. Some of these new developments, such as forward osmosis, membrane distillation, and dewvaporation, have the potential to increase the use of low-grade or waste heat, whereas others, including nanostructured/nanoenhanced membranes and capacitive deionization, have the potential to minimize total energy use.

4.3.2.1 Forward Osmosis (FO)

Forward osmosis (FO) is a membrane-based separation mechanism that drives water flux through a semipermeable membrane using a normal (rather than forced) osmotic pressure differential between a concentrated pull solution (or osmotic agent) and feed stream. The most difficult aspect of choosing an osmotic agent is finding a combination with enough osmotic ability to drive transmembrane conversion, which is especially difficult with high-salinity feed sources (DOE, 2014). The magnitude of water flux and degree of salt rejection will be competitive with RO if a reasonable differential in osmotic pressure is reached (McCutcheon et al., 2005). For seawater, FO recovery has been shown to be greater than 60%, and for brackish water, it has been shown to be greater than 90% (Chaudhry, 2014). There are several additional problems in FO systems that aren't present in RO systems. Since the product water is combined with the osmotic agent after the FO phase is completed, the FO procedure does not produce high-quality water for usage in a single step (Thompson & Nicoll, 2011). As a result, an osmotic agent must be chosen that is either suitable to include in the product water or that can be extracted quickly and economically (Li et al., 2013). For example, removing an osmotic agent dissolved in water that consists of a mixture of ammonia (NH_3) and carbon dioxide (CO_2) gases needs just a small amount of electrical power.

Switchable polarity solvents, which shift from miscible to immiscible with the addition or subtraction of CO_2, are currently being tested as alternative FO draw solutions (Stone et al., 2013). Control of membrane fouling, optimizing boron and arsenic removal, and solving capacitive polarization issues are all obstacles for FO (DOE, 2014). There are several compelling reasons to improve and commercialize this technique, including the fact that this method of extracting high-quality permeate uses relatively little resources, operates at naturally low pressures, and operates at ambient temperatures (Thompson and Nicoll, 2011). In Oman, on the Arabian Sea, a commercial system (capacity of 200 m³/day) is already in service. Although FO has seen some commercial traction, further study and testing are needed to confirm its effectiveness as a cost-effective desalination solution in a number of applications (Phuntsho et al., 2012).

4.3.2.2 Reverse Osmosis (RO)

Reverse osmosis (RO) is a form of pressurized filtration that involves pushing water through a semipermeable membrane filter with a solution/diffusion system under pressure. Some larger organic organisms (e.g., bacteria and viruses) may be removed via RO methods, but tiny uncharged species (e.g., hydrogen sulphide and carbon dioxide) may travel via the membrane. Spiral-wound and hollow fine fibre membranes are the two most widely used varieties of membranes on the market. Cellulose acetates, polyamides, polyetheramides, and polyethersulfones are all examples of RO membrane formulations. A thin-film hybrid polymer incorporating a microporous polysulfone support layer with a thin polyamide layer is currently the most commonly utilized membrane material (NRC, 2008). To achieve realistic product water fluxes via the membranes, today's best available RO membranes are worked at pressures well above the osmotic pressure. Increased salt rejection strength, greater surface area per unit volume, increased permeability, better recovery ratios, extended membrane life, and the ability to operate at higher pressures have all benefited from advances in membrane materials and efficiency.

These advancements also lowered membrane costs and reduced the strain needed to manufacture realistic water fluxes, lowering desalination plant operating and energy costs (NRC, 2008; Ghaffour et al., 2013). Membrane fouling, on the other hand, continues to be a significant problem. Pre-treatment (e.g., MF and UF) to extract suspended and colloidal waste, as well as dissolved organic matter, is now the most straightforward and efficient approach to defend against fouling. The creation of fouling-resistant membranes or membranes that can be conveniently washed with low-cost oxidants (e.g., chlorine) will be advantageous as an option. Current RO membranes for seawater and brackish water desalination, on the other hand, are not built to accept oxidants like free chlorine and need chlorine removal from the feedwater before the RO modules may filter it (NRC, 2008).

The most widely employed desalination method is RO. Its installed volume varies from 0.1 m³/day (for maritime and domestic uses) to almost 400,000 m³/day (for industrial uses) (Al-Karaghouli & Kazmerski, 2013). The amount of energy (i.e., osmotic pressure) used to force water through membranes for separation is proportional to the solution's salinity. The majority of RO operating costs are made up of energy, mainly in the form of electricity to power high-pressure pumps. Because of the poor net recoveries of the extremely pressurized feedwater, up to 60% of the added energy in the phase will be wasted if the concentrate is discharged to atmosphere with little effort to recover it. In the focus stream of a seawater RO plant, energy recovery devices will typically recover 75%–96% of the input energy (Sallangos, 2005; NRC, 2008).

Despite the fact that the RO mechanism is relatively advanced, there are possibilities to reduce its energy use by small yet important quantities. Improvements in process management and widely accessible state-of-the-art technology (e.g., more reliable energy recovery systems) have had a substantial and direct impact on lowering the energy and maintenance costs of RO systems over the last few decades (NRC, 2008). Although further advancements in pre-treatment (notably, NF) and membrane technology (e.g., fouling resistance) may allow further improvements, they are considered gradual, as RO technology is approaching its functional limits (DOE, 2014).

4.3.2.3 Electrodialysis and Electrodialysis Reversal

Electrodialysis (ED) is an electrochemical separation technique that uses two electrically charged electrodes to transfer ions across ion-selective semipermeable membranes, thus leaving freshwater behind. The ED works at ambient pressure and transfers ions across the membrane using direct current voltage. In certain cases, an ED unit will extract 50%–94% of TDS from the feedwater. ED membranes, like other membrane processes, are prone to fouling, so some pre-treatment of the feedwater is needed. The electrodialysis reversal (EDR) method is identical to the ED process, except that the electrode polarity is shifted on a regular basis to efficiently mitigate and eliminate scaling and fouling, enabling the device to work at higher recoveries (NRC, 2008). Desalination processes such as EDR and ED are deemed advanced. Since the expense of these methods is directly proportional to the TDS content and rises dramatically with higher salinity, they are commonly used to desalinate brackish water. Municipal ED/EDR applications have been recorded for brackish waters with TDS up to 7,500 ppm (Mickley, 2006). The capacity of an ED plant usually varies from 2 to 145,000 m³/day (Al-Karaghouli & Kazmerski, 2013).

In contrast to RO and thermal desalination, ED will only remove ionic components from a solution. As a consequence, relative to the RO procedure, fouling by uncharged species (e.g., silica) is less serious. Furthermore, new ED/EDR membranes are chlorine-resistant, rendering them more suitable for handling feedwaters containing higher amounts of organic matter, which would normally foul RO membranes (e.g., water reuse applications). These characteristics are essential factors that make ED/EDR more realistic than RO. EDR is now being used in hybrid applications as a focus reduction tool for RO processes due to its robust design (Kiernan & von Guttberg, 2008; Reahl, 2006).

4.3.2.4 Membrane Processes

Membranes are engineered to enable or prevent the movement of specific ions, such as salts. Membranes are essential in the natural processes of dialysis and osmosis since they separate salts. RO, NF, ED, and electrodialysis reversal (EDR) are commercially viable membrane methods. Pressure-controlled membrane processes, such as RO and NF, are categorized as pressure-driven membrane processes, while electrically driven membrane processes, such as ED and EDR, are classified as electrically driven membrane processes (NRC 2008). Important advancements in membrane technology have been made in recent years, and more new membrane desalination capability is installed per year across the world than new distillation capacity. In most cases, 35%–60% of the feed is recovered as product water in seawater; in brackish water, recovery will vary from 50% to 90%, based on initial salinity and the availability of sparingly soluble salts and silica (NRC, 2008).

Membrane processes will minimize salinity in the product water to less than 500 ppm TDS for both brackish and seawater. Membrane systems need the least amount of energy overall, but they are heavily affected by the salt content in the feedwater. Furthermore, since suspended and colloidal particles (e.g., iron, manganese, and sulphides), animals, and natural organic matter can trigger membrane scaling or fouling, pre-treatment is a critical phase in membrane desalination systems, especially those that use feedwater from surface water sources. Feedwater pre-treatment usually includes membrane filtration (e.g., microfiltration (MF) and ultrafiltration (UF)), sterilization, and the inclusion of chemicals to eliminate scaling and biofouling, in addition to traditional pre-treatment methods (e.g., coagulation and sand filtration). Membrane fouling will significantly reduce efficiency and salt rejection, as well as reduce tolerance to chlorine and other oxidants and increase energy consumption.

4.3.2.4.1 Membrane Distillation

Only water vapour molecules pass through a microporous hydrophobic membrane during membrane distillation, which is a thermally induced separation procedure. The vapour pressure differential caused by the temperature difference (which induces a pressure gradient) across the membrane is the guiding force in the MD operation (Alkhudhiri et al., 2012). This system uses both thermal and electrical resources, and it has the potential to reject 100% of salt.

This collection of innovations offers a low-temperature solution to conventional heating and cooling systems. Direct interaction, air distance, sweeping steam, and vacuum are some of the different settings. The hot solution (i.e., feed) is in close contact with the hot side of the membrane surface in the direct-contact configuration, resulting in evaporation at the feed-membrane surface.

This collection of innovations offers a low-temperature solution to conventional heating and cooling systems. Direct interaction, air distance, sweeping steam, and vacuum are some of the different settings. The hot solution (i.e., feed) is in close contact with the hot side of the membrane surface in the direct-contact configuration, resulting in evaporation at the feed-membrane surface (Alkhudhiri et al., 2012). The vapour is moved over the membrane by the pressure gap, and it condenses on the permeate side of the membrane module. The heat loss by conduction is the biggest disadvantage of direct contact MD. The feed fluid is only in close interaction with the hot side of the membrane surface in the air gap configuration. Between the membrane and the condensation surface, stagnant air is added, allowing the vapour to cross the air distance and condense across the cold surface within the membrane cell. The advantage of air-gap MD is that it reduces heat loss by conduction; nevertheless, it often creates additional resistance to mass transfer. In sweeping-gas MD, an inert gas is used to clean the vapour from the permeate membrane side of the membrane module outside the module. The gas serves to minimize heat loss in the same way as air-gap MD does, except when it is not stationary, the mass transfer coefficient is increased. The key drawback of sweeping-gas MD is that a tiny amount of permeate diffuses in a huge amount of sweep-gas, necessitating the usage of a large condenser. Finally, vacuum MD creates a vacuum in the permeate membrane side using a device. Condensation occurs beyond the membrane module, so heat loss by conduction is minimal (Alkhudhiri et al., 2012). MD rejects a large percentage of feed-stream solutes, which is similar to other distillation techniques (NRC, 2008).

MD systems, in general, have the same TDS-level insensitivity benefit as thermal processes, but they run at lower temperatures (i.e., 30°C–90°C) and energy intensities since the solution is not heated to the boiling point (DOE, 2014). This means that waste heat may potentially be used to fuel the desalination device. Furthermore, since these systems need a limited footprint, they have smaller capital costs. Furthermore, the hydrostatic pressure experienced in MD is smaller than that experienced in pressure-driven membrane processes such as RO. However, since it is a membrane operation, there are additional possible issues such as membrane deterioration and fouling of various kinds, which trigger problems with membrane hydrophobicity and lifetime. Other drawbacks include the high enthalpy of vaporization needed for the phase shift of water transferred through the membrane, as well as low rejection of volatile feed-stream pollutants that necessitate pretreatment (NRC, 2008).

This group of solutions is in the early stages of growth, but it has the ability to make substantial progress in the energy-efficient treatment of high-TDS waters (DOE, 2014). To construct an automated separation mechanism, the MD system may be paired with other separation processes such as UF or RO. While several larger-scale pilot plants have been designed to process freshwater, most existing MD applications are either in the laboratory or small-scale pilot plant stages (Alkhudhiri et al., 2012; Kullab & Martin, 2011).

4.4 GREEN TECHNOLOGIES FOR SALINE WATER TREATMENT AND MANAGEMENT

We look at a few key ways of utilizing environmental taxes or pollution penalties to encourage people to use advanced and "green" emissions-reduction technology. The use of saline water is an effective way to address the global lack of water for human use, as well as a vital supplement and strategic reservoir of water supplies in coastal areas. Saline water desalination has been a long-term sustainability policy in many countries due to its value for sustainable human development. Despite the fact that there are millions of desalination technologies available, the rate of clean water recovery remains unsatisfactory. Worse, the pollution generated by burning fossil fuels poses a danger to offshore habitats and the aquatic climate (Zhang et al., 2016).

As a result, scientists have made significant efforts to provide potable water through the use of green technology and have suggested several innovative desalination devices that have a strong probability of success (Ali et al., 2018; Fang et al., 2019; Khan et al., 2018). One of the most appealing and simple-to-implement desalination technologies is solar-driven water evaporation (SWE), which incorporates the two most available commodities on the planet (i.e., solar energy and water) (Tao et al., 2018). Antibiotics have been discovered in high salinity wastewater from aquatic aquaculture systems and pharmaceutical companies (Bôto et al., 2016), which has sparked widespread interest. High salinity in the supply water has a negative impact on soil productivity and plant nutrients, resulting in decreased crop growth and production. Pollutants must be degraded before being discharged into fresh water due to their harmful impacts on the environment and humans. There are many environmentally friendly techniques and strategies for removing salinity from water. Desalination is an essential technology for improving water quantity and efficiency.

4.4.1 Phytoremediation

Soil salinization is one of the main factors affecting farm production around the world. More than half a billion hectares of land are not suitable for crop development due to salinity. As a result, there is a need to find ways to develop saline soils so that they can sustain highly sustainable and meaningful land-use processes in order to face the emerging global food security challenges. While a sound drainage scheme is required to handle the increasing water table as a permanent solution to the soil salinity crisis, this alternative is too energy and cost-intensive to be used on a wide scale over large regions. A suitable alternative is phytoremediation or biological approaches, which are plant-based solutions for improving degraded soils. Plant roots' capacity to maximize dissolution and increase amounts of Ca in soil solution may be used to effectively strip Na from the soil cation exchange complex and leach it from the root zone in phytoremediation of saline soils. Stability of soil aggregates, root proliferation, soil hydraulic properties, and nutrient supply to plants are all increased during the amelioration period. This increase in soil properties allows for the planting of less resistant plants, as well as bettering the atmosphere and climatic conditions by increasing carbon sequestration (Ashraf et al., 2010).

Phytoremediation has been shown to be effective in a number of ways, including: (i) no need for chemical amendments; (ii) financial or other benefits from crops grown during amelioration; (iii) promotion of soil-aggregate stabilization and growth of macrospores, which enhance soil hydraulic properties and root proliferation; and (iv) increased plant-nutrient abundance in soil after phytoremediation. Phytoremediation is particularly effective when used on highly salty sodic and sodic soils (Qadir et al., 2007).

4.4.1.1 Using Plants

Contaminated soils and waters are a big environmental and human health issue that emerging phytoremediation technologies can help to solve in part. This low-cost plant-based remediation method takes advantage of plants' extraordinary capacity to concentrate elements and compounds from the atmosphere while still metabolizing different molecules in their tissues. Phytoremediation mostly addresses toxic heavy metals and chemical contaminants. Halophytic plants are of particular concern since they grow naturally in areas with high levels of harmful ions, mostly sodium and chloride. Furthermore, salt-tolerant plants have been found to be able to absorb metals. In comparison to salt-sensitive crop plants widely used for phytoextraction, halophytes have been proposed to be biologically better adapted to cope with environmental pressures, including heavy metals.

As a consequence, halophytes may be ideal candidates for heavy metal phytoextraction or phytostabilization in heavy metal-polluted soils, especially those affected by salinity. Toxic elements including cadmium, zinc, iron, and copper concentrate and are excreted through salt glands or trichomes on the leaf surface, an innovative phytoremediation mechanism known as phytoexcretion. Finally, halophytes have been proposed for soil desalination, with salt deposition in plant tissue or rhizosphere calcite dissolution producing Ca^{2+} that can be combined with Na^+ at places where cations are exchanged (Manousaki & Kalogerakis, 2011a).

Some halophytes in extraction of plants have salt excretion organs (salt glands, salt bladders, or trichomes) in their leaves that regulate plant tissue ion concentration as a secondary tolerance mechanism to excess salt (Lefèvre et al., 2009; Orcutt & Nilsen, 2000), and although not all halophytes have these regulating organs, salt accumulating glands are most common in the Poaceae, Tamaricaceae, Chenopodi, and some poisonous ions such as cadmium, zinc, copper, and lead are excreted through the salt glands from leaf tissues into the leaf surface in halophytes such as *Tamarix aphylla* L., 50 *Atriplex halimus* L., 12 mangroves, and a host of other estuarine and saltmarsh halophytes 51–55. Furthermore, it has been discovered that up to 50% of the salt entering the leaf of a salt-excreting halophyte may be excreted, suggesting that this innovative phytoremediation technique may be used for the remediation of salt-affected soils based on the same concept of collecting excreted salt until it returns to the soils, resulting in a decrease of soil salinity.

Suaeda maitima, Suaeda portulacastrum, Suaeda fruticosa, Suaeda salsa, Suaeda calceoliformis, Kalidium folium, SesuVium portulacastrum, Arthrocnemum indicum, Atriplex nummularia, and *Atriplex prostrata* have been recorded to accumulate high concentrations of salt in their above-ground parts (Rabhi et al., 2009; Keiffer & Ungar, 2002). For example, studies of *Suaeda fruticosa* found that a single harvest of the aerial parts in the fall per year will extract more than 2,400 lbs (1,088.6 kg) of salt from 1 acre. If the plants were harvested at the end of the growing season, studies with *Suaeda salsa* showed that a density of 15 plants/m² could theoretically extract 3,090–3,860 kg Na^+/ha (Zhao, 1991). After it was discovered that it shows no long-term growth inhibition under comparatively high salinity stress conditions, the above is the case (Manousaki & Kalogerakis, 2009). *Atriplex halimus* L., a salt-accumulating halophyte with functional salt bladders, warrants further research to determine the concentrations of salt absorption and accumulation in order to investigate its possible usage for phytodesalination. Phytodesalination has piqued interest as a method of desalinating soils by collecting salt-accumulating halophytes, which obtain and accumulate high concentrations of salt in their above-ground tissues, and thus salty soils may be strengthened (Manousaki & Kalogerakis, 2011b).

4.4.1.2 Using Algae

Established water supplies will no longer satisfy human needs due to population growth and city extension, as well as a global shortage of fresh water sources. Desalination, or the elimination of salinity of water using biological processes, entails the use of plant species, microorganisms, algae, or a mixture of these, both of which are capable of lowering water salinity. The aim of this research was to see whether *Dunaliella salina* algae could be used to desalinate salty water. Desalination of saline water is now done using a variety of methods all over the world, depending on the quantity and quality of water needed (El Nadi Shiva et al., 2012; Gorjian & Ghobadian, 2015). Desalination methods include membrane for RO, multiple-effect distillation process (MED), thermal evaporation, flash multi-stage (MSF), distillation vapour compression (VCD), and ion exchange (IX). Each has advantages and disadvantages that are applied based on regional conditions and needs (Sehn 2008; Al-Odwani et al., 2006; Toufic et al., 2011).

The usage of algae to remove salt from saline water to provide water for various uses is a novel and cost-effective concept. This is a novel approach to desalination that has the potential to be cost-effective (Kesaano & Sims, 2014; Nagy et al., 2019). Algae are basic plant organisms that lack a stem and root and contain chlorophyll. They range in size from single-celled organisms to massive, high-density species with varying abilities to thrive in aqueous and dry environments. The majority of algae organisms will thrive in salty waters with high water solubility (Munoz & Guieysse, 2006). Water salts, among all nutrients, are incorporated into biomass by living creatures (plants and animals), and the absorption level of salts from water salinity is decreased as a result. In this situation, certain species are more able to ingest water soluble than their dietary requirements, allowing them to lower water salinity more effectively. Desalination, or the reduction of water salinity using biological means, entails the use of various macrofite (aquatic plants), microfite (algae), and microorganism types, or their mixture, which may be useful in lowering water salinity (El Sergany et al., 2019).

Algae, unlike organic plants, can complete their life cycle at a variety of salinities. According to several latest reports, using algae to desalinate freshwater waters would cost the least amount of money. The findings revealed that algae have a beneficial impact on seawater desalination, which is a novel approach for desalinating salt water using modern methods. According to certain findings, sewage with a salinity of 35,000–10,900 mg/L has a content of 35,000–10,900 mg/L. Salinity was lowered by up to 2,500 times, according to the findings (El Nadi et al., 2011). According to Badawy et al. (2012), the microalgae *Dunaliella salina* is the most salinity-resistant eucaryote genus, which can be found in a variety of salty habitats such as lakes and saltwater lagoons. According to the findings, the algae's effectiveness in high salinity may be used to reduce the salinity in seawater, waste, and factory recycling water. Overall, the findings suggest that using green algae to reduce water salinity is advantageous. Variations in physical and chemical conditions, such as salinity, have a direct impact on green microalgae development and biochemical composition (Munoz & Guieysse, 2006).

Created waters are saline wastewater that contains both salt (mainly NaCl) and nutrients and are commonly produced in the food processing, leather, and petroleum industries (Lefebvre and Moletta, 2006). The use of seawater as a source of water in cases where high-quality water isn't needed encourages the production of vast volumes of salty wastewater. High salinity has been shown to impair nitrifying bacteria metabolism and result in nutrient reduction loss in saline wastewaters (Uygur & Kargı, 2004).

4.4.1.3   Using Biofilms and Membrane Bioreactors

The electrical conductivity of urban wastewater is estimated in the range of 0.7–3.0 mS/cm (Norton-Brandão et al., 2013). The wastewater generated by many industrial processes (e.g., seafood production, milk processing, petroleum refining, and textile manufacturing) has a high salinity (Yang et al., 2013). The usage of salt for cooking, snow melting, and toilet flushing with marine water, among other things, could result in high-salinity wastewater. Salinity has been shown to influence microbial populations in wastewater treatment by reducing compound bioavailability, causing microbial mortality due to differential osmotic pressure through cell membranes, or inhibiting biodegradation of potentially harmful intermediates, resulting in their aggregation in the bioreactor. For the treatment of salinity-amended urban wastewater, a pilot plant membrane bioreactor (MBR) and two pilot plant hybrid moving bed biofilm reactor–membrane bioreactors (MBBR–MBRs), divided into three aerobic and one anoxic chamber, were started up (Castillo-Carvajal et al., 2014).

Biological processes, on the other hand, are both economical and environmentally sustainable (Lefebvre & Moletta, 2006). In such situations, extremophilic bacteria's salinity resistance appears to be successful and allows for long-term waste care (Dopson et al., 2016). Microbial fuel cells (MFC) have gotten a lot of recognition from scientists all over the world because of their potential as waste-to-energy converters. MFCs have been shown to be capable of both generating electricity and treating waste (Choudhury et al., 2017). The electrode spacing, temperature, and ionic activity of the electrolyte all play a role in power generation in MFCs. Conductivity rises as the ionic pressure of the electrolyte rises, reducing the system's internal resistance and resulting in higher MFC power production (Dopson et al., 2016). Thus, if the correct collection of bacteria capable of decaying organic matter in saline wastes can be detected, salinity in wastes aids power output from MFC and is an asset rather than a drawback. The use of halophilic microorganisms to handle high-saline wastewater is an important method for dealing with the issue of saline waste disposal (Abou-Elela et al., 2010). According to Uygur and Kargı, chloride salts are commonly contained in wastewaters, but their impact on nitrification appears to be more prominent at high concentrations (1 percent m/v, NaCl) (Tramper & Van't Riet, 1991). Salt, on the other hand, can alter microbial metabolism, sludge settleability characteristics, and the architecture of sludge flocs and biofilms. Salinity may also interrupt biofilm composition/extracellular polymeric substances (EPS) and has a significant impact on oxygen maximal solubility and rate of transition to the liquid process (Uygur & Kargı, 2004).

4.4.2 Nanomaterials

Unconventional water supplies such as rain water, polluted groundwater, brackish water, industrial wastewater, and seawater are becoming more common as global water demand grows, particularly in traditionally water-stressed areas. Seawater is the most plentiful supply of fresh water for human demand on the planet. Seawater, on the other hand, is unsafe for household use due to its excessive salinity. Conventional water management systems are good at removing nutrients and biological matter, but not so good at lowering salinity. Desalination is a viable solution in this situation since it can bridge the difference between usable capacity and increasing demand. In reality, in many abandoned and remote places, as well as coastal regions, it has long been used to provide clean drinking water. The separation of salts and minerals from seawater to create freshwater is known as desalination (Misdan et al., 2012).

According to Mohamed's final science report from 2011, there are four types of energy used in seawater desalination processes: nuclear, mechanical, electrical, and chemical. Highly efficient, modular, and multifunctional nanotechnology-enabled water and wastewater treatment solutions are expected to provide high-performance, low-cost water, and wastewater treatment solutions that not only overcome major challenges faced by current water treatment technologies but also provide new treatment capabilities that could allow for the cost-effective use of unconventional water sources to expand the water supply. Nanomaterials are structures with a diameter in at least one dimension of less than 100 nm. At this rate, materials have unique size-dependent properties not found in their larger counterparts. Many of these nanomaterials have been studied and identified as possible items for water and wastewater treatment applications including adsorption, membrane filtration, photocatalysis, disinfection, and microbial control, as well as sensing and monitoring. Among the components are magnetic nanoparticles (NPs), dye-doped silica NPs, noble metal NPs, quantum dots, carbon nanotubes (CNTs), nanoscale metal oxide, nanofibers, zeolites, aquaporin (AQP), graphene, nano-magnetite, and a number of others. Carbon-based nanomaterials (CNMs) such as carbon nanotubes (CNTs) and graphene have piqued scientists' interest due to their peculiar properties such as large surface size, strong porosity range, excellent thermal and electrical conductivity, and high mechanical strength and stiffness (Li et al., 2009).

The introduction of carbon nanotubes and graphene with these novel properties has been suggested as a major possible advance in water desalination. Previous research on the transport properties of CNMs has discovered that smooth and frictionless graphitic walls of CNTs and graphene, as well as a rapid sorption-desorption mechanism, can allow rapid transport of water molecules, resulting in high flux separation efficiency (Goh et al., 2013). In contrast, zeolites, a modern flexible, special, and low-cost nanomaterial, have also been widely used in the desalination sector (Safarpour et al., 2017; Hosseini et al., 2014). Zeolites have a porous composition of several channels and cavities, making them a possible nanomaterial with a large surface region. The presence of zeolite pores between the diameters of hydrated salt ions and water molecules is advantageous for zeolites in separating ions from saline water, resulting in exceptional water purification efficiency.

The usage of nanomaterials will boost these dimensions of CDI, which would result in major improvements in desalination processes. The development of nanomaterials may theoretically play a significant role in averting possible water shortages. Improvements in the science and engineering of nanomaterials are predicted to accelerate as the field develops. To boost and preserve water quality and quantity, it is critical to take advantage of the leapfrogging possibilities presented by these nanomaterials. The threat to the environment and public health posed by nanomaterials released unintentionally or inadvertently remains a major concern. In order to monitor them at risk assessment and policies, these factors can be objectively evaluated and proactively analysed by environmental experts. A new policy for the use of nanomaterials in water purification and desalination should be established, and a protection guideline should be created (Teow & Mohammad, 2019).

4.4.3 Solar Desalination

Desalination is a process that removes a portion of pure water from a salt water supply using a lot of electricity. As salt water (feedwater) is pumped into the process, it produces two production streams: one of plain water and another of high-salt wastewater (Pugsley et al., 2016). Desalination is becoming extremely important as a highly reliable, non-conventional supply of fresh water in many places around the world that are experiencing extreme water scarcities. Markets for desalination have grown dramatically in recent decades and are expected to continue to grow in the coming years, especially in the Mediterranean, Middle East, and North African regions. Solar energy may be used to power water treatment processes, which may be a long-term solution to the world's water shortage issue. Seven main solar desalination processes, as well as the solar photocatalysis method and solar disinfection, are investigated in terms of recent advances and implementations. The resources used in the treatment phase must be taken into account if a sustainable water source is to be reached (Li et al., 2013).

Solar desalination technologies have received a lot of attention due to their applicability in arid or remote regions. For decades, large commercial desalination plants fuelled by fossil fuels have been installed in countries with water scarcity, especially in the oil-rich countries of the Middle East. Solar energy may be used to power desalination plants either directly or indirectly. In direct solar desalination systems, solar energy is harvested either by solar thermal collectors to provide heat or photovoltaic panels to generate electricity for thermal or membrane desalination technologies, with solar still being the most representative technology; while in indirect solar desalination systems, solar energy is harvested either by solar thermal collectors to provide heat or photovoltaic panels to generate electricity for thermal or membrane desalination technologies, with solar still being the most representative technology (Esfahani et al., 2016).

4.4.3.1   Direct Solar Desalination Still

Solar is also the most widely used direct solar desalination technology and is best suited to low-capacity water supply schemes in rural areas where pipeline installation or water distribution by truck are both uneconomical and unreliable (Tiwari et al., 2003). A single-basin solar still is made up of an airtight, slanting translucent cover that encloses a black-painted basin filled with saline water. After being melted by the solar radiation consumed by the basin, the water evaporates. The inner surface of the sloping cover condensed, and purified water was stored at the lower end of the cover. Solar is still not commonly used due to its poor efficiency per unit installation area, which is typically 1–5 L/m²/day for a single basin despite its technological versatility and reasonably low maintenance requirements. As a result, the construction of solar stills necessitates a considerable amount of room. Absorption area, water density, inlet water temperature, water-glass cover temperature differential, and other factors influence the efficiency of a basin-style solar still. Solar stills have undergone extensive modifications in order to increase their efficiency. Modifications are mostly intended to improve water evaporation in the basin, condensation on the lid, or the recovery of latent heat of evaporation (Al-Sulaiman et al., 2015).

4.4.3.2   Indirect Solar Desalination

4.4.3.2.1   Solar Humidification-Dehumidification

The isolation of salt water and pure water was achieved using humidification-dehumidification (HDH) technology, which took advantage of the moisture carrying ability of hot air (moisture). HDH is usually combined with external heaters such as solar collectors, unlike solar now, which is mainly passive. Furthermore, humidification and dehumidification are carried out in different modules, allowing each component to be planned and configured separately (Seifert et al., 2013). As a result, HDH could attain much higher thermal efficiency than solar even. Humidifier, dehumidifier, and external heater are the main components of the HDH operation. Solar HDH desalination is widely accepted as a viable option for low-capacity decentralized water production. Based on the cycle structure, the HDH mechanism can be divided into three categories: closed air, open water cycle (CAOW); closed water, open air cycle (CWOA); and open water, open air cycle (OWOA) (Al-Sulaiman et al., 2015).

4.4.3.2.2   Solar Disinfection of Water (SODIS)

Solar disinfection of drinking water (SODIS) is an easy, low-cost household water treatment technology promoted by the World Health Organization (WHO) that is ideal for developing countries that lack access to clean drinking water but obtain sufficient solar radiation. SODIS only demands that water be kept in clear tanks (typically PET bottles) that are subjected to intense sunlight over long periods of time in order for waterborne pathogens to be inactivated (Keogh et al., 2015). SODIS effectiveness has been well established. Clinical studies have shown that SODIS will decrease rates of childhood dysentery and baby diarrhoea by 45% when used against nearly all waterborne microbial organisms (McGuigan et al., 2012). Researchers are looking into a variety of factors that influence the SODIS operation, such as obtained UV radiation, container properties, water content, water temperature, and external configurations, among others. The annual cost of SODIS is calculated to be $0.63 per user (Vivar et al., 2015; McGuigan et al., 2012).

4.5 CONCLUDING REMARKS

Salt water is produced naturally as well as anthropogenically. The salt water can affect plant, soil, and human health adversely. To overcome the negative effects of salt water and reduce its associated consequences, different conventional and advanced treatment technologies are used throughout the world having salt-water problems. Alternatively, due to the side effects and associated problems of pollution, green or eco-friendly technologies are being used for the treatment and management of salt water. After the treatment, water can be used for irrigation or drinking water.

REFERENCES

Abdulsalam, A., Idris, A., Mohamed, T. A., & Ahsan, A. 2017. An integrated technique using solar and evaporation ponds for effective brine disposal management. *International Journal of Sustainable Energy* 36: 914–925.

Abou-Elela, S. I., Kamel, M. M., & Fawzy, M. E. 2010. Biological treatment of saline wastewater using a salt-tolerant microorganism. *Desalination* 250(1): 1–5.

Adviento-Borbe, M. A. A., Doran, J. W., Drijber, R. A., & Dobermann, A. 2006. Soil electrical conductivity and water content affect nitrous oxide and carbon dioxide emissions in intensively managed soils. *Journal of Environmental Quality* 35: 1999–2010.

Afrasiabi, N., & Shahbazali, E. 2011. RO brine treatment and disposal methods. *Desalinisation and Water Treatment Journal* 35: 39–53.

Ahmed, M., Shayya, W. H., Hoey, D., & Al-Handaly, J. 2001. Brine disposal from reverse osmosis desalination plants in Oman and the United Arab Emirates. *Desalination* 133: 135–147.

Akbarimoghaddam, H., Galavi, M., Ghanbari, A., & Panjehkeh, N. 2011. Salinity effects on seed germination and seedling growth of bread wheat cultivars. *Trakia Journal of Sciences* 9(1): 43–50.

Ali, A., Tufa, R. A., Macedonio, F., Curcio, E., & Drioli, E. 2018. Membrane technology in renewable-energy-driven desalination. *Renewable and Sustainable Energy Reviews* 81: 1–21.

Ali, M. 2010. Agriculture problems in Pakistan and their solutions. South Asia Partnership Pakistan. Retrieved from https://sappk.wordpress.com/2010/03/08/agriculture-problems-in-pakistan-and-their-solutions/.

Al-Karaghouli, A., & Kazmerski, L. L. 2013. Energy consumption and water production cost of conventional and renewable-energy-powered desalination processes. *Renewable and Sustainable Energy Reviews* 24: 343–356.

Alkhudhiri, A., Darwish, N., & Hilal, N. 2012. Membrane distillation: A comprehensive review. *Desalination* 287: 2–18.

Al-Odwani, A, El-Sayed, E. E. F., Al-Tabtabaei, M, & Safar, M. 2006. Corrosion resistance and performance of copper–nickel and titanium alloys in MSF distillation plants. *Desalination* 201: 46–57. doi:10.1016/j.desal.2006.01.034.

Al-Sulaiman, F. A., Zubair, M. I., Atif, M., Gandhidasan, P., Al-Dini, S. A., & Antar, M. A. 2015. Humidification dehumidification desalination system using parabolic trough solar air collector. *Applied Thermal Engineering* 75: 809–816.

Amato, M., & Ladd, J. N. 1994. Application of the ninhydrin-reactive N assay for microbial biomass in acid soils. *Soil Biology Biochemistry* 26: 1109–1115.

Anderson, J. P. E., & Domsch, K. H. 1980. Quantities of plant nutrients in the microbial bio-mass of selected soils. *Soil Science* 130: 211–216.

Andronov, E. E., Petrova, S. N., Pinaev, A. G., Pershina, E. V., Rakhimgalieva, S. Z., Akhmedenov, K. M., & Sergaliev, N. K. 2012. Analysis of the structure of microbial community in soils with different degrees of salinization using T-RFLP and real-time PCR techniques. *Eurasian Journal of Soil Science* 45: 147–156.

Ashraf, M. 2004. Some important physiological selection criteria for salt tolerance in plants. *Flora* 199: 361–376.

Ashraf, M. Y., Ashraf, M., Mahmood, K., Akhter, J., Hussain, F., & Arshad, M. 2010. Phytoremediation of saline soils for sustainable agricultural productivity. In: *Plant Adaptation and Phytoremediation*, pp. 335–355. Springer, Dordrecht.

Bano, A., & Fatima, M. 2009. Salt tolerance in *Zea mays (L.) following inoculation with Rhizobium and Pseudomonas*. *Biology and Fertility of Soils* 45: 405–413.

Batra, L., & Manna, M. C. 1997. Dehydrogenase activity and microbial biomass carbon in salt-affected soils of semiarid and arid regions. *Arid Land Research and Management* 11: 295–303.

Blaylock, A.D., 1994. Soil salinity, salt tolerance and growth potential of horticultural and landscape plants. Co-operative Extension Service, University of Wyoming, Department of Plant, Soil and Insect Sciences, College of Agriculture, Laramie, Wyoming.

Bôto, M., Almeida, C. M. R., & Mucha, A. P. 2016. Potential of constructed wetlands for removal of antibiotics from saline aquaculture effluents. *Water* 8(10): 465–479.

Brandt, H., & Tait, J. H. 1997. Process for Brine Disposal. United States patent US 5,695,643. Available from: http://www.freepatentsonline.com/5695643.html.

Castillo-Carvajal, L. C., Sanz-Martín, J. L., & Barragán-Huerta, B. E. 2014. Biodegradation of organic pollutants in saline wastewater by halophilic microorganisms: A review. *Environmental Science and Pollution Research* 21(16): 9578–9588.

Chaudhry, S. 2014. An Overview of Industrial Desalination Technologies, panel on American Society of Mechanical Engineers' (ASME's) New Industrial Desalination Handbook. *ASME 2014 Power Conference*, Baltimore, MD, July.

Cheng, R., Zhu, H., Cheng, X., Shutes, B., & Yan, B. 2020. Saline and Alkaline tolerance of wetland plants–what are the most representative evaluation indicators?. *Sustainability* 12(5): 1913–1930.

Chinnusamy, V., Zhu, J., & Zhu, J. -K. 2006. Gene regulation during cold acclimation in plants. *Plant Physiology* 126(1): 52–61.

Choudhury, P., Uday, U. S. P., Mahata, N., Tiwari, O. N., Ray, R. N., Bandyopadhyay, T. K., & Bhunia, B. 2017. Performance improvement of microbial fuel cells for waste water treatment along with value addition: A review on past achievements and recent perspectives. *Renewable and Sustainable Energy Reviews* 79: 372–389.

National Research Council. 1993. Soil and Water Quality: An Agenda for Agriculture. Washington, DC: The National Academies Press. https://doi.org/10.17226/2132.

DOE. 2014. "Best Management Practice #10: Cooling Tower Management," Federal Energy Management Program. http://energy.gov/eere/femp/best-management-practice-10-cooling-towermanagement. Accessed Oct. 7, 2014.

Dopson, M., Ni, G., & Sleutels, T. H. 2016. Possibilities for extremophilic microorganisms in microbial electrochemical systems. *FEMS Microbiology Reviews* 40(2): 164–181.

El Nadi, M. H. A., Waheb, I. S. A., & Saad, S. A. H. A. 2011. Using continuous flow algae ponds for water desalination. *El Azhar Univ., Faculty of Eng., CERM of Civil Eng.* 33(4).

El Nadi Shiva, M. H., Nasr, N. A. H., El Hosseiny, O. M. & Badawy, M. A. 2012. Algae Application for biological desalination. *2nd International Conference & Exhibition, Sustainable Water Supply & Sanitation, (SWSSC2012), Holding Company for Water and Wastewater*, Cairo, Egypt, December.

El Sergany, F. A. G. H., El Hosseiny, O. M., & El Nadi, M. H., 2019. The optimum algae dose in water desalination by algae ponds. *International Research Journal of Advanced Engineering and Science* 4: 152–154.

Esfahani, I. J., Rashidi, J., Ifaei, P., & Yoo, C. 2016. Efficient thermal desalination technologies with renewable energy systems: A state-of-the-art review. *Korean Journal of Chemical Engineering* 33(2): 351–387.

Fang, S., Tu, W., Mu, L., Sun, Z., Hu, Q., & Yang, Y. 2019. Saline alkali water desalination project in Southern Xinjiang of China: A review of desalination planning, desalination schemes and economic analysis. *Renewable and Sustainable Energy Reviews* 113: 109–268.

Farid, H. U., Bakhsh, A., Ali, M. U., Khan, Z. M., Shakoor, A., & Ali, I. 2018. Field investigation of aquifer storage recovery (ASR) technique to recharge groundwater a case study in Punjab province of Pakistan. *Water Supply* 18(1): 71–83.

Fernández-Torquemada, Y., Gónzalez-Correa, J. M., Loya, A., Ferrero, L. M., Díaz-Valdés, M., & Sánchez-Lizaso, J. L. 2009. Dispersion of brine discharge from seawater reverse osmosis desalination plants. *Desalinization of Water Treatment* 5: 137–145.

Ghaffour, N., Missimer, T. M., & Amy, G. L. 2013, Technical review and evaluation of the economics of water desalination: Current and future challenges for better water supply sustainability. *Desalination* 309: 197–207.

Gilron, J., Folkman, Y., Savliev, R., Waisman, M., & Kedem, O. 2003. WAIV – wind aided intensified evaporation for reduction of desalination brine volume. *Desalination* 158: 205–214.

Goh, P. S., Ismail, A. F., & Ng, B. C. 2013. Carbon nanotubes for desalination: Performance evaluation and current hurdles. *Desalination* 308: 2–14.

González, E., Arconada, B., Delgado, P., Alcaraz, A., Álvarez, A., Antequera, M., Fernández, Y., Garrote, A., de Rico, S., Salete, E., & Sánchez Lizaso, J. L. 2011. Environmental research on brine discharge optimization: A case study approach. *Desalinization of Water Treatment* 3: 197–205.

Hashim, A., & Hajjaj, M. 2005. Impact of desalination plants fluid effluents on the integrity of seawater, with the Arabian Gulf in perspective. *Desalination* 182: 373–393.

Hosseini, S. M., Rafiei, S., Hamidi, A. R., Moghadassi, A. R., & Madaeni, S. S. 2014. Preparation and electrochemical characterization of mixed matrix heterogeneous cation exchange membranes filled with zeolite nanoparticles: Ionic transport property in desalination. *Desalination* 351: 138–144.

Hu, Y., & Schmidhalter, U. 2002. Limitation of salt stress to plant growth. In: Hock, B., Elstner, C. F., editors. *Plant Toxicology*, pp. 91–224. Marcel Dekker Inc., New York.

Keiffer, C. H., & Ungar, I. A. 2002. Germination and establishment of halophytes on brine-affected soils. *Journal of Applied Ecology* 39(3): 402–415.

Keogh, M. B., Castro-Alférez, M., Polo-López, M. I., Calderero, I. F., Al-Eryani, Y. A., Joseph-Titus, C. & Fernandez-Ibanez, P. 2015. Capability of 19-L polycarbonate plastic water cooler containers for efficient solar water disinfection (SODIS): Field case studies in India, Bahrain and Spain. *Solar Energy* 116: 1–11.

Kesaano, M., & Sims, R. C. 2014. Algal biofilm based technology for wastewater treatment. *Algal Research* 5: 231–240. doi:10.1016/j.algal.2014.02.003

Khan, M. A., & Abdullah, Z. 2003. Salinity–sodicity induced changes in reproductive physiology of rice (Oryza sativa) under dense soil conditions. *Environmental and Experimental Botany* 49(2): 145–157.

Khan, M. A., Rehman, S., & Al-Sulaiman, F. A. 2018. A hybrid renewable energy system as a potential energy source for water desalination using reverse osmosis: A review. *Renewable and Sustainable Energy Reviews* 97: 456–477.

Kiernan, J., & von Guttberg, A. 2008. Selection of EDR Desalting Technology rather than MF/RO for the City of San Diego Water Reclamation Project, GE Water & Process Technologies Technical Paper.

Kullab, A., & Martin, A. 2011. Membrane Distillation and applications for water purification in thermal cogeneration plants. *Separation and Purification Technology* 76: 231–237.

Laura, R. D. 1974. Effects of neutral salts on carbon and nitrogen mineralization of organic-matter in soil. *Plant and Soil* 41: 113–127.

Lefèvre, I., Marchal, G., Meerts, P., Corréal, E., & Lutts, S. 2009. Chloride salinity reduces cadmium accumulation by the Mediterranean halophyte species Atriplex halimus L. *Environmental and Experimental Botany* 65(1): 142–152.

Lefebvre, O., & Moletta, R. 2006. Treatment of organic pollution in industrial saline wastewater: A literature review. *Water Research* 40(20): 3671–3682.

Li, D., Zhang, X., Simon, G. P., & Wang, H. 2013. Forward osmosis desalination using polymer hydrogels as a draw agent: Influence of draw agent, feed solution and membrane on process performance. *Water Research* 47(1): 209–215.

Li, X., Cai, W., An, J., Kim, S., Nah, J., Yang, D., & Ruoff, R. S. 2009. Large-area synthesis of high-quality and uniform graphene films on copper foils. *Science* 324(5932): 1312–1314.

Machado, R. M. A., & Serralheiro, R. P. 2017. Soil Salinity: Effect on vegetable crop growth. *Management Practices to Prevent and Mitigate Soil Salinization* 3(2): 30.

Manousaki, E., & Kalogerakis, N. 2009. Phytoextraction of Pb and Cd by the Mediterranean saltbush (Atriplex halimus L.): Metal uptake in relation to salinity. *Environmental Science and Pollution Research* 16(7): 844–854.

Manousaki, E., & Kalogerakis, N. 2011a. Halophytes an emerging trend in phytoremediation. *International Journal of Phytoremediation* 13(10): 959–969.

Manousaki, E., & Kalogerakis, N. 2011b. Halophytes present new opportunities in phytoremediation of heavy metals and saline soils. *Industrial & Engineering Chemistry Research* 50(2): 656–660.

McCutcheon, J. R., McGinnis, R. L., & Elimelech, M. 2005. A novel ammonia-carbon dioxide forward (direct) osmosis desalination process *Desalination* 174(1): 1–11.

McGuigan, K. G., Conroy, R. M., Mosler, H. J., du Preez, M., Ubomba-Jaswa, E., & Fernandez-Ibanez, P. 2012. Solar water disinfection (SODIS): A review from bench- top to roof-top. *Journal of Hazardous Materials* 235: 29–46.

Mickley, M. 2001. E.R.T. (U. S.). W.T.E. and R. Group, Membrane Concentrate Disposal: Practices and Regulation, Final Report. U.S. Department of the Interior, Bureau of Reclamation, Technical Service Center, Water Treatment Engineering and Research Group. Available from: https://books.google.co.in/books?id=GbmNOz1fywMC.

Mickley, M. C. 2006. *Membrane Concentrate Disposal: Practices and Regulation (Second Edition)*, Report 123. U.S. Bureau of Reclamation, Denver, CO.

Millán, A., Velasco, J., Gutiérrez-Cánovas, C., Arribas, P., Picazo, F., Sánchez-Fernández, D., & Abellán, P. 2011. Mediterranean saline streams in southeast Spain: What do we know? *Journal of Arid Environment* 75: 1352–1359.

Misdan, N., Lau, W. J., & Ismail, A. F. 2012. Seawater Reverse Osmosis (SWRO) desalination by thin-film composite membrane-Current development, challenges and future prospects. *Desalination* 287: 228–237.

Munns, R. 2002. Comparative physiology of salt and water stress. *Plant Cell and Environment* 25:239–250.

Unns R. (2009) Strategies for Crop Improvement in Saline Soils. In: Ashraf M., Ozturk M., Athar H. (eds) *Salinity and Water Stress*. Tasks for Vegetation Sciences, vol. 44. Springer, Dordrecht. https://doi.org/10.1007/978-1-4020-9065-3_11

Munns, R., & James, R. A. 2003. Screening methods for salinity tolerance: A case study with tetraploid wheat. *Plant and Soil* 253: 201–218.

Munoz, R., & Guieysse, B. 2006. Algal–bacterial processes for the treatment of hazardous contaminants: A review. *Water Research*: 2799–2815. doi:10.1016/j.watres.2006.06.011.

Nagy, A. M., El Nadi, M. H. & Hussein, H. M. 2019. Determination of the best water depth in desalination algae ponds. *El Azhar Univ., Faculty of Eng., CERM of Civil Eng.* 38(4) 1.

Netondo, G. W., Onyango, J.C., & Beck, E. 2004. Sorghum and salinity: II. Gas exchange and chlorophyll fluorescence of sorghum under salt stress. *Crop Science* 44: 806–811.

Nikiforakis, I. K., & Stamou, A. I. 2015. Integrated modelling for the discharge of brine from desalination plants into coastal waters. *Desalinisation and Water Treatment journal* 53:3214–3223.

Norton-Brandão, D., Scherrenberg, S. M., & van Lier, J. B. 2013. Reclamation of used urban waters for irrigation purposes–a review of treatment technologies. *Journal of Environmental Management* 122: 85–98.

Nouri, H., Chavoshi Borujeni, S., Nirola, R., Hassanli, A., Beecham, S., Alaghmand, S., Saint, C., & Mulcahy, D. 2017. Application of green remediation on soil salinity treatment: A review on halophytoremediation. *Process Safety and Environmental Protection* 107: 94–107.

NRC (National Research Council). 2008. *Desalination: A National Perspective.* National Research Council, Washington, DC.

Orcutt, D. M., & Nilsen, E. T. 2000. *Physiology of Plants Under Stress: Soil and Biotic Factors.* John Wiley & Sons, New Jersey, USA.

Panagopoulos, A., Haralambous, K.-J., & Loizidou, M. 2019. Desalination brine disposal methods and treatment technologies – a review. *Science of Total Environment* 693: 133545–133570.

Pathak, H., & Rao, D. L. N. 1998. Carbon and nitrogen mineralization from added organic matter in saline and alkali soils. *Soil Biology and Biochemistry* 30: 695–702.

Pearson, K.E. 2003. The Basic of Salinity and Sodicity. Effects on soil physical properties. Adapted by Krista E. Pearson. From a paper By Nikos, J. Warrence, Krista E. Pearson and James W. Bauder. Montana State University. MSU Extension Services.

Phuntsho, S., Vigneswaran, S., Kandasamy, J., Hong, S., Lee, S., & Shon, H. K. 2012. Influence of temperature and temperature difference in the performance of forward osmosis desalination process. *Journal of Membrane Science* 415: 734–744.

Pugsley, A., Zacharopoulos, A., Mondol, J. D., & Smyth, M. 2016. Global applicability of solar desalination. *Renewable Energy* 88: 200–219.

Qadir, M., Oster, J. D., Schubert, S., Noble, A. D., & Sahrawat, K. L. 2007. Phytoremediation of sodic and saline-sodic soils. *Advances in Agronomy* 96: 197–247.

Rabhi, M., Hafsi, C., Lakhdar, A., Hajji, S., Barhoumi, Z., Hamrouni, M. H., Abdelly, C., & Smaoui, A., 2009. Evaluation of the capacity of three halophytes to desalinize their rhizosphere as grown on saline soils under nonleaching conditions. *African Journal of Ecology* 47(4): 463–468.

Reahl, E. R. 2006. Half a Century of Desalination with Electrodialysis, GE Water & Process Technologies Technical Paper.

Reca, J., Trillo, C., Sánchez, J. A., Martínez, J., & Valera, D. 2018. Optimization model for on-farm irrigation management of Mediterranean greenhouse crops using desalinated and saline water from different sources. *Agricultural Systems* 166: 173–183.

Reeder, L. R. 1981. A review of three decades of deep-well injection and the present state of the art. *Journal of Environmental Sciences and Studies* 17: 373–383.

Reynolds, A. J. 1962. Observations of a liquid-into-liquid jet. *Journal of Fluid Mechanics* 14(4): 552–556.

Rietz, D. N., & Haynes, R. J. 2003. Effects of irrigation-induced salinity and sodicity on soil microbial activity. *Soil Biology and Biochemistry* 35: 845–854.

Rousk, J., Elyaagubi, F. K., Jones, D. L., & Godbold, D. L. 2011. Bacterial salt tolerance is unrelated to soil salinity across an arid agroecosystem salinity gradient. *Soil Biology and Biochemistry* 43: 1881–1887.

Saddiq, M. S., Iqbal, S., Afzal, I., Ibrahim, A. M., Bakhtavar, M. A., Hafeez, M. B., Jahanzaib, & Maqbool, M. M. 2019. Mitigation of salinity stress in wheat (Triticum aestivum L.) seedlings through physiological seed enhancements. *Journal of Plant Nutrition* 42(10): 1192–204.

Safarpour, M., Vatanpour, V., Khataee, A., Zarrabi, H., Gholami, P., & Yekavalangi, M. E. 2017. High flux and fouling resistant reverse osmosis membrane modified with plasma treated natural zeolite. *Desalination* 411: 89–100.

Saline Water and Salinity – USGS. https://www.usgs.gov › water- science-school › science. Accessed on 9 April, 2021.

Salomaa, S. 2004. Research projects of STUK 2003–2005.

Sarig, S., & Steinberger, Y. 1994. Microbial biomass response to seasonal fluctuation in soil–salinity under the canopy of desert halophytes. *Soil Biology and Biochemistry* 26: 1405–1408.

Sarig, S., Fliessbach, A., & Steinberger, Y. 1996. Microbial biomass reflects a nitrogen and phosphorous economy of halophytes grown in salty desert soil. *Biology and Fertility of Soils* 21: 128–130.

Schmitz, W. 1956. Salzgehaltschwankungen in der Werra und ihre fischereilichen Auswirkungen. *Vom Wasser* 23: 113–136.

Sehn, P. 2008. Fluoride removal with extra low energy reverse osmosis membranes: Three years of large scale field experience in Finland. *Desalination* 223(1–3): 73–84. doi:10.1016/j.desal.2007.02.077.

Seifert, B., Kroiss, A., Spinnler, M., & Sattelmayer, T. 2013. About the history of humidification-dehumidification desalination systems. In *The IDA World Congress on Desalination and Water Reuse*.

Setia, R., Marschner, P., Baldock, J., & Chittleborough, D. 2010. Is CO_2 evolution in saline soils affected by an osmotic effect and calcium carbonate? *Biology and Fertility of Soils* 46: 781–792.

Setia, R., Marschner, P., Baldock, J., Chittleborough, D., & Verma, V. 2011. Relationships between carbon dioxide emission and soil properties in salt-affected landscapes. *Soil Biology and Biochemistry* 43: 667–674.

Shahid, S. A., Zaman, M., & Heng, L. 2018. Soil salinity: Historical perspectives and a world overview of the problem. In: Zaman, M., Shahid, S. A., Heng, L., editors. *Guideline for Salinity Assessment, Mitigation and Adaptation Using Nuclear and Related Techniques*, pp. 43–53. Springer International Publishing, Switzerland AG.

Shammas, N., Sever, C., & Wang, L. 2010. Deep-well injection for waste management. In: *Advanced Biological Treatment Processes*, pp. 521–582. doi:10.1007/978-1-60327-170-7_14.

Shao, D. D., Law, A. W. K., & Li, H. Y. 2008. Brine discharges into shallow coastal waters with mean and oscillatory tidal currents. *Journal of Hydro-Environmental Research* 2: 91–97.

Singh, K. N., & Chatrath, R. 2001. Salinity tolerance. In: Reynolds, M. P., Monasterio, J. I. O., McNab, A., editors. *Application of Physiology in Wheat Breeding*, pp. 101–110. CIMMYT, Mexico, DF.

Siyal, A. A., Siyal, A. G., & Abro, Z. A. 2002. Salt affected soils and their reclamation. *Pakistan Journal of Applied Sciences* 2(5): 537–540.

Stone, M. L., Rae, C., Stewart, F. F., & Wilson, A. D. 2013. Switchable polarity solvents as draw solutes for forward osmosis. *Desalination* 312: 124–129.

Olga L. Villa Sallangos, Operating experience of the Dhekelia seawater desalination plant using an innovative energy recovery system. March 2005. *Desalination* 173(1): 91–102. doi:10.1016/j.desal.2004.07.045

Tao, P., Ni, G., Song, C., Shang, W., Wu, J., Zhu, J., Chen, G., & Deng, T. 2018. Solar-driven interfacial evaporation. *Nature Energy* 3(12): 1031–1041.

Tate, R. L. 2000. *Soil Microbiology.* John Wiley & Sons, New York.

Teow, Y. H., & Mohammad, A. W. 2019. New generation nanomaterials for water desalination: A review. *Desalination* 451: 2–17.

Tester, M., & Davenport, R. 2003. Na$^+$ tolerance and Na$^+$ transport in higher plants. *Annals of Botany* 91: 503–507.

Thompson, N. A., & Nicoll, P.G. 2011. Forward Osmosis Desalination—A Commercial Reality. *International Desalination Association (IDA) World Congress*, Perth, Western Australia, September 4–9. http://www.modernwater.com/assets/pdfs/PERTH%20 Sept11%20-%20FO%20Desal%20A%20Commercial%20Reality.pdf.

Tiwari, G. N., Singh, H. N., & Tripathi, R. 2003. Present status of solar distillation. *Solar Energy* 75(5): 367–373.

Toufic, M, Hasan, F., Zeina, A., & Khaled, A. 2011. Techno-economic assessment and environmental impacts of desalination technologies. *Desalination* 2011: 263–273. doi:10.1016/j.desal.2010.08.035.

Tramper, J., & Van't Riet, K. 1991. *Basic Bioreactor Design.* Marce lDekker Inc, New York, USA.

Uygur, A., & Kargı, F. 2004. Salt inhibition on biological nutrient removal from saline wastewater in a sequencing batch reactor. *Enzyme and Microbial Technology* 34(3–4): 313–318.

Velmurugan, V., & Srithar, K. 2008. Prospects and scopes of solar pond: A detailed review. *Renewable and Sustainable Energy Review* 12: 2253–2263.

Vivar, M., Fuentes, M., Castro, J., & García-Pacheco, R. 2015. Effect of common rooftop materials as support base for solar disinfection (SODIS) in rural areas under temperate climates. *Solar Energy* 115: 204–216.

Weller, T. F. 1993. Produced brine underground disposal injection cost model. *Society of petroleum Engineering* 10. doi:10.2118/26887-MS.

Williams, W. 2001. Anthropogenic salinisation of inland waters. *Hydrobiologia* 466: 329–337. https://doi.org/10.1023/A:1014598509028

Williams, W. D. 2002. Environmental threats to salt lakes and the likely status of inland saline ecosystems in 2025. *Environmental Conservation* 29: 154–167. doi:10.1017/S0376892902000103.

Wong, V. N. L., Dalal, R. C., & Greene, R. S. B. 2009. Carbon dynamics of sodic and saline soils following gypsum and organic material additions: A laboratory incubation. *Applied Soil Ecology* 41: 29–40.

Yang, J., Spanjers, H., Jeison, D., & Van Lier, J. B. 2013. Impact of Na$^+$ on biological wastewater treatment and the potential of anaerobic membrane bioreactors: A review. *Critical Reviews in Environmental Science and Technology* 43(24): 2722–2746.

Yuan, B. C., Xu, X. G., Li, Z. Z., Gao, T. P., Gao, M., Fan, X. W., & Deng, H. M. 2007. Microbial biomass and activity in alkalized magnesic soils under arid conditions, *Soil Biology and Biochemistry* 39: 3004–3013.

Zaman, S. B., & Ahmad, S. 2009. Salinity and waterlogging in the Indus Basin of Pakistan: Economic loss to agricultural economy. *Managing Natural Resources for Sustaining Future Agriculture* 1:4–8.

Zhang, A., Zhao, S., Wang, L., Yang, X., Zhao, Q., Fan, J., & Yuan, X. 2016. Polycyclic aromatic hydrocarbons (PAHs) in seawater and sediments from the northern Liaodong Bay, China. *Marine Pollution Bulletin* 113(1–2): 592–599.

Zhao, K. -F. 1991. Desalinization of saline soils by Suaeda salsa. *Plant Soil* 135: 303.

Zhu, J. K. 2002. Salt and drought stress signal transduction in plants. *Annual Review of Plant Biology* 53: 247–273.

5 Advanced Oxidation Processes for the Treatment of Pollutants from Saline Water

Mohammed Nayeemuddin and Palaniandy Puganeshwary
Universiti Sains Malaysia

Shaik Feroz
Prince Mohammad Bin Fahd University

CONTENTS

5.1 INTRODUCTION

Sustainable supply of pure water is a primary concern all over the world due to continuous increase in consumption, industrial activity and population growth. The exploitation of freshwater resources for domestic and industrial use led to stress and scarcity. The extinction rates of freshwater resources were among the highest worldwide [1]. Various research works have been reported on the treatment of pollutants present in saline water before being fed to the desalination systems. Advanced oxidation processes (AOPs) are one of the most common and effective methods for the

DOI: 10.1201/9781003185437-6

degradation of recalcitrant pollutants. By using AOPs, it is possible to oxidize many pollutants directly to water and carbon dioxide or at least to transform them into simpler forms that can be removed by other, more traditional methods of treatment. Many different types of oxidizing agents such as chlorine, chlorine dioxide, oxygen, ozone, potassium permanganate and hydrogen peroxide can be used in the advanced oxidation treatment processes. To increase the efficiency of the oxidation process, a combination of oxidation agents and different process modifications are carried out in the treatment of pollutants from the saline water systems.

AOPs are methods of removing pollutants from saline water using highly reactive hydroxyl free radicals that are extremely effective oxidizing agents. There are three main methods of generating free radicals, which are chemical reactions (H_2O_2/O_3, O_3/OH^-, H_2O_2/Fe^{2+} or Fe^{3+}), photochemical reactions (H_2O_2/UV, O_3/UV), and photocatalytic reactions (semiconductor/UV). One of the most important AOPs is a semiconductor-mediated photocatalysis, which is a widely used technique for pollutant degradation in aqueous and nonaqueous medium [2].

Photocatalysis emerged as a proven technology to treat the recalcitrant trace pollutants present in water streams including saline water. This technique is cost effective and offers total mineralization of trace organic pollutants into harmless carbon dioxide and water. Photocatalysis depends on the idea of generating free radicals with the application of photonic energy and successive attack of the same on the pollutant molecules. Semiconductor nanomaterials are a promising option for eco-friendly and low-cost remediation systems in which chemicals, energy sources, and catalysts are nontoxic and economical and produce no secondary product. Compared to other semiconductors, TiO_2 acts as a promising photocatalyst used in the photocatalysis system due to its chemical and biological inertness, economical, and long-term stability against chemical and photo corrosion [3].

Researchers are doing tremendous work in developing novel photocatalysts for the remediation of pollutants from an aqueous medium. The workhorse photocatalyst TiO_2, which dominates the global market in photocatalysis, has limitations of its activity to UV region. The photocatalytic activity of TiO_2 is intrinsically lower due to the limited specific surface area. Nevertheless, the photo-generated electrons and holes in TiO_2 tend to a reunion, thereby restraining remediation performance. Therefore, it is highly necessary to develop a suitable photocatalytic degradation system to prevent the recombination rate of electrons and holes by modifying TiO_2 into other well-crystalline morphology.

A lot of efforts are being made to develop novel photocatalytic materials that can work efficiently in the visible region. However, the overall efficiencies of the novel photocatalytic materials are very far from industrial applications. Various approaches are followed in the design and manufacturing of novel photocatalysts such as coupling with plasmonic nanomaterial, surface engineering, and heterostructuring with other materials. The main focus was to boost light harvesting, photo capture, charge separation, and mass transfer that play a significant role in photocatalysis applications [4]. Visible light-responsive photocatalysis, which can utilize natural sunlight is a promising technology for saline water treatment with high purification efficiency [5]. Industrial wastewater containing highly toxic pollutants was generally released into rivers and streams in developing countries. These pollutants eventually enter into seawater, accumulates to increase their concentration, thereby creating an environmental hazard especially for marine life. These pollutants get decomposed under natural sunlight, but the natural process of degrading the pollutants becoming ineffective to ever increasing in the amount of toxic pollutants. The natural sunlight radiation in combination with the catalyst will accelerate the degrading process of pollutants effectively. Homogeneous and heterogeneous photocatalysis under natural and artificial light source have been successfully employed for the degradation of pollutants present in saline water and industrial saline wastewater.

5.2 SOLAR ENERGY-BASED AOPS FOR SALINE WATER TREATMENT

5.2.1 SOLAR SEMICONDUCTOR PHOTOCATALYTIC SYSTEMS

The solar photocatalytic mechanism was well defined in many research works. When a photocatalyst usually semiconductor exposed to the sunlight, it will absorb photonic energy. Electrons get excited from the excess energy and promote the conduction band from the valence band, therefore creating the negative-electron (e^-) and positive-hole (h^+) pair. The positive hole can oxidize the pollutant directly or oxidize water to form ($HO^\bullet$) radicals. At the same time, the electron reduces the oxygen adsorbed to the photocatalyst, which prevents the combination of electrons and the positive hole [2]. Photocatalytic reaction breaks down the pollutant molecules without any residue. The photocatalyst lasts for a longer time and the process does not need any additional chemicals, which makes the operation simple and economical.

The general mechanism of solar photocatalysis is shown in Equations 5.1–5.6 [2, 6].

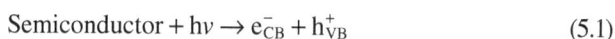

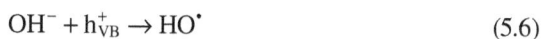

$$\text{Semiconductor} + h\nu \rightarrow e_{CB}^- + h_{VB}^+ \tag{5.1}$$

$$H_2O + h_{VB}^+ \rightarrow H^+ + HO^\bullet \tag{5.2}$$

$$O_2 + e_{CB}^- \rightarrow O_2^{\bullet-} \tag{5.3}$$

$$O_2^{\bullet-} + H_2O \rightarrow 2H_2O_2 \tag{5.4}$$

$$H_2O_2 + e_{CB}^- \rightarrow OH^- + HO^\bullet \tag{5.5}$$

$$OH^- + h_{VB}^+ \rightarrow HO^\bullet \tag{5.6}$$

Investigations were carried out with continuous recirculation to treat the pollutants present in the saline water under natural sunlight. The photocatalysts TiO_2 and ZnO along with an antimicrobial agent were used in suspension as well as immobilized inside the glass tube. A substantial reduction in total organic and inorganic carbon was observed. The performance of TiO_2 to degrade the pollutants in the presence of an antimicrobial agent was slightly better than ZnO with an antimicrobial agent [7]. A novel design reactor, which circulates the flow inside of it using a peristaltic pump, was used to investigate the degradation of the pollutants present in seawater. This novel reactor system shows a better approach for effective utilization of photocatalyst surface without an agitator to mix the content. TiO_2 photocatalyst along with antibacterial agent was employed in the batch recirculation process under direct natural sunlight. The treatment efficiency was evaluated by measuring total organic and inorganic carbon, chemical oxygen demand (COD), total dissolved solids (TDS), and pH [8,9].

Many marine toxins have very complex structures and are very difficult to remove by conventional treatment methods. The marine toxic okadaic acid present in seawater was found to be causing major socioeconomic, environmental, and health issues. This acid showed little change in methanolic solutions or deionized water at different temperatures. Heating to 45°C also showed no reduction in toxin content of the okadaic acid. Shellfish getting contaminated with okadaic acid, which consumed creating gastrointestinal symptoms in human beings. UV/TiO$_2$ photocatalysis has been utilized to degrade this toxic pollutant present in seawater. Complete degradation of the pollutant was observed after 30 minutes in seawater and 7.5 minutes in deionized water on exposure to UV/TiO$_2$ photocatalysis. The data were well fitted with a pseudo first- order kinetic model. The slower rate of acid degradation in seawater, may be due to the presence of other organic matter and competitive degradation under photocatalysis mechanism [10]. Photocatalytic degradation of organotin pesticide (tributyltin) present in seawater was evaluated. It has been observed a percentage removal of 41% of the pesticide present in seawater. The surface of photocatalyst gets inactivated and the hydroxyl radicals get scavenged by the salinity and other organic matter present in the seawater [11]. A photocatalytic membrane reactor with photocatalyst immobilized in/on a membrane and photocatalytic membrane reactor with catalyst in suspension has been employed for the removal of organics from the saline water. Membrane fouling was observed in suspension mode, and the same could be avoided and photocatalytic reaction can be conducted either in feed or in permeate in the case of the photocatalytic membrane immobilized reactor. However, the efficiency was found to be lower compared to a photocatalyst in suspension [12].

5.2.2 Solar Photo-Fenton Systems

The photo-Fenton process was another advanced oxidation technique that was widely employed for the treatment of pollutants present in an aqueous medium. Some research studies reported on the application of photo-Fenton process for the treatment of pollutants present in saline water/wastewater. However, very limited work has been done by applying this technique for the treatment of saline water/wastewater. Heterogeneous solar-Fenton process has been the most promising for water treatment applications with low amounts of sludge production. It is less sensitive to deactivation of catalyst at neutral and near-neutral pH and solid catalyst inherently more stable for reuse. The reaction mechanism involved in solar photo-Fenton process is represented by Equations 5.7–5.9 [13,14].

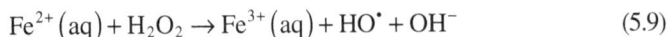

$$H_2O_2 + h\nu \rightarrow 2HO \tag{5.7}$$

$$Fe^{3+}(aq) + H_2O \rightarrow Fe^{2+}(aq) + H^+ + HO^{\bullet} \tag{5.8}$$

$$Fe^{2+}(aq) + H_2O_2 \rightarrow Fe^{3+}(aq) + HO^{\bullet} + OH^- \tag{5.9}$$

TiO_2 and H_2O_2 were used in suspension as well as thin-film immobilization to treat the inorganic pollutants present in seawater under direct sunlight irradiation. Suspension process will be having a uniform photocatalyst distribution in the reactor system. Higher efficiency can be achieved through the suspension process due to the availability of a larger surface area and low-pressure drop as the suspension particles are well mixed. The suspension process minimizes the catalyst fouling due to continuous removal; however, the removal requires nanofilter and adds additional cost to the process. The thin-film method is continuous and does not require the separation of catalyst. The film method has a lower light utilization efficiency due to the less availability of the catalyst surface area. The decrease in inorganic carbon value was more when the seawater was treated with a combination of TiO_2 and H_2O_2 compared to TiO_2 coating alone [15,16].

The presence of chloride ions is a major challenging aspect in the photocatalysis treatment of the pollutants present in saline water. To overcome this chloride ion effect, a nanosecond laser flash technique was employed in the photo-Fenton reaction under UV light source. This research work suggests that inhibition of the photocatalytic step of the photo-Fenton process in the presence of chloride ions can be circumvented by maintaining the pH at or slightly above 3.0 throughout the reaction [17]. A novel solar light-driven two-chamber cell has been used to degrade the organic pollutants present in seawater. Chloride ions are used as a medium in the process with WO_3 photoanode as a radical initiator. Under solar light, photoanode promotes chloride ions to convert to chlorine species, which could oxidize the organic compounds more rapidly [18].

5.2.3 Solar Photoelectrocatalytic Systems

Solar photoelectrocatalytic oxidation is another recent development technique adopted for water treatment. This technique uses both photons and electrical energy to generate hydroxyl radicals. Total mineralization of recalcitrant pollutants to water and carbon dioxide can be achieved in a short period of time through this technique compared to conventional photocatalysis. The use of bias potential in this technique significantly reduces the rapid and spontaneous recombination of charge carriers [19]. A microfluidic was employed to investigate the photoelectrocatalytic effect for the decontamination of seawater without chlorine production. The production of chlorine during the photocatalytic process was considered as highly toxic, and this research work was very significant. The high efficiency and elimination of chlorine generation suggest that the photoelectrocatalytic micro-reactor device has high potential to be scaled up for industrial applications. This research work provides a way forward for simultaneous treatment of pollutants and renewable energy generation [20].

The photoelectrocatalysis technique was employed for the removal of organics from municipal reverse osmosis concentrate. The participation of primary oxidants and the formation of disinfection byproducts were revealed in this research work at different anodic potentials and at different pHs. The highest organics removal rate of 63% was achieved, with a total disinfection byproduct formation of 5.41 μmol/L. At lower pH between 4 and 6, the mineralization rates enhanced by 20%. An increase in anodic potential slightly increases the production of hydroxyl radicals and benefitted the oxidation of bulk organics. Hydrophilic organics humic substances with larger molecular weights are transformed to smaller molecular weights and degrade within 30 minutes. However, the operational conditions should be carefully optimized to control the formation of disinfection byproducts [21].

Photoelectrochemical (PEC) reactor module consisting of nine photoelectro-chemical cells equipped with spray-deposited TiO_2 catalysts has been employed for the removal of pollutants present in the seawater under solar irradiation. TOC and COD analysis of treated seawater confirms the complete mineralization of pollut-ants. This novel approach of using photoelectrochemical cells shows a promising technique for the removal of pollutants present in saline water systems. However, cost analysis needs to be carried out to check the feasibility for large-scale industrial applications [22]. Photocatalytic and electrochemical degradation (ECD) technolo-gies have been evaluated for saline water purification. In the chlorine-mediated ECD approach, significant quantities of free chlorine (hypochlorite, Cl_2) and chlorinated hydrocarbons were formed in the solution. No chlorine-based compounds formed in the photocatalytic degradation and resulted in better removal of non-purgeable hydrocarbons than ECD. Non-formation of chlorine compounds in photocatalytic process was a good development indicating less scavenging of hydroxyl radicals by chloride ions [23].

5.3 CASE STUDY USING SURFACE RESPONSE METHODOLOGY (RSM)

A case study was carried out to optimize the solar photocatalytic process parameters for the removal of pollutants from seawater using response surface methodology (RSM). RSM is a statistical tool that utilizes experimental data generated from cen-tral composite design (CCD) to optimize and model any system. RSM is a promising technique for process optimization, strategically calculating optimum conditions for a multivariate process. This technique establishes the relationships between response variables and controlled input variables or process factors. CCD is well suited for fit-ting quadratic curve in order to optimize significant process parameters by conduct-ing a minimum number of experiments [24–26].

5.3.1 Design of Experiments (DoE)

RSM for experimental design and optimization was carried out using Design-Expert software (version 11, Stat-Ease, USA). CCD-RSM model generated using data from 17 experimental runs. The total experimental runs conducted was calculated based on Equation 5.2,

$$\text{Number of experiments} = 2^k + 2k + n \qquad (5.10)$$

where 2^k = factorial runs, $2k$ = axial runs, and n = number of factors/central runs.

Photocatalyst dosage, reaction time, and pH were considered as input variables. The response variables were the percentage removal efficiencies of TOC and COD.

Table 5.1 shows the design of experiments (DoE). Each independent process variable was varied between −1(low), 0 (center), and +1(high). Table 5.2 shows the ranges and levels assigned to the variables under study.

5.3.2 Batch Experiments

Batch experimental studies were carried by considering 1,000 mL of seawater in a glass beaker as shown in Figure 5.1. Calculated amount of TiO$_2$ was added and

TABLE 5.1

Design of Experiments (DoE)

Run	Space Type	Factor 1 A:A: Dosage	Factor 2 B:B: Time	Factor 3 C:C: pH	Response 1 R1:TOC	Response 2 R2: COD
		g/L	min		%	%
1	Factorial	1	60	6	2.9304	25
2	Axial	2.5	180	6	44.3223	69.6273
3	Axial	2.5	300	7.5	60.4396	75.9091
4	Axial	2.5	180	9	50.5495	73.1818
5	Factorial	1	60	9	8.42491	30.4545
6	Factorial	1	300	9	45.7875	44.0909
7	Factorial	4	300	6	82.0513	84.7
8	Factorial	4	60	9	65.9341	83.8182
9	Center	2.5	180	7.5	44	71.5455
10	Axial	4	180	7.5	73.6264	84.6364
11	Axial	2.5	60	7.5	15.7509	69.8182
12	Factorial	1	300	6	36.63	40
13	Factorial	4	60	6	64.4689	81.8364
14	Factorial	4	300	9	83.1502	85.7273
15	Center	2.5	180	7.5	48	73
16	Axial	1	180	7.5	10.989	34.9091
17	Center	2.5	180	7.5	47	72.54

TABLE 5.2

Ranges and Levels of the Independent and Dependent Variables

Name	Units	Type	Changes	Std. Dev.	Low	High
A: Dosage	g/L	Factor	Easy	0	1	4
B: Time	min	Factor	Easy	0	60	300
C: pH		Factor	Easy	0	6	9
R1:TOC	%	Response		6.16389	2.9304	83.1502
R2: COD	%	Response		0.695712	25	85.7273

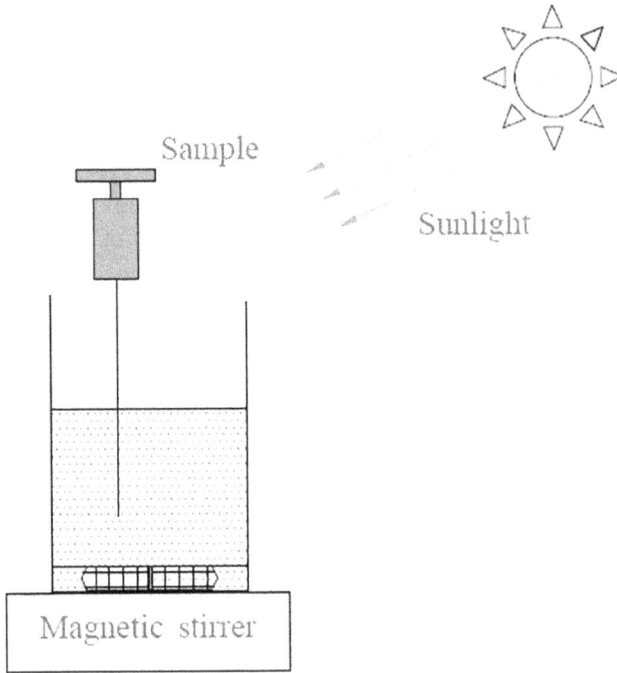

FIGURE 5.1 Experimental set-up.

constant stirring was provided using a magnetic stirrer. The photocatalysis process was carried out under the open sunlight between 10.00 AM and 2.00 PM, so that maximum UV radiation can be utilized. Samples were taken at equal intervals, and the same was analyzed for different parameters.

The percentage removal efficiency was evaluated using Equation 5.1 [20].

$$\text{Percentage removal} = \frac{\left(C_0 - C_f\right)}{C_0} \times 100 \tag{5.11}$$

where C_0 and C_f are the initial and final concentrations in mg/L, respectively.

5.3.3 ANALYSIS OF VARIANCE (ANOVA)

The analysis of variance (ANOVA) was carried out to estimate the significance of model and variables. The percentage removal of parameters and the biodegradability were fitted to a quadratic model that includes the interaction terms. Tables 5.3–5.5 show the results of ANOVA for each response.

The model F-value of 28.23 for response 1 (TOC) implies the model is significant. There is only a 0.01% chance that an F-value this large could occur due to noise. P-values less than 0.05 indicate model terms are significant. In this case A, B is significant model terms. Values greater than 0.1 indicate that the model terms are not significant. The Lack of Fit F-value of 11.87 implies there is a 7.95% chance that a

TABLE 5.3
ANOVA Results for Response 1 (R1: TOC)

Source	Sum of Squares	df	Mean Square	F-value	P-value	
Model	9,651.82	9	1,072.42	28.23	0.0001	Significant
A-A: Dosage	6,994.38	1	6,994.38	184.09	< 0.0001	
B-B: Time	2,266.51	1	2,266.51	59.66	0.0001	
C-C: pH	54.96	1	54.96	1.45	0.2682	
AB	164.38	1	164.38	4.33	0.0761	
AC	18.26	1	18.26	0.4807	0.5104	
BC	1.36	1	1.36	0.0358	0.8554	
A^2	4.52	1	4.52	0.1189	0.7404	
B^2	22.75	1	22.75	0.5989	0.4643	
C^2	110.65	1	110.65	2.91	0.1317	
Residual	265.95	7	37.99			
Lack of Fit	257.29	5	51.46	11.87	0.0795	Not significant
Pure Error	8.67	2	4.33			
Cor Total	9,917.77	16				

TABLE 5.4
ANOVA Results for Response 2 (R2: COD)

Source	Sum of Squares	df	Mean Square	F-value	P-value	
Model	6,991.77	9	776.86	1,605.04	< 0.0001	Significant
A-A: Dosage	6,064.58	1	6,064.58	12,529.74	< 0.0001	
B-B: Time	156.02	1	156.02	322.36	< 0.0001	
C-C: pH	25.95	1	25.95	53.61	0.0002	
AB	71.18	1	71.18	147.07	< 0.0001	
AC	5.34	1	5.34	11.03	0.0127	
BC	0.6717	1	0.6717	1.39	0.2773	
A^2	421.01	1	421.01	869.82	< 0.0001	
B^2	0.8268	1	0.8268	1.71	0.2325	
C^2	2.19	1	2.19	4.52	0.0711	
Residual	3.39	7	0.4840			
Lack of Fit	2.28	5	0.4565	0.8259	0.6274	not significant
Pure Error	1.11	2	0.5527			
Cor Total	6,995.16	16				

Lack of Fit F-value this large could occur due to noise. The **Predicted R^2** of 0.8108 is in reasonable agreement with the **Adjusted R^2** of 0.9387; i.e. the difference is less than 0.2. **Adeq Precision** measures the signal-to-noise ratio. A ratio greater than 4 is desirable. The obtained ratio of 18.549 indicates an adequate signal. The percentage removal efficiency of TOC (response 1) was fitted to a quadratic model as shown in Equation 5.12.

TABLE 5.5
Experimnetal and Predicted Values

S.No	Experimental Values		RSM Predicted Values	
	%TOC	%COD	%TOC	%COD
1	2.93	25	1.38	25.17
2	44.32	69.63	47.37	69.82
3	60.44	75.91	55.43	76.84
4	50.55	73.18	52.06	73.04
5	8.42	30.45	5.51	30.61
6	45.79	44.09	45.51	43.89
7	82.05	84.70	83.83	84.54
8	65.93	83.82	64.45	84.19
9	44.00	71.55	43.29	72.33
10	73.63	84.64	71.04	84.42
11	15.75	69.82	25.32	68.94
12	36.63	40.00	36.98	39.62
13	64.47	81.84	63.61	82.02
14	83.15	85.73	86.32	85.55
15	48.00	73.00	43.29	72.33
16	10.99	34.91	18.14	35.17
17	47.00	72.54	43.29	72.33

$$R1 = 101.78782 + 24.31574A + 0.244099B - 40.01371C$$
$$-0.025183AB - 0.671551AC + 0.002289BC$$
$$+0.577023A^2 - 0.000202B^2 + 2.85623C^2. \tag{5.12}$$

where 'A' is the dosage, 'B' is the time, and 'C' is pH.

The **Model F-value** of 1,605.04 for response 2 (COD) implies the model is significant. There is only a 0.01% chance that an F-value this large could occur due to noise. **P-values** less than 0.05 indicate model terms are significant. In this case, A, B, C, AB, AC, and A^2 are significant model terms. Values greater than 0.1 indicate the model terms are not significant. The **Lack of Fit F-value** of 0.83 implies the Lack of Fit is not significant relative to the pure error. There is a 62.74% chance that a Lack of Fit F-value this large could occur due to noise. The **Predicted R^2** of 0.9971 is in reasonable agreement with the **Adjusted R^2** of 0.9989; i.e. the difference is less than 0.2. **Adeq Precision** measures the signal-to-noise ratio. A ratio greater than 4 is desirable. The ratio of 113.149 indicates an adequate signal. The percentage removal efficiency of COD (response 2) was fitted to a quadratic model as shown in Equation 5.13.

$$R2 = -55.29204 + 49.98047A + 0.072533B + 8.29544C$$
$$-0.016572AB - 0.363131AC - 0.001610BC \tag{5.13}$$
$$-5.57129A^2 + 0.000039B^2 - 0.401593C^2$$

5.3.4 Effect of Process Variables as 3D Response Surfaces

Figures 5.2 and 5.3 show the 3D response surface plots for the percentage removal efficiency of TOC and COD that are obtained by the statistical analysis through Design-Expert software. The plots demonstrate the effects of reaction time and photocatalyst dosage on responses at constant pH. The quadratic effect in the plots was similar to the terms in the respective model equations.

5.4 THE WAY FORWARD

Semiconductor photocatalytic technology using artificial light or natural light source has become a prominent viable alternative technology for commercialization in the near future. Different recalcitrant water pollutants ranging from pesticides, herbicides, detergents, pathogens, viruses, coliforms, and spores are effectively removed by this technology. A lot of efforts are being dedicated to the development of visible light-responsive photocatalysts. The treatment of the pollutants present in saline water under natural light is still in the immature stage. The major problem of chloride ions scavenging the hydroxyl radicals needs to be addressed. Cost-effective process for immobilization of photocatalyst and its stability under high concentration of salts are yet to be addressed. The accumulation of a large amount of theoretical and modeling work is also useful and imperative in the quest to foster a deep understanding

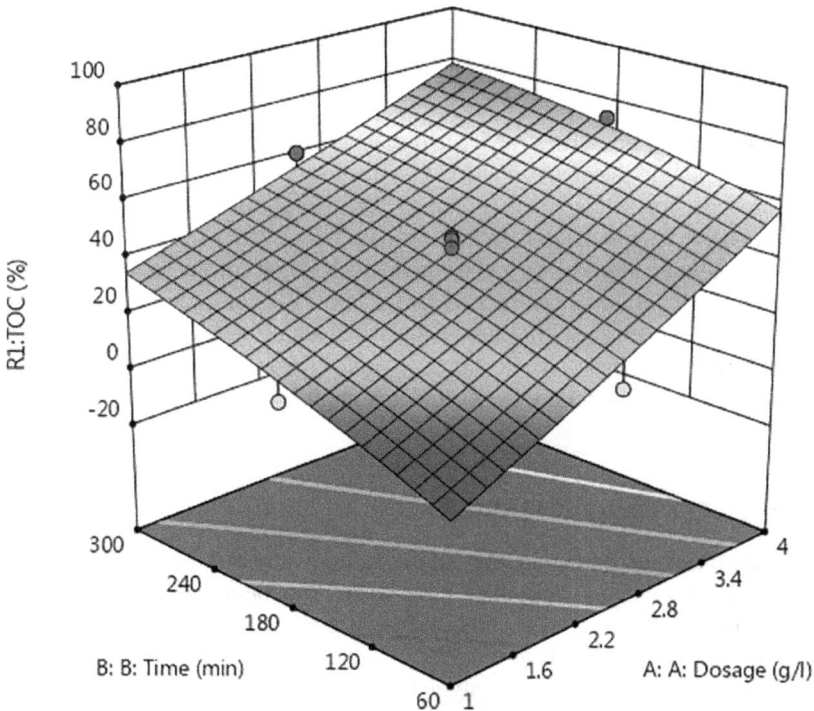

FIGURE 5.2 3-D plot for % removal efficiency of TOC.

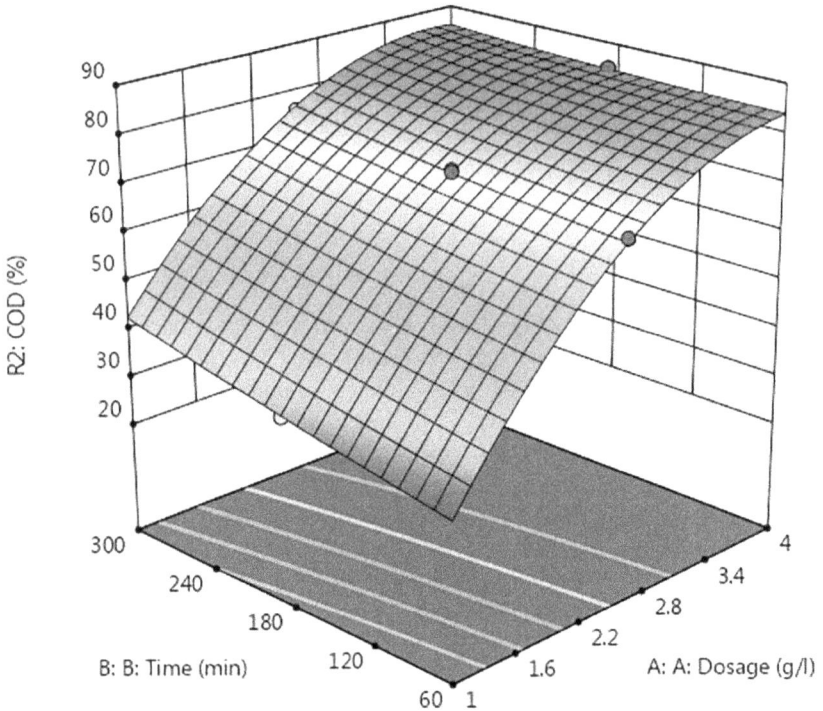

FIGURE 5.3   3-D plot for % removal efficiency of COD.

of the preparation, properties, and performances of photocatalysts and their optimization for saline water treatment. Multi-technology integration will provide a
bright prospect for saline water treatment and energy-related issues using visible
light-responsive photocatalysts with advanced efficiency and good robustness. More
efforts on scientific innovations and technological advancements are required for the
commercial applications of solar photocatalytic treatment of saline water. More demonstration and modeling studies are needed to examine the viability of the system
upscaling and its applications. Technology transfer from laboratory to commercial
scale is challenging especially in developing countries due to financial and technological aspects.

5.5   CONCLUDING REMARKS

The treatment of the pollutants presents in saline water using photocatalysis technique is still considered to be at developing stage. A lot of efforts need to be done for
the commercialization of solar photocatalytic treatment of saline water. The scavenging of radicals by chloride ions is still a major issue and needs to be overcome for
the effective utilization of photocatalytic technique to remove pollutants from saline
water. There is a wide scope for more tests under visible light with novel photocatalysts. Wider studies are needed on the optimization of photocatalytic parameters in

the treatment of saline water and industrial saline wastewater. Focus needs to be given on the economic aspects and feasibility of technology for commercial treatment aspects. A case study to optimize the process parameters in seawater treatment using RSM techniques was presented, which can be further explored for commercialization purposes.

REFERENCES

1. Nayeemuddin, M., Puganeshwary, P., Feroz, S: Pollutants removal from saline water by solar photocatalysis: A review of experimental and theoretical approaches. *International Journal of Environmental Analytical Chemistry.* doi:10.1080/03067319. 2021.1924160.
2. Feroz, S., Raut, N., Maimani, R.: Utilization of solar energy in degrading organic pollutant-A case study. *International Journal of COMADEM.* 14(3), 33–37 (2011).
3. Mohammed, M.R., Al Zobai, M.K.M.: Photocatalytic degradation of organic pollutants in petroleum refinery using TiO_2/UV and ZnO/UV by batch and continuous process. *Solid Sate Technology.* 63(3), 5390–5404 (2020).
4. Likodimos, V.: Advanced photocatalytic materials. *Materials.* 13, 821–824 (2020).
5. Dong, S., Feng, J., Fan, M., Pi, Y., Hu, L., Han, X., Liu, M., Sun, J.: Recent developments in heterogeneous photocatalytic water treatment using visible light responsive photocatalysts: A review. *Journal of Royal Society of Chemistry.* 5, 14610–14630 (2015).
6. Cambrussi, A.N.C.O., Morais, A.I.S., Neris, A.D.M., Osajima, J.A., Filho, E.C.D.S., Ribeiro, A.B.: Photo degradation study of TiO_2 and ZnO in suspension using miniaturized tests. *Revista Materia.* 24(4), 1–11 (2019).
7. Feroz, S., Jabri, H.: The effect of combining TiO_2 and ZnO in the pretreatment of seawater reverse osmosis process. *International Journal of Environmental Science and Development.* 6(5), 348–351 (2015).
8. Feroz, S., Harthy, W., Baawain, M., Saadi, S., Varghese, M., Rao, L.: Experimental studies for treatment of seawater in a re-circulation batch reactor using TiO_2 P25 and polyamide. *International Journal of Applied Engineering Research.* 10, 259–266 (2015).
9. Araimi, M., Feroz, S., Sangeetha, B., Saadi, S., Rao, L.: Treatment of seawater using solar energy in a re-circulation pipe reactor. *International Journal for Innovative Research in Science and Technology.* 2, 422–426 (2016).
10. Munoz, D. C., Lawton, L. A., Edwards, C.: Degradation of okadaic acid in seawater by UV/TiO_2 photocatalysis-proof of concept. *Science of the Total Environment.* 733, 1–8 (2020).
11. Muff, J., Simonsen, M.E., Sogaard, E.G.: Removal of tributyltin from contaminated seawater by combination of photolytic and TiO_2 mediated photocatalytic processes. *Journal of Environmental Chemical Engineering.* 5, 3201–3206 (2017).
12. Mozia, S.: Photocatalytic membrane reactors (PMRs) in water and wastewater treatment-A review. *Separation and Purification Technology.* 73(2), 71–91 (2010).
13. Yousif Mohamed Salih, F., Sakhile, K., Shaik, F., Lakkimsetty, N.R.: Treatment of petroleum wastewater using synthesized Hematite (α-Fe_2O_3) photocatalyst and optimization with response surface methodology. *International Journal of Environmental Analytical Chemistry.* 101, 1–20 (2020).
14. Laftani, Y., Boussaoud, A., Chatib, B., Hachkar, M., El-Makhfouk, M., Khayar, M.: Comparison of advanced oxidation processes for degrading ponceau S dye: Application of photo-Fenton process. *Macedonian Journal of Chemistry and Chemical Engineering.* 38(2), 197–205 (2019).

15. Jabri, H., Hudaifi, A., Feroz, S., Marikar, F., Baawain, M.: Investigation on the effect of TiO_2 and H_2O_2 for the treatment of inorganic carbon present in seawater. *International Journal of Engineering and Science*. 5, 50–55 (2015).
16. Jabri, H., Rao, L., Feroz, S.: Application of solar nano-photocatalysis in the pretreatment of reverse osmosis seawater desalination. *International Journal of Applied Nanotechnology*. 3, 11–17 (2017).
17. Machulek, A., Giongo, C., Moraes, J., Nascimento, C., Quina, F.: Laser flash photolysis study of the photocatalytic step of the photo-Fento reaction in saline solution. *Photochemistry and Photobiology*. 82, 208–212 (2006).
18. Xiao, K., Liang, H., Chen, S., Yang, B., Zhang, J., Li, J.: Enhanced photoelectrocatalytic degradation of bisphenol A and simultaneous production of hydrogen peroxide in saline wastewater treatment. *Chemosphere*. 222, 141–148 (2019).
19. Orimolade, O.B., Arotiba, A.O.: Towards visible light driven photoelectrocatalysis for water treatment: Application of a $FTO/BiVO_4/Ag_2S$ heterojunction anode for the removal of emerging pharmaceutical pollutants. *Scientific Reports*. 10, 5348–5361 (2020).
20. Wang, N., Tan, F., Tsoi, C., Zhang, X.: Photoelectrocatalytic micro-reactor for seawater decontamination with negligible chlorine generation. *Microsystem Technologies*. 23, 4495–4500 (2016).
21. Chen, Y., Li, S., Hu, J.: Photoelectrocatalytic degradation of organics and formation of disinfection byproducts in reverse osmosis concentrate. *Water Research*. 168, 105–115 (2019).
22. Shinde, S., Bhosale, C., Rajpure, K.: Photocatalytic activity of sea water using TiO_2 catalyst under solar light. *Journal of Photochemistry and Photobiology*. 103(2), 111–7 (2011).
23. Bruninghoff, R., Duijne, A., Braakhuis, L., Saha, P., Jeremiasse, A., Mei, B., Mul, G.: Comparative analysis of photocatalytic and electrochemical degradation of 4-ethylphenol in saline conditions. *Environmental Science and Technology*. 53, 8725–8735 (2019).
24. Varghese, M.J., Feroz, S., Dutta, S.: Solar nano-photocatalytic pretreatment of seawater: Process optimization and performance evaluation using response surface methodology. *Journal of Indian Chemical Society*. (In press.)
25. Varghese, M.J., Feroz, S., Dutta, S.: TiO_2/photo-Fenton process for seawater pretreatment: Modeling and optimization using response surface methodology (RSM) and artificial neural networks (ANN) coupled genetic algorithm (GA). *Journal of Indian Chemical Society*. (In press.)
26. Varghese, M.J., Feroz, S., Dutta, S.: Solar nano-photocatalytic pretreatment of seawater: Process optimization and performance evaluation using response surface methodology and genetic algorithm. *Applied Water Science*. 11(18), 1–15 (2021).

6 Application of Artificial Intelligence in the Treatment of Saline Water

Varghese Manappallil Joy
National Institute of Technology Durgapur
National University of Science and Technology

Shaik Feroz
Prince Mohammad Bin Fahd University

Susmita Dutta
National Institute of Technology Durgapur

Ahmed Yousuf Khalfan Al-Busaidi and
Lakkimsetty Nageswara Rao
National University of Science and Technology

CONTENTS

DOI: 10.1201/9781003185437-7

6.1   INTRODUCTION

Seawater is becoming a highly significant water resource globally due to the lack of freshwater availability and environmental pollution [1]. Ninety-seven per cent of total water available on earth is composed of seawater and brackish water [2]. Seawater reverse osmosis (RO) technology is widely employed in the desalination process and dominates other technologies like multistage flash and multi-effect distillation processes. RO membrane technology is highly energy-intensive, and efforts are being made to link renewable solar energy with desalination processes. However, the performance of RO membranes depends on effective fouling control and proper pre-treatment feed water [3]. Fouling caused by organic pollutants in RO feed seawater significantly lowers the permeate flux and damages the membranes resulting in high-pressure drop and increased energy consumption [4]. Many laboratory experimental studies have reported that biofilms deposited on membranes consist of about 50% organic matter and remaining organic acids.

Renewable solar energy is abundant in many parts of the world, especially in the Middle East regions where desalination plays a crucial role in the potable water supply, and it can be efficiently utilized in RO pre-treatment processes [5]. A UV-driven photocatalytic treatment can be employed to degrade high molecular organics, humic substances and microorganisms present in the RO feed seawater [6]. Hydroxyl radicals (OH$^•$) developed during the advanced oxidation process can degrade potential organic contaminants and recalcitrant pollutants present in an aquatic environment [7]. Many research studies reported the synergistic effect of catalysts (TiO$_2$/ZnO), combined photocatalytic treatment (Solar/Fenton/TiO$_2$), the addition of oxidants (H$_2$O$_2$) to enhance the production of hydroxyl radicals and thereby accelerate the organic contaminant degradation.

Jiménez et al. [8] reported a 79% of TOC removal from cleaning water used in RO desalination membranes by the application of solar photo-Fenton treatment. The optimal operating conditions reported as: solar irradiation time = 60 minutes, H$_2$O$_2$ dosage = 1.4 g/L and Fe^{2+} dosage = 70 mg/L. It was reported that the addition of Degussa P25 TiO$_2$ enhanced the photo-Fenton process. One of the previous research works [9] reported enhanced degradation of organic pollutants from feed water when ZnO modified TiO_2 nanocomposites were used instead of pure TiO_2. The photocatalytic efficiency of TiO_2 can be improved by the addition of ZnO since it prevents electron-hole recombination and thereby enhancing the photodegradation process [10]. A high production rate of hydroxyl radicals was observed when TiO_2 integrated with ZnO photocatalytic system. However, the treatment of seawater is challenging due to the interference of chloride ions and hydroxyl radical scavenging issues.

The synergistic effect of TiO_2/Fenton and TiO_2/ZnO solar photocatalytic system for reverse osmosis pre-treatment of seawater is presented here. The organic foulant removal from the RO feed water enhances the membrane life and reduces the energy requirement through improved permeate flux and lowered pressure drop. Organic degradation is measured in terms of Total Organic Carbon (TOC) removal and Chemical Oxygen Demand (COD) removal from the RO feed seawater. Machine-learning models were developed using computational intelligence methods. Process optimization using evolutionary algorithms was performed to determine the optimum operating conditions that result in maximum organic removal from RO feed seawater. The electron and hydroxyl radical scavenging issues can be surpassed by operating the photocatalytic process at optimum settings of input parameters.

6.2 TIO₂/FENTON FOR SOLAR NANO-PHOTOCATALYTIC TREATMENT OF SEAWATER

6.2.1 PHOTOCATALYSIS HARDWARE AND STATISTICAL DESIGN OF EXPERIMENTS (DoE)

The photocatalytic reactor set-up used for the pre-treatment of seawater using the TiO_2/Fenton process is shown in Figure 6.1.

The reactor system consists of nine borosilicate glass tubes (650 mm long, 22 mm inner diameter with 2 mm wall thickness) which are configured in parallel but connected in series to ensure continuous recirculation of seawater through the system. The seawater fed with TiO_2/Fenton dosage is recirculated through the photoreactor using a peristaltic pump (0–1,100 L/min, 0.5–350 rpm) at a flow rate of 1.1 L/min. An air pump connected with a diffusor ensures a sufficient supply of oxygen required for the photocatalytic reaction. The seawater required for the experimentation was collected in sterile bags from a location near one of the desalination plants operating in Muscat, Oman (nearly 1.5 km away from the shore). The average TOC and COD values of the seawater samples were determined as 3.15 and 5.13 mg/L, respectively.

For photocatalytic experimentation, a central composite design (CCD) of experiments matrix (Table 6.1) consists of four input factors (TiO_2 dosage (g/L), Fenton dosage (g/L), pH and solar irradiation time (min)), and two response variables (TOC removal (%) and COD removal (%)) are developed. The coded range of +1 (high) level and −1 (low) level of each input factors were set at 1.6 and 0.6 g/L (TiO_2 dosage), 3.57 and 0.71 g/L (Fenton dosage), 7 and 3 (pH) and 180 minutes and 60 (solar

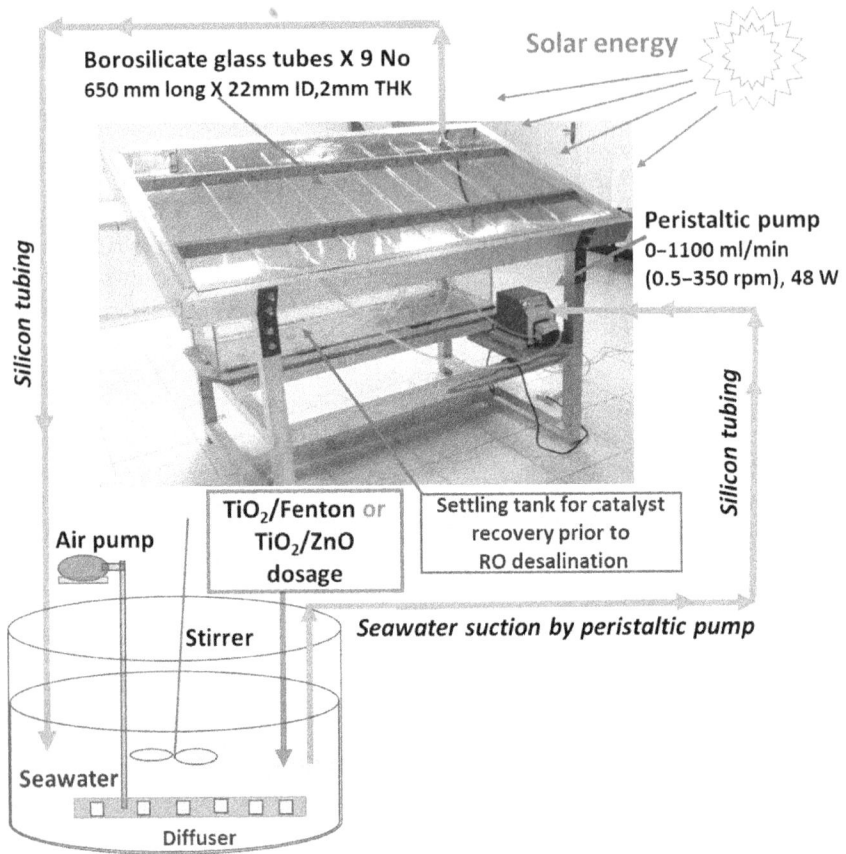

FIGURE 6.1 Experimental set-up of solar photocatalytic reactor system for reverse osmosis pre-treatment.

irradiation time), respectively. As per the average photo-Fenton dosage suggested in previous work [11], in all the experimental runs, the H_2O_2/Fe^{2+} dosage ratio was maintained as 100: 1.

Overall, 25 experimental runs ($= 2^n + 2n + n_c$, where n is the number of factors which is equal to 4) were conducted as per the face-centred hyper-cube version of CCD. Thus, the types of design space points in this research work consist of 2^n factorial runs ($=16$), $2n$ axial runs ($=8$) and a single ($n_c = 1$) run at the centre of the input design space. The distribution of experimental settings at ideal locations using CCD enables to cover the entire design space efficiently and leads to high-fidelity modelling-optimization performance. The TOC and COD removal efficiency after each experimental run was calculated as per Equation 6.1.

$$[\text{TOC or COD}]\,\text{removal}\,(\%)$$
$$= \frac{[\text{TOC or COD}]_{\text{before treatment}} - [\text{TOC or COD}]_{\text{after treatment}}}{[\text{TOC or COD}]_{\text{before treratment}}} \times 100 \quad (6.1)$$

TABLE 6.1
Central Composite Design of Experiments (DoE) Matrix with Experimental and ANN Predicted Values

Run	Space Point Type	TiO$_2$ (g/L)	Fenton (g/L)	pH	Time (min)	TOC Experimental	TOC ANN Predicted	COD Experimental	COD ANN Predicted
1	Axial	1.1	2.14	5	60	59.25	60.35	54.55	53.83
2	Axial	1.1	3.57	5	120	65.48	67.89	59.85	63.63
3	Axial	1.1	0.71	5	120	47.38	48.07	42.86	41.04
4	Axial	0.6	2.14	5	120	69.47	71.19	64.63	63.55
5	Axial	1.1	2.14	3	120	48.95	48.43	44.31	45.78
6	Axial	1.1	2.14	7	120	32.82	34.51	27.12	26.08
7	Axial	1.1	2.14	5	180	67.93	69.38	62.49	64.06
8	Axial	1.6	2.14	5	120	73.15	70.67	66.82	66.54
9	Centre	1.1	2.14	5	120	77.46	71.12	66.59	66.66
10	Factorial	1.6	3.57	7	180	40.83	38.92	34.85	32.72
11	Factorial	0.6	0.71	3	180	44.14	48.47	38.94	39.62
12	Factorial	1.6	3.57	3	60	35.69	35.45	30.14	31.94
13	Factorial	1.6	0.71	7	180	23.58	17.71	17.84	21.86
14	Factorial	1.6	0.71	3	180	43.64	48.50	37.98	38.12
15	Factorial	0.6	0.71	7	60	30.89	29.72	26.46	24.44
16	Factorial	0.6	3.57	7	180	35.59	29.27	29.85	33.74
17	Factorial	0.6	0.71	7	180	17.49	17.52	12.15	17.64
18	Factorial	0.6	0.71	3	60	49.48	48.53	44.93	43.79
19	Factorial	1.6	0.71	3	60	52.41	48.52	47.69	49.95
20	Factorial	0.6	3.57	3	60	42.96	43.13	36.12	35.86
21	Factorial	0.6	3.57	3	180	42.35	42.21	35.99	38.14
22	Factorial	1.6	3.57	3	180	33.65	33.73	27.58	33.07
23	Factorial	0.6	3.57	7	60	27.47	27.44	23.78	22.76
24	Factorial	1.6	0.71	7	60	19.78	19.79	13.68	10.05
25	Factorial	1.6	3.57	7	60	22.58	22.54	16.95	21.96

6.2.2 ANN MODELLING OF TiO$_2$/FENTON TREATMENT SYSTEM

Advanced oxidation processes (AOPs) constitute a promising technology for the removal of organic pollutants from wastewater. Fenton's oxidation is one of the well-regarded metal catalyzed oxidation reactions of aquatic-based organic compounds. Fenton process is encompassing hydrogen peroxide (H$_2$O$_2$) with iron ions reaction to form active oxygen species that oxidize organic or inorganic compounds. The mixture of FeSO$_4$ or any other ferrous complex and H$_2$O$_2$ (Fenton's reagent) at low pH results in Fe^{2+} catalytic decomposition of H$_2$O$_2$. The reaction proceeds via a free radical chain process that generates hydroxyl radicals which have extremely high oxidizing ability and could oxidize recalcitrant organic compounds in a short time. The efficiency of Fenton process depends on the enhanced oxidation power of hydroxyl radicals (OH$^{\bullet}$) formed from Fenton's reaction. In a Fenton reaction system,

(Equation 6.2–6.4) hydrogen peroxide and ferrous ions are simultaneously generated by the $2e^-$ reduction of the oxygen molecule which is dissolved and $1e^-$ reduction of ferric ions in the aquatic medium.

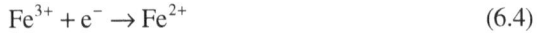

$$Fe^{2+} + H_2O_2 + H^+ \rightarrow Fe^{3+} + OH^{\bullet} + H_2O \tag{6.2}$$

$$O_2 + 2H^+ + 2e^- \rightarrow H_2O_2 \tag{6.3}$$

$$Fe^{3+} + e^- \rightarrow Fe^{2+} \tag{6.4}$$

ANN is a nature-inspired algorithm that mimics the operation of the human nervous system [12]. A neural network consists of computational units called artificial neurons. A typical ANN architecture includes an input layer, one or more hidden layers and one output layer. The number of neurons in the input and output layers is equal to the number of input and response variables, respectively. The hyperparameters like the number of neurons in the hidden layers and the number of hidden layers are optimized using validation run data during training. Each neuron is connected to other neurons in the next layer with a synaptic weight known as connection weights. Artificial intelligence is created by embedding the network with optimum bias and connection weights determined using procedures like backpropagation training algorithms. In this research work, both feedforward and cascade forward (which includes a weight transfer from the input to each layer and from each layer to successive ones) networks were used, and the one that gives the least root mean squared error (RMSE) was selected for multi-criterion optimization paradigm.

The photocatalytic reactions and the organic pollutant degradation kinetics are very complex. Due to the presence of various reactive oxidation species and the intricate spatial distribution of solar radiation within the photoreactor, the modelling of the system using conventional methods is very cumbersome [13]. Hence in this research work, the complex mapping between input variables (TiO_2 dosage, Fenton dosage, pH and solar irradiation time) and output variables (TOC and COD removal) was performed using the ANN technique. Eighty per cent of the data were taken for training the ANN model, and the remaining 20% were used for validation purpose. While training using training data, the network converges to the optimum synaptic weights (connection weights) between neurons, whereas the validation runs are used to determine the hyperparameters like the number of hidden layers and number of neurons in the hidden layers.

Overall, 160 ANN models were developed for each output variables using *nested for-loop program codes* written in MATLAB® R2019b environment. To ensure that all the inputs have the same weights and to accelerate the training process, all the inputs are normalized by using Equation 6.5. Since the output data were skewed to the left side, a log transformation of the response matrix was performed. The training of the network was performed using four different algorithms viz. (i) Levenberg-Marquardt ("trainlm"), (ii) Variable Learning Rate Gradient Descent ("traingdx"), (iii) Bayesian Regularization ("trainbr") and (iv) Scaled Conjugate Gradient ("trainscg"). In the feedforward net architecture, two types of transfer functions (activation functions) were used in the hidden layer neurons viz. (i) Tangent Sigmoid ("tansig") and (ii) Log Sigmoid ("logsig"). The network with least validation

RMSE was selected as the final model for incorporating in the multi-objective optimization process.

$$A_i = \frac{X_i - \min(X_i)}{\max(X_i) - \min(X_i)} \tag{6.5}$$

6.2.2.1 Garson's Method for Analysing the Influence of Input Parameters on TOC and COD Removal

The relative significance of each input factors on TOC and COD removal can be estimated by using Garson's algorithm given in Equation 6.6.

$$I_j = \frac{\displaystyle\sum_{m=1}^{m=N_h}\left(\left(\left|W_{jm}^{ih}\right|\middle/\displaystyle\sum_{k=1}^{N_i}\left|W_{km}^{ih}\right|\right)\times\left|W_{mn}^{ho}\right|\right)}{\displaystyle\sum_{k=1}^{k=N_i}\left\{\displaystyle\sum_{m=1}^{m=N_h}\left(\left|W_{km}^{ih}\right|\middle/\displaystyle\sum_{k=1}^{N_i}\left|W_{km}^{ih}\right|\right)\times\left|W_{mn}^{ho}\right|\right\}} \tag{5.6}$$

where I_j is the relative significance of the jth input factor on TOC or COD removal; N_i and N_h are the number of neurons in the hidden and output layers, respectively; W_s are connection weights; the superscripts "i", "h" and "o" indicate input, hidden and output layers, respectively; and subscripts "k", "m" and "n" refer to neurons in input, hidden and output layers, respectively. The relative importance of input factors on the removal of TOC was also estimated by using the connection-weight method.

6.2.3 MULTI-OBJECTIVE OPTIMIZATION USING NON-DOMINATED SORTING GENETIC ALGORITHM (NSGA-II)

The multi-objective optimization paradigm is a stochastic search process used to estimate the vector of input variables within the feasible region to generate optimum values for objective functions. Instead of developing a single solution, a typical multi-objective optimization algorithm will generate a series of optimal solutions known as the Pareto-optimal set [14]. Evolutionary algorithms like Genetic Algorithm (GA) mimic natural evolutionary processes occurring in nature. NSGA-II proposed by Deb et al. [15] is one of the most well-regarded algorithms used for multi-objective optimization. In this research work, the TOC and COD ANN models were used as objective functions of NSGA-II to evaluate the cost and to subsequently converge at Pareto-optimal solutions. In complex advanced oxidation processes like seawater photocatalysis, estimating the global optimum solutions using conventional methods is a tedious task. Hence evolutionary stochastic metaheuristics like NSGA-II are promising and better tools due to their fast and elitist nature.

NSGA-II performs multi-criterion optimization using operators employed for non-dominant sorting, ranking based on the dominance and crowding distance calculations. Apart from operators like selection, crossover (reproduction) and mutation operators used in GA, NSGA-II employs parameter-less niching operators and techniques like elitist preservation and non-dominated sorting. Like GA, the evolution (generation or iteration) starts with a random population consists of individuals known as chromosomes. Each individual or solution is represented by a vector of input variables known as genes. The niching operator (crowding distance calculator) ensures isolated solution has a preference for selection to next generation. The niching operator enhances the exploration capability of the algorithm and ensures the spread of solutions in the entire search space (input variable space). The Pareto-optimal front, which consists of non-dominated solutions (no other decision vector dominates it), is calculated on the basis of dominance of one decision vector over the other. A case study was carried out for the removal of TOC from seawater using the developed ANN model code.

6.3   CASE STUDY-1 OUTCOMES

6.3.1   ANN MODEL DEVELOPMENT FOR TOC REMOVAL EFFICIENCY

The CCD matrix, along with experimental and ANN predicted results, is shown in Table 6.1. Using the input data and experimental results for output responses, two ANN models (TOC and COD removal) were created separately using program codes written in MATLAB® R2019b. For TOC modelling, overall 160 ANN models were trained, and the model (Model no:26) with the least root mean square value (RMSE = 4.594) for validation was selected. The selected TOC model was utilized as one of the objective functions in multi-objective optimization using the NSGA-II algorithm. The distribution of RMSE for prediction and validation of each trained models for TOC removal (%) is depicted in Figure 6.2.

The performance of the selected TOC-ANN model in predicting the output is shown in Figure 6.3. It can be observed that the selected ANN model successfully

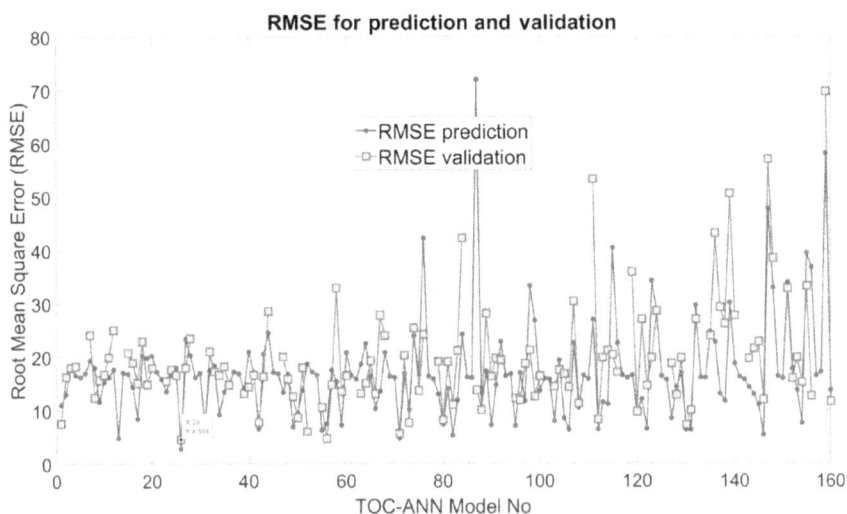

FIGURE 6.2   Distribution of RMSE for prediction and validation of 160 number of trained TOC-ANN models.

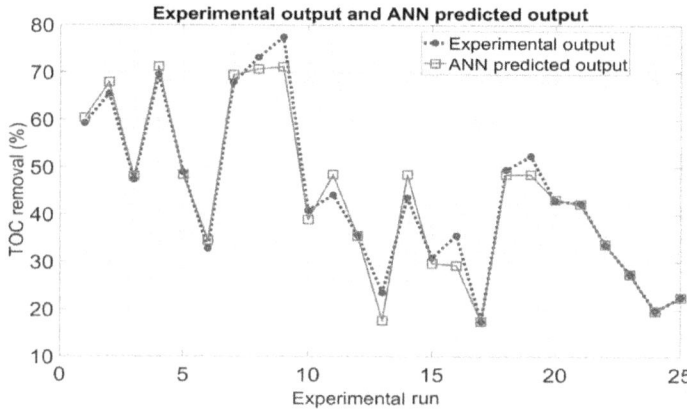

FIGURE 6.3 Experimental output versus ANN model predicted output for TOC removal using TiO$_2$-Fenton solar photocatalysis.

TABLE 6.2
The ANN Model Weight Matrix for TOC Removal Using TiO$_2$-Fenton Solar Photocatalysis

	W1					W2	
	Factor				Bias		
Hidden Layer Neuron	TiO$_2$ (g/L)	Fenton (g/L)	pH	Time (min)		Hidden Layer Neuron	W
1	−0.303	2.250	−0.506	0.611	8.468	1	2.031
2	−5.102	2.683	−2.432	−7.130	−8.065	2	−1.552
3	2.360	4.695	6.198	−10.029	−6.220	3	−1.902
4	−0.706	−4.463	2.490	−3.736	−4.292	4	−0.532
						Bias	−1.136

maps the input-output relation and captures the complex relation between TOC removal efficiency and input parameters (dosage of TiO$_2$ and Fenton, pH and solar irradiation time)

6.3.1.1 The Relative Importance of Input Variables on TOC Removal

The selected ANN model for TOC removal was having an RMSE of 4.594 with a feedforward (4-4-1) architecture (four neurons in the hidden layer). Table 6.2 shows the matrix of bias and weights (connection weights from input neurons to hidden layer neurons, and connection weights from the hidden layer neuron to the output neuron) of the final selected TOC-ANN model. The impact of input parameters (TiO$_2$ and Fenton dosage, pH and solar irradiation time) on output response (TOC removal) was calculated using Garson's algorithm and connection weight methodology.

TABLE 6.3
The Relative Importance of Input Factors on TOC Removal

Input Factors	Garson's Algorithm (%)		Connection Weight Approach	
	Relative Importance (%)	Rank	Importance	Rank
TiO_2 (g/L)	33.68	1	2.37	1
Fenton (g/L)	28.46	2	−1.98	2
Time (min)	22.35	3	0.96	4
pH	15.5	4	1.4	3

The results from both Garson's approach and the connection-weight method for the TOC removal model are shown in Table 6.3. The connection-weight approach estimates the product of the connection weights from input to hidden and from hidden to output neurons. The products from each input to output across all hidden neurons are then added and ranked. The relative influence of each input factor on TOC removal is then determined based on their ranks. The discrepancy between the results of Garson's and connection weight method is logical. In Garson's method, the absolute values of connection weights are considered, but in the connection-weight approach, raw data is taken. It can be observed that TiO_2 dosage (relative influence = 33.68%) and Fenton dosage (relative influence = 28.46%) have more impact on the photocatalytic degradation of organic pollutants from seawater. This result is similar to the outcome reported in a previous work [16] where TiO_2 dosage is obtained as the most influencing parameter for TOC removal from petroleum wastewater.

6.3.2 ANN MODEL DEVELOPMENT FOR COD REMOVAL EFFICIENCY

For COD modelling, overall 160 ANN models were trained, and the model (Model no:90) with the least root mean square value (RMSE = 3.581) for validation was selected. The selected COD model was utilized as one of the objective functions in multi-criterion optimization using the NSGA-II algorithm. The distribution of RMSE for prediction and validation of each trained models for COD removal (%) is depicted in Figure 6.4.

The performance of the selected COD-ANN model in predicting the output is shown as a regression curve in Figure 6.5. Both training and validation data fall on or close to the 45° regression line. Hence it can be concluded that the selected ANN model successfully maps the input-output relation and captures the complex relation between COD removal efficiency and input parameters (dosage of TiO_2 and Fenton, pH and solar irradiation time).

6.3.3 MULTI-OBJECTIVE OPTIMIZATION OF TiO₂-FENTON SOLAR PHOTOCATALYTIC SYSTEM USING NSGA-II

The Pareto-optimal front (non-dominated solutions) derived from the NSGA-II algorithm is shown in Figure 6.6. The NSGA-II codes created in MATLAB®

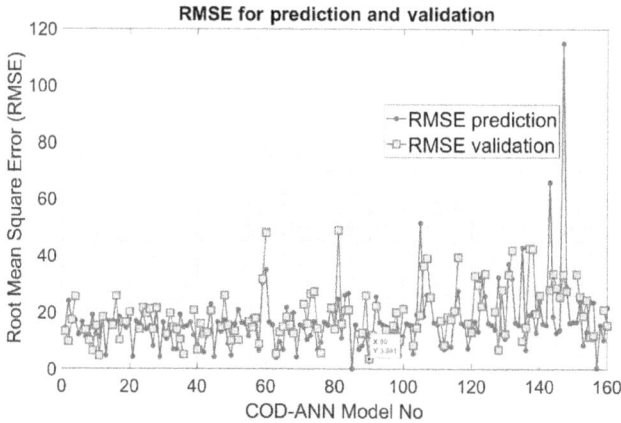

FIGURE 6.4 Distribution of RMSE for prediction and validation of trained COD-ANN models.

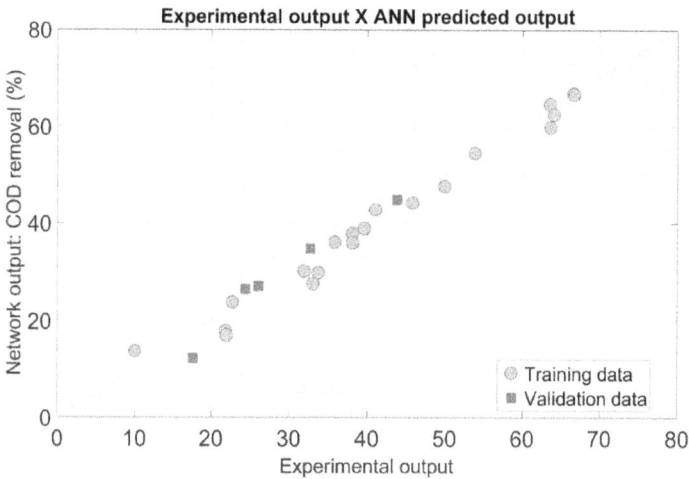

FIGURE 6.5 Experimental output versus ANN model predicted output for COD removal using TiO$_2$-Fenton solar photocatalysis.

were essentially for minimizing the objective space, and hence the cost functions (TOC/COD-ANN models) were negated to achieve maximum cost values in each iteration. For maximizing both TOC removal and COD removal simultaneously, both should have smaller values along both abscissa and ordinate. Each chromosome or solution in the Pareto-optimal front corresponds to a unique set of decision vector, i.e. an individual decision variable list in the Pareto-optimal set. The selected solution (chromosome) suggests the optimal TOC and COD removal as 85.13% and 70.42%, respectively. The optimal values of input factors (genes) at this position were obtained as: TiO$_2$ dosage = 0.91 g/L, Fenton dosage = 3.1 g/L, pH= 5.3 and Solar irradiation time = 162.7 minutes.

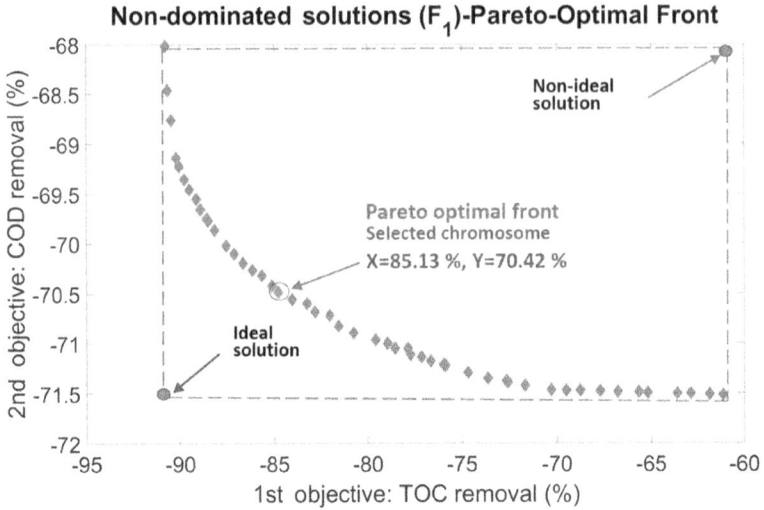

FIGURE 6.6 Non-dominated solutions: Pareto-optimal front derived by NSGA-II algorithm for TiO$_2$-Fenton solar photocatalytic treatment of seawater.

6.4 TIO$_2$/ZNO FOR SOLAR NANO-PHOTOCATALYTIC TREATMENT OF SEAWATER

6.4.1 CENTRAL COMPOSITE DESIGN (CCD) OF EXPERIMENTS FOR TiO$_2$/ZnO TREATMENT SYSTEM

For photocatalytic experimentation, a central composite design (CCD) of experiments matrix (Table 6.4) consists of four input factors (TiO$_2$ dosage (g/L), ZnO dosage (g/L), pH and solar irradiation time (min)), and one response variables (TOC removal (%)) is developed. The coded range of $+1$ (high) level and -1 (low) level of each input factors were set at 0.6 and 0.4 g/L (TiO$_2$ dosage), 0.6 and 0.4 g/L (ZnO dosage), 7 and 3 (pH) and 180 minutes and 60 (solar irradiation time), respectively. The same experimental set-up shown in Figure 6.1 was used for TiO$_2$-Fenton photocatalytic treatment seawater.

6.4.2 ANN MODELLING OF TiO$_2$/ZnO TREATMENT SYSTEM

Heterogeneous photocatalysis based on TiO$_2$ suspensions has been extensively reported as one of the most promising advanced oxidation technology for degrading organic pollutants under natural or artificial irradiation. Even though in the literature the ZnO application is less reported, it presents promising results especially when using in combination with TiO$_2$ as drawbacks on one constituent will be taken care of by the other.

Overall, 320 ANN models were developed for TOC removal (%) efficiency using *nested for-loop* program codes written in MATLAB® R2019b environment. All the input factors were normalized between zero and one, and the output is

TABLE 6.4

The Central Composite Design of Experiments (DoE) Matrix with Experimental and ANN Predicted Values for TiO_2-ZnO Solar Photocatalytic System

Run	Space Type	Factor 1 A:TiO_2 Dosage	Factor 2 B:ZnO Dosage	Factor 3 C:pH	Factor 4 D:RT	Response 1: TOC Removal (%) Experimental	ANN Predicted
1	Center	0.5	0.5	5	120	63.79	64.78
2	Axial	0.5	0.5	5	180	71.23	69.38
3	Axial	0.5	0.6	5	120	59.37	57.38
4	Axial	0.5	0.4	5	120	58.22	56.89
5	Axial	0.5	0.5	3	120	53.76	54.78
6	Axial	0.6	0.5	5	120	69.34	69.67
7	Axial	0.5	0.5	7	120	66.79	66.84
8	Axial	0.4	0.5	5	120	59.44	59.73
9	Axial	0.5	0.5	5	60	62.21	61.64
10	Factorial	0.4	0.4	3	60	36.09	36.76
11	Factorial	0.4	0.6	7	180	53.44	55.78
12	Factorial	0.6	0.4	7	60	55.86	57.15
13	Factorial	0.4	0.4	7	180	55.87	55.65
14	Factorial	0.4	0.4	7	60	53.2	51.26
15	Factorial	0.6	0.6	3	60	47.07	48.76
16	Factorial	0.4	0.6	3	60	38.39	37.73
17	Factorial	0.6	0.4	7	180	64.65	65.41
18	Factorial	0.6	0.6	7	180	66.07	66.39
19	Factorial	0.6	0.6	7	60	59.39	60.18
20	Factorial	0.6	0.4	3	60	42.31	43.26
21	Factorial	0.6	0.6	3	180	57.11	55.28
22	Factorial	0.6	0.4	3	180	54.47	52.73
23	Factorial	0.4	0.6	3	180	40.92	41.58
24	Factorial	0.4	0.6	7	60	50.58	50.40
25	Factorial	0.4	0.4	3	180	43.06	42.95

log-transformed. The training of the network was performed using four different algorithms viz. (i) Levenberg-Marquardt ("trainlm"), (ii) Variable Learning Rate Gradient Descent ("traingdx"), (iii) Bayesian Regularization ("trainbr") and (iv) Scaled Conjugate Gradient ("trainscg"). In each of the Feedforward net and Cascade forward network architecture, two types of transfer functions were used in the hidden layer neurons, namely: (i) Tangent Sigmoid ("tansig") and (ii) Log Sigmoid ('logsig'). The network with the least validation RMSE was selected as the final model for optimization using Grey Wolf Optimizer (GWO) algorithm.

6.4.3 TOC REMOVAL OPTIMISATION USING GREY WOLF OPTIMIZER (GWO)

The selected TOC-ANN model for TiO$_2$-ZnO solar photocatalytic degradation system was optimized using GWO algorithm. The GWO algorithm proposed by Mirjalili et. al.[17] is one of the most well-regarded swarm intelligence techniques used for single-objective optimization problems. "This algorithm mimics the dominance hierarchy and hunting behaviour of grey wolves in nature. Grey wolves live in one of the most organised natural groups called a pack. Wolves in a pack are divided into four classes: alpha, beta, delta, and omega. The alpha is normally the strongest wolf that leads the pack in navigation and hunting. All wolves should follow the alpha's order. In the next dominance lever, beta wolves help alpha in decision making and leadership. Omega wolves are the least powerful" [14].

The GWO algorithm initially starts the optimization process using a set of random wolves (solutions) in the search space. In each iteration or "hunt," the fitness of each solution is evaluated using an objective function. The top three wolves are selected as alpha, beta and delta according to the quantity of each solution. The position of each wolf is updated iteratively, and if any of the solutions becomes better than alpha, beta or delta, it will be replaced with a new solution. The GWO algorithm terminates after reaching the stopping criterion.

6.4.4 CASE STUDY-2 OUTCOMES

6.4.4.1 ANN Model Development for TOC Removal Efficiency in TiO$_2$-ZnO Solar Photocatalysis of Seawater

The face-centred-CCD matrix, along with experimental and ANN predicted results, is shown in Table 6.4. Using the input data and experimental results, output response (TOC removal) was developed using program codes written in MATLAB® R2019b. Overall, 320 ANN models were trained, and the model (Model no:34) with the least root mean square error value (RMSE = 0.9728) for validation was selected. The selected TOC model was optimized using the GWO algorithm to determine the optimum operating parameter settings for TiO$_2$-ZnO photocatalytic treatment of seawater. The distribution of RMSE for prediction and validation of each trained models for TOC removal (%) is depicted in Figure 6.7.

The performance of the selected TOC-ANN model in predicting the output is shown as a regression plot in Figure 6.8. It can be observed that the selected ANN model with a regression coefficient (=0.99213) successfully maps the input-output relation and captures the complex relation between TOC removal efficiency and input parameters (dosage of TiO$_2$ and ZnO, pH and solar irradiation time).

6.4.4.2 Single-Objective Optimization of TiO$_2$-ZnO Solar Photocatalytic System Using GWO Algorithm

Figure 6.9 shows the convergence curve generated by GWO. Maximum TOC removal converged in 1,000 iterations was found to be 78.5% at the following optimum input settings: TiO$_2$ dosage = 0.6 g/L, ZnO dosage = 0.5164 g/L, pH = 6.0795 and solar irradiation time = 180 minutes. The average of three validation runs performed at

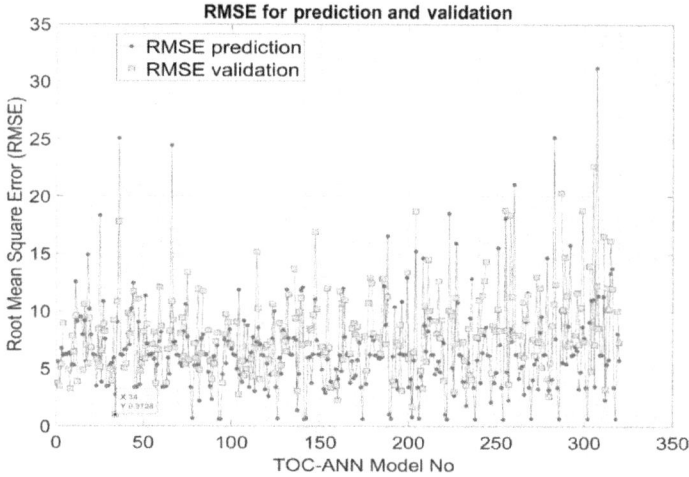

FIGURE 6.7 Distribution of RMSE for prediction and validation of 320 number of trained TOC-ANN models using TiO_2-ZnO solar photocatalysis of seawater.

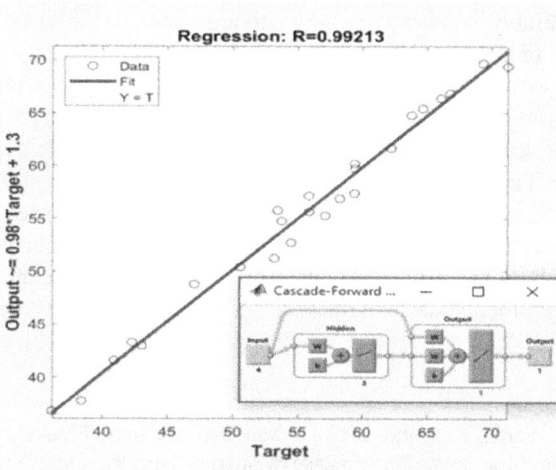

FIGURE 6.8 Regression plot generated by Cascade Forward Neural Network (architecture 4-3-1) for TOC-ANN model of TiO_2-ZnO solar photocatalysis.

these optimum settings was found to be TOC removal $= 73.64\%$. Similar results were reported in a previous study [18] where the optimum TOC removal was 74% in TiO_2-ZnO-aeration-solar photocatalytic treatment of petroleum refinery wastewater.

6.5 CONCLUDING REMARKS

RO membrane fouling can be mitigated by solar photocatalytic pre-treatment of feed saline water. The integration of renewable solar energy in the organic degradation

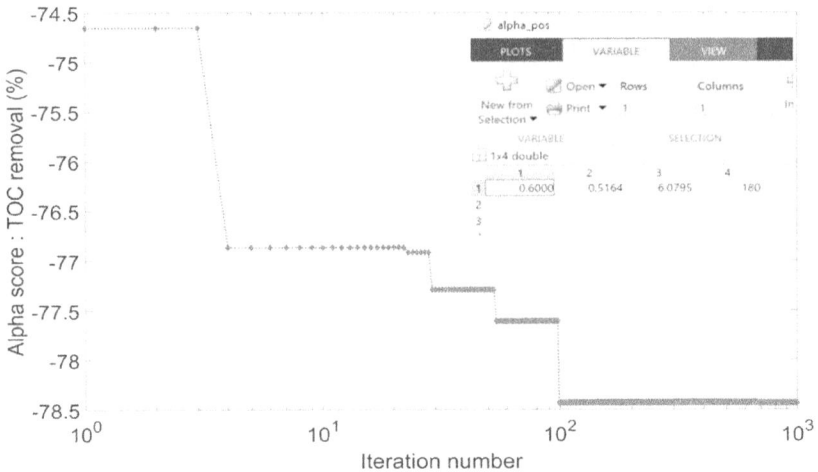

FIGURE 6.9 Convergence curve of GWO (maximum TOC removal = 78.5% at optimum input settings: TiO_2 dosage = 0.6 g/L, ZnO dosage = 0.5164 g/L, pH = 6.0795 and solar irradiation time = 180 minutes).

process considerably reduces the energy requirement in the desalination process. The application of computational intelligence and evolutionary algorithms in modelling and multi-variate optimization of the photocatalytic system enables the determination of optimum values of operational parameters. Running the photocatalytic reactors at these settings enhances the mass transfer, photon-pollutant-hydroxyl radical contact and the photodegradation of organic contaminants from saline water. This surpasses the electron and hydroxyl radical scavenging issues and negates the interference of chloride ions in the pre-treatment process. Effective utilization of nano-photocatalysts and renewable solar energy enables this technique to be implemented at the commercial level in the solar photocatalytic RO pre-treatment process.

REFERENCES

1. Joy, V. M.; Feroz, S.; Dutta, S. Solar Nanophotocatalytic Pretreatment of Seawater : Process Optimisation and Performance Evaluation Using Response Surface Methodology and Genetic Algorithm. *Appl. Water Sci.*, **2021**. doi:10.1007/s13201-020-01353-6.
2. Feroz, S. Application of Solar Nanophotocatalysis in Reverse Osmosis Pretreatment Processes. *Renew. Energy Technol. Water Desalin.*, **2017**, 43–56. doi:10.1201/9781315643915.
3. Amy, G.; Ghaffour, N.; Li, Z.; Francis, L.; Linares, R. V.; Missimer, T.; Lattemann, S. Membrane-Based Seawater Desalination: Present and Future Prospects. *Desalination*, **2017**, *401*, 16–21. doi:10.1016/j.desal.2016.10.002.
4. Weinrich, L.; LeChevallier, M.; Haas, C. N. Contribution of Assimilable Organic Carbon to Biological Fouling in Seawater Reverse Osmosis Membrane Treatment. *Water Res.*, **2016**, *101*, 203–213. doi:10.1016/j.watres.2016.05.075.
5. Ghaffour, N.; Bundschuh, J.; Mahmoudi, H.; Goosen, M. F. A. Renewable Energy-Driven Desalination Technologies: A Comprehensive Review on Challenges and Potential Applications of Integrated Systems. *Desalination*, **2015**, *356*, 94–114. doi:10.1016/j.desal.2014.10.024.

6. Malato, S.; Fernández-Ibáñez, P.; Maldonado, M. I.; Blanco, J.; Gernjak, W. Decontamination and Disinfection of Water by Solar Photocatalysis: Recent Overview and Trends. *Catal. Today*, **2009**, *147* (1), 1–59. doi:10.1016/j.cattod.2009.06.018.

7. Malato, S.; Maldonado, M. I.; Fernández-Ibáñez, P.; Oller, I.; Polo, I.; Sánchez-Moreno, R. Decontamination and Disinfection of Water by Solar Photocatalysis: The Pilot Plants of the Plataforma Solar de Almeria. *Mater. Sci. Semicond. Process.*, **2016**, *42*, 15–23. doi:10.1016/j.mssp.2015.07.017.

8. Jiménez, S.; Micó, M. M.; Arnaldos, M.; Corzo, B.; Contreras, S. Treatment of Cleaning Waters from Seawater Desalination Reverse Osmosis Membranes for Reutilization Purposes. Part I: Application of Fenton Processes. *J. Water Process Eng.*, **2017**, *19* (August), 283–290. doi:10.1016/j.jwpe.2017.08.008.

9. Yang, T.; Peng, J.; Zheng, Y.; He, X.; Hou, Y.; Wu, L.; Fu, X. Enhanced Photocatalytic Ozonation Degradation of Organic Pollutants by ZnO Modified TiO$_2$ Nanocomposites. *Appl. Catal. B Environ.*, **2018**, *221*, 223–234. doi:10.1016/j.apcatb.2017.09.025.

10. Qin, R.; Meng, F.; Khan, M. W.; Yu, B.; Li, H.; Fan, Z.; Gong, J. Fabrication and Enhanced Photocatalytic Property of TiO$_2$-ZnO Composite Photocatalysts. *Mater. Lett.*, **2019**, *240*, 84–87. doi:10.1016/j.matlet.2018.12.139.

11. Varghese, M. J.; Dutta, S.; Feroz, S. Modeling and Optimisation of TiO$_2$ /Photo-Fenton Process for Seawater Pretreatment with Artificial Neural Networks (ANN) and Response Surface Methodology (RSM). *J. Indian Chem. Soc.*, **2020**, *97* (10), 1–8.

12. Nwobi-Okoye, C. C.; Uzochukwu, C. U. RSM and ANN Modeling for Production of Al 6351/ Egg Shell Reinforced Composite: Multi Objective Optimization Using Genetic Algorithm. *Mater. Today Commun.*, **2020**, *22* (August), 100674. doi:10.1016/j.mtcomm.2019.100674.

13. Shargh, M.; Behnajady, M. A. A High-Efficient Batch-Recirculated Photoreactor Packed with Immobilized TiO$_2$-P25 Nanoparticles onto Glass Beads for Photocatalytic Degradation of Phenazopyridine as a Pharmaceutical Contaminant: Artificial Neural Network Modeling. *Water Sci. Technol.*, **2016**, *73* (11), 2804–2814. doi:10.2166/wst.2016.132.

14. Mirjalili, S.; Dong, J. S. *Multi-Objective Optimization Using Artificial Intelligence Techniques*; Springer, 2020. doi:10.1007/978-3-030-24835-2.

15. Deb, K.; Pratap, A.; Agarwal, S.; Meyarivan, T. A. Fast and Elitist Multiobjective Genetic Algorithm: NSGA-II. *IEEE Trans. Evol. Comput.*, **2002**, *6* (2), 182–197. https://doi.org/10.1109/4235.996017.

16. Aljuboury, D. A. D. A.; Palaniandy, P.; Aziz, H. B. A.; Feroz, S. Treatment of Petroleum Wastewater Using Combination of Solar Photo-Two Catalyst TiO$_2$ and Photo-Fenton Process. *J. Environ. Chem. Eng.*, **2015**, *3* (2), 1117–1124. doi:10.1016/j.jece.2015.04.012.

17. Mirjalili, S.; Mirjalili, S. M.; Lewis, A. Grey Wolf Optimizer. *Adv. Eng. Softw.*, **2014**, *69*, 46–61. doi:10.1016/j.advengsoft.2013.12.007.

18. Aljuboury, D. A. D. A.; Palaniandy, P.; Aziz, H. B. A.; Feroz, S.; Amr, S. S. A. Evaluating Photo-Degradation of COD and TOC in Petroleum Refinery Wastewater by Using TiO2/ZnO Photo-Catalyst. *Water Sci. Technol.*, **2016**, *74* (6), 1312–1325. doi:10.2166/wst.2016.293.

7 Emerging Membrane Technologies for Saline Water Treatment
Challenges and Future Perspectives

Afzal Husain Khan
Jazan University

Mohammad Saood Manzar
Imam Abdurrahman University

Nadeem Ahmad Khan
Jamia Millia Islamia Central University

CONTENTS

DOI: 10.1201/9781003185437-8

7.1 INTRODUCTION

Water scarcity around the sphere is becoming a noteworthy problem because of inadequate obtainability of freshwater resources and the high cost of conveying fresh water from distant bases to water demand areas. This condition has led to a renewed attention on emerging salt-water and saline water as alternate bases of clean water. Abundant seawater and brackish water sources and fast-developing salt removal methods and investigation will offer noteworthy prospects to discourse existing and future water scarceness glitches. Water is abundant on the earth's surface; about 97% of it is saline, with just 3% being fresh water ideal for humans, vegetations, and wildlife. The usage of clean water has augmented because of increasing population, suburbanization, and modernization [1,2].

A lot of drought-hit areas have a lot of saline water which further needs treatment for use in different modes. In order to overcome water scarcity problems in arid regions, the approaches, like water conservation and recycling are implemented, but these approaches are also prohibitively expensive [3]. Water desalination is the primary method for obtaining drinking water from a number of sources [4]. The desalination method is typically used in places where there is an abundance of saline water. The derivative of the desalination process is generally brine water. Desalination technique is also practiced on numerous maritime craft and warships. The majority of modern interest in desalination is centered on providing worthwhile fresh water for human drinking. Reverse osmosis (RO) is a desalination technique that applies the osmosis technique to eliminate salt and extra filths by transporting water via a sequence of semipermeable membranes [5]. As per Edward Jones et al., about 16,000 operative desalination units are placed athwart 177 nations, which produce a valued 95 million m^3/day of drinking water [6]. Desalination is frequently applied in countries in the Middle East and North Africa, like Saudi Arabia, the United Arab Emirates, and Kuwait [7]. Approximately 21,000 desalination units are in process around the world of which the major ones are operative in nations like United Arab Emirates, Saudi Arabia, and Israel [8]. Desalination is not so cost-effective when associated with other viable sources of water treatment, as it generates a very small portion of fresh water which can be used by human whereas it is economical for household and industrial application in dry zones [9]. Despite the probabilities of having higher costs, desalination when coupled with technological advancement can perform better than other conventional water treatment methods. The aim of this chapter is to explore the various methods for purification of saline water, as well as the management and technical approaches for implementing these methods, and to provide insight into future recommendations, trials, and opportunities involving sustainable water sources reliant on desalination technologies.

7.2 GLOBAL DESALINATION PRACTICE

Desalination has grown into a feasible water supply substitute permitting to tap the major water reservoir present in the sea. Brine salt removal technology, accessible for decades, made great strides in several dry parts of the global platform like the Middle East, the Mediterranean, Australia, and the Caribbean [6]. Desalination plants work in countries like Saudi Arabia, Oman, United Arab Emirates, Cape Verde, Portugal, Italy, India, China, Japan, and Australia. Universal desalination plants yield around 3.5 billion gallons of potable water per day. Water quality as a relationship of salinity is stated by the concentration of total dissolved solids (TDS), the amount of entire minerals, metals, cations, and anions present in the aquatic medium [10]. Plot drawn below displays numerous categories of water and likely arrays of TDS concentration. The more amount of TDS in aquatic medium distresses human well-being and leads to scaling in piping and erosion of piping and fittings. The U.S. Environmental Protection Agency (USEPA) has fixed the Secondary Maximum Contaminant Level (SMCL) standard for TDS in potable water lower than 500 mg/L; in drinking water, a TDS concentration of <200 mg/L is required. The main goal of desalination is to remove salt from the water so that the resultant water should be suitable for fit to drink use or other planned usages. RO plants yield high TDS concentrates (>65,000 mg/L) that might also comprise nearly noxious chemicals used during raw water pretreatment and posttreatment (cleaning) methods. Figures 7.1 and 7.2 show the global distribution of large desalination plants by capacity, feedwater type desalination technology, and feed water sources based on TDS for desalination in the United States, Saudi Arabia, Australia, China, Israel, and others.

7.3 MEMBRANE SHAPES AND MODULES

Plate-and-frame, tubular, spiral wound, and hollow fiber are the four most common types of modules used nowadays in desalination technologies. The detailed description and illustration (Figure 7.3) are given below:

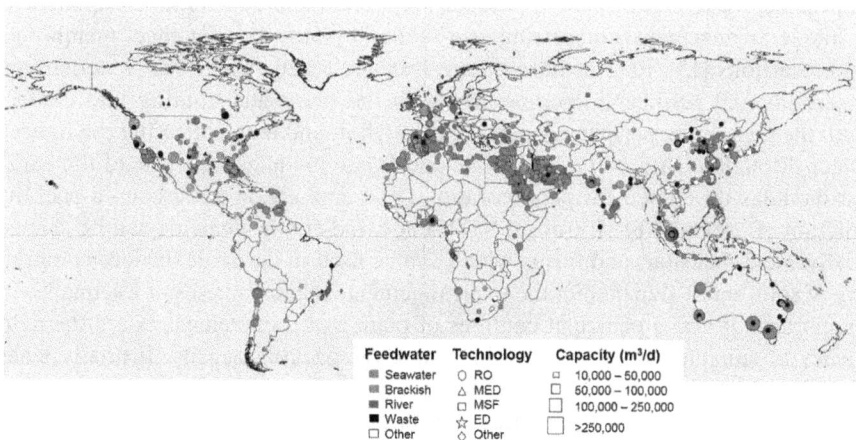

FIGURE 7.1 Global distribution of large desalination plants by capacity, feed water type, and desalination technology.

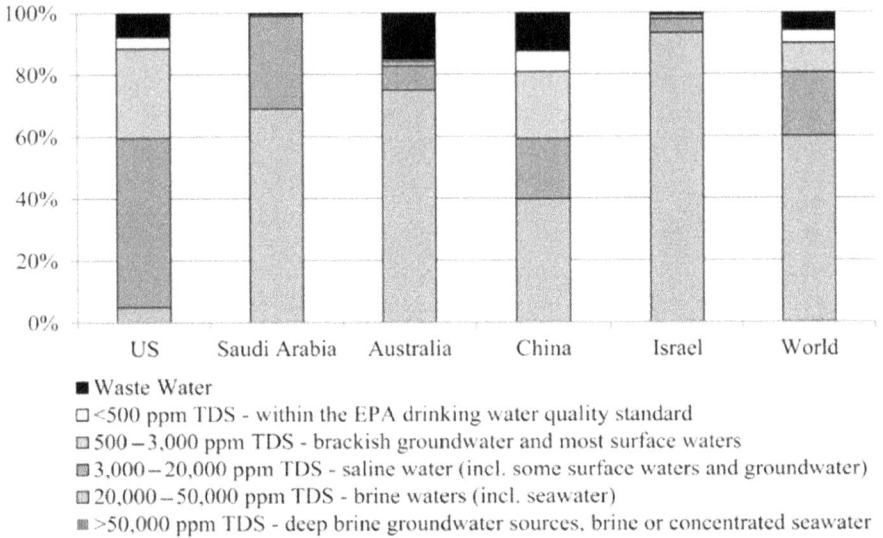

100%

80%

60%

40%

20%

0%

US   Saudi Arabia   Australia   China   Israel   World

■ Waste Water
□ <500 ppm TDS - within the EPA drinking water quality standard
▨ 500 – 3,000 ppm TDS - brackish groundwater and most surface waters
▨ 3,000 – 20,000 ppm TDS - saline water (incl. some surface waters and groundwater)
▨ 20,000 – 50,000 ppm TDS - brine waters (incl. seawater)
▣ >50,000 ppm TDS - deep brine groundwater sources, brine or concentrated seawater

FIGURE 7.2   Feed water sources based on TDS for desalination in the United States, Saudi Arabia, Australia, China, Israel, and others. (See [11].)

7.3.1 PLATE AND FRAME

Membrane processes are generally effective when compared to separation approaches like sedimentation and filtration [13]. Exertions have been done to augment RO procedures and lessen both the investment and functioning prices, mostly via enhancements in membrane expertise. Plate and frame membrane modules are applied in numerous water treatment techniques which contain contaminated waste streams for treatment [14]. The plate-and-frame component is a simple arrangement, comprising two-fold end plates, the flat sheet membrane, and insertions [15]. Flat membranes are fastened together with raw water stream insertions and permeable membranes, with the raw water coming into contact with the membrane peripheral, passing through it, and then collecting the filtered water through a main outlet [16]. Plate and frame modules were one of the earliest modules developed for pervaporation (PV) application possessing a packing fraction of around 100–400 m^2/m^3 [17]. Plate-and-frame elements can be shaped in diverse dimensions and forms which can be used in lab-scale devices comprising of solo, small-size membrane to arrangements that comprises of a sequence of membranes [9]. The principal confines of plate-and-frame elements are the deficiency of suitable membrane provision and low packing density. It finally leads to truncated hydraulic pressure or process at comparable pressures on together borders of the membrane which finally needs comparatively higher process regulation [18]. Low-packing density results in an elevated investment and working charges. Additional confines of the plate-and-frame arrangement comprise difficulties with interior and exterior fastening and difficulty in monitoring membrane for a confined range of flow speeds and pressures.

(a) Plate and frame module

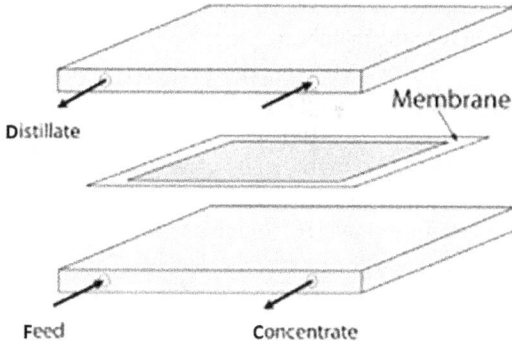

(b) Tubular module

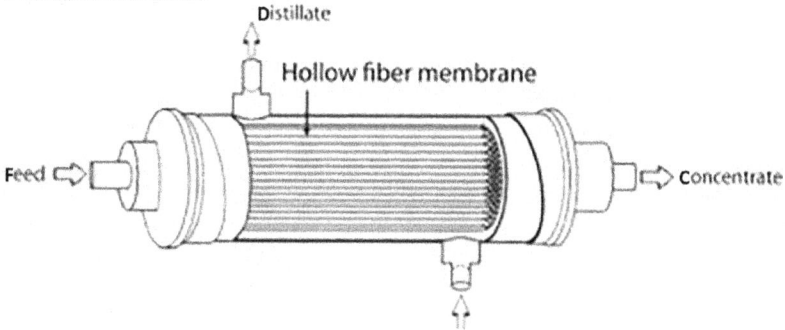

(c) Spiral wound module

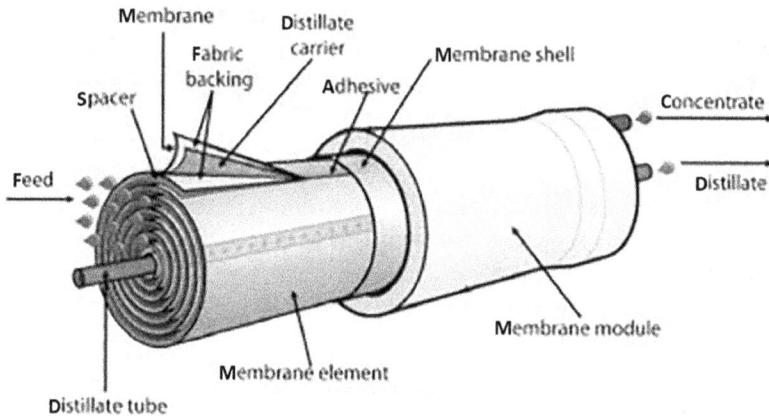

FIGURE 7.3 Membrane distillation module arrangements: (a) plate and frame module; (b) tubular module-based hollow fiber membranes; and (c) spiral wound components. (See [12].)

7.3.2   Tubular Membrane

Tubular membranes function in lateral, or against the stream, plan where process liquid is driven laterally across the membrane exterior. In tubular modules, the membrane is regularly on the inner of a conduit, and the raw water is propelled via the conduit [12]. This strategy upholds steady purification rates for process watercourses with more loads of salts or precipitates. Tubular membranes possess an uneven structure comprised of robust polymeric ingredients; consequently, they can effortlessly treat a higher amount of suspended solids and distillate product [19]. Tubular membranes are accessible in ultrafiltration (UF) and microfiltration skills.

7.3.3   Spiral Wound Membrane

Spiral-wound membrane (SWM) modules are the elementary gadgets of modern desalination treatment expertise. A SWM is applied for detaching a primary constituent from a raw water blend of the primary constituent and the subsequent constituent. This module has a flat sheet membrane enfolded around a pricked permeate assemblage conduit. The treated water (permeate) passes on single sideways of the membrane. A membrane cover made of binary panes is pasted at the three boundaries of an archetypal SWM component, with a fabric filling the permeate station. This envelope's showing permeate side is attached to a pricked internal station, which amasses the treated water [20]. The SWM element design variables also comprise raw water and treated water passage physiognomies and number of membrane coverings for continuous entire zone with membrane surface properties. SWM modules have a compact design, stuffing a bulky membrane surface area per unit volume which results in a thin space of gap less than 1 mm, which finally leads to worsening working difficulties like abrasion loss, contamination of the membrane, and scale formation [21].

7.3.4   Hollow Fiber

These units can have a shell-side feed configuration where the feed passes along the outside of the fibers and exits the fiber ends as illustrated in Figure 7.3. Hollow fiber modules can also be used in a bore side feed configuration where the feed is circulated through the fibers. Hollow fiber membranes are more promising for the production of membrane modules commercially due its high membrane area to component capacity proportions [22]. Hollow fibers employed for wastewater treatment and in membrane bioreactors are not always used in pressure vessels. Bundles of fibers can be suspended in the feed solution, and the permeate is collected from one end of the fibers. Hollow fiber membranes having a diameter of approximately 1 mm are promising for the production of membrane modules since they lead to higher membrane zone to module volume proportions [23].

7.4   MEMBRANE MATERIALS

Microfiltration (MF), UF, reverse osmosis (RO), and nanofiltration (NF) membranes are non-natural organic polymers. MF and UF membranes are regularly fabricated

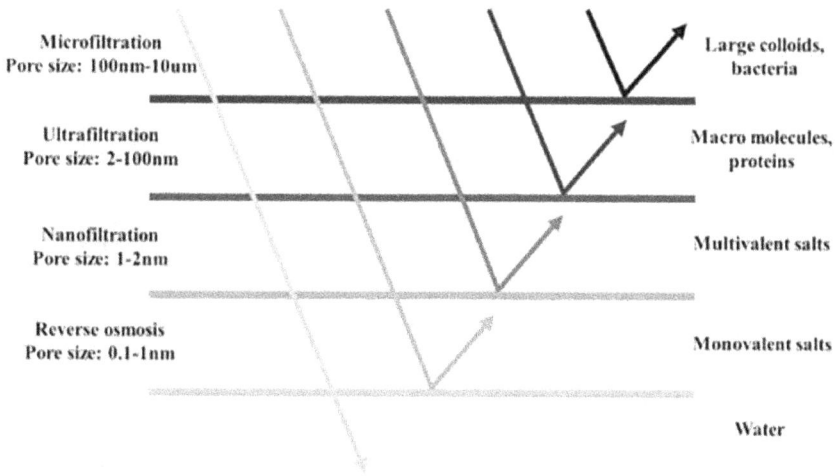

FIGURE 7.4 Categorization of membranes for water purification. (See [25].)

using similar resources, but employing various membrane creation conditions in order to attain different pore dimensions is shown in Figure 7.4 [24].

7.4.1 ORGANIC MEMBRANES

Polymeric or organic membranes are frequently used salt removal and wastewater treatment membranes today, because of their established and better usage archives [26]. Currently, the engineering market built on polymeric or organic membranes for water decontamination is remarkably vast. Cay-Durgun and Lind discussed the usage of permeable ingredients in polymeric membranes for salt removal. The author validated that the addition of thin film nanocomposites in polymeric membranes has improved perm-selectivity, antifouling, and chemical tolerance. There are other kinds of organic membranes which are made up of cellulose which is a polymer originated from plants like fiber. Cellulose being linear, rod-like ingredient support cellulose-based membranes in a strong way [25]. Cellulose-based membranes have a lot of compensation as compared to other RO membranes like simple to fabricate and possess exceptional mechanical features [27]. The other compensations of organic-based membranes are comparatively resilient to attack via chlorine and can also stand up to 5 mg/L of free chlorine, which is much higher than the tolerance revealed by various other membranes like aromatic polyamides [28]. Organic-based membranes decompose in the presence of water with the passage of time, thus cuts their selection [29]. These types of membranes are also enormously subtle to variations in pH, and the salt rejection rate of these types of membranes declines as the temperature surges [30,31].

Although polymeric membranes endure to grow at a fixed percentage, inorganic membranes with excellent chemical and physical features like high adjustability and recyclability are also gaining acceptance. The inorganic membranes comprise ceramic membranes [32,33] and carbon derivative membranes [34,35]. Oxides like alumina [36,37], silica [38,39], or other inorganic elements are normally used publicized membranes. Figure 7.4 depicts the carbon templated existence in a mesoporous silica matrix which can avert the movement of silica clusters from hydrolysis and can hinder micropore contraction. As of safeguarding and cost-effective point of assessment, inorganic membranes are perfect for in situ chemical washing at higher temperatures [40]. Advantages of inorganic membranes are that they are less vulnerable to erosion by microbes, which finally leads to the fouling decay of majority of the polymer-based membranes [41]. Inorganic membranes can be inserted into adequately solid and porous thin film and multilayered-based structures, and tubes if they are suitably pervious [40]. The other types of inorganic membranes like ceramic membranes are comparatively novel for desalination processes and possess numerous important rewards over polymeric membranes because of their exceptional thermal and chemical solidities [42,43]. Ceramic membranes can withstand at higher temperatures and in the occurrence of solvent deprived of apprehension that it may deteriorate further [44]. Other than ceramic-based membranes, membranes fabricated by graphene oxide (GO), carbon nanotubes (CNTs), etc. have involved huge consideration because they hold required features like adjustable pervious morphology, outstanding chemical, mechanical, and thermal properties, decent salt rejection rate, and higher water absorptivity [45]. Furthermore, inorganic membranes can endure high operational pressure conditions, high mechanical stability, and no chances of fouling [45]. Other than this, the drawbacks of inorganic membranes are their high cost and higher energy consumed in the manufacturing methods [46].

7.4.2 Hybrid Organic–Inorganic Membrane

Metal-organic frameworks (MOFs) are a type of hybrid organic-inorganic membrane mixture crystal-like permeable ingredients that comprise an unvarying collection of positive-charged metal ions bounded by carbon-based connecting elements to produce a reiterating, cage-like unit. The development of these innovative nanomaterials like carbon nanotubes (CNTs), graphene, and its offshoots has offered an innovative era for aiding membrane science and technology with proficient separation recital. The exclusive characteristics of these ingredients have forced the researchers to plan a novel type of inorganic membrane. The high porousness and choosiness of these membranes may perhaps permit higher water fluidity and display enhanced energy savings at high recapture. In addition, the novel structural concepts of these membrane materials are anticipated to upgrade some existing problems found in the contemporarily used ingredients and achieve the practically effective, high efficiency, and energy-saving desalination and wastewater treatment methods.

7.5 MEMBRANE TECHNOLOGIES FOR SALINE WATER PURIFICATION

World oil production has increased since the late 1850s and is expected to exceed 106.6 million barrels annually in 2030. The demand for petroleum has increased. In addition to the importance of the petrochemical industry, the operation added one of the most important waste sources, including more than 80% of liquid waste and 95% of aged oils. Water is a dynamic crude oil byproduct (oil and saline generated wastewater). Oil and saline water products can be widely categorized as organic and inorganic compounds based on dissolved and/or dispersed oil and saline, heavy metals, and dissolved salts. In order to prevent significant environmental harm as a result of its toxic properties and the need to comply with the 10–15 part per million release requirements, oil-generated wastewater must be treated. Increased clean water consumption is the result of population growth, urbanization, industrialization, and increased demand. Two-thirds of the world's population, according to the UN (2007), will face water deficits by 2025 because of the unparalleled safe supply of water. The "Food, joint duty" study shows that nearly 1.2 billion people have no access to clean drinking water, 2.6 billion are left without or insufficient sanitation, and millions will die from infected water. Along with appropriate water store-taking, wastewater recuperation, and desalination, sustainable solutions to water stress problems were thoroughly explored. The disposal waste wastewater treatment of the oil and saline seems to be another feasible solution in the light of the huge extraction and wastewater release to clean water in order to satisfy water requirements and to save the atmosphere for future generations.

The statistical analysis presented in Figure 7.5 shows a rising interest in the recovery of oil-generated waste water as shown by the growing number of publications on the subject between 2003 and 2020. According to data, different technologies have been applied in the oil and saline and gas industry for oil-generated wastewater treatment. Because of its low-energy consumption and simple and well-controlled process management without the use of expensive chemical compounds, membrane technology is an appealing replacement technology. Studies during 2003–2020 for oil and saline wastewater reclamation for the membrane technologies, including RO, FO, and MD, are shown in Figure 7.5. Due to its benefits, membrane distillation and forward osmosis trials overcame conventional RO in the last decade. The future osmotic and membrane distillation earn significant attention as a medium for energy-efficient membrane desalination technologies. In general, the overall objective of all wastewater desalination technologies is to eliminate and use treated drinking water to dissolved components while still reducing environmental and water scarcity issues.

Oil- and saline-generated waste water has many salinities and requires particular consideration in the conception and management, rather than the use of municipal oily waste water, of the special treatment plant. Thanks to its high-quality salt extraction and product water, advanced membrane treatment is the perfect alternative to traditional wastewater treatment from oil and saline generation. RO, front osmosis, and distillation of the membrane are both strategies for diaphragms dependent on oil and saline-produced wastewater. These approaches have been extensively researched for freshwater supplies for medicinal and industrial applications due to their ability to remove a broad range of molecules and ions. Efforts to address membrane-based separation issues, especially membrane foulation and energy consumption, continue despite the extensive use of membrane treatment in order to desalt oil-generated waste water. At present, high-performance membrane manufacture and process optimization, such as hybrid technology deployment to improve the performance and use of whole processes in the energy sector, are two major fields of membrane technology.

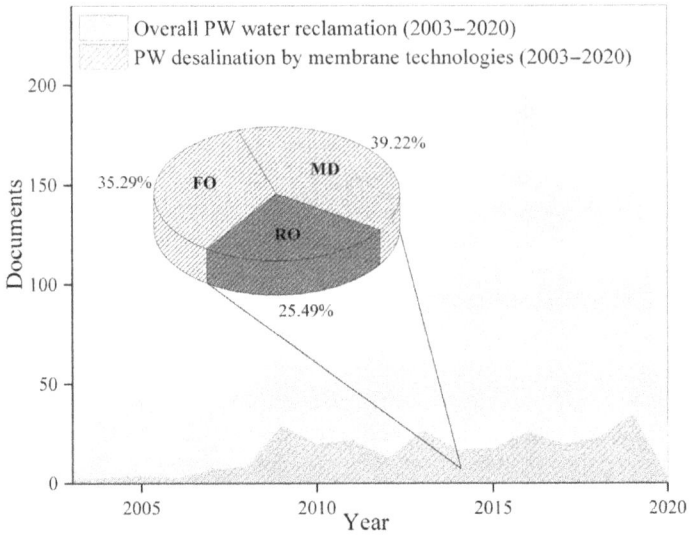

FIGURE 7.5 Increase in oil and saline generated wastewater reclamation studies between 2003 and 2020. (See [47].)

The advancement of membranes, including the selection of materials, advanced membrane design, and increased membrane performance, is making much progress. Among these advances is the combination of nanotechnology and membranes to alter physical-chemical properties, including pores, surface loading, hydrophilicity, and surface rugging. Optimizing membrane performance and working conditions has multiple benefits including increased performance of liquid water, energy conservation, and resistance to fouling. A feasible approach to realize these advantages would seem to be combining two or more membrane or non-diaphragm technologies. The method of incorporation covers both the pre- and posttreatment processes. Numerous tests of advanced membrane technology in oil-generated wastewater treatment have shown the promise. Table 7.1 shows the efficacy of membrane desalination. Tong et al. illustrated the membrane technologies for the demulsification and RO generated by oil and saline water, forward osmosis, and desalination membrane distillation. In contrast, Shaffer et al. conducted a crucial assessment of oil-borne wastewater desalination and reuse with various membrane technologies such as forward osmosis and membrane distillation. However, in terms of treatment efficiency and basic methods/improvements in membrane architectures, progress in membrane technologies for waste water handling is inadequate to solve existing shortcomings. The overall feasibility and effectiveness of oil-generated wastewater desalination technologies remain uncertain.

Fouling phenomena represent the greatest challenge to the creation of consistently advanced desalination membranes such as organic foulings, inorganic scaling, colloidal fouling, and bio-fouling. It first examines the complexity and treatment by conventional technologies of the oil-generated wastewater composition and its advantages and drawbacks. A comprehensive summary of the state-of-the-art wastewater desalination

TABLE 7.1
Efficiency of Desalination Using Membrane Desalination

Use of Membrane Desalination Configuration	Type of Membrane	Input Waste	Rejection
Direct contact membrane distillation	Polyvinylidene fluoride (80.77% porosity; 0.4 nm thickness; contact angle, 82°)	Synthetic oil and saline generated wastewater	>99% of TOC
		18.10 mg/L of TOC plus 247,900 ppm of TDS	>90% of salt waste
	Polyvinylidene fluoride incorporated with SiO_2	Synthetic oil and saline generated wastewater	–
		1 M NaCl plus 1,000 mg/L of oil produced	
	Polyvinylidene fluoride functionalized with KOH	Synthetic oil and saline generated wastewater	99% of salt waste
		0.05 v.% canola oil plus 0.6 M of NaCl	
	Polyvinylidene fluoride electrospray coated with ZnO	Synthetic oil and saline generated wastewater	>99% of salt waste
		saline oil solution (0.015 v/v) plus 3.5 wt% NaCl solution plus dissolved solute (0.4 mM) feed	
	Polyvinylidene fluoride incorporated with Silica	Synthetic oil and saline generated wastewater	~100% of salt waste
		500 mg/L of crude oil plus 1 M NaCl	
	Polyvinylidene fluoride + polyvinyl alcohol	Real oil generated wastewater from the Wattenberg field in northeast Colorado	>99% of salt waste
	Marketable Polyvinylidene fluoride (Merck-Millipore)	Synthetic oil and saline generated wastewater	–
		10,000 mg/L NaCl plus 1,000 mg/L of oil plus 0.5 mM dissolved solute	
	Polyvinylidene fluoride-coated with SiO_2-forward osmosis	Synthetic oil and saline generated wastewater	> 99% of salt waste
		1 M NaCl plus 1,000 mg/L of oil Plus 0.4 mM dissolved solute	
Vacuum membrane distillation	Marketable polyvinylidene fluoride mean pore size 0.1 mm, membranes porosity 62%	Artificial coal shale gas	99.9%
	Marketable polypropylene with pore size, inner diameter, and membrane porosity 0.2 μm, 450 μm, 1,800 μm, and 73% respectively	Artificial oil and saline generated wastewater	–
		NaCl, KCl, $MgCl_2$, $CaCl_2$ Na_2SO_4, $NaHCO_3$ Na_2CO_3 SiO_2 were 30,750 mg/L, 370 mg/L, 260 mg/L, 420 mg/L, 3,460 mg/L, and 1,250 mg/L, respectively	

(Continued)

TABLE 7.1 (*Continued*)
Efficiency of Desalination Using Membrane Desalination

Use of Membrane Desalination Configuration	Type of Membrane	Input Waste	Rejection
	Marketable (membrane) porosity = 73.9%, average pore sizes = 0.238 µm, and contact angle of outer surface were 147.7°	Artificial oil and saline generated wastewater	99.5% of salt waste
Air gap membrane distillation	Marketable polyvinylidene fluoride pore size of 0.22 µm	Synthetic oil and saline generated wastewater 3.5 wt% NaCl solution	99.2% of salt waste
	Marketable microporous	Real oil and saline generated wastewater from Saudi Arabian Oil Company, 63,372 mg/L of NaCl	> 99% of TOC > 99.9% of Salt waste

with forward osmosis, RO, membrane distillation, and their hybrid approach is given in the following paragraph. Significant issues such as membrane expansion, operating conditions optimization, and membrane cleaning procedures are discussed in these processes. Pre-handling was also widely used in practical applications to reduce the strain on the following desalination of the membrane. Fresh developments into the ongoing membrane desalination technology Zero Liquid Discharge (ZLD) in the treatment of concentrations of saline solutions have also been investigated.

7.5.1 FORWARD OSMOSIS (FO)

The membrane processes used for mass transfer are similar in the desalination of waste water produced by oil and saline and involve moving along the membrane interface of the component. On the other hand, the membranes necessary for the separation procedure differ in terms of pores, surface load, and hydrophilicity. The popular properties of polymers also include their lightweight, easy manufacture, and lower costs than ceramic membranes in membranes manufacturing. The use of polymeric membranes is also challenging, including low permittiveness, low fouling resistance, and limited service life. It is essential to establish membrane types of oil-generated wastewater treatment with desirable features like reliability and performance. The membrane of thin-film composite (TFC) is the most often used for forward osmosis and RO. A standard TFC is made up of a layer of interfacial polymerization (IP), consisting of a permanent selective polyamide (PA) layer, modelled using the techniques of reverse phase on a porous substratum membrane support. Commercial TFC membranes are generally characterized by a robust, low hydrophilic PA sheet.

Dipping oil and saline-generated waste water often benefits from membrane changes, as they strengthen the hydraulic surface and minimize surface roughness. Membrane modifications using novel functional nanomaterials, such as the processing of waste water generated by oil and saline for a range of applications, are currently at the forefront of membrane development. Oil-generated wastewater membranes have a series of attractive properties, such as preliminary measuring, high flux, and anti-fouling. Nanomaterials including carbon nanomaterials, nanoscale metal oxides, and zwitterions have been used to enhance the membrane properties in many studies. Nanomaterials can alter their structures and characteristics by adding a protective hydration layer to polymers. Nanomaterials have been integrated into the PA layer or substratum through combining or infusion in each of these techniques. Additional hydrophilic materials were applied to adjust membrane membranes to achieve favorable properties such as high hydrophilicity and low roughness. Many studies have shown that high-performance TFC membranes with hydrophilic surface material coverings and new forms of usable IP monomers are efficiently manufactured and modified.

Typical membrane distillation membranes include polyvinyl-fluoride (PVDF), polyethylene (PE), and polypropylene (PP) due to their low energy level and high water repellence. Several mechanisms have been used to enhance membrane distillation characteristics of the hydrophobic membrane. The inclusion of functional groups that favorably modify the surface properties of the membrane usually leads to surface changes. One such change is the incorporation into the traditional method of hydrophobic membrane distillation of the hydrophobic/omniphobic membrane or the hydrophobic surface membrane called the Janus membrane. The superhydrophobic/omniphobic membrane is produced with low-energy materials such as spherical and fluoroalkylsilane, which are surface coated. Depending on the omniphobic membrane, the re-entering arrangement of the sphere nanoparticles provides a kinetic barrier to enable the particles to migrate from a solid metal condition of Cassie-Baxter to a completely wetted Wenzel with low surface pressure fluids. On the other side, a uniform durable water-repellent coating is generated by the use of hydrolyzed silane cups such as 1H, 1H, 2H, and 2H, perfluorodecyl, and fluorinated silane. Fluorosilane is fluorsilanized with the alkyl silane. These routes will combine surfaces with low free surface energies, which are superhydrophobic/omniphobic. Those amendments show promising weather tolerance when used in combination with organic solvent solutions with low surfactants and low surface tension. Later studies identified a hydrophilic JM sequence to improve the anti-weathering and anti-fouling properties of the membrane. The latest JM can be realized in two ways: by integrating a coating layer or by incorporating a coating layer superhydrophobic. On hydrophobic and/or omniphobic substrates, the use of self-installation or graft is performed on the hydrophobic substances such as zwitterionic polymer polyvinyl alcohol (PVA), potassium hydroxide (SiO_2), and silica nanoparticles. These hydrophilic modifications inhibit the adsorption of organic compounds by creating a hydration layer at the membrane surface. Though the adhesion to omniphobic substrates is much harder than hydrophobic substrates, omniphobic JM is still superior to oil-generated wastewater treatment because, contrary to the hydrophobic Janus membranes, it incorporates anti-wrenching and anti-fouling roles, usually just antifouling. Although surface alteration by the surface cover usually produces a further film, reducing the flow, the addition of a hydrophilic layer will minimize the conductor thermal loss by the membrane, thus increasing the flow of the transmembrane.

7.5.2   Reverse Osmosis (RO)

RO is a well-known membrane process, suitable for the treatment of wastewater generated by oil and saline. Commercial RO membranes have been commonly used during this phase, and as a result of processing of the oil as well as saline and gas byproducts, significant quantities of oils and saline are produced by wastewater, as shown in Table 7.2. Since the primary reversal osmosis restriction for desalination of waste water generated by oil and saline is still fouling, the majority of studies are based on membrane changes and cleaning techniques. The polydopamine coating of the commercial membrane has been changed by McCloskey et al., and there is a notable increase in flows during the oil and saline wastewater treatment. Polydopamine was used to produce a very thin, clean, and hydrophilic top layer of the PA board. Flux has been improved by 30%–50% over unmodified membrane due to the fact that polydopamine aggregation efficiently improves membrane surface hydrophilicity. Although water could enter previously inaccessible canals by means of the hydrophilic polydopamine coating, the hydrophobic shape of the membrane was normal. As Pei and colleagues applied an advanced form of functional monomer for hyperbranched polyamidoamine (PAMAM)/TMC RO membrane, a similar RO membrane modification was observed. Due to its strongly radiated three-dimensional structure and self-inhibition, PAMAM developed a thin, smooth, and hydrophilic active layer during IP. Experiments with fouling indicated a continuation of the rejection, while the flow declined around five percent in 24 hours.

A RO membrane used in industry was more resistant to modified oil and saline adsorption than was used here. To restore its effectiveness, membrane cleaning is necessary. In conjunction with hydraulic washing, Xu and Drewes examined both the fouling of RO and NF devices and the application of some acid and metal chelating agents. To retrieve the stream of a commercial RO membrane Piemonte et al. study the performance of VSEP cleaning processing. Vibrating membranes can reduce by up to 100% the TSS content of oil-produced wastewater while reducing bio-oxygen (BOD), chemical oxygen (COD), and TOC content. Lin et al. have used RO to treat wastewater formed with high salinity oils and saline, and analyzed the relationship between pressure and permeate water and the connection between stability and

TABLE 7.2
Characteristic of Oil and Saline Generated Wastewater Content

Characteristics	Max. Limit Range (mg/L)
Total suspended solid	116
Total organic carbon	70
Chlorine	9,218
Potassium	13
Calcium	371
Total nitrogen	25
Sulfate radical	69

TABLE 7.3
Efficiency of RO

Type of Membrane	Input Water	Refusal Rate
Marketable RO	Real oil and saline generated wastewater 722 mg/L of Total dissolved solids (1,448 µS/cm) + 68.8 mg/L of Total organic carbon	87% of total dissolved solids
Marketable RO double filtration	Synthetic oil and saline generated wastewater 37,500 mg/L Total dissolved solids + 565 mg/L of oil	99.8% of total dissolved solids
Marketable RO	Synthetic oil and saline generated wastewater 4,800–5,500 mg/L Salts +9–13 mg/L petrolic	~99.9% of total organic carbon 96.6% of salt wasted
Marketable RO incorporated with polydopamine	Synthetic oil and saline generated wastewater 4.5 g surfactant plus 40.5 g soybean oil in 3 L of water using 2,000 mg/L salt	>99.1% of salt wasted
Hyperbranched polyamidoamine	Synthetic oil and saline generated wastewater 2.5 or 5-mL hexadecane plus 2,000 mg/L NaCl aqueous solution plus 250 mg	99% of total organic carbon 89.3% of salt wasted
Marketable RO (thin film composite)	Real oil and saline generated wastewater 5,243 ± 561 mg/L of total dissolved solids conductivity (9,647 ± 652 µS/cm)	80.4% of total organic carbon 91.4% of salt wasted

cleaning time of the membrane module. The integrity of the membrane was calculated at an operating pressure of 1.2 MPa for a 5-hour duration and during this time its flux continued to be constant. In the last step, flow is reduced and the module is removed from the osmosis membranes for the extraction of organic salt and hydrocarbons. The flow can be returned to its original state after a fourth cleaning with the detergent. The efficacy of the RO membrane depends heavily on feed levels according to these studies. The degree of membrane fouling also often has a direct impact on the frequency and expense of cleaning processes. However, RO can also be used to treat oil-producing waste water, especially by means of suitable membrane changes on the basis of RO water quality permeates.

7.5.3 MEMBRANE DISTILLATION (MD)

Membrane distillation (MD) was used as an energy-low desalination procedure. However, as solar panels, nanotechnology, and membrane technology have progressed, it is a new technology of considerable academic and business interest. Recent systems proposed include Forward osmosis-dipsy, freeze distillation (FD-C), forward osmosis-diaphragm desalination membrane (VDM), FD-desalination-C-diaphragm, VM-diaphragm desalination-C, and diaphragm desalination-free-fiber hollow cooling crystallation (Membrane desalination-SHFCC). Forward osmosis is an example of a high-power mechanism requiring the drawing solution to regenerate. In addition, regeneration problems in conventional pressure systems can be caused by a drawer solution of higher osmotic pressure. The drawer solution can be commercially recycled by using forward osmosis-membrane desalination. Advanced osmosis-membrane desalination is more cost efficient to recycle the solution using solar energy and residual gas than RO or NF. In addition, revitalization is done with the use of complex membrane and distillation methods for desalination of the membrane. It runs at moderate temperatures and pressures (atmospheric or vacuum pressure) and theoretically experiences 100% salt repulsion. It is modular and room savings. The separating potential is less influenced by feed concentration compared with other pressure-controlled membrane methods and is produced by solar or waste heat instead of electrical power.

Seawater must be converted into pure water without producing a stream of waste and precious salts to create a solution which is both sustainable and fluid free. The ZLD principle becomes more feasible today by integrating membrane desalination into its operation.

In addition to pure water, the membrane desalination unit produced a high concentration of savory for transparent crystallization in the following crystallizer. The desalination of the membrane concentrates the salmon to almost supersaturated depths in ZLD schemes. Anything about the scaling problem with such a high salt content is seriously concerned, despite the marginal effect that the high salinity has on membrane desalination, compared with other membrane processes induced by pressure, such as RO. Concentrations temperature and polarization also speed up crystal growth on the surface of the membrane. Inorganic scales can hinder the flight, deplete energy in the surface, and ultimately wet a membrane. We need a groundbreaking membrane desalination membrane and module designed to relieve the scaling problem and new techniques for restoring membrane desalination membranes to sustain the viability of ZLD systems. The desalination membrane used has traditionally been microporous membranes, such as PP, PVDF, and PTFE, as well as suitable derivatives of these materials. They perform well in the short term, but the scaling, moisture, and competitiveness of feed supplies pose the challenge of their long-term reliability.

The leakage of oil tanks and industrial runoff, both marine and waste water produce trace quantities of material with low surface voltage, and potential membrane desalination membranes should now be omniphobic, which means they must be robust and be capable of repelling both high and low surface tension. The reintegration structure and low surface energy or surface omniphobicity are two significant factors. The first is mostly developed by nanoparticles, while the second is produced by means of a coating, combining, or plasma processing of fluorinated materials. In various configurations, omniphobic membranes were developed with various supporting materials such as glass fiber and polyvinyl difluoride, including flat sheets,

TABLE 7.4
Combined Treatment Strategies

Combination Treatments	Process Used	Pressure Applied	Characteristics
Enhanced dewatering RO	RO	Osmotic pressure	RO used in pretreatment shows higher water recovery over forward osmosis
			Hybrid enhanced dewatering RO-RO produces more water than one phase RO with comparable energy efficacy
Adsorption-double staged Nanofiltration	Nanofiltration	Osmotic pressure	Organic substances adsorbed on activated carbon
			Less fouling on membrane surface
			Double stages nanofiltration membrane development of more flux permeation
			Rejection of total solids by 72%, turbidity by 6%, and salinity upto 90%.
Micro filters–Membrane desalination	Membrane desalination	Thermal gradient	Membrane desalination employed as pretreatment reduced total organic carbon %
			It can recover calcite and halite from oil wastewater
			Employed low-energy desalination
Micro filters–Forward osmosis–Membrane desalination	Membrane desalination	Thermal gradient	Purified by membrane desalination process are around ~99%
Membrane desalination	Membrane desalination	Thermal gradient	Crystallization is used to reduce temperature as well as remove solvent.
			Treatment of high saline wastewater
UltraFiltration–Nanofiltration–RO	RO	Osmotic pressure	Coagulant–ultrafiltration hybrid used for suspended and colloids removal
Biologically active filtration–Ultra filtration–Nanofiltration	Nanofiltration	Osmotic pressure	Biologically active filtration used as pretreatment removes organic matter under saline condition
			Ultrafiltration best for turbidity
			Nanofiltration process with > 94% salt rejection
Fenton oxidation–Membrane desalination	Membrane desalination	Thermal gradient	Fenton oxidation can degrade >95% of total organic carbon
			Membrane wetting during membrane desalination

(Continued)

TABLE 7.4 (*Continued*)
Combined Treatment Strategies

Combination Treatments	Process Used	Pressure Applied	Characteristics
Forward osmosis–Membrane desalination	Membrane desalination	Thermal gradient	Forward osmosis as a pretreatment Achieved 99.9% for oil and saline and salt
Micro filters–Membrane desalination	Membrane desalination	Thermal gradient	Microfilters as pretreatment Microfilters reduce fouling for further treatment
Forward osmosis–RO	RO	Osmotic pressure	Water flux steadily declined throughout the system RO maintained >99% ion rejection and >95% hydrocarbon rejection

hollow, and nanofibers (PVDF). Membrane diameter membranes with a cross-section sandwich construction consist of:

1. two porous sponge-like inner and outer layers and
2. a dense, thin, central, vacuum-resistant layer of small macro-voids.

In addition, membrane desalination diaphragms made of solar or photo-thermal components may be used as a way to minimize demand for electricity, thereby reducing the purchase of expensive solar panels.

7.6 ROLE OF NANOTECHNOLOGY IN MEMBRANE SALINE WATER PURIFICATION

Water safety and access to clean potable water are basic rights to the life of humans. Sadly, population increase, climate change, deforestation, and economic progress contribute, as seen by the freshwater decline, to the extraordinary crisis of the 21st century. Today, almost 1.2 billion people suffer from water shortages and their repercussions. By 2025, almost 1.8 billion people are expected to be affected by water shortages. Water desalination and agricultural processing of waste water have been developed in decades of development as two successful means to increase and reduce freshwater sources. Alter the fact that seawater accounts for more than 97% (about $1.4–109\,km^3$) of worldwide water supplies constitutes a crucial step in addressing this issue. It is also used in areas with serious water shortages and poor precipitation, such as the Middle East and Arab Peninsula. Alongside the hydrological cycle, the reuse of waste water is another good means of increasing the availability of fresh water. In recent years, municipal wastewater treatment has been a major obstacle to a safe potable water source. Wastewater is currently used to provide fresh water as a natural resource.

Due to its superiority to classical ones this drew great interest. Research on MD (membrane distillation) was extremely active, especially since the 1980s, in a variety of technologies and design techniques, including direct membrane distillation (DCMD) and membrane air gap distillation (AGMD). DCMD is the most often used research center, accounting for over 60% of MD research, mostly because of its high efficiency and ease of use. MD is a hybrid thermal/membrane system, powered by the heat pressure gradient, which carries steam through the hydrophobic membrane from the hot side (320–350 kg) to the cold side. MD's advantages over RO and MED are higher product performance, reduced energy use, high feed solution salinity, and low-grade waste heat. MD has a variety of advantages. MD is a theoretically clean wastewater ZLD production system. In ensuring a desalination and/or waste water treatment supply, MD is crucial. The principal downside of MD is its present industrial-scale, temperature polarisation (TP) effect. TP is an incident that's natural. Water evaporation at a feeder/membrane interface will reduce the temperature of the membrane surface compared to the feed mass, reducing the heat draw. The substance itself could also lead to the TP effect through lack of heat. TP has a major adverse effect on MD's results by minimizing mass transfers. At that point, the flow of permeates would decrease. The use of such new technologies, such as auto heat nanocomposites, is also far from satisfied. Furthermore, membrane tampering and wetting will reduce MD performance. The membrane cleaning costs are increasing, and the MD unit is disconnected. In order to promote this exciting green technology, attempts are also extensively examined to improve the MD through the use of membrane modules, modular structures, and process architecture.

The optimization of membrane parameters including hydrophobicity, porosity, and porous distribution is essential to determine MD efficiency by controlling mass/heat transfer and membrane wetting. MD membrane is mainly manufactured by the use of a range of substances, including polypropylene, polytetrafluoroethylene, polyvinylchloride, using traditional MF and UF membranes formed by various processes such as reversal, detention, sintering, and thermal-induced separation (PTFE). The low penetrating rate, low hydrophobicity, and a tendency toward fouling and weeding are the characteristics of these membranes. In the study of new membranes, it is essential to improve MD's efficiency, especially in high permeations.

Because of its special properties, which include a high surface to volumetric ratio, a tunable fiber diameter, and three-dimensional interconnecting structure, ENMs are used extensively in adsorption and filtration processes such as MF, UF, NF, RO, and water/oil, and saline separation. Apart from a number of advances, ENMs continue to be imported on a broad scale, and current laboratory deployment is largely marginal. ENMs are still unaware of their vast capacities in a number of areas though. Thanks to its advantages of a strong surface/volume ratio (100 times that of MF membranes), high porosity (up to 90 percent), interconnected structures (such as superhydrophobic relationships), and decent mechanical efficiency, ENM is an excellent MD membrane configuration substrate.

7.7 MEMBRANE TECHNOLOGIES FOR POLLUTANT REMOVAL FROM SALINE WATER

Membrane-based technologies are well-known separation technologies providing a variety of applications in water desalination, toxic metal separation, and the recovery of valuables from saline waters. Membrane technology is a word that refers to a variety of different exclusion procedures [48]. Membrane technology has become a popular solution for recuperating water for reprocessing from numerous industrial saline wastewater foundations [49,50]. The application of membranes to create process water from groundwater, surface water, or wastewater is becoming gradually common. A membrane, in its most basic form, is a blockade that splits two segments by averting designated section flow across it as shown in the above figure [51]. Subsequently, then, significant advancements have been made to make membranes more suitable for a variety of applications especially in the removal of recalcitrant pollutants from saline water. The membrane physiognomies also show a leading part in quickening the elimination of contaminants from the saline water. Figure 7.6 illustrates the solute elimination mechanisms conferring to dissimilar sorts of membranes used in salty water treatment technologies. Factually, the membrane separation processes are controlled through changes in concentration, temperature, and changes in electrical potential [52]. The factors that control the processes are grouped based on pore size of the recycled membrane to four operations viz. MF, UF, NF, and RO [53]. Forward osmosis (FO) has been effectively combined with membrane distillation (MD) to remove highly saline landfill leachate [54]. The combined technique could effectively remove NH_4^+-N, Hg, As, and Sb ions. Membrane technology can provide ample denunciation of nonvolatile ingredients in feed solution, and its efficacy is fairly not dependent on the amount of salt in the feed solution [55]. The collective effect of NF and RO technique was

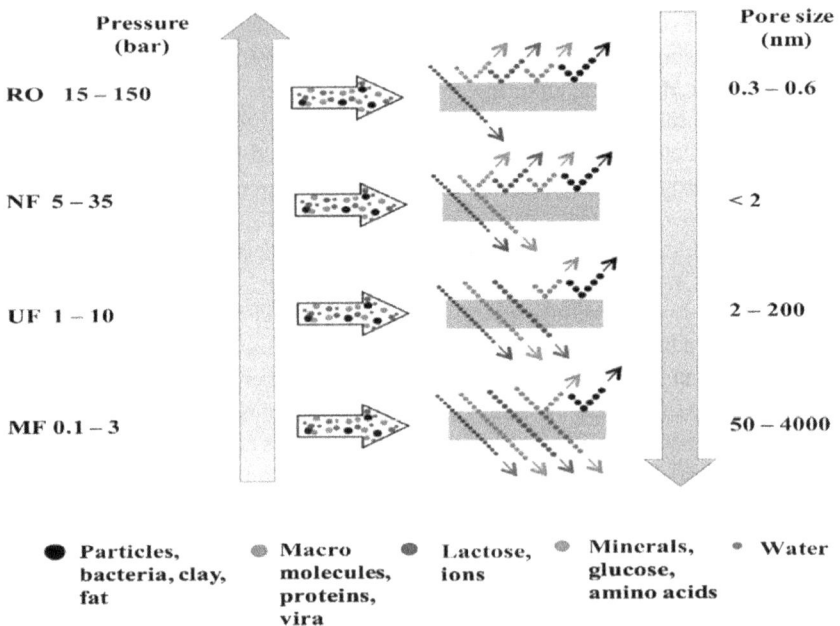

FIGURE 7.6 Membrane separation technologies for saline water pollutant removal its separation characteristic. (See [53].)

applied to purify saline distillery wastewater which comprised of contaminants like total dissolved solids and potassium which were finally eliminated effectively [56]. Innovative salt removal skills provide lesser feed pressure necessities whereas preserving rejection of the salt [57,58]. Graphene-based membrane showed 100% rejection of common ions present in seawater with a better water perviousness than other accessible RO membranes [59]. In another study, in terms of energy consumption, the membrane-based process consumed 2.93 kWh/m^3 at 50% recapture for making water which can be used for farming application needing small boron levels from saline water resources [60]. Membrane-based technology provides the benefits of maneuver at ambient pressure and low temperatures (30°C–90°C), with the skill to achieve 100% salt rejection [61]. Similarly, in another study microbial-based desalination cell was employed with water at diverse preliminary salt concentrations. The fabricated membrane could effectively remove 90% of the salt and other pollutants at a specific energy of 31 Wh/m^3 [62]. In another research, the desalination technology with specific energy consumption of 3.5 kWh/m^3 could attain 50% recovery of salt [63]. These revolutionary researches produced optimistic outcomes in removing pollutants from saline water along with salts through membrane technology.

7.8 CONCLUDING REMARKS

This chapter discusses the potential membrane technologies for the treatment of saline water. Among membrane desalination technologies, RO, FO, and MD are the predominant processes of membrane desalination. For the properties of the membrane, it is essential to control the efficiency of these processes. In order to achieve long-term permeability, selectivity, energy conservation, and anti-fouling/anti-weeding properties different strategies were used to enhance membrane structures and waste control features. In view of the complexity of waste, the probability of membrane failure has increased, and much work has already been done to recuperate the membrane production. An integrated device of two membrane processes or hybrid membrane processes and traditional technology often becomes common as an improved desalting technique. The selection of adequate pre- and posttreatments will improve the efficiency of energy usage and desalination separation. More detailed studies and scientific experiments are also important, despite recent progress in this area, to increase production and encourage the industrialization of technology. Desalination methods must be selected intentionally on the basis of efficiency and results. The research should also concentrate on the economic side and the possibility to sustain a sector-specific development therapy. Lastly, a continuous attempt to reduce the barriers to desalt productivity offers good prospects for reduced demand on water sources and the environmental effect of the oil and saline and gas industry.

REFERENCES

1. H. March and D. Sauri, *Local Environ.* **22**, 523 (2017).
2. M. Christaki, G. Stournaras, P.T. Nastos, and N. Mamassis, *Water Hist.* **9**, 411 (2017).

3. G.B. Beekman, *Int. J. Water Resour. Dev.* **14**, 353 (1998).
4. A.S.A. Mohamed, M.S. Ahmed, H.M. Maghrabie, and A.G. Shahdy, *Int. J. Energy Res.* **45**, 3698 (2021).
5. L.F. Greenlee, D.F. Lawler, B.D. Freeman, B. Marrot, and P. Moulin, *Water Res.* **43**, 2317 (2009).
6. E. Jones, M. Qadir, M.T.H. van Vliet, V. Smakhtin, and S. Kang, *Sci. Total Environ.* **657**, 1343 (2019).
7. M. Hameed, H. Moradkhani, A. Ahmadalipour, H. Moftakhari, P. Abbaszadeh, and A. Alipour, *Water* **11**, (2019).
8. U. Caldera, D. Bogdanov, S. Afanasyeva, and C. Breyer, *Water* **10(1)**, **3**, (2018).
9. K. Elsaid, E.T. Sayed, M.A. Abdelkareem, M.S. Mahmoud, M. Ramadan, and A.G. Olabi, *J. Environ. Chem. Eng.* **8**, 104099 (2020).
10. J.D. Rhoades, in *Methods Soil Anal.* (John Wiley & Sons, Ltd, New Jersey, USA 1996), pp. 417–435.
11. J.R. Ziolkowska, *Water Supply* **16**, 563 (2015).
12. L.M. Camacho, L. Dumée, J. Zhang, J. Li, M. Duke, J. Gomez, and S. Gray, *Water* **5**, 94 (2013).
13. P.S. Goh, T. Matsuura, A.F. Ismail, and N. Hilal, *Desalination* **391**, 43 (2016).
14. W. Suwaileh, N. Pathak, H. Shon, and N. Hilal, *Desalination* **485**, 114455 (2020).
15. A Figoli, S Santoro, F Galiano, A Basile Pervaporation membranes: preparation, characterization, and application. In book: *Pervaporation, vapour permeation and membrane distillation*, 1st Ed. Woodhead Publishing, Cambridge, 19–63 (2015).
16. R. Schwantes, J. Seger, L. Bauer, D. Winter, T. Hogen, J. Koschikowski, and S.-U. Geißen, *Membranes (Basel).* **9**, (2019).
17. E. Guillén-Burrieza, D.-C. Alarcón-Padilla, P. Palenzuela, and G. Zaragoza, *Desalination* **374**, 70 (2015).
18. A. Eliseus and M.R. Bilad, *AIP Conf. Proc.* **1891**, 20039 (2017).
19. M. Sule, J. Jiang, M. Templeton, E. Huth, J. Brant, and T. Bond, *Environ. Technol.* **34**, 1329 (2013).
20. J. Johnson and M. Busch, *Desalin. Water Treat.* **15**, 236 (2010).
21. A.A. Alsarayreh, M.A. Al-Obaidi, R. Patel, and I.M. Mujtaba, *Processes* **8**(5), 573 (2020).
22. S. Chou, L. Shi, R. Wang, C.Y. Tang, C. Qiu, and A.G. Fane, *Desalination* **261**, 365 (2010).
23. L.-Z. Zhang and G.-P. Li, *Desalination* **404**, 200 (2017).
24. K.P. Lee, T.C. Arnot, and D. Mattia, *J. Memb. Sci.* **370**, 1 (2011).
25. P. Cay-Durgun and M.L. Lind, *Curr. Opin. Chem. Eng.* **20**, 19 (2018).
26. L. Xu, B. Shan, C. Gao, and J. Xu, *J. Memb. Sci.* **593**, 117398 (2020).
27. G. Li, J. Wang, D. Hou, Y. Bai, and H. Liu, *J. Environ. Sci.* **45**, 7 (2016).
28. T. Shintani, H. Matsuyama, and N. Kurata, *Desalination* **207**, 340 (2007).
29. M. Safarpour, A. Khataee, and V. Vatanpour, *J. Memb. Sci.* **489**, 43 (2015).
30. D.H.N. Perera, S.K. Nataraj, N.M. Thomson, A. Sepe, S. Hüttner, U. Steiner, H. Qiblawey, and E. Sivaniah, *J. Memb. Sci.* **453**, 212 (2014).
31. S.M. Ghaseminezhad, M. Barikani, and M. Salehirad, *Compos. Part B Eng.* **161**, 320 (2019).
32. Z.S. Tai, M.H.A. Aziz, M.H.D. Othman, M.I.H.M. Dzahir, N.A. Hashim, K.N. Koo, S.K. Hubadillah, A.F. Ismail, M.A. Rahman, and J. Jaafar, *Sep. Purif. Rev.* **49**, 317 (2020).
33. K. Bin Bandar, M.D. Alsubei, S.A. Aljlil, N. Bin Darwish, and N. Hilal, *Desalination* **500**, 114906 (2021).
34. S. Ali, S.A.U. Rehman, H.-Y. Luan, M.U. Farid, and H. Huang, *Sci. Total Environ.* **646**, 1126 (2019).

35. H. Kitano, K. Takeuchi, J. Ortiz-Medina, I. Ito, A. Morelos-Gomez, R. Cruz-Silva, T. Yokokawa, M. Terrones, A. Yamaguchi, T. Hayashi, and M. Endo, *Nanoscale Adv.* **2**, 3444 (2020).
36. D.D. Athayde, J. Motuzas, J.C. Diniz da Costa, and W.L. Vasconcelos, *Desalination* **500**, 114862 (2021).
37. N.M. Mahpoz, S.N.N.M. Makhtar, M.Z.M. Pauzi, N. Abdullah, M.A. Rahman, K.H. Abas, A.F. Ismail, M.H.D. Othman, and J. Jaafar, *Desalination* **496**, 114697 (2020).
38. M. Elma, E.L.A. Rampun, A. Rahma, Z.L. Assyaifi, A. Sumardi, A.E. Lestari, G.S. Saputro, M.R. Bilad, and A. Darmawan, *J. Water Process Eng.* **38**, 101520 (2020).
39. D.A. Reino Olegário da Silva, L.C. Bosmuler Zuge, and A. de Paula Scheer, *Sep. Purif. Technol.* **247**, 116852 (2020).
40. P.S. Goh and A.F. Ismail, *Desalination* **434**, 60 (2018).
41. D.H. Seo, S. Pineda, Y.C. Woo, M. Xie, A.T. Murdock, E.Y.M. Ang, Y. Jiao, M.J. Park, S. Il Lim, M. Lawn, F.F. Borghi, Z.J. Han, S. Gray, G. Millar, A. Du, H.K. Shon, T.Y. Ng, and K. (Ken) Ostrikov, *Nat. Commun.* **9**, 683 (2018).
42. S.K. Hubadillah, M.H.D. Othman, T. Matsuura, A.F. Ismail, M.A. Rahman, Z. Harun, J. Jaafar, and M. Nomura, *Ceram. Int.* **44**, 4538 (2018).
43. C. Li, W. Sun, Z. Lu, X. Ao, and S. Li, *Water Res.* **175**, 115674 (2020).
44. K.P. Goswami and G. Pugazhenthi, *J. Environ. Manage.* **268**, 110583 (2020).
45. M.T. Alresheedi, B. Barbeau, and O.D. Basu, *Sep. Purif. Technol.* **209**, 452 (2019).
46. S.H. Park, Y.G. Park, J.-L. Lim, and S. Kim, *Desalin. Water Treat.* **54**, 973 (2015).
47. N.A. Ahmad, P.S. Goh, L.T. Yogarathinam, A.K. Zulhairun, and A.F. Ismail, *Desalination* **493**, 114643 (2020).
48. R.W. Baker, in *Encyclopedia of Polymer Science and Technology*. (American Cancer Society, USA 2001).
49. C.Y. Tang, Z. Yang, H. Guo, J.J. Wen, L.D. Nghiem, and E. Cornelissen, *Environ. Sci. Technol.* **52**, 10215 (2018).
50. A. Asad, D. Sameoto, and M. Sadrzadeh, in *Nanocomposite Membranes for Water and Gas Separation*, edited by M. Sadrzadeh and T. Mohammadi (Elsevier: Amsterdam, The Netherlands, 2020), pp. 1–28.
51. C.A. Quist-Jensen, F. Macedonio, and E. Drioli, *Desalination* **364**, 17 (2015).
52. S. Al-Obaidani, E. Curcio, F. Macedonio, G. Di Profio, H. Al-Hinai, and E. Drioli, *J. Memb. Sci.* **323**, 85 (2008).
53. M. Tawalbeh, A. Al Mojjly, A. Al-Othman, and N. Hilal, *Desalination* **447**, 182 (2018).
54. Y. Zhou, M. Huang, Q. Deng, and T. Cai, *Desalination* **420**, 99 (2017).
55. Q. Liu, C. Liu, L. Zhao, W. Ma, H. Liu, and J. Ma, *Water Res.* **91**, 45 (2016).
56. S. Šostar-Turk, M. Simonič, and I. Petrinić, *Dye. Pigment.* **64**, 147 (2005).
57. A. Subramani, M. Badruzzaman, J. Oppenheimer, and J.G. Jacangelo, *Water Res.* **45**, 1907 (2011).
58. S.S. Shenvi, A.M. Isloor, and A.F. Ismail, *Desalination* **368**, 10 (2015).
59. M. Xue, H. Qiu, and W. Guo, *Nanotechnology* **24**, 505720 (2013).
60. D.L. Shaffer, N.Y. Yip, J. Gilron, and M. Elimelech, *J. Memb. Sci.* **415–416**, 1 (2012).
61. A. Subramani and J.G. Jacangelo, *Water Res.* **75**, 164 (2015).
62. X. Cao, X. Huang, P. Liang, K. Xiao, Y. Zhou, X. Zhang, and B.E. Logan, *Environ. Sci. Technol.* **43**, 7148 (2009).
63. J. MacHarg, T.F. Seacord, and B. Sessions, *Desalin. Water Reuse* **18**, 30 (2008).

8 Emerging Membrane Technologies for Saline Water Purification Process

An Overview of the Efficiency Based on Structure–Property Relationship

Poosalaya Sangadi and Chandrasekar Kuppan
Vignan's Foundation for Science, Technology and Research

Murthy Chavali
NTRC-MCETRC and Aarshanano
Composites Technologies Pvt. Ltd.

CONTENTS

DOI: 10.1201/9781003185437-9

8.1 INTRODUCTION

Present-day global water scarcity can be mitigated through different purification processes which are thermally driven, membrane assisted, pressure driven and electrically driven or their combination with different variations in the module, membrane types, processing conditions, etc. Among the numerous process to reclaim clean potable water from wastewater, saline water and brackish water, membrane technology has proven to be the clear winner in terms of scale up to cost-effectiveness. Two types of membranes exist:

    a. symmetric or homogeneous and
    b. asymmetric or heterogeneous membranes which vary in pore homogeneity.[1]

The homogeneous membrane is made of single polymer material with uniform pore size, whereas heterogeneous or asymmetric membranes have two layers: one thin outer layer controlling the selectivity and a comparatively thicker, porous supporting inner layer.[2] When compared to the symmetric membranes, the latter exhibits higher flux arising due to their thinness which makes them largely applicable for large-scale water purification. The third type of membrane that is gaining popularity is composite membranes, where the selective layer is a composite of different inorganic metal oxides. On a comparative scale, asymmetric and composite membranes exhibit better performance as they have a thinner selective layer, which increases the water flux.

In this book chapter, authors have extensively discussed different membranes for saline water treatment, alongside biomimetic, ceramic, membranes made of nanocomposites, nanofibres. Membrane technologies include forward osmosis, membrane distillation, pressure driven and solar driven. They have also presented an overview of the efficiency based on structure–property relationship towards emerging membrane technologies for saline water purification process. Finally, conclusions and future perspectives regarding the membranes for water treatment applications are provided.

The choice of the membrane raw material is very much important in deciding the membrane property and efficiency. Some of the common materials used for membrane synthesis are cellulose acetate, cellulose diacetate, cellulose triacetate, polyamide, sulfonated polysulfone, aromatic polyamide (PA), polyacrylonitrile (PAN) and polyvinylidene fluoride (PVF).[3] The choice is particularly made to impart membrane properties like hydrophilicity, hydrophobicity, surface charge, chemical inertness, fouling and mechanical strength.

Though a plethora of membranes can be designed, the ultimate commercialization of the material depends on its overall cost and efficiency. The cost of the membrane is dependent on two parameters, i.e., permeate flux and simple recycling process. To have a higher flux, the thickness of the membrane can be compromised which apart from reducing the cost it will also improve the permeate flux. And for recyclability, the adsorbed foulants should be removed efficiently using any of the chemical or hydraulic process or surface modification of membranes with anti-fouling materials. The efficient and easiest way to prevent fouling in recent days is to functionalize the polymeric membrane material with a wide arrays of materials in the market along with other treatment techniques like ultraviolet, plasma and surface irradiation. Based on the size of the filtrate, membrane technology can be broadly classified into microfiltration with a thickness of 100 nm, ultrafiltration with a thickness of 10 nm, nanofiltration with a thickness of 1 nm and last reverse osmosis membranes with the thickness of <1 nm.[4] The size range in which the solutes that can be filtered is given in the schematic diagram (Figure 8.1).

Membrane technology has grown up vastly where some of the membrane processes though not used directly for saline water purification but can be modified to utilize them for saline water treatment. Such type of processes is mostly pressure-driven membrane process, forward osmosis, electrodialysis and electrodialysis reversal, pervaporation, hybrid membrane process; forward osmosis-reverse osmosis hybridization, solar-driven and membrane[5] distillation process. Table 1 provides different types of membranes their advantages and disadvantages.

Among the mentioned process, membrane distillation and hybrid membrane processes have huge scope for saline water system. Some applications of membrane process in saline water and brine treatment are explained in Table 8.2.

Another important material that was added to modern membranes is nanomaterials.[10] Nanomaterials because of their numerous advantages impart better efficiency in terms of enhancing the water flux and increasing the percentage of salt rejection. Some of the recent studies have used stimuli-responsive materials, bioproteins and zwitterions to enhance the purification and anti-fouling process. In this chapter, we will see some of the new and modern processes that can be improvised to have better flux, low fouling and high rejection. So, the discussion will not be based on the type

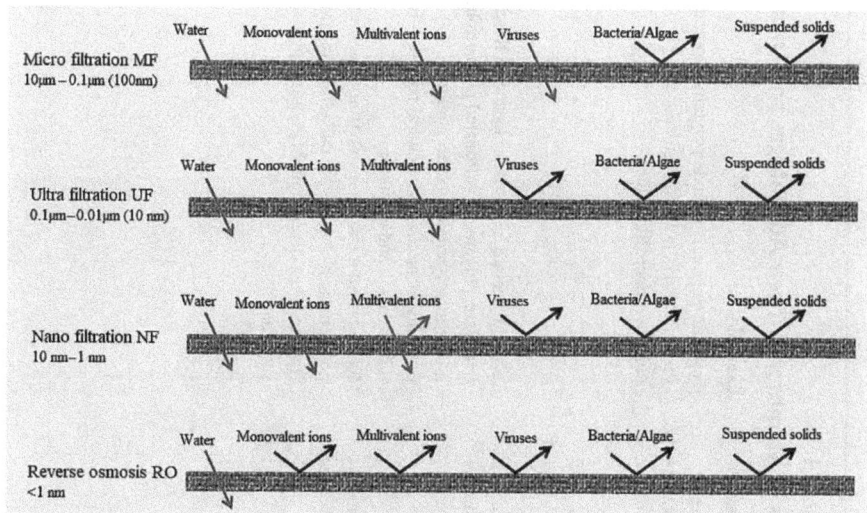

FIGURE 8.1   Different filtration methods with various filtration ranges.

TABLE 8.1

Types of Membranes Their Advantages and Disadvantages

Membranes	Examples	Advantages	Disadvantages
Polymer (nanocomposite) membranes	Synthetic polymers include polytetrafluoroethylene (Teflon PTFE), polyamide-imide, and polyvinylidene difluoride	Production costs are very low; High mechanical stability; Easy synthesis and fabrication; Easy for upscaling is easy; Making variations in module form; **Separation mechanism**: Solution diffusion	Low chemical and thermal stability; Plasticization; Pore size not controllable; Cannot withstand harsh cleaning conditions; Membrane fouling; Inherent hydrophobicity; Low permeability; Low selectivity; Follows the trade-off between; permeability and selectivity; Difficulty in cleaning after fouling
Organic membranes	Cellulose acetate, cellulose diacetate, cellulose triacetate, polyamide, sulfonated polysulfone, aromatic polyamide, polyacrylonitrile, polyvinylidene fluoride; Natural polymers include rubber, wool, and cellulose	High performance; Simple operation; Membranes available can be used to separate many kinds of contaminants; Disinfection can be performed without chemicals	Membrane fouling; Production of polluted water; Membranes have to be replaced regularly; Difficulty in cleaning after fouling
Inorganic membranes	Metallic membranes are made from sintering metal powders such as tungsten, palladium, or stainless steel and then depositing them onto a porous substrate	Tunable pore size; Can operate in harsh conditions; A moderate trade-off between permeability and selectivity; Inertness to microbiological degradation; Superior chemical resistance; High mechanical stability; Ease of cleaning after fouling; Good thermal stability; **Separation mechanism**: Molecular sieving (<6Å), surface diffusion (<10–20Å), capillary condensation (<30Å), and Knudsen diffusion (<0.1 µm)	Brittle; Higher costs due to specific thickness requirements; Difficulty to scale-up: Surface poisoning effect; Cannot withstand pressure drop differences

(Continued)

TABLE 8.1 (Continued)
Types of Membranes Their Advantages and Disadvantages

Membranes	Examples	Advantages	Disadvantages
Zeolite membranes	Analcime; Chabasite; Clinoptilolite; Erionite; Heulandite; Laumontite; Mordenite	Highly uniform pore size; Beneficial for catalytic membrane reactor applications	Relatively low gas flux and thicker layer requirements to prevent cracks and pinholes
Ceramic membranes	Metal (aluminium or titanium) Non-metal (oxides, nitride, or carbide)	Highly acidic environments; Highly basic environments due to inertness	High sensitivity to a temperature gradient, which leads to membrane cracking
Mixed matrix membranes	Inorganic material in the form of micro or nano-particles (discrete phase) incorporated into a polymeric matrix (continuous phase)	Enhanced mechanical strength; Improved thermal stability; Reduced plasticization; Low-energy requirement; Compacting at high pressure; Surpasses the trade-off between permeability and selectivity; Enhanced separation performance over native polymer membranes; **Separation mechanism**: combined polymeric and inorganic membrane principle	Brittle at a high fraction of fillers in a polymeric matrix; Chemical and thermal stabilities depend on the polymeric matrix
Nanofibre membranes	Chitosan, gelatin (which consists of a mixture of peptides and proteins produced by partial hydrolysis of collagen extracted from the skin, bones, and connective tissues of animals)	Unique characteristics; Simple and cost-effective; Surface area; High aspect ratio; Extreme porous structure; Small pore size; High salt rejection; Range of shapes and sizes; Low-operating temperature; Low-energy consumption; Low-operating pressure requirement; Fewer requirements of mechanical properties	Uniform porosity is difficult; High costs; Low flux; Low permeability; Pore wetting; Water loss; Higher machinery cost; Large volume scaffolds not possible

TABLE 8.2
Various Applications of Membrane Process in Saline Water and Brine Treatment

Process type	Feed	Membrane type	Performance	Reference
Forward osmosis (FO)	Saline water with activated sludge	Polyamide thin film composite membrane (FO)	96% COD rejection	[6]
	Produced water	Polyamide thin film composite membrane (FO)	90% rejection of neutral hydrophobic compounds	[7]
	Synthetic wastewater	Cellulose triacetate membrane (FO)	>97% Pollutants rejection	[8]
Membrane distillation (MD)	Synthetic produced water	NA	70% freshwater recovery. 88.8% TOC removed	[9]

NA= Not Applicable

of membranes and process, and it will be on the recent developments for a particular type of membrane to achieve the above-said targets. Some of the recent innovation, which was not commercialized, but successfully implemented in small units, is also discussed.

8.2 CONVENTIONAL MEMBRANES OPERATED BY PRESSURE

Pressure-driven separation processes are by far the most exhaustively employed membrane process in water purification. Among the other processes reverse osmosis (RO) has been the forerunner since its inception in the industrial water treatment process. In RO, an excess pressure higher than the natural osmotic pressure is generated on the feed side to stop the transport of salt from the feed to the draw side. The adsorbed water molecules due to the pressure difference get diffused into the membrane and get transported to the permeate side and get desorbed.[11] This principle of RO has been used and implemented in various water treatment processes as the pressure-driven separation process is undoubtedly a superior process for reclaiming pure water from the saline feed. Though successful in purifying, still challenge remains in terms of energy consumption to maintain desired pressure for a longer duration.

Though RO was found to be the most energy-efficient and environmentally friendly technique, it suffers from low permeability and membrane fouling. Since the permeability is related to thickness and to overcome it, ultra-thin nanoporous membranes were designed using nanofillers like CNT, boron nitride NT, graphene, graphene oxide and MoS^{12} nanomaterials. Graphene was found to show better performance with respect to flux rate when compared to conventional membranes. Similar 2D nanomaterials with and without modification are proposed and theoretically studied for the seawater demineralization process. Hexagonal boron nitride nano-tubes (h-BN)[13] are one such material analogous to graphene[14] because of their similar thickness. They are commercially called white graphene, and they are composed of boron and nitrogen atoms placed alternatively in the honeycomb structure. Because of its remarkable mechanical property, it can be exfoliated in liquid and can be used as molecular sieves. Water transport property of h-BN and graphene was compared, and boron nitride was found to show the superior property as the adhesiveness of water to the h-BN monolayer was higher (low ST on the surface of BN), when compared to graphene. Though the efficiency of BN as a membrane for demineralization was not studied to date, it is expected to show enhanced salt rejection along with resistance to oxidation which makes them highly stable. Molecular dynamics simulation studies were performed by simulating a boron nitride structure with triangular nanopores having[15] free nitrogen and boron edges as a protrusion for brackish water salt removal. The results revealed that the efficiency of separation is dependent on the porosity, edge chemistry and applied external pressure for removal. With this theoretical prediction, it is possible to have the complete rejection of dissolved salt by having control on the pore size and hence the permeation flux.

8.3   FORWARD OSMOSIS

Forward osmosis (FO), unlike the pressure-driven RO process, uses the natural osmotic pressure to draw only the water molecules from the feed to the draw side using a semi-permeable membrane. Here, the draw side is made highly concentrated to induce a concentration flux to draw the water molecule from the feed side. The pressure because of the concentration gradient is enough to drive the water from feed to the draw side and proceeds until equilibrium is reached.[16] But the issue associated with this process requires an additional recovery process to recover freshwater from the collected draw water.

The forward osmosis process utilizes an asymmetric semi-permeable membrane, where the osmotic gradient across the membranes forces the water to permeate by rejecting the salt and impurities. FO process is advantageous over other pressure-driven techniques in terms of an easy cleaning process, comparatively less or no pressure conditions and chances of less fouling propensity.[17] The mechanism of water purification either uses a concentrative internal concentration polarization (ICP) process or a dilutive internal concentration polarization process, which depends on the asymmetric nature of the membranes and also on the position of the active thin layer facing either the feed or draw solution side. Though both the process competes to have an effective driving force with decreasing flux, the concentrative ICP often exhibits higher severity as more foulants start filling up the porous support of the membrane and decreasing the permeate flux drastically with time in real-life applications.[18] The schematic diagram of FO is explained in Figure 8.2.

Double-skinned FO membranes are designed to reduce the fouling characteristics, by introducing a primary dense layer that rejects the draw solute efficiently and a relatively less dense layer to reject the foulants from the feed. This advanced process has been pioneered by experts in the field with varying techniques and compositions ranging from polymers to polyelectrolytes to nanoparticles, depending on the nature

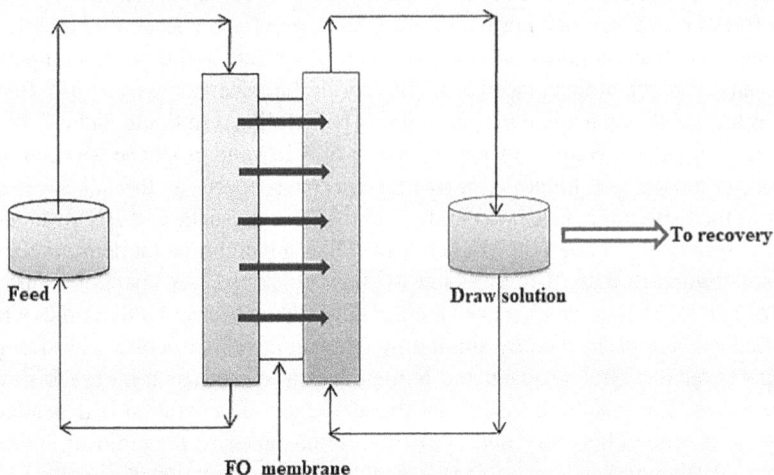

FIGURE 8.2   Schematic diagram of forward osmosis (FO).

of the membrane required.[19] Studies were experimentally and mathematically demonstrated to have reduced fouling characteristics. FO membranes with a double skin of different density were fabricated using techniques like phase inversion polymerization, interfacial polymerization and layer-by-layer deposition[20] for primary dense skin, and the secondary skin was preferably made of nanofiltration membranes which have the maximum rejection ability. Phuoc et al.[21] designed double-skinned FO membranes using cellulose acetate by phase inversion process which was found to have a less ICP and fouling probability when compared to a single-skinned membrane. Their studies on double-skinned membranes made of polyamide primary skin layer, and Nexar copolymer (a sulfonated penta-block copolymer) as secondary skin layer also supported their earlier results of better water flux and reduced foulant contamination over the membranes.

8.4   THIN-FILM COMPOSITES

Thin-film composite membranes use a selective active layer mostly made of polyamide or other polymers, with control of water permeability and rejection by modifying the structure.[22] To reduce the fouling behaviour of the membranes, various classes of nanomaterials were used like CNT, alumina silicates, bio-inspired aquaporin, MOF and graphene to improve the permeability and chemical resistance in comparison to the conventional membranes. Since the permeability flux is inversely proportional to the membrane thickness, graphene and its chemically modified GO sheets in combination with other membrane materials will be efficient in rejecting the salt from saline water. Graphene is unique in the purification process as it allows only the water to penetrate and rejects all the solutes which are >1 nm, making the permeate flux higher and rejection high. Chemical modification over the GO with hydroxyl groups makes the surface more hydrophilic,[12] enhances the wetting characteristics and increases further the flux of permeate with the expense of selectivity. It is demonstrated that there is a strong correlation between the hydration radius of permeate and the membrane porosity. Coating the thin-film composite with material resisting fouling on the collector side was proven to be one of the most effective approaches to reduce the fouling tendency.

Anti-fouling surfaces are designed with hydrophilic coatings like sorbitol, peptoids and zwitterions as they were efficient in resisting oil, grease and organic molecules.[23] Though a wide range of anti-fouling chemistries exists, none of the methods was not good enough to resist fouling that has salinity greater than seawater. Zwitter-ionic coatings were found to be resistant to chlorine oxidation and also mitigate fouling in water purification.. A recent work[24] showed that pyridine-based zwitter-ionic coating material was found to resist fouling to a higher extent when compared to other commercial materials, and surprisingly, the anti-fouling property was observed even in higher saline conditions. One such study attempted to graft zwitter-ionic polymeric brushes[25] on the backside of the membranes which imparted strong hydrophilicity and high resistance to foulants because of electrostatic repulsion of the ionic structures. The percentage of coverage behind the membranes was not sufficient to cover the entire large pores on the membranes, which looked for other surface modification techniques at the same time retaining the flux and selectivity. CNT was one of the membrane materials which can be used to design novel anti-fouling membrane. Chemical modification of the CNT surface using polymeric zwitter-ionic brush was found to be highly active in resisting fouling by designing double-skinned membranes with conventional membrane-making materials like polyamide with salt-rejecting capability.

8.5 MEMBRANE DISTILLATION

Membrane distillation works on the concept of varying hydrophobicity of the volatilities across the membrane. Water vapour's comparatively higher hydrophobicity than liquid water and higher vapour pressure makes the transport of water vapour towards the draw side.[26] This heat-assisted process can be highly beneficial in saline water purification. Since lesser heat is used to just convert the water to vapour, low-grade thermal energy is sufficient to run the whole process efficiently and cost-effectively compared to other techniques. Membrane distillation (MD) is a novel technique where the transport of water vapour is enhanced over a hallow membrane at the same time blocking the transport of liquid water, which operates on the vapour flux between the warmer and cooler section of the membrane. Another major advantage of MD is its energy efficiency, which is much lower than other conventional distillation processes, where the operating temperature is between 60°C and 90°C for MD.

Among all the water purification processes, MD is gaining popularity for its low greenhouse gas emission. The main criteria in designing an MD membrane is, it should have high vapour permeability, appropriate pore size and high porosity.[27] The efficiency of purification can be improved if the membranes are superhydrophobic as they will avoid membrane pore wetting and reduce fouling because of the low water-membrane contact area. The superhydrophobic surface was made by many advanced techniques like template dissolving, chemical etching and glass fibre drawing.[28] The nanocomposite was found to be an area that can be fine-tuned to get the desired property, likewise for membrane demineralization, nanocomposite can be a helpful tool to get the maximum superhydrophobicity. Silica-PVDF composite will be a promising material as the support and strength will be given by PVDF which can be made into a membrane by electrospinning technique with high porosity and desired thickness.[29] The presence of silica makes the membranes hydrophobic giving a CA of >150°.

In a recent study,[30] CNT was immobilized in membranes and their efficiency of purification by membrane distillation process was studied, where CNTs acted as sorbents providing an additional pathway for the transport. Because of the high thermal conductivity of CNT, the temperature gradient across the membrane will be minimized preventing condensation of vapour and permeating faster through the pores. On comparing the membranes without CNT and with CNT, it is observed that the rejection of salt increased along with permeate flux with increasing temperature up to a maximum of 80°C, after which both get decreased. Whereas at all temperature, both the flux and salt rejection were maintained higher making the superiority of the CNT immobilized membranes for saline water purification. It is proved in a normal RO process that water clusters of four or more molecules can get transported via the pores of the membrane entrapping salt molecules with it as the pore size is much bigger than the individual ions in saline water. Since the transport of liquid water is not allowed in the CNT membranes for MD, 100% rejection of salt can be achieved.

MD efficiency can be improved by changing the configurations. Four different configurations which are in current use are direct contact MD (DCMD), air gap MD (AGMD), vacuum MD (VMD) and sweeping gas MD (SGMD). A few other configurations with hybrid combinations are also in use like thermostatic SGMD and the liquid GMD process. Among the other configuration, the direct contact membrane distillation (DCMD) process is the most used process. Similar to conventional distillation, a hydrophobic microporous membrane maintains a direct contact between the feed and permeate. The schematic diagram of DCMD is illustrated in Figure 8.3a. AGMD is designed with a stationary air gap between the membrane and the condensation surface maintained at low temperature. This module is primarily designed to reduce the loss of heat due to conduction at the loss of permeate flux.[31] The schematic diagram of AGMD is explained in Figure 8.3b. In the case of vacuum membrane distillation, the process is fastened by using a vacuum at the permeate side which constantly provides a low pressure lower than the partition pressure of the volatile components at the feed side. The low pressure enhances the permeate flux, and with this process, the drawback of heat loss is completely neglected. The schematic diagram of VMD is explained in Figure 8.3c. In sweeping gas MD, an inert gas is swept along the membrane to transport the vapour and gets condensed at an external condensation chamber preventing the heat loss completely. The schematic diagram of SGMD is illustrated in Figure 8.3d. Thermostatic sweeping gas membrane distillation is a modified version of sweep gas and air gap MD. Here a cold wall is designed inside the permeate side, where the temperature of the sweeping gas is reduced before the vapour reaches the external condenser where the water permeates water condenses. Liquid gap MD introduces the third channel to induce condensation

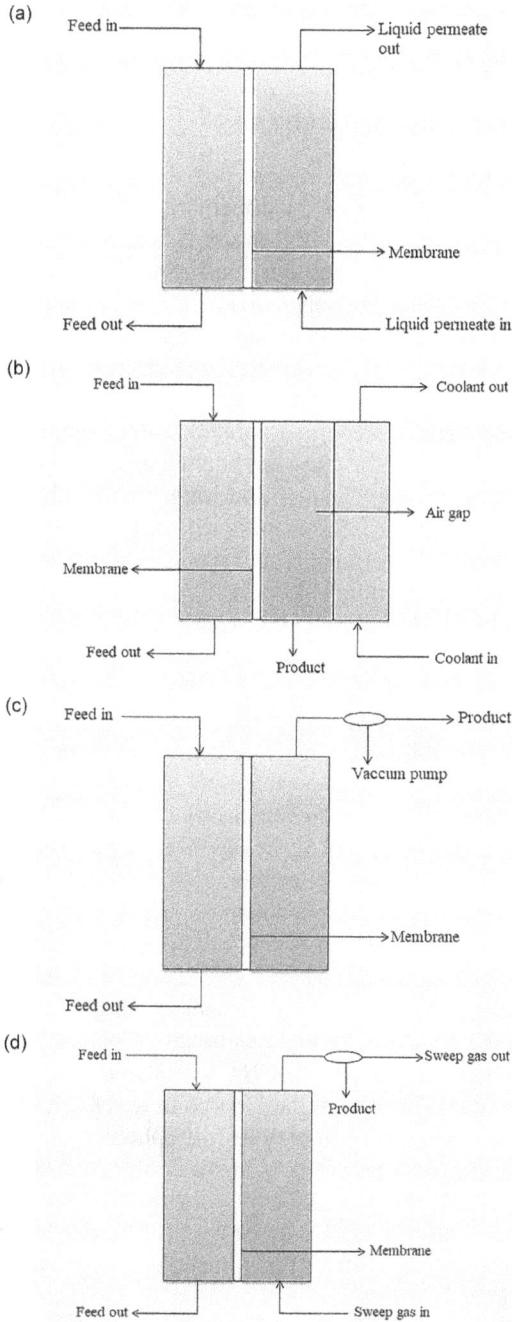

FIGURE 8.3 Schematic MD-based configurations of (a) Direct Contact Membrane Distillation (DCMD); (b) Air Gap Membrane Distillation (AGMD); (c) Vacuum Membrane Distillation (VMD); and (d) Sweeping Gas Membrane Distillation (SGMD).

of permeate. The channel contains a non-permeable foil which separates permeate and reduces the temperature across the membrane.

8.6   NANOCOMPOSITE MEMBRANES

RO membranes are highly efficient in removing small-sized ions like sodium and chloride ions in saline water. Polyamide membranes are thought to be efficient in the RO process because of their high-water flux and salt rejection,[32] but their major disadvantage comes from their low fouling ability and chemical resistance towards chlorine. This makes the polyamide membranes not that suitable for salt-water filtration, but modifications or compositing with other efficient materials will make the polyamide one of the cheapest and efficient materials for saltwater purification.

TiO$_2$ and Ag nanoparticles were incorporated for anti-fouling property, zeolites to improve the water flux, CNT for gas and liquid permeability, antibacterial property and mechanical stability.[33] Though the water flux in presence of CNT may rise upto 1,000 L/m^{2}h bar if CNTs are aligned uniformly throughout the membrane, because of increased hydrophobicity, synthesis of aligned CNT membranes is still under lab scale as aligning CNT is not easy. In the case of randomly dispersed CNT membranes, the salt rejection was less, and the water flux was high, whereas if the CNT is functionalized with Zwitter ion, then salt rejection increased to around 98% and water flux got decreased drastically. For a membrane to be effective for any application, it should be selective in removing the pollutant and it should have high permeability, a win-win situation between these two makes the process effective.

Nanomaterials introduction into membrane science has revolutionized the area to a different level making them applicable in wide areas which helped to overcome the tradeoff in properties. Nanocomposite membranes can be either mixed matrix membranes or thin-film nanocomposite membranes, where a thin selective membrane embedded with nanomaterials is supported over a porous substrate in the latter case and nanomaterials are dispersed randomly in the former. Though the efficiency of filtration is high, depending on the size of the nanomaterials and the charge on them the performance varies.[34] CNT exhibits unique properties when it comes to membrane filtrations, where the channels of CNT will also help in permeating the permeate selectively and efficiently. But the incorporation of CNT into the membrane structure is not that compatible because of its hydrophobicity, which can be successfully done by modifying the surface accordingly. To understand the role of CNT for filtration, CNT modified thin-film nanocomposite membranes[35] were designed and their efficiency was compared. CNT modification can be done on both the selective and supportive material. Four different combinations of membranes are likely to be possible like

   a. composite of thin selective layer,
   b. the nanocomposite of supportive layer,
   c. thin-film nanocomposite and
   d. nanocomposite supported thin-film nanocomposite.

8.7 NANOCOMPOSITE NANOFILTRATION

Polymer matrix membranes embedded with nanomaterials are new emerging materials that can find application in gas separation and purification techniques. Nanomaterials in the form of sheets, tubes and particles foster low resistance channels which enhance water permeability and salt rejection.[36] Stability of the nanomaterials in the polymer matrix is the main factor that controls the selectivity in separation and permeability. The instability may arise due to the spontaneous agglomeration of nanomaterials in the polymer matrix, their non-uniform dispersion and a large gap between the matrix and nanosheets. The adhesion and wettability characteristic of the nanomaterial with the polymer matrix control the stability and membrane performance to a greater extent. The weak non-covalent interactions between the nanomaterial and the matrix impart non-selective defects like large gaps which decrease the salt rejection and hence the performance or efficiency.[37]

One way of improving the adhesion characteristics of the nanomaterials with the polymer matrix is to functionalize the surface of nanomaterials with the required surface energy. Dopamine from the catechol-amines group can undergo polymerization under alkaline conditions on any type of surfaces with covalent bonds, which will be further enhanced if the surface has thiol or amine groups because of strong interaction.[38] Dispersing MWCNT, in presence of dopamine, PEI in polyamide matrix to form positively charged nanofiltration membranes was expected to have better permeability and rejection. These membranes work under the principle of electrostatic repulsion and steric hindrance to have better selectivity. The selectivity of divalent ions was found to be higher than the monovalent cations, which might be due to lesser charge and smaller size.

The order of rejection using these filtration membranes for different divalent and monovalent cations is

$$Zn^{2+} > Mg^{2+} > Cu^{2+} > Ca^{2+} > Na^+$$

By modifying the charges over the CNT dispersed polymer matrix, the selectivity can be improved and can be used effectively for saline water filtration.

8.8 NANOSHEET MEMBRANES

For a membrane to be selective to specific ions or molecules, porosity of the membrane is the controlling factor and is one of the prime criteria for the purification process. Graphene nanosheets have the advantage of functionalizing with various groups like carboxyl, hydroxyl, carbonyl and epoxy towards a specific application which is termed graphene oxide.[39] These nanosheets are layered or stacked together where the functionalized regions maintain the interlayer space forming nanochannels for the transport of selective molecules or ions. The propensity of the stacked layered structures as membranes for water purification was retarded because of low permeation flux and the relative instability of the nanosheets to get peeled of due to electrostatic repulsion. To address both the issues, the interlayer distance can be broadened by incorporating other nanomaterials like Titania, MOF, CNT and other similar materials.[40]

Though these modifications increase the flux, the selectivity of the membranes is compromised to achieve the former. Modification over the GO like creating pores via oxidation/surface activation, reducing some of the surface oxygen to enhance stability and crosslinking using multivalent metal ions to improve the stability over different acidic and basic conditions ultimately resulted in increased flux compared to pristine GO but couldn't meet the expectation as the membranes are unstable after a few hours of operation or with no selectivity, etc. One approach was to incorporate these nanosheets over porous covalent organic frameworks by chemical bonding. These mixed nanosheets reduce the disadvantage arising due to inter-lamellar space, further the porous nature of CNT increases the selectivity and flux for saline water purification and other nanofiltration techniques.

8.9 NANOFILTRATION MEMBRANES

The properties of nanofiltration membranes fall between the reverse osmosis membranes and ultrafiltration membranes, and the filtration process follows the principle of the Donnan, dielectric and size exclusion process.[41] Modern-day nanofiltration process uses thin-film composites over dense porous support prepared by interfacial polymerization. Another comparatively easier approach to prepare NF composites is by self-assembling oppositely charged polyelectrolytes by Layer-by-layer (LbL) approach. This type permits the fabrication of hollow fibre nanofiltration membranes with a selective layer inside the hollow fibre of the membranes. Further, this technique allows for easy fine-tuning of the pores by controlling the charges of the chosen polyelectrolytes. Membranes designed using 2D materials like graphene and its surface-modified graphene oxide are a novel class of materials utilized mainly because of their well-defined nanostructures.

Laminates of graphene oxide sheets were found to be useful for gas separation, pervaporation, and water treatment. These laminates were relatively effective towards rejection up to >85% of organic molecules and ineffective[42] for salt rejection because of their comparatively bigger pore as the channels of GO get swollen due to electrostatic repulsion. The swelling characteristics can be controlled by crosslinking the GO laminates which help in controlling the permeation of solute molecules by varying the sizes of intercalating cross-linker molecules either by covalent or electrostatic interaction. Smaller molecules make the inter-lamellar distance smaller thereby making the channel size smaller and reject smaller ions efficiently, whereas if the crosslinker is large, then the efficiency of rejection goes down as the channel sizes are growing bigger. Porphyrin molecules are 2D molecules that can assemble spontaneously due to pi-pi interaction between GO sheets.[43] Tetrakis(1-methyl-pyridinium-4-yl)porphyrin (TMPyP) a porphyrin derivative is a cationic molecule that can be efficiently assembled on the GO sheets and can exist as a single molecule and the channel size was found to be 1 nm, which is expected to reject high concentration of salt from the feed.

One recent approach to mitigate the drawbacks of membrane technology is to use stimuli-responsive materials for membrane design. These stimuli responsiveness can be activated externally and can be grouped into direct stimuli having localized contact with the responsive group, indirect stimuli which is responsive against the general thermodynamic environment and field responsiveness, which respond instantaneously because of an external magnetic field.[44] One successful observation with the magnetically activated micromixers is their effective suppression of concentration polarization occurring on the membrane surfaces. Concentration polarization often results when the salt concentration is higher at the membrane surface, which will lead to increased osmotic back pressure which automatically decreases the permeate flux across the membrane. The same principle was used by Azmi et al.[45] to suppress the polarization concentration during filtration of humic acid. The magnetic responsiveness of the nanofiltration membranes was effective in rejecting high salt and increasing permeate flux across the membrane.

Nanofiltration membranes with porosity of <10 nm can successfully stop the permeation of molecules with a molecular weight range of 200–1,500 Da which is energy efficient with a low carbon footprint are found to be ideal material for saline water purification. Despite their advantages, some of the features which hinder their activity are their thickness, porosity, pore distribution, tortuosity and water adhesion property (wettability). To achieve higher water flux and selectivity in separating ions, thin-film composites made with mesoporous silica with controllable pore size were designed by stacking the active layer over alumina support. The membranes designed like the above-said process were found to be super hydrophilic in nature with a uniform pore distribution and low tortuosity.

The nanofiltration process shows performance between ultrafiltration and reverses osmosis for separating mono and divalent ions. Key features which are expected from nanofiltration is, it should have high water flux which is possible by reducing the thickness, it should have high efficiency to remove divalent and multivalent ions, it should show low rejection to monovalent ions which will reduce the operating pressure and energy consumption and it should show high stability and resistance against fouling for a long term operation.[46] Despite intense research, scientists are still in the process of designing membranes that can show less fouling and less operating pressure. Layer-by-layer (LbL) self-assembly of polyelectrolytes with opposite charges was designed on a porous substrate to have high water flux and high salt rejection. Ninety per cent rejection of divalent ions by polyallylamine hydrochloride and polysodium styrene sulfonate-based hollow fibre nanofiltration membranes was impressive, but these membranes showed only 10% of rejection for monovalent ions.[47]

Apart from other drawbacks of LbL membranes, the membranes cannot withstand harsh conditions at varying pH. Dopamine a well-known bio-glue, with the potential to autoxidize and self-polymerize on any surface via both covalent and non-covalent interaction, can be used to design membranes with homogeneously dispersed nanoparticles. Apart from increasing the wettability, these polymers are selective and have anti-fouling property, which can be utilized in nanofiltration membranes to increase the efficiency of water flux with less rejection and fouling.

8.10 NANOFIBRE MEMBRANES

Electrospun nanofibre-based membranes have a unique advantage because of their nanometre thickness, which when cast into membranes impart voids of micron/submicron on the surface and in bulk. These submicron voids can be utilized for removing solutes in that size range. When these nanofibres were designed with pores of nanometres, then the permeate of higher size is inhibited thereby enhancing the filtration processes.[48] Moreover, the fouling characteristics will be drastically reduced without sacrificing the particle rejection and the usage capacity will be increased. This shows that electrospun membranes were found to be superior to the conventional membranes as the former shows a higher water flux and higher rejection with less fouling. To further enhance the filtration characteristics, the surface of the nanofibre was functionalized with different polymers and their behaviour in saline conditions was measured.[49] Modified cellulose-based membranes were designed with hydrophobic (pHEMA) and hydrophilic (PAA) polymers surfaces, and their permeation and fouling behaviour showed that hydrophobic chains restricted the permeation of water despite higher anti-fouling characteristics. Whereas the PAA-modified membranes showed excellent permeate flux and less biofouling characteristics as the acrylic acid chains were water soluble which allows the water permeate to flow easily.

8.11 SOLAR-DRIVEN DISTILLATION

Commercial water treatment processes often utilize huge amounts of non-renewable energy resources making the process less cost-effective to be implemented in economically poor regions. The solar-driven water purification process was a green approach that gets enhanced by using photothermal conversion materials like graphene, carbon black, polypyrrole and nanoparticles. Porous microstructure-based membranes derived by impregnating fibrous silver carbon nitride and membranous GO (SCG) have been used for solar-driven water purification process as the material is less dense and thermally conductive with high absorption capacity, which makes them absorb maximum solar heat and efficiently transfer to the water surface thereby improving the distillation process. The designed membranes were found to eliminate or filter soluble ions to an extent of 99.86%, which can be operated continuously for 5 days and can be recycled efficiently with minimal fouling and salt accumulation.

Conventional technologies to produce membranes mostly use a large sum of fossil fuels and emit CO_2 which makes solar-based technologies highly promising. One such process for water purification is solar-driven photothermal evaporation, which is not possible in all regions. The photothermal layer is one of the main components which acts as an absorbent of solar energy and at the same time acts as a converter of solar energy to heat energy. To date, there are two viable strategies implemented, where metallic and carbonaceous nanomaterials[50] are used as the active absorbent material by homogeneously suspending over the membrane surface. Though this process is not industrially applicable, with less cost-effective materials, the process can be scalable with enhanced salt rejection capability.

One such efficient material designed for solar evaporator was a hybrid gel material made of CuS and bacterial-derived cellulose as photothermal material. These gel materials are porous in nature and diffuse[51] the evaporated water into the air. Some of the active materials which have been used for the solar evaporator are carbon-based nanomaterials, plasmonic metal particles, black metal oxides and semi-metallic nanoparticles. Because of their light-trapping efficiency over a wide range of wavelength, non-stoichiometric compounds like tungsten oxide, molybdenum oxide and titanium oxide are some of the potential materials which can be used for this type of water purification process. $CoWO_{4-x}$ NP a non-stoichiometric compound that has a full spectral absorption range of 450–2,500 nm[52] can be fabricated easily with cost-effective raw materials and can be used for the water purification process.

8.12 BIOMIMETIC MEMBRANES

Water channel protein "Aquaporin" was expected to deliver better selectivity against the permeable with energy efficiency by the formation of layer by layer self-assembled biomimetic membranes. This protein has the narrowest pore channels of 2.8 Å, which restricts the transport of solute molecules except for water molecules.[53] Combination of Aquaporin with other lipids and copolymers were found to exhibit excellent demineralization property by supported bilayer structures despite their low stability. The efficiency of the protein when embedded into the membrane material was studied as this will increase the stability of the material. One approach to overcome the stability was by forming AQP-laden vesicles which are immobilized over a dense polymer layer. Biomimetic membranes prepared by the LBL process over a polyelectrolyte membrane show an enhanced performance of high flux and rejection. The rejection of ions was mainly due to the charged layers of the lipid bilayer and polyelectrolyte substrate which apart from attracting the charged ions gave excellent stability and durability.

8.13 CERAMIC MEMBRANES

Ceramic membranes are a class of materials, which were found to be superior to the commonly employed polymeric membranes where they exhibit excellent chemical resistance and stable under harsh conditions like high salinity, high pressure and high temperature.[54] despite their low-cost effectiveness. Silicon oxycarbide was one potential ceramic material that can overcome the cost-effectiveness and commercialization issue of ceramic membranes. The incorporation of silicon oxycarbide improves resistance against high temperature and chemicals at the same time imparting good mechanical strength, which makes this material sustain harsh conditions. Tuning the porosity of ceramic materials is always challenging as the pores formed are in the microscale range making them less utilized for the purification process. PDMS was used to control the size, and the synthesized membranes were found to show a high surface area as a function of molecular weight.

8.14 SUMMARY

For the past few decades, diverse aspects of membrane technology application have grown to different extents. Other areas such as cosmetics, pharmaceutical, fuel processing and production and food industries are all taking benefits from various range of membrane processes. Yet, as the applications are more limited and the processes are fairly complicated, the growth rate is not comparable to the water treatment industry. More specific application of thin-film filters in association with biomedical areas (artificial organs) more specifically blood purification is also experiencing continuous improvements.

Membrane and membrane-based technologies have grown up vastly and are widely applied in various water purification and waste treatment applications, where some of the membrane processes though not used directly for saline water purification but can be modified to utilize them for saline water treatment. Such type of processes is mostly pressure-driven membrane process, forward osmosis, electrodialysis and electrodialysis reversal, pervaporation, hybrid membrane process; forward osmosis-reverse osmosis hybridization, solar-driven and membrane distillation process. Developing sustainable technologies for saline water treatment using membranes is a major challenge, in addition to membrane-based challenges, other common challenges like reducing energy consumption, lowering the environmental impact; and decreasing water production costs. Currently, membrane processes such as RO, FO, UF, NF and MD are being used in treating brackish water and various applications of wastewater treatment.

The 21st century's greatest challenge is to manage the evergrowing water demand. With decreasing amount of potable water, there is a need for sustainable techniques to purify contaminated waters and saline water. Nanotechnology could provide solutions while developing novel materials of extraordinary properties, which can demineralize seawater at a rapid pace with limited costs, making it potable. This work reviewed recent developments in nanotechnology for saline water treatment using emerging efficient membrane materials and technologies.

8.15   FUTURE PROSPECTS

Membrane technologies with effective deployment can become a sustainable solution for different applications and even can emerge and evolve for future filtration applications. The course may require novel materials for membrane preparation with unique customized characteristics. Recent developments in a large assortment of polymers alongside the arrival of novel nanomaterials lead the way for more membrane-based processes for a wide range of prospective industrial applications and also the development of a specific set of features for a customized application while preparing these membranes. Such highly tailor-made functionalities and dynamic texture controlled membrane syntheses could enhance purification, ultrafiltration and even segregation and recovery of resources, fulfilling the sustainable goals.

Membrane processes play a major role in wastewater treatment water purification and sustainable energy generation. Regardless of its advantages, some of the key concerns with parameters like chemical resistance, mechanical stability, thermal stability, selectivity in separation, fouling, flocculation and blockage of membrane pores are becoming more common and have to mitigate immediately as per the requirement of the application, which means more and more customization. With these existing and upcoming challenges, material scientists and membrane technologists have to find sustainable ways to advance robust materials and economic hybrid processes. However, membrane-based approaches open up new avenues for the development of a sustainable environment concerning efficient saline water treatment either separately or together with other membrane-based or non-membrane-based techniques, while utilization of low-cost resources making the process highly economic and viable towards various industrial applications.

REFERENCES

1. Cheryan, M. *Ultra-filtration Handbook*: Technomic Publishing: Lancaster, PA, 1986.
2. Matsuyama, H., Teramoto, M., Uesaka, T. (1997), "Membrane formation and structure development by dry-cast process, *J. Membrane Sci.* 135) 271–288.
3. Gullinkala, T., Escobar, I.C. (2011), "Membranes for water treatment applications–an overview," *Mod. Appl. Membr. Sci. Technol.* 155–170. doi: 10.1021/bk-2011-1078.ch010.
4. Adnan, S., Hoang, M., Bolto, H.T.W.B., Xie, Z.L. (2010), "Recent trends in research, development and application of membrane technology in the pulp and paper industry," *Appita* 63, 235–241.
5. Jhaveri, J.H., Murthy, Z.V.P. (2016), "A comprehensive review on anti-fouling nanocomposite membranes for pressure-driven membrane separation processes." *Desalination* 379, 137–154.
6. Bell, E., Holloway, R., Cath, T. (2016), "Evaluation of forward osmosis membrane performance and fouling during long-term osmotic membrane bioreactor study," *J. Membr. Sci.* 517, 1–13.
7. Bell, E., Poynor, T., Newhart, K., et al. (2017), "Produced water treatment using forward osmosis membranes: Evaluation of extended-time performance and fouling," *J. Membr. Sci.* 52577–52588.
8. Li, S., Kim, Y., Chekli, L., et al. (2017), "Impact of reverse nutrient diffusion on membrane bio-fouling in fertilizer-drawn forward osmosis," *J. Membr. Sci.* 539, 108–115.
9. Francesco, R. (2019), "Desalination of produced water by membrane distillation: Effect of the feed components and a pretreatment," *Sci. Rep.* 9, 14964.
10. Savage, N., Diallo, M.S. (2005), "Nanomaterials and water purification: Opportunities and challenges," *J. Nanoparticle Res.* 7, 4–5, 331–342.
11. Greenlee, L.F., Lawler, D.F., Freeman, B.D, et al. (2009), "Reverse osmosis desalination: Water sources, technology, and today's challenges," *Water Res.* 43, 2317–2348.
12. Cohen-Tanugi, D., Lin, L.C., Grossman, J.C. (2016), "Multilayer nanoporous graphene membranes for water desalination," *Nano Lett.* 16, 1027–1033.
13. Lin, Y. Williams, T.V. Connell, J.W. (2010), "Soluble, exfoliated hexagonal boron nitride nanosheets," *J. Phys. Chem. Lett.* 1, 277–283.
14. Nair, R.R., Wu, H.A., Jaya ram, P.N., et al. (2019), "Unimpeded permeation of water through helium-leak–tight graphene-based membranes," *Science* 335, 442–444.
15. Hashido, M., Kidera, A., Ikeguchi, M. (2007), "Water transport in aquaporins: Osmotic permeability matrix analysis of molecular dynamics simulations," *Bio Phys. J.* 93, 373–385.
16. Suwaileh, W.A., Johnson, D.J., Sarp, S., et al. (2018), "Advances in forward osmosis membranes: Altering the sub-layer structure via recent fabrication and chemical modification approaches," *Desalination* 436, 176–201.
17. Klaysom, C., Cath, T.Y., Depuydt, T., et al. (2013), "Forward and pressure retarded osmosis: Potential solutions for global challenges in energy and water supply," *Chem. Soc. Rev.* 42, 6959–6989.

18. Su, J., Chung, T.S. (2011), "Sub layer structure and reflection coefficient and their effects on concentration polarization and membrane performance in FO processes," *J. Membr. Sci.* 376, 214–224.

19. Su, J., Chung, T.S., Helmer, B.J., et al. (2012), "Enhanced double-skinned FO membranes with an inner dense layer for wastewater treatment and macromolecule recycle using Sucrose as draw solute," *J. Membr. Sci.* 396, 92–100.

20. Saren, Q., Qiu, C.Q., Tang, C.Y. (2011), "Synthesis and characterization of novel forward osmosis membranes based on layer-by-layer assembly," *Environ. Sci. Technol.* 45, 5201–5208.

21. Phuoc, H., Duong, T., Chung, S., et al. (2014), "Highly permeable double-skinned forward osmosis membranes for anti-fouling in the emulsified oil–water separation process," *Environ. Sci. Technol.* 48, 4537–4545. doi:10.1021/es405644u.

22. Bassyouni, M., Abdel-Aziz, M.H., Zoromba, M.S., et al. (2019), "A review of polymeric nanocomposite membranes for water purification," *J. Ind. Eng. Chem.* 73, 19–46.

23. Dimitriou, M.D., Zhou, Z., Yoo, H.-S., et al. (2011), "A general approach to controlling the surface composition of poly (ethylene oxide)-based block copolymers for antifouling coatings," *Langmuir*, 27, 13762–13772.

24. Shao, Q., Jiang, S. (2015), "Molecular understanding and design of Zwitter ionic materials," *Adv. Mater.* 27, 15–26.

25. Le, N.L., Quilitzsch, M., Cheng, H., et al. (2017), "Hollow fiber membrane lumen modified by poly Zwitter ionic grafting," *J. Membr. Sci.* 522, 1–11.

26. Nagy, E. Chapter 19-Membrane distillation. In Nagy, E., Ed. *Basic Equations of Mass Transport Through a Membrane Layer*, 2nd ed, pp. 483–496, 2019, Elsevier: Amsterdam, The Netherlands.

27. Alkhudhiri, A., Hilal, N. Membrane distillation-Principles, applications, configurations, design, and implementation. In Gude, V.G., Ed. *Emerging Technologies for Sustainable Desalination*, 2018, Butterworth- Heinemann: Oxford, UK.

28. Ma, Z., Hong, Y., Ma, L., et al. (2009), "Superhydrophobic membranes with ordered arrays of nano spiked microchannels for water desalination," *Langmuir* 25 (10), 5446–5450.

29. Jingling, Y., Geng-Sheng, L., Chung, Y., et al. (2020), "Mesoporous silica thin membrane with tunable pore size for ultra high permeation and precise molecular separation," *ACS Appl. Mater. Interfaces*, 12, 7459–7465, doi:10.1021/acsami.9b21042.

30. Hone, J., Whitney, M., Piskoti, C. (1999), "Thermal conductivity of single-walled carbon nanotubes," *Phys. Rev. B: Condens. Matter Mater. Phys.* 15, 2514–2516.

31. Ken, G., Ornthida, S., Somenath, M. (2011), "Water desalination using carbon-nanotube-enhanced membrane distillation," *ACS Appl. Mater. Interfaces*, 3, 110–114. doi:10.1021/am100981s.

32. Li, D., Wang, H.T. (2010), "Recent developments in reverse osmosis desalination membranes," *J. Mater. Chem.* 20, 4551–4556.

33. Peng, F.B., Hu, C.L., Jiang, Z.Y. (2007), "Novel poly (vinyl alcohol)/carbon nanotube hybrid membranes for pervaporation separation of benzene/cyclohexane mixtures," *J. Membr. Sci.* 297, 236–242.

34. Lau, W.J., Gray, S., Matsuura, T., et al. (2015), "A review on polyamide thin-film nanocomposite (TFN) membranes: History, applications, challenges and approaches," *Water Res.* 80, 306–324.

35. Xiangju, S., Li, W., Lili, M. (2016), "Nanocomposite membrane with different carbon nanotubes location for nanofiltration and forward osmosis applications," *ACS Sustain. Chem. Eng.* 4(6), 2990–2997. doi: 10.1021/acssuschemeng.5b01575.

36. Wang, X., Yeh, T.M., Wang, Z. (2014), "Nanofiltration membranes prepared by interfacial polymerization on thin-film nanofibrous composite scaffold," *Polymer* 55, 1358–1366.

37. Ghaemi, N., Madaeni, S.S., Daraei, P. (2015), "PES mixed matrix nanofiltration membrane embedded with polymer wrapped MWCNT: Fabrication and performance optimization in dye removal by RSM," *J. Hazard. Mater.* 298, 111–121.

38. Hong, S.K., Na, Y.S., Choi, S.H. et al. (2012), "Non-covalent self-assembly and covalent polymerization co-contribute to poly dopamine formation," *Adv. Funct. Mater.* 22, 4711–4717.

39. He, R., Cong, S., Wang, J. et al. (2019), "Porous graphene oxide/porous organic polymer hybrid nanosheets functionalized mixed matrix membrane for efficient CO_2 capture," *ACS Appl. Mater. Interfaces*, 11, 4338–4344.

40. Zhu, J., Tian, M., Hou, J., et al. (2016), "Surface zwitter ionic functionalized grapheme oxide for a novel loose nanofiltration membrane," *J. Mater. Chem. A.* 4, 1980–1990.

41. Szymczyk, A., Fievet, P. (2006), "Ion transport through nanofiltration membranes: The steric, electric and dielectric exclusion model," *Desalination* 200, 122–124.

42. Xiao-Ling, X., Fu-Wen, L., Yong, D., et al. (2016), "Graphene oxide nanofiltration membranes stabilized by cationic porphyrin for high salt rejection," *ACS Appl. Mater. Interfaces* 8(20), 12588–12593. doi: 10.1021/acsami.6b03693.

43. Xu, X.-L., Lin, F.-W., Xu, W., et al. (2015), "Highly sensitive INHIBIT and XOR logic gates based on ICT and ACQ emission switching of a porphyrin derivative," *Chem. Eur. J.* 21, 984–987.

44. Qian, X., Yang, Q., Vu, A., et al. (2016), "Localized heat generation from magnetically responsive membranes," *Ind. Eng. Chem. Res.* 55(33), 9015–9027.

45. Azmi, N.A., Ng, Q.H., Low, S.C. (2015), "Ultrafiltration of aquatic humic substances through magnetically responsive polysulfone membranes." *J. Appl. Polym. Sci.* 132 (21), 41874.

46. Liang, Y., Lin, S. (2020), "Intercalation of Zwitter ionic surfactants dramatically enhances the performance of low-pressure nanofiltration membrane," *J. Membr. Sci.* 596, 117726. doi:10.1016/j.memsci.2019.117726.

47. Liu, C., Shi, L., Wang, R. (2015), "Cross-linked layer-by-layer polyelectrolyte nanofiltration hollow fiber membrane for low-pressure water softening with the presence of SO4 2− in fee water," *J. Membr. Sci.* 486, 169–176. doi:10.1016/j.memsci.2015.03.050.

48. Gautam, A.K., Lai, C., Fong, H., et al. (2014), "Electrospun polyimide nanofiber membranes for high flux and low fouling microfiltration applications," *J. Membr. Sci.* 466, 142–150.

49. Sahadevan, R., Zhao, Y., Fong, H., et al. (2016), "Polyacrylonitrile nanofiber membranes modified with ionically cross-linked polyelectrolyte multilayers for the separation of ionic impurities," *Nanoscale* 8, 18376–18389.

50. Ding, T., Zhu, L., Wang, X.-Q., et al. (2018), "Hybrid photothermal pyroelectric and thermo galvanic generator for multi situation low-grade heat harvesting," *Adv. Energy. Mater.* 8 (33), 1802397. doi: 10.1002/aenm.201802397.

51. Jiang, Q., Tian, L., Liu, K.-K., et al. (2016), "Bi layered bio foam for highly efficient solar steam generation," *Adv. Mater.* 28(42), 9400–9407. doi: 10.1002/adma.201601819.

52. Liu, H., Yang, Q., Guo, W., et al. (2020), "CoWO4-x-based nanoplatform for multimode imaging and enhanced photothermal/ photodynamic therapy," *Chem. Eng. J.* 385, 123979.

53. Alleva, K., Chara, O., Amodeo, G. (2012), "Aquaporins: Another piece in the osmotic puzzle," *FEBS Lett.* 586, 2991–2999.

54. Dong, B., Yang, M., Wang, F., (2019), "Porous Al_2O_3 plates prepared by co (sMBR) performances for Oil & Gas wastewater treatment," *J. Membr. Sci.* 594, 117459.

9 Micro- and Nanoplastics (MNPs) Pollutant Removal from Saline Water

Parul Sahu

CSIR – Central Salt and Marine Chemicals Research Institute

CONTENTS

9.1 INTRODUCTION

Microplastics (MPs) are defined as plastic particles of a regular or irregular shape with a size range between 1 and 500 μm, while nanoplastics (NPs) are defined as plastic particles below 100 nm.[1] In 1994, for the first time, Shaw and Day[2] had predicted the possibility of microplastic formation of floating plastic debris present in the sea due to mechanical fragmentation. However, the physical or chemical risks of MPs were then unknown, and a little emphasis was given on MNPs research. Later, the detrimental effects of MNPs on aquatic life[3] and their transport to the food chain were discovered, and MNPs were identified as a new class of pollutants globally.[4] These plastic debris mainly consists of polyvinyl chloride (PVC), polyethylene (PE), polyamide, (PA), polypropylene (PP), polystyrene (PS), and polyethylene terephthalate (PET).[5] MNPs also act as the absorber, transporter, and releaser of various organic contaminants thus can have a more severe impact on the environment.[6]

DOI: 10.1201/9781003185437-10

9.1.1 Sources, Occurrence, and Impacts of MNPs

The sources of MNPs can be classified into two categories:

1. Primary source: MNPs polymers in the form of microbeads, capsules, fibers, or pellets produced for specific applications like optical fiber, capillary films, injection molding, 3D printing, cosmetics, etc., fall under the primary source. MNPs are accidentally discharged to waterways during the transport, packaging, and processing of these products.[7] The wastewater from these industries may also possess a significant amount of MNPs contamination. Figure 9.1: A and B show primary MPs (microbeads derived from personal care products).[8]

2. Secondary source: MNPs generated from the degradation and fragmentation of macroplastic debris from land and sea-based human activities fall under secondary source. UV and chemically degraded terrestrial plastic wastes from domestic and industrial effluents also contribute as the secondary source of MNPs. Figure 9.1: C and D show secondary MPs fragments from breakdown of larger plastics and synthetic textile fibers.[8]

With the proliferating use of various plastics in domestic and industrial settings, the stress caused by MNPs is increasing in standing and running water as well as wastewater and drinking water.[7] Wastewater treatment plants have also been reported as one of the sources of untreated MNPs into rivers and oceans.[9] The harmful effects

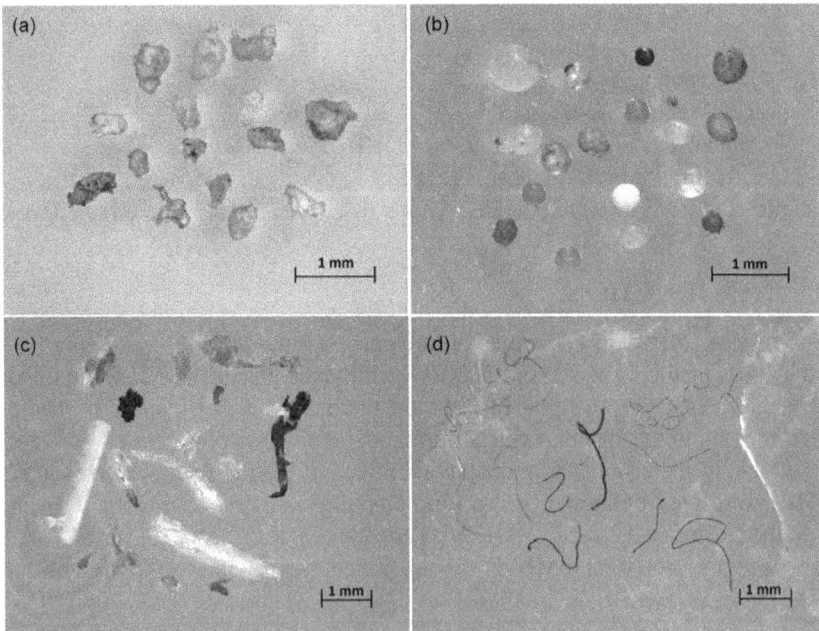

FIGURE 9.1 (a)-(b): Primary MPs; (c)-(d): Secondary MPs. Images of primary and secondary MPs in wastewater samples. (See Ref. 8.)

of plastic debris, especially MNPs found in various saline water bodies, are evident in humans and the environment.

Several physical, chemical, and biological methods have been discussed by researchers to treat and dispose of saline and industrial wastewater. However, the presence of MNPs (that can absorb and transport chemical pollutants or be toxic) in saline/industrial wastewater and their removal is not much discussed.[10]

9.1.2 Method of Detection, Identification, and Quantification of MNPs

Water and sediment samples collected to detect MNPs need pretreatment to remove any adhered organic/inorganic material in the sample. Different cleansing/oxidizing agents, e.g., 30% H_2O_2,[11] NaClO, and Fenton's reagent,[9] are reported to pretreat the MNPs containing samples. The density separation method is generally used to separate MNPs wherein a brine solution is added to recover plastic particles. Subsequently, the supernatant containing MNPs is filtered using filter paper and dried for further observations (Figure 9.2).

Morphological and physical forms (fibers, strands, filaments, films, sheets, foams pellet) of MNPs can be determined using optical techniques, namely, stereoscope, microscope, and scanning electron microscopy (SEM).[12] SEM images provide high magnification of surface textures of plastics (e.g., grooves, flakes, cracks, fractures, or pits). Nanoplastic particles (<400 nm) characterization using SEM have also been reported recently in the context of water and wastewater treatment.[13] While SEM analysis reveals the physical features, energy-dispersive X-ray spectroscopy (EDS) during SEM observations can establish chemical

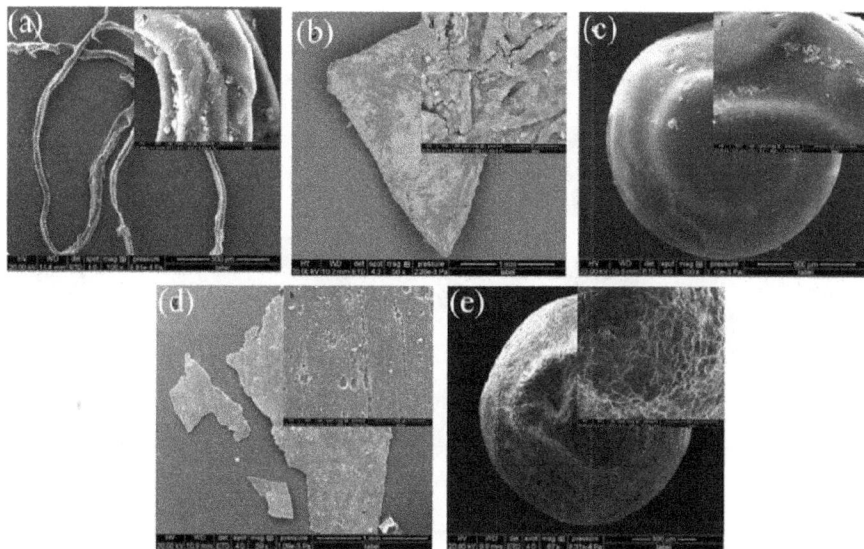

FIGURE 9.2 SEM image of various MPs, classified by their shape (a) fiber, (b) fragment, (c) pellet, (d) film, and (e) Styrofoam. (See Ref. 12.)

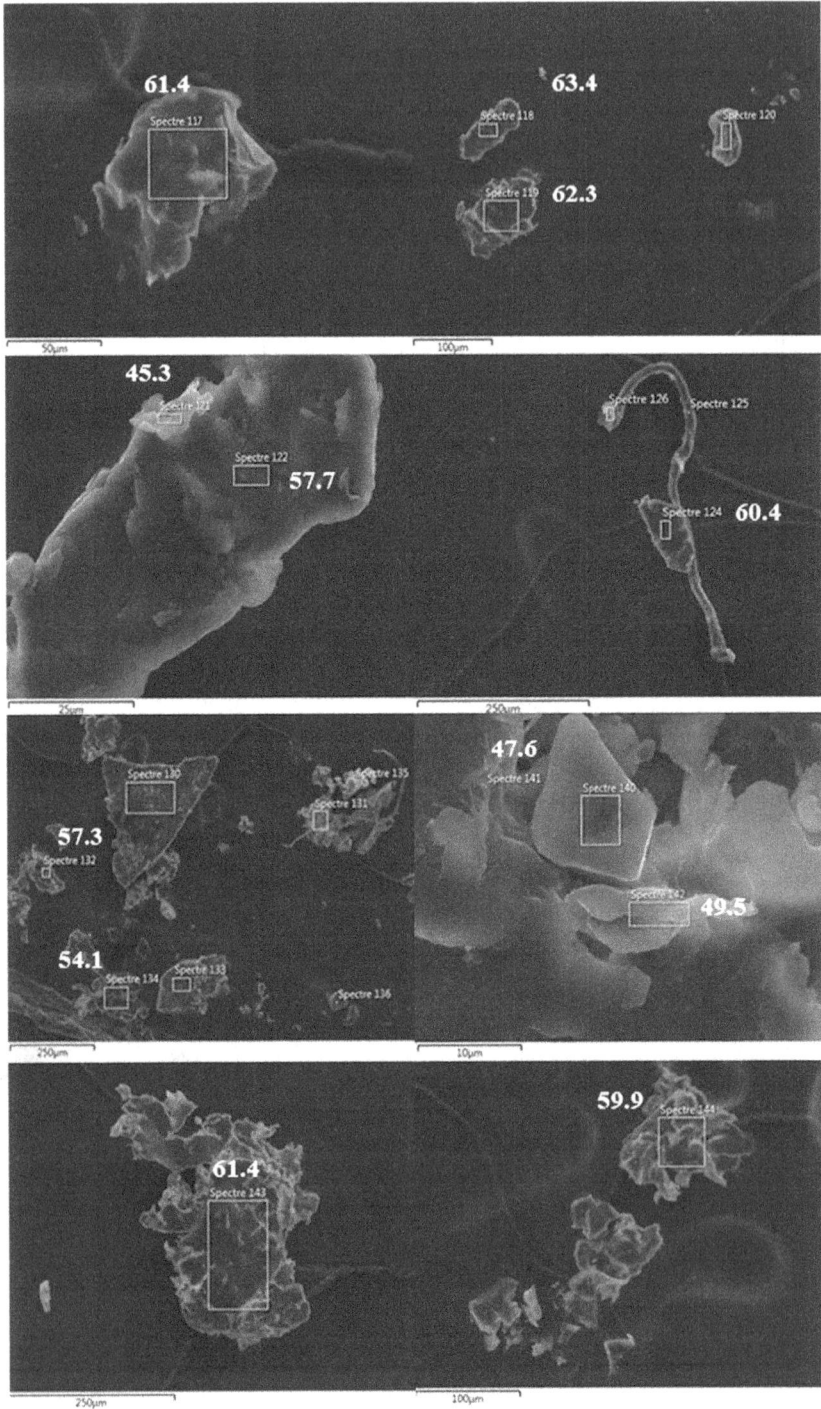

FIGURE 9.3   EDS analysis of SEM images of MPs. (See Ref. 14.)

components of particles. Figure 9.3 shows the SEM/EDS of fluoro-polymers in the MPS of the marine samples. These plastics contain fluoride and carbon elements with molar ratios F/C in the range 0.82–1.73 that agree with the structure of polyfluoride compounds like polyvinylidene fluoride (PVFD) or polytetrafluoroethylene (PTFE).[14]

Spectroscopic methods like Fourier Transform Infrared (FTIR) spectroscopy and Raman microspectroscopy[15,16] are proven to help analyze the chemical composition of large and small MPs. For example, Di and Wang[15] have performed a compositional study of MPs present in surface waters and sediments using micro-Raman spectroscopy (Raman spectra shown in Figure 9.4). Among the various MPs detected by Raman spectra scanned in the range of 50–3,500 cm^{-1}, PS, PP, and PE were found predominant. The chemical composition of the particles is determined by comparing the obtained spectrum with the database.[15]

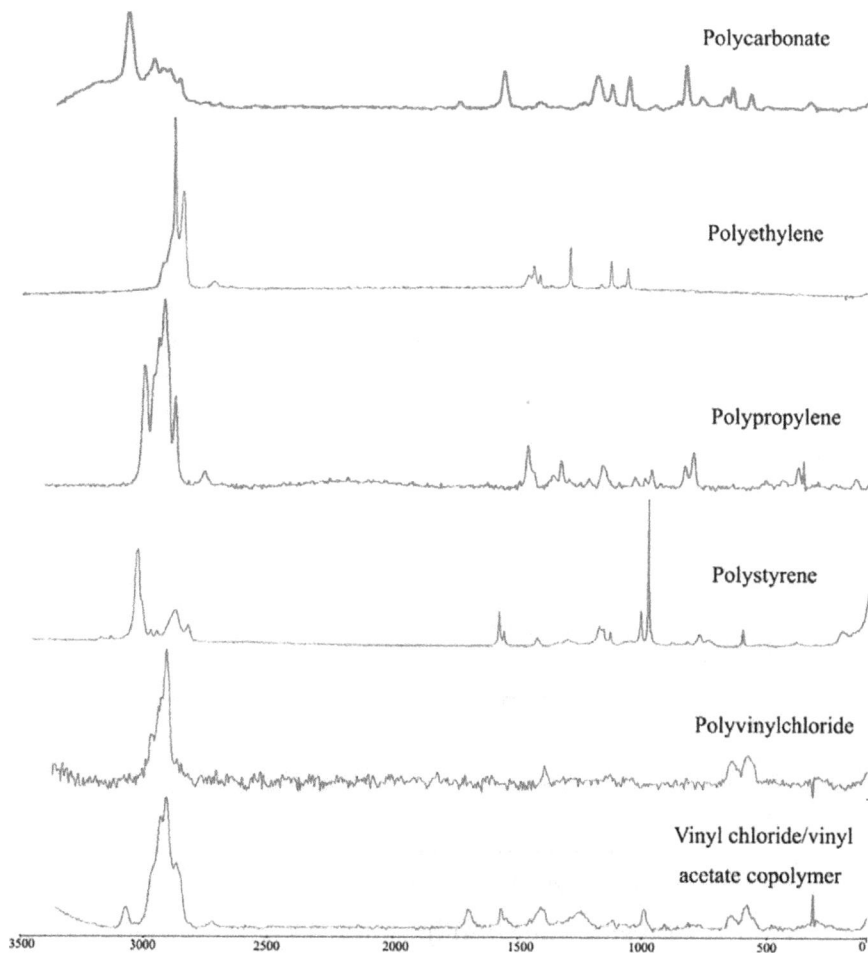

FIGURE 9.4   Raman spectra of the MPs. (See Ref. 15.)

Seth and Shriwastav[17] reported the presence of MPs in Indian sea salt samples using micro-FTIR with a spectrum range of 4,000–400 cm^{-1}. The presence of four types of polymers (shown in Figure 9.5) was determined by comparing the obtained spectra with the database.

Analytical techniques for the detection, size determination, identification, and quantification of sub-20-μm MNPs are still limited, and mass spectroscopy (MS) based analytical methods, namely GC-MS and TD-PTR-MS, MALDI-TOF-MS, are reported promising for the selective quantification of NPs.[5]

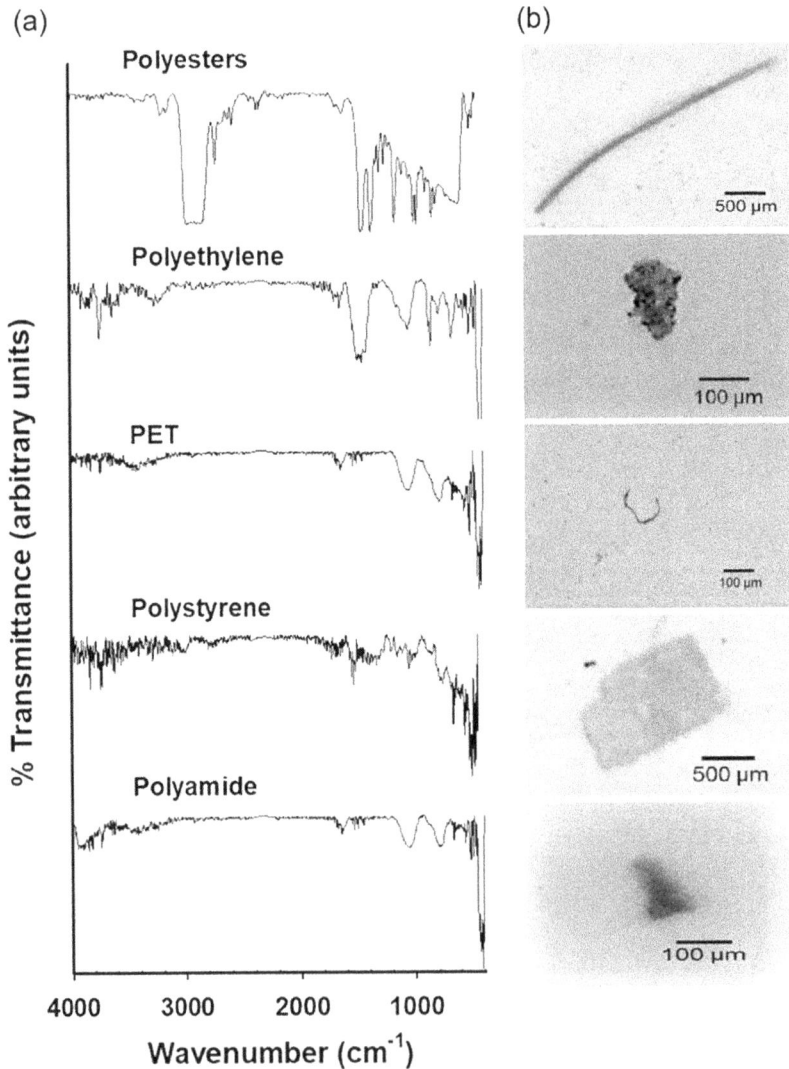

FIGURE 9.5 (a) Shows FTIR spectra of the MPs, and (b) shows the corresponding particle.

9.2 TREATMENT METHODS

9.2.1 DENSITY SEPARATION

The density difference of the MPs particles and liquid in which they are suspended can isolate MPs from solution. For example, as part of MPs quantification in Singapore's coastal environment, Ng and Obbard[18] have separate microplastic particulates from sediment samples via floatation in a 1.2 kg/L hypersaturated saline solution. However, for plastic particles with a broader range of specific density (e.g., 0.8–1.43 kg/L), the liquid's density needs to be higher than MPs' density. This can be achieved by adding the mineral salt (NaCl, $ZnCl_2$, NaI, etc.) or solvents heavier than the feed solution to the sample, wherein particles become lighter, float to the surface, and can be separated.[19]

A Munich Plastic Sediment Separator (MPSS) for separation of small microplastic particles (<1 mm) from solution under density difference is shown in Figure 9.6.[19] The MPSS consists of a stirred sediment-container, standpipe, and dividing chamber. MPs laden solution is introduced through the sediment inlet flange, and additives are mixed to adjust solution density. By adjusting the flow rate, the solution passes through the tapered standpipe. The reduced diameter at the top of the standpipe allows a high concentration of plastic particles in the extracted sample volume.

A rotor mounted at the bottom revolves and assists flotation of plastic particles to the top layer of liquid, separated in a dividing chamber-using filter. The bottom valve drains the additive solution and enables the liquid inflow to MPSS in further separation. As the microplastic separation in MPSS is based on the density difference, continuous monitoring and adjustment of the additive solution density are essential. Knowledge of the specific density of the microplastic particles of interest may be necessary to utilize this technique. Density-based separation may not be efficient for nanoplastic removal.

9.2.2 SAND FILTRATION

Application of conventional sand filters (disc filters and rapid sand filters) is reported by researchers for the removal of MPs present in saline water and wastewater streams.[17,20,21] Sand filters use quartz-grade fine sand (effective size 0.63–0.85 mm) with a specific gravity range of 2.6–2.7 (bulk density ~1.56 g/cc). Besides, other filter media such as anthracite, granular activated carbon, garnet, pumice, expanded clay particles, and glass can also be used for filtration.[22] The mechanism of such filtration is based on bed formation and pore blocking through a layer of sand. The bed is subsequently cleaned by a backwash operation to put back into service. Microplastic particles and fibers ranging from 100 to 500 μm were removed efficiently using simple sand filters. However, the MPs of <100 μm could not be prevented using sand filters.

9.2.3 ELECTROCOAGULATION

Electrocoagulation (EC) involves an electrolysis process using a metal electrode to produce coagulants in situ that eliminate micro-plastics. EC offers advantages of

To vacuum pump

Filter holder

Sample
chamber

(III) Dividing chamber
with ball valve and
filter holder

Vent screw
Sediment inlet flange

(II) Standpipe

(I) Stirred sediment container

Bottom valve

To electromotor

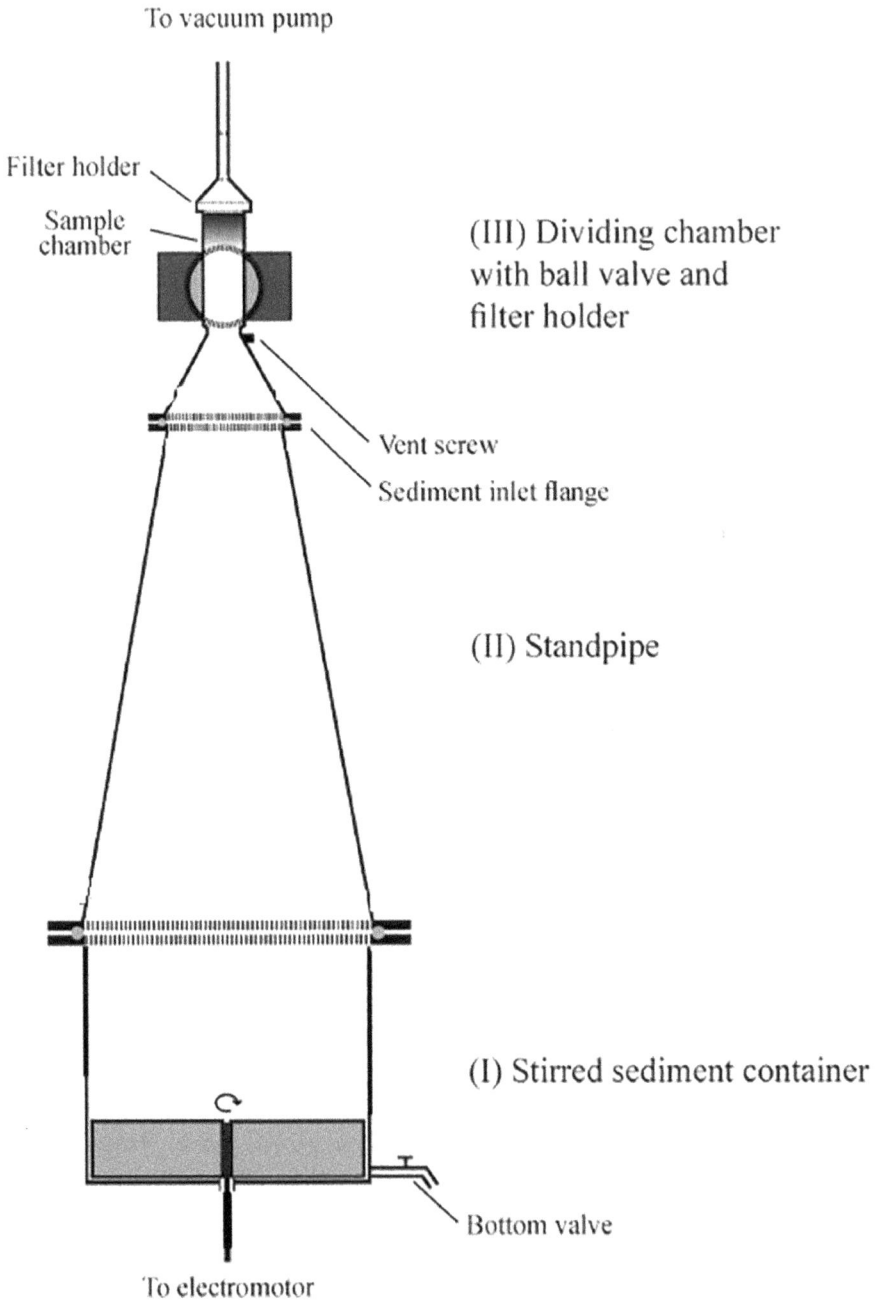

FIGURE 9.6   Schematic of Munich plastic sediment separator (MPSS). (See Ref. 19.)

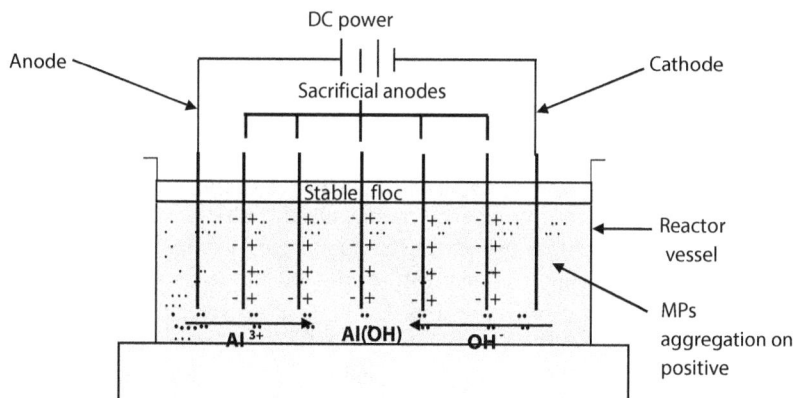

FIGURE 9.7 Electrocoagulation setup. (Adapted from Perren, W., Wojtasik, A. & Cai, Q., *ACS Omega*, 3(3), 3357–3364, 2018.)

environmental compatibility, low capital cost, energy efficiency, sludge minimization, and cost effectiveness.

Commonly used coagulants produced by EC comprise Fe^{2+} or Al^{3+} metal ions with OH^- ions resulting in metal hydroxide coagulants. Schematic of an electrocoagulation setup comprising of a reactor vessel and parallel placed electrodes connected to a DC power supply is shown in Figure 9.7. The voltage and current density of the EC are controlled using a DC power supply to outermost electrodes, one acting as the primary anode and the other as the primary cathode. The unconnected electrodes in between the outer electrodes act as bipolar sacrificial electrodes. Under the application of current, the metal ions are liberated from the electrodes to form microcoagulants. The suspended MPs lose their stability under the influence of microcoagulant and form aggregates on positive electrode faces. Perren and coworkers[23] have demonstrated successful removal (up to 99%) of polyethylene microbeads (300–355 μm) from simulated wastewater using EC at a lab scale. Various factors like pH, conductivity, operation time, power, and electrode consumption should be considered for the design and operation of the EC process for large-scale applications.

9.2.4 Membrane Filtration

Membrane-based processes are commercially being used for the treatment of saline wastewater worldwide. Membrane processes (microfiltration, ultrafiltration, reverse osmosis, electrodialysis, etc.) involve the passage of the contaminated water stream through a semipermeable membrane such that the separation occurs based on the molecular size or charge of the interaction.[24] Poerio and coworkers[25] reviewed various membrane technologies applied as tertiary treatment of effluents containing MNPs with a smaller size (<100 nm). Ultrafiltration (UF), a low-pressure process (1–10 bar) using UF membranes with pore size between 1 and 100 nm, in combination with coagulation, is reported to remove MPs for the production of drinking water.[26] Addition of suitable coagulation agents like $FeCl_3.6H_2O$ and anionic polyacrylamide (PAM) was found to enhance the coagulation performance of polyethylene (PE) microplastic (<500 μm) in a subsequent study.[27] Apart from microplastic filtration, membrane filtration or membrane ultrafiltration can be used to filter nanoplastics as low as 5 nm in size by choosing a suitable pore size of the membrane. As filters with small sizes have relatively lower flow rates, the filtration of NPs is expected to be very tedious and time-consuming. For samples with a varying range of plastic/suspended particles, the application of sequential filtration stages having higher to lower pore size filters can avoid pore-clogging and separate plastic particles on the basis of size.[1]

Reverse osmosis (RO) involves using nonporous or nanofiltration membranes (pore size >2 nm) at high application pressure (10–100 bar) to remove salts, contaminants, heavy metals, and other impurities. RO being an advanced tertiary treatment method removes the MPs effectively. However, RO suffers from fouling phenomena while handling the nanoplastics (<100 nm) and may require adequate pretreatment and improved membrane material.[28]

Another variant of membrane system: membrane bioreactors (MBR), a combination of membrane filtration processes with suspended growth biological reactors (Figure 9.8), are being used as advanced wastewater systems for the effective removal of organic-laden saline effluents.[29–31] MBR can handle organic pollutants and TSS, including MPs present in high-salinity effluents.

MBR consists of a bioreactor and a membrane unit, as shown in Figure 9.8. In the bioreactor, the solution complexity is reduced by the biodegradation of the organic matter, allowing the purification and treatment of the MPs. The mixed liquor from the bioreactor is pumped along with the membrane filtration system for the separation process. Treated water is collected as permeate stream, and MPs concentrate in the retentate stream.[25]

Another variant of membrane technique, i.e., dynamic membrane (DM) technique, has attracted attention for MPs removal in wastewater treatment. In the DM technique, the particles and other foulants (present in the wastewater being filtered) form a cake

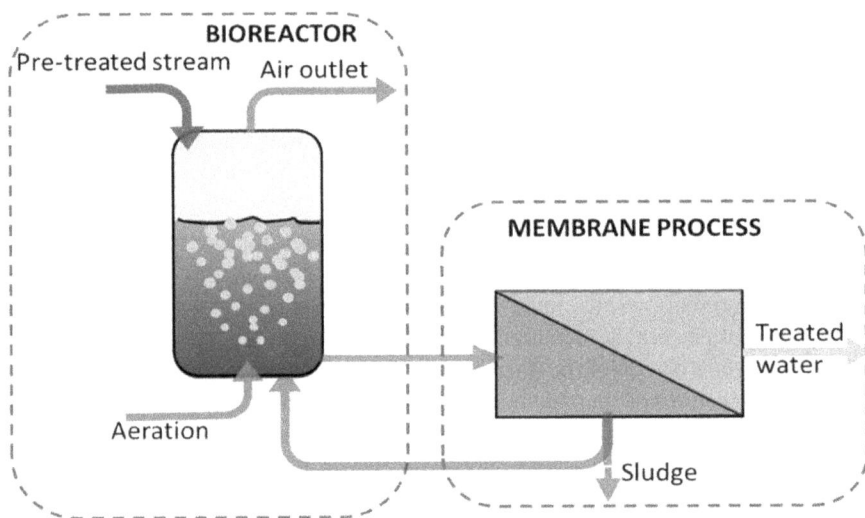

FIGURE 9.8 Schematic of a membrane bioreactor (MBR) process. (See Ref. 25.)

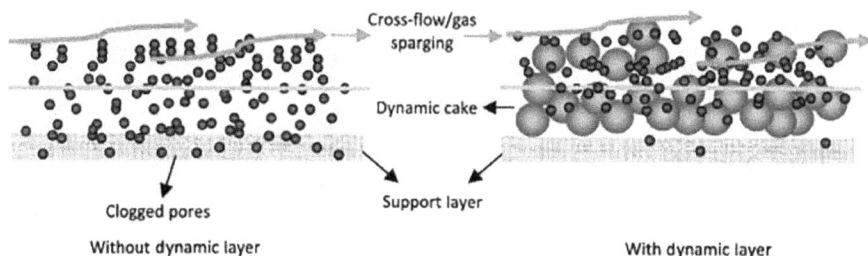

FIGURE 9.9 Formation of the dynamic cake layer on membrane support. (See Ref. 32.)

(filtration) layer on a supporting membrane surface. The resistance to filtration is caused by this cake layer, and the thickness and density of the layer affect membrane performance.[25] As illustrated in Figure 9.9, the deposited cake layer acts as a "secondary" membrane before actual membrane/supporting material (e.g., mesh, woven filter-cloth, etc.). The performance of the DM technique mainly affects by supporting material, sludge properties, operation conditions, configuration, and operation mode.[32]

Li et al.[33], in a lab-scale study, have demonstrated the potential of DM for removing the microparticles (ranging from 1.65 to 516 µm) from synthetic wastewater on a 90 µm supporting mesh. The effluent turbidity was reported to be <1 NTU from 200 NTU after 20 minutes of filtration, verifying the effective removal of MPs by DM. With an increase in influent flux and influent particle concentration, transmembrane pressure (TMP) and filtration resistance increased, resulting in quick cake formation on the supporting mesh. DM offers advantages of low filtration resistance, low transmembrane pressure, and filtration under gravity. Also, it offers easy cleaning with surface brushing, air scouring and/or water backwashing, and low-energy consumption.[33] While DM was found effective at removing microparticles from wastewater, its effectiveness in the presence of high salt concentration may require further investigations.

9.2.5 Magnetic Extraction

Magnetic extraction-based separation takes advantage of the hydrophobic surface of MNPs to magnetize the plastics.[34] A magnetically active binding agent (i.e., Fe nanoparticles) is used to magnetize MNPs, allowing for recovery under the external magnetic field. Fe nanoparticles offer the advantages of low cost, high-specific surface area, and ferromagnetic properties (Figure 9.10).[35]

Grbic et al.[34] investigated the magnetic extraction of MPs of three sizes range large (1–8 mm), medium (200 µm–1 mm), and small (<20 µm) from freshwater, seawater, and sediments. Using surface modified (coated with hexadecyltrimethoxysilane) Fe nanoparticles, recovery of 92% of 10–20 µm microbeads and 93% of MPs above 1 mm from seawater was achieved. Their proof-of-concept experiments established magnetic extraction as a potential technique for plastic removal, particularly small MPs < 20 µm. Recovering and reusing bound Fe nanoparticles from MNPs for a cyclic process will be the key driver for applying this technique in the future.

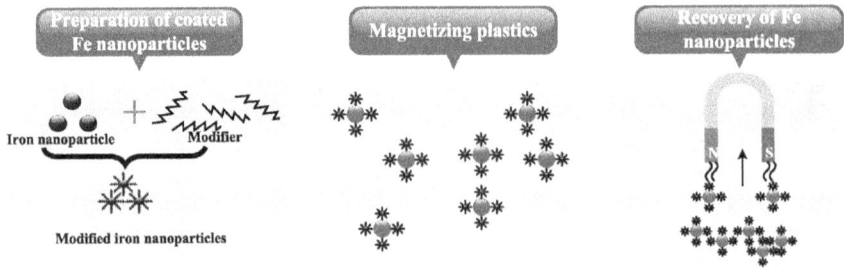

FIGURE 9.10   Magnetic extraction of MPs using coated Fe nanoparticles. (See Ref. 35.)

Apart from the methods discussed above, techniques, namely centrifugation and ballasted flocculation for the removal of NPs (333 ± 76 nm), have also been reported by researchers.[13] While centrifugation (at 10,000 rpm) was found highly energy-intensive, ballasted flocculation method required additional chemicals (aluminum sulfate, polymer, and micro-sand), making the process unattractive.

9.3   TECHNICAL AND ECONOMIC ASSESSMENT OF TREATMENT METHODS

It is evident that the treatment methods discussed in Section 9.2 are capable of handling MNPs in saline water. A technical and economic assessment of the treatment methods for MNPs removal is presented in Table 9.1.

9.4   CONCLUDING REMARKS

This chapter discussed a new class of pollutants, i.e., micro- and nanoplastics (MNPs), and their occurrence in saline and industrial wastewater. Various conventional and emerging treatment technologies for MNPs removal have been discussed in detail. While density separation and sand filtration techniques can be implemented to remove small to large MPs, membrane-based MBR can be used to recover plastic particles with smaller sizes (<100 nm). Emerging techniques, namely, electrocoagulation, dynamic membrane technique, and magnetic extraction, can be potential alternatives subject to scale-up studies with practical MNPs laden samples.

REFERENCES

1. Barbosa F, Adeyemi JA, Bocato MZ, Comas A, Campiglia A. A critical viewpoint on current issues, limitations, and future research needs on micro- and nanoplastic studies: From the detection to the toxicological assessment. *Environ Res.* 2020;182:109089. doi:10.1016/j.envres.2019.109089
2. Shaw DG, Day RH. Colour- and form-dependent loss of plastic micro-debris from the North Pacific Ocean. *Mar Pollut Bull.* 1994;28(1):39–43. doi:10.1016/0025-326X (94)90184-8
3. Thompson RC, Olsen Y, Mitchell RP, et al. Lost at Sea: Where is all the plastic? *Science.* 2004;304(5672):838. doi:10.1126/science.1094559
4. News. *Mar Pollut Bull.* 2013;75(1):4–7. doi:10.1016/j.marpolbul.2013.09.022

TABLE 9.1

A Technical and Economic Assessment of Various Treatment Methods for MNPs Recovery

Treatment Method	Advantages/ Applicability	Status of Technology/ Economic Aspects	Limitations	References
Density separation	Applicable for separating MPs from sediments and water samples Small (<1 mm) to large MPs (5–1 mm) can be removed using this technique	As density/gravity separation systems are prevalent in wastewater treatment, further intensification can suit MPs removal	Required additional solution (high density) for MPs having a broad range of density	Imhof et al.[19]
Sand filtration	Shown superior filtration of larger MPs (>500 μm) Effective for MPs particulate over microfibers	Adequate size of filtration media (quartz sand) in conventional sand filters (disc filter, RSF, etc.) can remove large MPs	Not applicable to small MPs and NPs	Seth and Shriwastav[17], Bayo et al.,[20] Wolff et al.[21]
Electrocoagulation	Shown potential for the removal of polyethylene microbeads ranging 300–355 μm	Lab-scale potential is demonstrated (99.24% removal at pH 7.5) An operating cost of £0.05/m³ of synthetic wastewater is established	pH control (i.e., 7.5) is required	Perren et al.[23]
Membrane technology	Among the tertiary membrane technologies, membrane bioreactor (MBR) proves to be promising for plastic particles with smaller size (<100 nm) and fibers	MBR is an established technology for organic and saline wastewater treatment and can be adopted for MNPs removal	MNPs in sludge needs proper management Membrane fouling and recycling are persistent challenges	Poerio et al.[25] and Bayo et al.[20]
Magnetic extraction	>90% Recovery of 10–20 μm MPs is achieved The method is efficient for various sizes, polymer types, and sample matrices	Proof-of-concept experiments have shown recovery of small to large MPs (10–1,000 μm) recovery Scale-up studies and economic evaluation are needed	Recycling of bonded Fe nanoparticles is not addressed	Grbic et al.[34]

5. Velimirovic M, Tirez K, Voorspoels S, Vanhaecke F. Recent developments in mass spectrometry for the characterization of micro- and nanoscale plastic debris in the environment. *Anal Bioanal Chem.* 2021;413(1):7–15. doi:10.1007/s00216-020-02898-w

6. Koelmans AA, Bakir A, Burton GA, Janssen CR. Microplastic as a vector for chemicals in the aquatic environment: Critical review and model-supported reinterpretation of empirical studies. *Environ Sci Technol.* 2016;50(7):3315–3326. doi:10.1021/acs.est.5b06069

7. Herbort AF, Schuhen K. A concept for the removal of microplastics from the marine environment with innovative host-guest relationships. *Environ Sci Pollut Res.* 2017;24(12):11061–11065. doi:10.1007/s11356-016-7216-x

8. Talvitie J, Mikola A, Koistinen A, Setälä O. Solutions to microplastic pollution – Removal of microplastics from wastewater effluent with advanced wastewater treatment technologies. *Water Res.* 2017;123:401–407. doi:10.1016/j.watres.2017.07.005

9. Sol D, Laca A, Laca A, Díaz M. Approaching the environmental problem of microplastics: Importance of WWTP treatments. *Sci Total Environ.* 2020;740:140016. doi:10.1016/j.scitotenv.2020.140016

10. Baztan J, Carrasco A, Chouinard O, et al. Protected areas in the Atlantic facing the hazards of micro-plastic pollution: First diagnosis of three islands in the Canary Current. *Mar Pollut Bull.* 2014;80(1):302–311. doi:10.1016/j.marpolbul.2013.12.052

11. Vetrimurugan E, Jonathan MP, Sarkar SK, et al. Occurrence, distribution and provenance of micro plastics: A large scale quantitative analysis of beach sediments from southeastern coast of South Africa. *Sci Total Environ.* 2020;746:141103. doi:10.1016/j.scitotenv.2020.141103

12. Strungaru S-A, Jijie R, Nicoara M, Plavan G, Faggio C. Micro- (nano) plastics in freshwater ecosystems: Abundance, toxicological impact and quantification methodology. *TrAC Trends Anal Chem.* 2019;110:116–128. doi:10.1016/j.trac.2018.10.025

13. Murray A, Örmeci B. Removal effectiveness of nanoplastics (<400nm) with separation processes used for water and wastewater treatment. *Water.* 2020;12(3). doi:10.3390/w12030635

14. Sunitha TG, Monisha V, Sivanesan S, et al. Micro-plastic pollution along the Bay of Bengal coastal stretch of Tamil Nadu, South India. *Sci Total Environ.* 2021;756:144073. doi:10.1016/j.scitotenv.2020.144073

15. Di M, Wang J. Microplastics in surface waters and sediments of the Three Gorges Reservoir, China. *Sci Total Environ.* 2018;616–617:1620–1627. doi:10.1016/j.scitotenv.2017.10.150

16. Yaranal NA, Subbiah S, Mohanty K. Identification, extraction of microplastics from edible salts and its removal from contaminated seawater. *Environ Technol Innov.* 2021;21:101253. doi:10.1016/j.eti.2020.101253

17. Seth CK, Shriwastav A. Contamination of Indian sea salts with microplastics and a potential prevention strategy. *Environ Sci Pollut Res.* 2018;25(30):30122–30131. doi:10.1007/s11356-018-3028-5

18. Ng KL, Obbard JP. Prevalence of microplastics in Singapore's coastal marine environment. *Mar Pollut Bull.* 2006;52(7):761–767. doi:10.1016/j.marpolbul.2005.11.017

19. Imhof HK, Schmid J, Niessner R, Ivleva NP, Laforsch C. A novel, highly efficient method for the separation and quantification of plastic particles in sediments of aquatic environments. *Limnol Oceanogr Methods.* 2012;10(7):524–537. doi:10.4319/lom.2012.10.524

20. Bayo J, López-Castellanos J, Olmos S. Membrane bioreactor and rapid sand filtration for the removal of microplastics in an urban wastewater treatment plant. *Mar Pollut Bull.* 2020;156:111211. doi:10.1016/j.marpolbul.2020.111211

21. Wolff S, Weber F, Kerpen J, Winklhofer M, Engelhart M, Barkmann L. Elimination of microplastics by downstream sand filters in wastewater treatment. *Water.* 2021;13(1). doi:10.3390/w13010033

22. Brandt MJ, Johnson KM, Elphinston AJ, Ratnayaka DD. Chapter 9 – Water filtration. In: Brandt MJ, Johnson KM, Elphinston AJ, Ratnayaka DD (eds.) *Twort's Water Supply*, Seventh Edn. Butterworth-Heinemann; 2017:367–406. doi:10.1016/B978-0-08-100025-0.00009-0

23. Perren W, Wojtasik A, Cai Q. Removal of microbeads from wastewater using electrocoagulation. *ACS Omega.* 2018;3(3):3357–3364. doi:10.1021/acsomega.7b02037

24. Maddah HA, Alzhrani AS, Bassyouni M, Abdel-Aziz MH, Zoromba M, Almalki AM. Evaluation of various membrane filtration modules for the treatment of seawater. *Appl Water Sci.* 2018;8(6):150. doi:10.1007/s13201-018-0793-8

25. Poerio T, Piacentini E, Mazzei R. Membrane processes for microplastic removal. *Mol.* 2019;24(22). doi:10.3390/molecules24224148

26. Ma B, Xue W, Hu C, Liu H, Qu J, Li L. Characteristics of microplastic removal via coagulation and ultrafiltration during drinking water treatment. *Chem Eng J.* 2019;359:159–167. doi:10.1016/j.cej.2018.11.155

27. Ma B, Xue W, Ding Y, Hu C, Liu H, Qu J. Removal characteristics of microplastics by Fe-based coagulants during drinking water treatment. *J Environ Sci.* 2019;78:267–275. doi:10.1016/j.jes.2018.10.006

28. Malankowska M, Echaide-Gorriz C, Coronas J. Microplastics in marine environment: a review on sources, classification, and potential remediation by membrane technology. *Environ Sci Water Res Technol.* 2021;7(2):243–258. doi:10.1039/D0EW00802H

29. Deowan SA, Korejba W, Hoinkis J, et al. Design and testing of a pilot-scale submerged membrane bioreactor (MBR) for textile wastewater treatment. *Appl Water Sci.* 2019;9(3):59. doi:10.1007/s13201-019-0934-8

30. Luo G, Wang Z, Li Y, Li J, Li A-M. Salinity stresses make a difference in the start-up of membrane bioreactor: Performance, microbial community and membrane fouling. *Bioprocess Biosyst Eng.* 2019;42(3):445–454. doi:10.1007/s00449-018-2048-3

31. Abdollahzadeh Sharghi E, Shourgashti A, Bonakdarpour B. Considering a membrane bioreactor for the treatment of vegetable oil refinery wastewaters at industrially relevant organic loading rates. *Bioprocess Biosyst Eng.* Published online 2020. doi:10.1007/s00449-020-02294-9

32. Ersahin ME, Ozgun H, Dereli RK, Ozturk I, Roest K, van Lier JB. A review on dynamic membrane filtration: Materials, applications and future perspectives. *Bioresour Technol.* 2012;122:196–206. doi:10.1016/j.biortech.2012.03.086

33. Li L, Xu G, Yu H, Xing J. Dynamic membrane for micro-particle removal in wastewater treatment: Performance and influencing factors. *Sci Total Environ.* 2018;627:332–340. doi:10.1016/j.scitotenv.2018.01.239

34. Grbic J, Nguyen B, Guo E, You JB, Sinton D, Rochman CM. Magnetic extraction of microplastics from environmental samples. *Environ Sci Technol Lett.* 2019;6(2):68–72. doi:10.1021/acs.estlett.8b00671

35. Shen M, Song B, Zhu Y, et al. Removal of microplastics via drinking water treatment: Current knowledge and future directions. *Chemosphere.* 2020;251:126612. doi:10.1016/j.chemosphere.2020.126612

Section II

Brackish Water

10 A Qualitative Examination of Applications and Treatment Technologies of Brackish Water

Saisanthosh Vamshi Harsha Madiraju
The University of Toledo

Abhiram Siva Prasad Pamula
Oklahoma State University

CONTENTS

DOI: 10.1201/9781003185437-12

10.1   WHAT IS BRACKISH WATER?

Water contains dissolved solids which can make surface or groundwater brackish if they are present in sufficient concentrations. Brackish water is typically defined as salty in taste, and salt concentration ranges from 500 to 30,000 mg/L (Nthunya et al. 2018). Furthermore, brackish water is more saline than freshwater and not as salty as freshwater. Varying salt concentrations in brackish water are mostly because of seawater mixing with freshwater, such as in estuaries. Moreover, certain human activities produce brackish water, including civil infrastructure (dikes) and prawn farming by flooding coastal marshlands (Nguyen, Dargusch, and Moss 2016). The main concern of brackish water is that it is hostile to the growth process of most terrestrial plants (Cloern, Canuel, and Harris 2002). Therefore, managing brackish water is important where proper management can be utilized for irrigation without affecting crop yields and soil properties (Figures 10.1 and 10.2; Table 10.1).

The dissolved solids concentration of brackish water is greater than that of freshwater and less than seawater, ranging from 1,000 to 10,000 mg/L. Typically, the term saline refers to water when dissolved solids concentration is more significant than 1,000 mg/L (USGS 2021). Therefore, brackish water comes in the saline range.

10.2   TYPES OF BRACKISH WATER

Classification of brackish water can be done based on the hydrology of brackish water availability on the earth (Den Hartog 1974). The brackish water is classified as surface brackish water and underground brackish water.

10.2.1   SURFACE BRACKISH WATER

The source of surface brackish water is commonly located in transitional points where freshwater meets seawater. These transitional areas of water bodies are commonly known as estuaries. Most of the estuaries are sensitive to change in salinity due to changes in tidal cycles (Ahuja et al. 2014). Besides the seasonal increase in the influx of freshwater due to precipitation, snowmelt results in varying salinity concentrations in estuaries. This change in salinity makes the brackish water environment interesting to study from an ecology and microbiology point of view

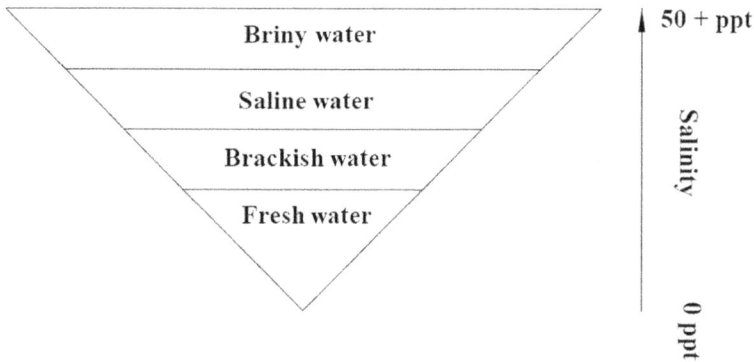

FIGURE 10.1 Classification of water based on salinity levels.

FIGURE 10.2 Bird's eye view of the estuary and its surrounding lush landscape.

(Rheinheimer 2012). In addition to naturally formed surface brackish water bodies, human activities result in their formation by consumptive use and the addition of salts to them.

10.2.2 UNDERGROUND BRACKISH WATER

The high demand for water resources in the agriculture and mining industry makes underground brackish water an important natural resource. Based on a study by the United States Geological Survey (USGS), most western hemispheres are bound with brackish water. It can act as a potential water resource for domestic, agricultural, industrial, mining, and environmental needs (USGS 2021). Recently, brackish groundwater became a significant component in the water supply where an estimated 3,920 Mgal/d of saline groundwater was used in the United States in 2010

TABLE 10.1

Classification of Water Based on Source, Salinity, Conductivity, and Application (US EPA, OW 2014)

Type	Source	Salinity (ppt)	TDS (mg/L)	EC (mS/cm)	Application
Briny water	Brine pool water	<50	>50,000	>42	Industrial applications like mining, food preservation, de-icing on roads, refrigerating fluid, auxiliary agent in water purification, and softening
Saline water	Ocean, Sea	30–50	30,000–50,000	2.5–42	Thermoelectric power-plant cooling, minerals like Na, Mg, Ca, and K are extracted
Brackish water	Estuaries, deep inland and coastal aquifers	0.5–30	>500–30,000	0.7–2.5	Agricultural (Limited), Livestock (TDS < 1,000 mg/L only), aquaculture, power plants, oil and gas industry
Freshwater	Aquifers, rivers, lakes	<0.5	<500	<0.7	Potable water and food manufacturing, livestock

(Stanton et al. 2017). Most saline groundwater use is reported in California, Florida, Oklahoma, Texas, Utah, and Wyoming for mining, oil, and gas production. Although the reported data of saline groundwater usage is based on the compilation of scientific reports, the actual use might be more than reported (Dieter et al. 2018). In addition to inland brackish groundwater, the discovery of an expansive 15,000 square miles of brackish water aquifer under the seabed from New Jersey to Massachusetts indicates the availability of water resources under the oceans (Gustafson, Key, and Evans 2019). With improvement in desalination technologies, this vast amount of brackish water resources is viable during times of water stress due to unpredictable weather patterns.

10.3   COMPOSITION OF BRACKISH WATER

The salt content in brackish water makes it essential due to dissolved basic minerals compared to freshwater which is the closest to neutral pH of 7. The brackish water contains ions including sodium (Na^+), potassium (K^+), nitrate (NO_3^-), magnesium (Mg^{2+}), calcium (Ca^{2+}), chloride (Cl^-), sulfate (SO_4^{2-}), carbonate (CO_3^{2-}), and bicarbonate (HCO_3^-). These dissolved solids in the brackish water are in the concentration ranging from 1,000 to 2,000 ppm. Moreover, the average salinity of the brackish water ranges between 0.5 and 30 ppt. This increased salt content and dissolved solids make the brackish water denser than the freshwater such that the specific gravity ranges from 1.005 to 1.010 (Ahuja et al. 2014) (Figure 10.3).

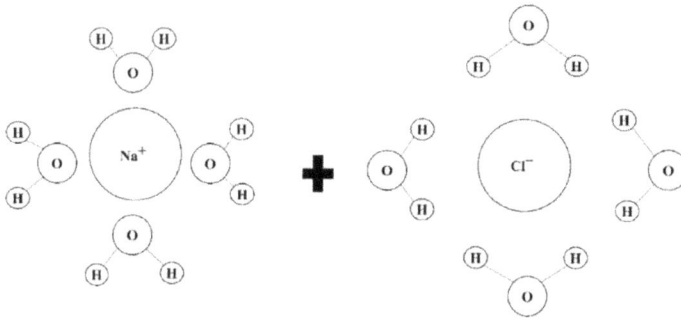

FIGURE 10.3 Sodium and chloride ions surrounded by water molecules structure in brackish water.

10.4 IS THE TREATMENT OF BRACKISH WATER NECESSARY?

Increased water demand in recent years paved the way for exploring water resources and treatment technologies (Vliet et al. 2021). Brackish water is an unconventional water source that might offer a partial solution to water scarcity issues (Mohsen and Al-Jayyousi 1999). Some of the modern applications of brackish water include thermoelectric power cooling, mining operations in oil and gas, and other industrial purposes. If the salinity and salt concentrations are higher, then it is imperative to address the issue of water quality by water treatment methods (McMahon et al. 2016). Once treated, the water can be used for irrigation, livestock, and potable water purposes.

Utilizing brackish water has been broadly classified into four categories.

10.4.1 DIRECT CONSUMPTION OR DRINKING

Brackish water is an alternative source to freshwater during times of water stress. Unpolluted brackish water can be easily treated to potable water standards using desalination techniques. Removing salinity from brackish water and using it for ecosystem services has been a challenge because of the lack of public acceptance due to lack of awareness and negative perceptions of water treatment and reuse methods (Scruggs and Thomson 2017). As far as technology and treatment capacity is concerned, there is a rapid increase in brackish water treatment wherein California 17 brackish water desalination plants with a combined capacity of 252,000 m³/d were in the development of 2017 (Ahdab and Lienhard 2021). Moreover, brackish water has dissolved solids and salinity, which is easier to treat, and the effluent can reach potable water standards.

Fewer salts and dissolved solids in brackish water allow the desalination treatment to be energy and cost-efficient. For example, NIROBOX smart desalination technology owned by Fluencecorp is an energy-efficient treatment technology without a significant capital investment required to build the infrastructure and maintain it. These NIROBOX units are ideal for use in remote regions because of the size of the treatment unit being equivalent to a shipping container and low cost and energy requirements. Moreover, another small desalination treatment option from Fluencecorp includes NIROFLEX where the infrastructure can be established in a short span of time (preferably 3 months) for brackish water treatment. Besides the NIROFLEX plant has the capability to be flexible because of the customizable options to treat different pollutants based on the water source ("Global Water & Wastewater Solutions | Fluence" 2021) (Figure 10.4; Table 10.2).

FIGURE 10.4   A sample figure displaying the application of treated water for drinking water.

TABLE 10.2
Major Chemical Parameters and Their Standards According to WHO (Khadsan and Kadu 2004; WHO 2021)

Chemical Parameter	WHO Standard (ppm except pH)	
	Permissible Limit	Excessive Limit
pH	7–8.5	6.5–9.2
TDS	500	1,500
Ca	75	200
Mg	50	150
Cu	1.0	1.5
Fe	0.3	1
Mn	0.1	0.5
Cl^-	200	600
SO_4^{2-}	200	400
NO_3^-	50	100
F^-	0.5	1–1.5
As	0.01	0.2
Cd	<0.005	0.05
CN^-	<0.01	0.05
Pb	–	0.1
Se	–	0.01

10.4.2 INDUSTRIAL PURPOSE

Treated brackish water can be used for industrial purposes in addition to potable use. Traditionally, saltwater has been used for cooling in thermoelectric plants in the United States. In 2005, more than 95% of saline groundwater pumped was used as a coolant in the electricity generation industry, and the rest 5% was used in mining (oil and gas production) industries (World Energy Council 2016). In the area of the petrochemical industry, an oilfield in Puerto Gaitán in Colombia is a case study where brackish water was used for irrigation purposes after reverse osmosis (RO). The initial feedwater from the oil field contains high salinity, residual oils, hydrocarbons, and other contaminants. All these contaminants are needed to be removed initially from the feedwater before desalination and then the treated water can be used for irrigation ("Global Water, Wastewater & Reuse Solutions for the Municipal, Industrial and Commercial Markets – Texas Desal Association" 2016) (Figure 10.5).

10.4.3 IRRIGATION

Brackish water can be used for irrigation purposes if freshwater availability is scarce in the region and the salt concentration should not exceed the tolerance limit for the crops. Apart from surface water, wells also provide brackish water for irrigation. In arid coastal regions, farmers use water from bays, rivers, and streams which mix with saltwater from the sea. Moreover, the amount of ionic species, dissolved solids, and salt concentration are varying constantly such that this water needs consistent quality monitoring, especially before irrigation (Xiaobin, Yaohu, and Wahba 2019).

The use of brackish water during irrigation might cause a consistent accumulation of salts in the soil (Zhang et al. 2020). In humid areas, irrigation is performed using sprinkler systems. Usually, sprinkling irrigation will not penetrate the soil beyond 6–9 in. If the sprinkler application of water on the field is confined to the surface of the soil, then there shall be minimal leaching of salt into the ground (Wang, Zhao, and Duan 2021; Sun et al. 2011).

Using brackish water to maintain surface moisture is an application where salts are most likely confined to the surface foot of the soil. Besides, it is important to keep track of precipitation before irrigation because second irrigation might affect the salinity levels in soil that can impede crop growth. For instance, with little or no precipitation, second irrigation will double the salt content in the surface foot, and more than saturation rainfall can cause leaching of the salt to the subsurface. Therefore, the intensity of precipitation shall determine the concentration of salt content and movement in the soil profile. Moreover, good drainage in the subsurface is essential and good in

FIGURE 10.5   Usage of brackish water in power plants, petroleum, and other oil and gas.

removing salt accumulation, and leaching of salts out of the root zone makes a significant improvement in crop growth (Lunin 1960; Mu et al. 2018; Zhang et al. 2020).

Brackish water can also be used for the greenhouse production of vegetables. Research conducted by M. Abdal and M. Suleiman at Kuwait Institute for Scientific Research in 2003 suggest vegetables like cabbage, garlic, cauliflower, and beetroot can be cultivated using brackish water (Abdal and Suleiman 2003). Their study also suggests a particular category of these vegetables which has more yield capacity (Figure 10.6).

10.4.4 RECREATION

Recreational activities cater to increasing ecotourism in the inland waterbodies and coastal areas. In addition, coastal water bodies provide opportunities for aesthetic enjoyment, recreation, sense of place, and collective identity (Conlan, White, and Hawkins 1992). Besides natural surface brackish water, treated brackish water can be used for recreation purposes such as in swimming pools, water theme parks, etc. It is imperative that the water quality standard should meet at least the World Health Organization (WHO) guidelines for safe recreational water environments (Helmer and Hespanhol 1997). According to the US EPA, the water should not have more than 235 most probable number (MPN)/100 mL for primary contact for all recreational activities that involve water (Jin et al. 2004). Similarly, the brackish water used for recreation activities should have *E. coli* with log samples of five or more in the range of around 200 MPN/100 mL and no single draw sample should ever exceed 400 MPN/100 mL (Figure 10.7).

10.4.5 SECONDARY APPLICATIONS

The direct applications of brackish water are limited. The following are some of the secondary applications of brackish water:

- Livestock watering (Only if TDS < 3000) (Anderson and Hawkes 1985)
- Amenity use (like aesthetic view purposes where salt scaling is not accountable) (Jury and Vaux 2005)

FIGURE 10.6   Usage of brackish water in agriculture/irrigation, gardening, greenhouse farming.

FIGURE 10.7   Usage of brackish water for recreational activities like swimming, diving, scuba diving, kayaking.

- Floriculture/floral industry (to grow salt-tolerant plants like daylilies and moss rose) (Cassaniti et al. 2013)
- Aquaculture and saltwater/brackish aquariums (Boyd et al. 1994)

10.5 CURRENT TREATMENT TECHNOLOGIES

Current desalination treatment methods include removing dissolved salts from brackish water, seawater, and treated wastewater. In the contemporary scenario, various treatment technologies have been developed for reducing salinity including membrane treatment, ion exchange, distillation, and so on. In this section, the major treatment technologies used in removing salts from water are discussed (Kabeel et al. 2013; Eyvaz and Yüksel 2018) (Figure 10.8).

10.5.1 THERMAL TREATMENT PROCESSES

10.5.1.1 Multistage Flash Distillation (MSFD)

MSFD is a thermal process for desalination of brackish/saline water at low pressures. The process involves the basic principle of "decrease in air pressure decreases the boiling temperature of water" (El-Dessouky and Ettouney 2002).

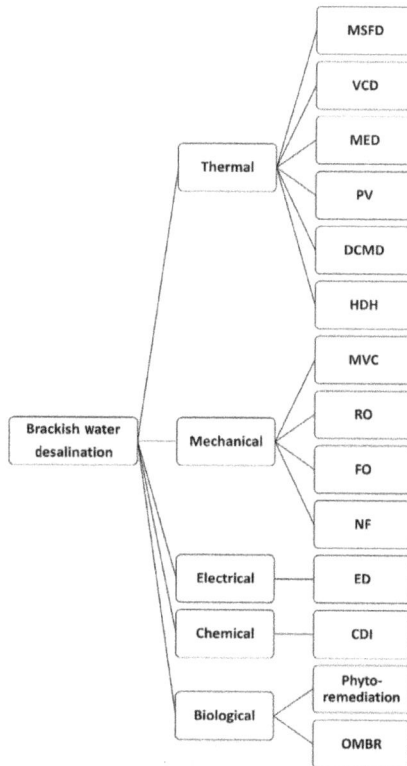

FIGURE 10.8 Technologies for brackish water desalination.

Feedwater is preheated and sent into a tank for condensation. The heat exchange process occurs in the same tank, and the condensed vapor is collected. This collected water is called the condensed treated distillate. The process is continued similarly for the multiple stages to collect condensed steam. The brine solution is collected from the untreated water. The process flow diagram of MSFD is presented in Figure 10.9 (Buschow 2001).

A multistage flash distillation system is located in Al-Jubayl, Saudi Arabia. It can produce 200×10^6 gallons/d of desalinated water (Toth 2020).

10.5.1.2 Vacuum Membrane Distillation (VCD)

VCD is a membrane-based separation process to remove volatile compounds, extraction of dissolved gasses, and also remove salts from saline water. This process has been characterized into three processes including single, binary, and multiple component transport processes. It is a promising technology in removing salts from aqueous solutions (Obotey Ezugbe and Rathilal 2020).

The preheated feedwater is evaporated and sent through a membrane that is microporous and hydrophobic. The vapor is collected from the other side of the membrane, separated, and stored in a vacuum. The mass transfer inside the membrane occurs in the vapor phase. The permeated vapor stream from the membrane apparatus is then condensed under vacuum (if any volatile organic compounds are present in the liquid, then a two-step process is preferred with recirculation) (Hussain et al. 2021). The process flow diagram of VCD is presented in Figure 10.10.

The applications of these processes consist of desalination, concentration, and extraction of organics and dissolved gases from water.

10.5.1.3 Multi-Effect Distillation (MED)

MED is a process involving repeated series of evaporator units where brackish/saline water is evaporated in multiple stages at low temperatures to generate clean distillate

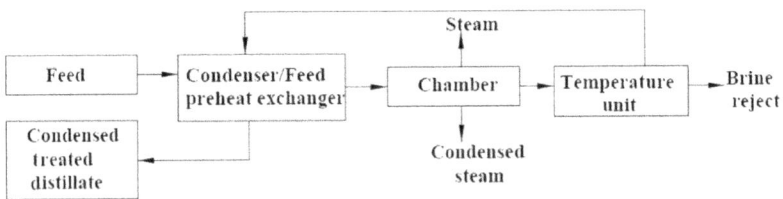

FIGURE 10.9 A simple process flow diagram of MSFD.

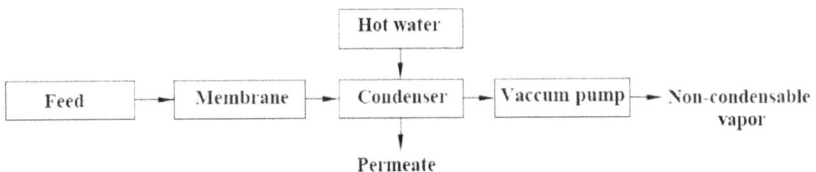

FIGURE 10.10 A simple process flow diagram of VCD.

water. Generally, temperatures are lower than 343.15 K ("Multiple Effect Distillation (MED)" 2021).

An MED evaporator consists of several consecutive units with reduced temperatures and pressures. Initially, the preheated feed brackish water is transferred into the condenser, where the vapor stage is condensed to the liquid state. The steam produced in the preheating stage is cooled using saline water (heat exchanger tubes). The exchange tubes consist of an internal heating system and external cooling system to collect the treated distilled water after condensation. Saltwater is added at the outlet of the condenser as a make-up unit. Brine solution and vapor are collected at each stage with the use of pumps (centrifugal). The vapor is transferred into the next unit until the maximum condensed distillate is collected. Finally, the untreated water is sent back to the source (Vane 2017). The process flow diagram of MED is presented in Figure 10.11. MED process can be designed to produce distilled water with steam or waste heat from power production or chemical processes, and/or to produce potable water.

There are mainly two types of MED:

- Multiple Effect Distillation with Thermal Vapor Compression (MED-TVC)
- Multiple Effect Distillation with Mechanical Vapor Compression (MED-MVC)

10.5.1.4   Pervaporative Distillation (PV)

PV is a hybrid distillation process that involves the separation of constituents in the feed based on relative solubilities in the membrane and relative boiling points. PV has a wide range of applications including desalination and is conventionally used for the separation of azeotropic liquid mixtures, bioethanol dehydration, reactive separations, etc. (Wang et al. 2020).

Initially, the feed preheated and transferred through a distillation tank where the water temperature is regulated to separate vapor. Then the vapor is transferred to a condensing tank where the vapor is cooled and transferred through a PV membrane unit (usually a cylindrical tank). The rejected water after condensing was sent to the feed tank to reuse and recirculate in the system. The temperature of the feed mixture is regularly monitored. In the PV membrane unit, at high temperature and atmospheric pressure, the feed mixture is sent directly to the hydrophilic organo-silane coated side of the membrane. A vacuum pump is used to collect the permeate in the vapor state from the other side of the membrane. The permeate in the liquid state is stored in a tank by condensing the collected vapor. Later stream analyzers were used to identify the ionic composition of the permeate (Sridhar and Moulik 2018). The process flow diagram of PV is presented in Figure 10.12.

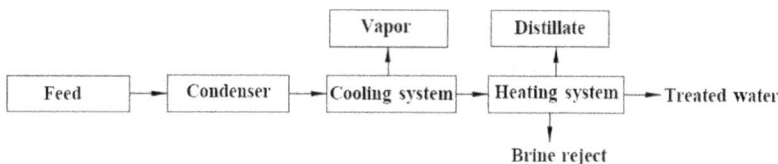

FIGURE 10.11   A simple process flow diagram of MED.

10.5.1.5 Direct Contact Membrane Distillation (DCMD)

DCMD is a thermally driven membrane process where the water vapor transfer is transferred through a microscopic membrane due to pressure and temperature difference.

The feed is heated and maintained at high temperatures. Stripping of permeate is performed using cold water. The water vapor formed from the heater water is transferred to the DCMD. The vapor passes through a membrane inside the DCMD. The vapor is condensed at the other side of the membrane (Chakraborty 2006). The process flow diagram of DCMD is presented in Figure 10.13.

10.5.1.6 Humidification and Dehumidification (HDH)

HDH is a process that involves water heating to generate the water vapor and then condensing the vapor to produce distilled water. HDH is mostly suited for water with high salinity like sea or ocean water. But brackish water can also be treated using the HDH.

The process of HDH is described as follows: Initially, the feedwater is heated by an electric heater and stored in a water tank. Hot water is pumped from the water tank to the top of the humidifier. The water is pumped into the humidifier in such a way that water is distributed throughout the humidifier. A cellulose pad is placed in the humidifier to maintain efficiency in wastewater generation. The rejected water (retentate) is collected at the bottom of the tank and sent to the electric heater to reuse in the system. The hot humid air is passed through an air duct into the dehumidifier. There is a cooling system in the dehumidifier to condense the humid air. A storage tank is set up below the dehumidifier to collect the water (Kabeel et al. 2013). The process flow diagram of HDH is presented in Figure 10.14.

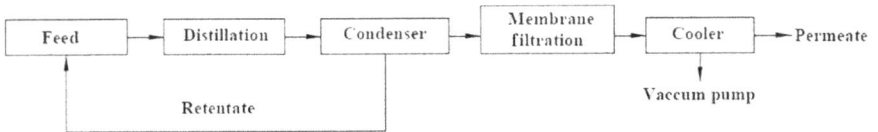

FIGURE 10.12 A simple process flow diagram of PV.

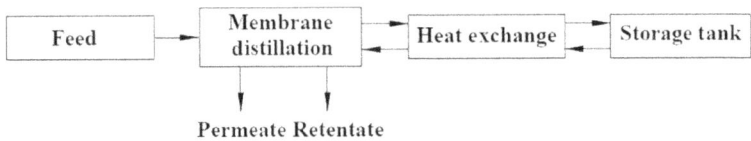

FIGURE 10.13 A simple process flow diagram of DCMD.

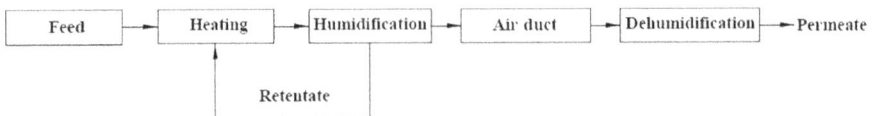

FIGURE 10.14 A simple process flow diagram of HDH.

10.5.2 Mechanical Treatment Processes

10.5.2.1 Mechanical Vapor Compression (MVC)

MVC is a process of thermal desalination involving the continuous exchange of latent heat by simultaneous evaporation and condensation. This process is similar to MED.

The vapor is generated from the feed in the evaporation unit. The water vapor generated in the evaporation unit is compressed and then condensed in heat exchange. The temperature is increased while water vapor is compressed. The evaporation unit still contains liquid while the condensation process is happening in the heat exchange unit. This is a cyclic process where the treated water is collected after multiple cycles depending on the requirement (Lara, Osunsan, and Holtzapple 2011). The process flow diagram of MVC is presented in Figure 10.15.

MVC units can be more efficient when operated combined with MED and MSFD in series.

10.5.2.2 Reverse Osmosis (RO)

The development of high-quality membranes that handles transmembrane pressure while maintaining sufficient treated water flux made the membrane treatment feasible for treating large volumes of water. Currently, membrane treatment processes are used to extract freshwater from saline ocean water.

RO separation is a method where sufficient pressure is applied against feedwater to mechanically force it through a semipermeable barrier to separate water molecules from other chemical constituents. This process separates the feedwater into desired permeate and high concentrated retentate (reject). Retentate characteristics depend on various factors including ionic charge, molecular weight, the shape of the molecule, and hydrophobicity (Alghoul et al. 2009). The process flow diagram of RO is presented in Figure 10.16. In the water quality management and treatment industry,

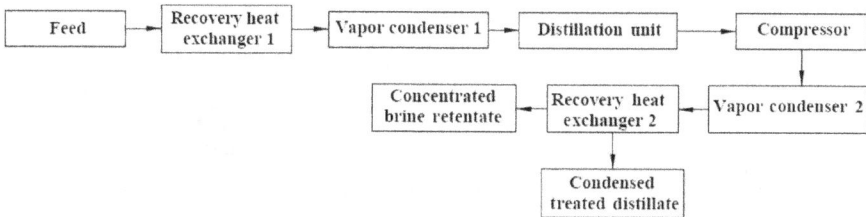

FIGURE 10.15 A simple process flow diagram of MVC.

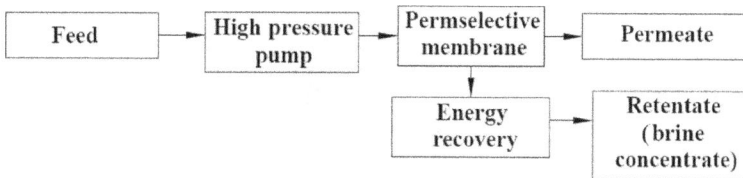

FIGURE 10.16 A simple process flow diagram of RO.

RO process is suitable for producing potable (drinking) water from brackish and salt waters. Moreover, feedwaters which are used in the RO process can vary including surface water, well water, industrially processed water, seawater, etc.

10.5.2.3 Forward Osmosis (FO)

Forward osmosis (FO) is the exact opposite of the RO process where water is moved across the semipermeable membrane from low to high concentration. Removing water from salt is easier when the salt content is at a higher concentration. In addition, maintaining flux of water for salt separation application does not require much energy when compared to reverse osmosis. Also, the draw solution can be recycled in the forward osmosis process. The process flow diagram of FO is presented in Figure 10.17. Moreover, FO plays a complementary role in the treatment of brackish water. Since the pressure required in the FO treatment process is significantly less than the RO process the rate of fouling in membrane technology can be reduced. Therefore, FO is a good alternative in areas where RO cannot produce high water quality with a limited amount of electricity and energy requirements (Cath, Childress, and Elimelech 2006). Furthermore, FO is gaining prominence in the field of wastewater treatment, water reuse, and other environmental applications.

Although FO is a promising technology, more work needs to be done in creating reliable, cost-effective solutions that require minimum treatment after recovery. Since the treatment requires less energy requirement, then renewable technologies like solar and thermal can be used during treatment.

10.5.2.4 Nanofiltration (NF)

NF is a pressure-driven membrane filtration method that uses an NF membrane (contains pores usually sized between 1 and 10 nm) that rejects salts and contaminants and allows the water to penetrate. This method has mixed properties of ultrafiltration and reverse osmosis. The order of pore sizes in membranes is given as:

Reverse osmosis < Nanofiltration < Ultrafiltration < Microfiltration.

Initially, the feedwater is pretreated and sent to the NF membrane system through a membrane system pump. The water is forced through the NF membrane with external pressure from the pump. The NF membrane has maximum treatment efficiency and provides a high-water flux by the rejection of salts and other contaminants like organic matter at low operating pressure. The permeate is collected from

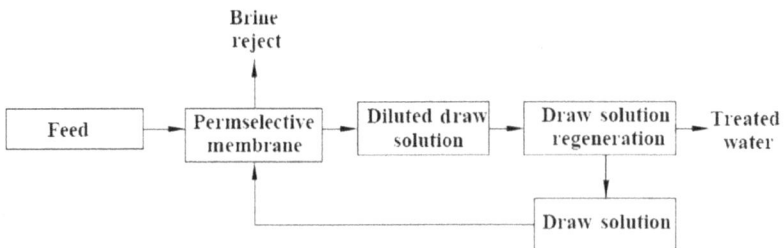

FIGURE 10.17   A simple process flow diagram of FO.

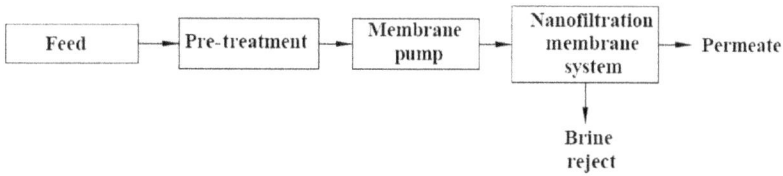

FIGURE 10.18 A simple process flow diagram of NF.

the NF membrane system and stored which can be used for culinary applications (Mohsen, Jaber, and Afonso 2003). The process flow diagram of NF is presented in Figure 10.18. The retentate brine solution is rejected and can be used for some non-culinary applications like toilets, etc. This process prevents salt scale formation in other desalination technologies.

Some of the examples of nanomembranes are NF90, NF270, and N30F (Hilal et al. 2007).

Recent studies show that NF membranes reject salts like NaCl, Na_2SO_4, $MgCl_2$, $MgSO_4$, and $CaSO_4$. Surface characterization (to determine properties like pore size, pore size distribution, and surface roughness) of the membranes can be done using atomic force microscopy (Ndlwana, Motsa, and Mamba 2020).

10.5.3 ELECTRICAL TREATMENT PROCESSES

10.5.3.1 Electrodialysis (ED)

Recently, electrodialysis (ED) has become a priority in desalination applications, especially in industrial wastewater treatment and recovery. Since the 1950s, ED has been used to remove ionic materials from aqueous solutions. In treating brackish water, ED is considered energy-intensive when compared to the ED treatment process. Besides high-energy treatment requirements, ED needs less pretreatment of feedwater compared to pressure-driven membrane processes such as ultrafiltration, nanofiltration, and RO (Tsiakis and Papageorgiou 2005). The process flow diagram of ED is presented in Figure 10.19.

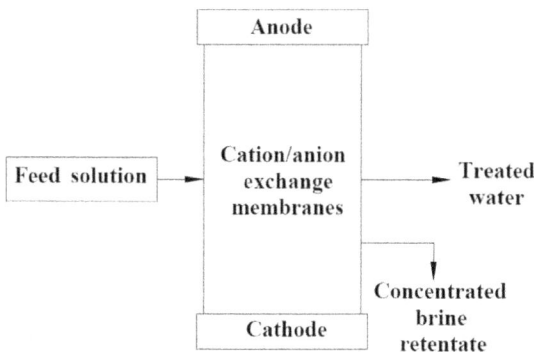

FIGURE 10.19 A simple process flow diagram of ED.

The ED treatment is able to process difficult feedwater solutions because the electrode polarity is more tolerant to concentrated solutions. Therefore, reducing pretreatment of feedwater and the option of polarity reversal have made the ED process competitive when compared to other membrane treatment processes.

10.5.4 CHEMICAL TREATMENT PROCESS

10.5.4.1 Capacitive Deionization Technology (CDI)

CDI is a process that removes charged species from water with an electrical potential difference between a pair of porous carbon electrodes. Cathode adsorbs positively charged ions and anode adsorbs negatively charged ions.

Initially, the feedwater is pretreated and sent to the mixing tank. Then the feed is transferred into the CDI system through membrane filtration. The CDI system of brackish water desalination mainly involves two phases. (i) Adsorption: In this phase, a potential difference is applied between the two carbon porous electrodes. Then there exists an attraction between the ions present in the water and electrodes. The conductivity of the cell effluent is monitored during this process. Once the maximum salt adsorption capacity (mSAC) is achieved, the conductivity of the cell effluent stabilizes. At this stage, the salts are completely removed from the feed brackish water. The change in salt concentration is measured by the difference between the inflow and outflow. This method is repeated in the batch mode until the maximum decrease in the salt concentration. (ii) Regeneration: Once the electrodes are saturated then they are regenerated. The potential difference between the electrodes is decreased to zero during regeneration. All the salt ions then leave the electrodes and exit CDI as brine solution. The treated water is collected after reaching the mSAC as the permeate (Ahmad et al. 2016). The process flow diagram of CDI is presented in Figure 10.20.

10.5.5 BIOLOGICAL TREATMENT PROCESSES

10.5.5.1 Phytoremediation

Phytoremediation is generally a process used to remove impurities like organic compounds and toxic metals in the water and soil with the direct use of cost-effective higher plants. Pollutants with very low to moderate levels of contamination can be treated with the use of phytoremediation. A selected variety of plants used in phytoremediation can also remove salts from the water. This treatment method can be used for saline water as well as brackish water (Pilon-Smits 2005). A sample phytoremediation treatment in a controlled boundary is presented in Figure 10.21.

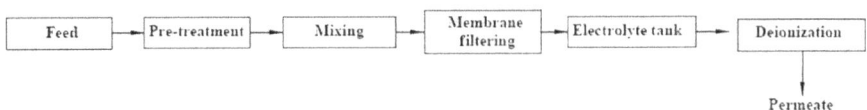

FIGURE 10.20 A simple process flow diagram of CDI.

Volatile pollutants
release through
plant

Salts/Contaminants
accumulation area
(Any harvestable
area like leaves)

Salts/Contaminants
mobilized in soil

Degradation of
contaminants occurs
in roots by
microorganisms
in soil

Salts/Contaminants
absorption area

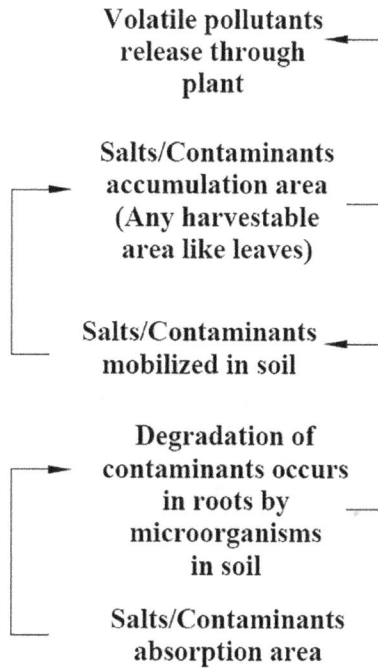

FIGURE 10.21   A sample phytoremediation treatment in a controlled boundary.

A particular variety of plants was selected to grow near a brackish water stor-age area. The plants are grown in such a way that the entire water to be treated is exposed to the roots of these plants. The plants absorb heavy metals, salts, and other organic pollutants from the roots and circulate them through the plant. Then these salts and contaminants are accumulated in plant tissues and leaves. During this process, plants convert salts and contaminants into less toxic and volatile forms and also destroy all the organic pollutants. Some of the salts and contaminants are escaped into the ambient air through atmospheric deposition (Arthur et al. 2005). Other salts and contaminants still are in plants which can be collected through fly ash after combustion of dried leaves and branches. This fly ash can be used in the manufacturing of bricks and other construction materials. The water exposed to these roots will be purified and desalinated. This treated brackish water can be used for irrigation and industrial purposes and can be used as potable water after post-treatment.

10.5.5.2   Osmotic Membrane Bioreactor (OMBR)

OMBR is a combination of FO and the biological treatment process. It is initially used for desalination. A membrane bioreactor (MBR) was created as a substitute for the clarifier with ultrafiltration/microfiltration membrane units. The membranes in MBR are the microbial barrier that apprehends the biomass to restrict its recircula-tion inside the bioreactor. An OMBR is a revised version of MBR by incorporation of FO into it.

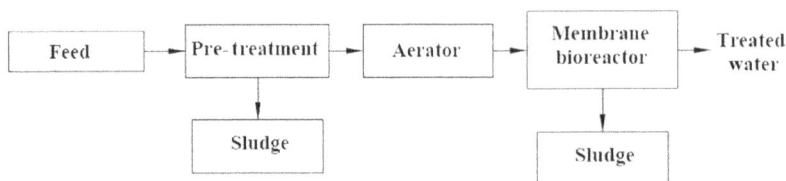

FIGURE 10.22 A simple process flow diagram of OMBR.

The feedwater is directly transferred into the OMBR membrane unit which consists of FO membranes. The saline water is added to the membrane unit. The pressure-driven FO membranes filter the feed and are transferred to the system which consists of a bioreactor. The bioreactor contains microorganisms that degrade the contaminants. Various approaches like aerobic and anerobic processes are available to enhance the microbial degradation depending on the quality of the feed (Holloway, Achilli, and Cath 2015). This is the single-step OMBR process. The treated solution is sent to the secondary membrane process (like membrane distillation) to completely desalinate the water. The process flow diagram of OMBR is presented in Figure 10.22.

10.5.6 Additional Treatment Technologies

Emerging treatment technologies make the desalination of brackish water more reliable. Some of the desalination treatment technologies that under research/partially implemented are stated as follows:

- Hybrid coagulation–ultrafiltration/NF (Ang et al. 2016)
- NF–FO–RO hybrid system (Altaee and Hilal 2015)
- Closed circuit RO (Stover 2013)
- Graphene-based filtration (Almomani et al. 2021)
- Carbon aerogel based capacitive deionization technology (Xu et al. 2008)
- Microbial desalination cell technology (Zhang and He 2013)
- Mineral recovery enhanced desalination (Thomson et al. 2020)
- Electro-sorptive de-ionization (Li et al. 2019)
- Vibratory shear enhanced process (Subramani et al. 2012)
- Carbon nanotubes (Das et al. 2014)
- Short-circuited closed-cycle operation of flow-electrode CDI (He et al. 2018)
- Photobiological treatment process (Ikehata et al. 2018)
- Electrokinetic desalination (Pan et al. 2018)
- An integrated fertilizer-driven forward osmosis-renewables powered membrane distillation system (Suwaileh et al. 2019)
- Hybrid ion-exchange–pressure-driven membrane processes (Al Abdulgader, Kochkodan, and Hilal 2013)
- Desalination using *Chlorella vulgaris* microalgae (Barahoei, Hatamipour, and Afsharzadeh 2021)
- Nuclear desalination (Al-Mutaz 2001)

10.6 SUMMARY

In the current scenario, more than 17 countries around the world in the Middle East, Africa, and South Asia are facing high-water stress problems ("International Decade for Action 'Water for Life' 2005–2015. Focus Areas: Water Scarcity" 2021). Continuous research is being conducted to identify the best alternatives to satisfy human life. A qualitative examination is performed on the current applications and treatment technologies of brackish water. Brackish water similar to seawater is a current and future alternative resource to freshwater in many areas around the world. The current treatment technologies to desalinate the seawater can be used for brackish water as well. The reason is that the salinity of the brackish water is more than the freshwater but less than the seawater.

The brackish water is undervalued, and no major applications were discovered except for limited agricultural irrigation purposes. But the severity of the freshwater depletion leads to an increase in the significance of brackish water treatment. Desalination is the fundamental treatment to increase the applications of brackish water.

Many treatment technologies to desalinate the brackish water have arrived starting initially from the basic water evaporation techniques. Every technology has its significance based on the geographic location and the availability of technical, financial, and natural resources. Some of the noteworthy processes in membrane treatment include reverse osmosis (RO) and forward osmosis (FO). RO is a widely used technology and comprises up to 60% of desalination techniques around the world (Esmaeilion 2020). The thermal desalination process including MED, MVC, and MSFD is next popular in many areas of desalination. Other membrane technologies are also viable options for the desalination of brackish water. The scope toward the improvement of CDI to make it efficient is increasing. Incorporation of renewable energy concepts like solar, tidal, and wind into the existing treatment technologies to make it energy-efficient and sustainable. A recent study conducted by Texas A&M AgriLife extension showed that a 7.5 sq ft of solar distillation unit can desalination enough water for a person to survive ("Desalination Methods for Producing Drinking Water – Source Waters" 2021). Further advancements such as water reclamation projects increase the scope of membrane technology.

The authors conclude that brackish water can be an alternative source to overcome the growing water availability concern. Besides, the future of alternative water treatment to meet the increasing demand is investigated using hybrid treatment processes (exploring synergy of thermal, mechanical, and electrical, chemical, and biological treatment processes), which is an area of interest both from scientific and engineering perspective. Based on the examination of the emerging desalination technologies and the necessity of brackish water, if the treated water meets the local and federal regulations then the applications are endless.

ACKNOWLEDGMENTS

The authors thank the University of Toledo and Oklahoma State University for providing the facilities to conduct the study. We sincerely thank icon developers at Flaticon (www.flaticon.com) for their contribution to the icons used in this book chapter. Besides, we are thankful to photographers at Pexels (www.pexels.com) for their contribution of Figure 10.2 in this book chapter.

REFERENCES

Abdal, M., and Suleiman, M. 2003. "Vegetables Production in Greenhouse Using Brackish Water." https://pubag.nal.usda.gov/catalog/302255.

Ahdab, Yvana D. and John H. Lienhard. "Desalination of brackish groundwater to improve water quality and water supply." *Global Groundwater: Source, Scarcity, Sustainability, Security and Solutions*, edited by Abhijit Mukherjee et al, Elsevier (in press).

Ahmad, Fawad, Sher Jamal Khan, Yousuf Jamal, Hussain Kamran, Aitzaz Ahsan, Muhammad Ahmad, and Amir Khan. 2016. "Desalination of Brackish Water Using Capacitive Deionization (CDI) Technology." *Desalination and Water Treatment* 57 (17). Taylor & Francis: 7659–66.

Ahuja, Satinder, Matthew C. Larsen, Jo Leslie Eimers, Craig L. Patterson, Sukalyan Sengupta, and Jerald L. Schnoor. 2014. *Comprehensive Water Quality and Purification*. Elsevier, Amsterdam.

Al Abdulgader, Hasan, Victor Kochkodan, and Nidal Hilal. 2013. "Hybrid Ion Exchange – Pressure Driven Membrane Processes in Water Treatment: A Review." *Separation and Purification Technology* 116 (September): 253–64. doi:10.1016/j.seppur.2013.05.052.

Alghoul, M.A., Poovanaesvaran, P., Sopian, K. and Sulaiman, M.Y. 2009. "Review of Brackish Water Reverse Osmosis (BWRO) System Designs." *Renewable and Sustainable Energy Reviews* 13 (9). Elsevier: 2661–67.

Almomani, Fares, Yasser Vasseghian, Jéssica Andrade Vilas-Boas, and Elena-Niculina Dragoi. 2021. "Graphene-Based Nanomaterial for Desalination of Water: A Systematic Review and Meta-Analysis." *Food and Chemical Toxicology*. 148. Elsevier: 111964.

Al-Mutaz, Ibrahim S. 2001. "Potential of Nuclear Desalination in the Arabian Gulf Countries." *Desalination* 135 (1): 187–94. doi:10.1016/S0011-9164(01)80001-1.

Altaee, Ali, and Nidal Hilal. 2015. "High Recovery Rate NF–FO–RO Hybrid System for Inland Brackish Water Treatment." *Desalination* 363. Elsevier: 19–25.

Anderson, Mark T., and Clifford L. Hawkes. 1985. "Water Chemistry of Northern Great Plains Strip Mine and Livestock Water Impoundments 1." *JAWRA Journal of the American Water Resources Association* 21 (3): 499–505. doi:10.1111/j.1752-1688.1985.tb00162.x.

Ang, W. L., Mohammad, A. W., Benamor, A., and Hilal, N. 2016. "Hybrid Coagulation–NF Membrane Processes for Brackish Water Treatment: Effect of PH and Salt/Calcium Concentration." *Desalination* 390. Elsevier: 25–32.

Arthur, Ellen L., Pamela J. Rice, Patricia J. Rice, Todd A. Anderson, Sadika M. Baladi, Keri L. D. Henderson, and Joel R. Coats. 2005. "Phytoremediation – An Overview." *Critical Reviews in Plant Sciences* 24 (2). Taylor & Francis: 109–22.

Barahoei, Malihe, Mohammad Sadegh Hatamipour, and Saeed Afsharzadeh. 2021. "Direct Brackish Water Desalination Using Chlorella Vulgaris Microalgae." *Process Safety and Environmental Protection* 148 (April): 237–48. doi:10.1016/j.psep.2020.10.006.

Boyd, Claude E., Margaret E. Tanner, Mahmoud Madkour, and Kiyoshi Masuda. 1994. "Chemical Characteristics of Bottom Soils from Freshwater and Brackishwater Aquaculture Ponds." *Journal of the World Aquaculture Society* 25 (4): 517–34. doi:10.1111/j.1749-7345.1994.tb00821.x.

Buschow, K. H. J. 2001. *Encyclopedia of Materials: Science and Technology*. Elsevier, Amsterdam.

Cassaniti, C., Romano, D., Hop, M. E. C. M., and Flowers, T. J. 2013. "Growing Floricultural Crops with Brackish Water." *Environmental and Experimental Botany* 92 (August): 165–75. doi:10.1016/j.envexpbot.2012.08.006.

Cath, Tzahi Y., Amy E. Childress, and Menachem Elimelech. 2006. "Forward Osmosis: Principles, Applications, and Recent Developments." *Journal of Membrane Science* 281 (1–2). Elsevier: 70–87.

Chakraborty, Amlan. 2006. *Direct Contact Membrane Distillation for Brackish Water Desalination: Comparison of Flat Sheet Membrane and Hollow Fiber Membrane Modules.* New Mexico State University, Las Cruces, NM.

Cloern, James E., Elizabeth A. Canuel, and David Harris. 2002. "Stable Carbon and Nitrogen Isotope Composition of Aquatic and Terrestrial Plants of the San Francisco Bay Estuarine System." *Limnology and Oceanography* 47 (3). Wiley Online Library: 713–29.

Conlan, K. White, K.N., and Hawkins, S.J. 1992. "The Hydrography and Ecology of a Re-Developed Brackish-Water Dock." *Estuarine, Coastal and Shelf Science* 35 (5). Elsevier: 435–52.

Das, Rasel, Md. Eaqub Ali, Sharifah Bee Abd Hamid, Seeram Ramakrishna, and Zaira Zaman Chowdhury. 2014. "Carbon Nanotube Membranes for Water Purification: A Bright Future in Water Desalination." *Desalination* 336 (March): 97–109. doi:10.1016/j.desal.2013.12.026.

Den Hartog, C. 1974. "Brackish-Water Classification, Its Development and Problems." *Hydrobiological Bulletin* 8 (1): 15–28. doi:10.1007/BF02254902.

"Desalination Methods for Producing Drinking Water – Source Waters." 2021. Texas A&M AgriLife Extension Service. Accessed May 1, 2021. https://agrilifeextension.tamu.edu/library/water/desalination-methods-for-producing-drinking-water/.

Dieter, Cheryl A., Molly A. Maupin, Rodney R. Caldwell, Melissa A. Harris, Tamara I. Ivahnenko, John K. Lovelace, Nancy L. Barber, and Kristin S. Linsey. 2018. *Estimated Use of Water in the United States in 2015.* US Geological Survey. Reston, VA.

El-Dessouky, Hisham T., and Hisham Mohamed Ettouney. 2002. *Fundamentals of Salt Water Desalination.* Elsevier, New York.

Esmaeilion, Farbod. 2020. "Hybrid Renewable Energy Systems for Desalination." *Applied Water Science* 10 (3): 84. doi:10.1007/s13201-020-1168-5.

Eyvaz, Murat, and Ebubekir Yüksel. 2018. *Desalination and Water Treatment.* BoD – Books on Demand, Norderstedt.

"Global Water & Wastewater Solutions | Fluence." 2021. Accessed April 30, 2021. https://www.fluencecorp.com/.

"Global Water, Wastewater & Reuse Solutions for the Municipal, Industrial and Commercial Markets – Texas Desal Association." 2016. Accessed August 19, 2020. https://www.texasdesal.com/fluence-global-water-wastewater-solutions/.

Gustafson, Chloe, Kerry Key, and Rob L. Evans. 2019. "Aquifer Systems Extending Far Offshore on the U.S. Atlantic Margin." *Scientific Reports* 9 (1). Nature Publishing Group: 8709. doi:10.1038/s41598-019-44611-7.

He, Calvin, Jinxing Ma, Changyong Zhang, Jingke Song, and T. David Waite. 2018. "Short-Circuited Closed-Cycle Operation of Flow-Electrode CDI for Brackish Water Softening." *Environmental Science & Technology* 52 (16): 9350–60. doi:10.1021/acs.est.8b02807.

Helmer, Richard, and Ivanildo Hespanhol. 1997. *Water Pollution Control: A Guide to the Use of Water Quality Management Principles.* CRC Press, Boca Raton, FL.

Hilal, Nidal, Habis Al-Zoubi, Naif A. Darwish, and Abdul Wahab Mohammad. 2007. "Performance of Nanofiltration Membranes in the Treatment of Synthetic and Real Seawater." *Separation Science and Technology* 42 (3). Taylor & Francis: 493–515.

Holloway, Ryan W., Andrea Achilli, and Tzahi Y. Cath. 2015. "The Osmotic Membrane Bioreactor: A Critical Review." *Environmental Science: Water Research & Technology* 1 (5). Royal Society of Chemistry: 581–605.

Hussain, Altaf, Arnie Janson, Joel Minier Matar, and Samer Adham. 2021. "Membrane Distillation: Recent Technological Developments and Advancements in Membrane Materials." *Emergent Materials*. Springer: 1–21. doi:10.1007/s42247-020-00152-8.

Ikehata, Keisuke, Yuanyuan Zhao, Harshad V. Kulkarni, Yuan Li, Shane A. Snyder, Kenneth P. Ishida, and Michael A. Anderson. 2018. "Water Recovery from Advanced Water Purification Facility Reverse Osmosis Concentrate by Photobiological Treatment Followed by Secondary Reverse Osmosis." *Environmental Science & Technology* 52 (15): 8588–95. doi:10.1021/acs.est.8b00951.

"International Decade for Action 'Water for Life' 2005–2015. Focus Areas: Water Scarcity." 2021. Accessed May 1, 2021. https://www.un.org/waterforlifedecade/scarcity.shtml.

Jin, Guang, Andrew J. Englande, Henry Bradford, and Huei-Wang Jeng. 2004. "Comparison of E. Coli, Enterococci, and Fecal Coliform as Indicators for Brackish Water Quality Assessment." *Water Environment Research* 76 (3). Wiley Online Library: 245–55.

Jury, William A., and Henry Vaux. 2005. "The Role of Science in Solving the World's Emerging Water Problems." *Proceedings of the National Academy of Sciences* 102 (44): 15715–20. doi:10.1073/pnas.0506467102.

Kabeel, A. E., Mofreh, H. Hamed, Z., Omara, M., and Sharshir, S.W. 2013. *Water Desalination Using a Humidification-Dehumidification Technique – A Detailed Review*. Scientific Research Publishing, Wuhan.

Khadsan, R. E., and Mangesh V. Kadu. 2004. "Drinking Water Quality Analysis of Some Bore-Wells Water of Chikhli Town, Maharashtra." *I Control Pollution* 20 (1). Research and Reviews. https://www.icontrolpollution.com/peer-reviewed/drinking-water-quality-analysis-of-someborewells-water-ofchikhli-town-maharashtra-45461.html.

Lara, Jorge R., Omorinsola Osunsan, and Mark T. Holtzapple. 2011. "Advanced Mechanical Vapor-Compression Desalination System." In Schorr, Michael (ed.). *Desalination, Trends and Technologies*. IntechOpen, London: 129–48.

Li, Guiju, Wenbo Cai, Ruihua Zhao, and Linlin Hao. 2019. "Electrosorptive Removal of Salt Ions from Water by Membrane Capacitive Deionization (MCDI): Characterization, Adsorption Equilibrium, and Kinetics." *Environmental Science and Pollution Research* 26 (17). Springer: 17787–96.

Lunin, Jesse. 1960. *Use of Brackish Water for Irrigation in Humid Regions*. Agricultural Research Service, US Department of Agriculture, Washington, DC.

McMahon, Peter B., John Karl Böhlke, Katharine Dahm, David L. Parkhurst, David W. Anning, and Jennifer S. Stanton. 2016. "Chemical Considerations for an Updated National Assessment of Brackish Groundwater Resources." *Groundwater*. doi:10.1111/gwat.12367.

Mohsen, Mousa S., and Odeh R. Al-Jayyousi. 1999. "Brackish Water Desalination: An Alternative for Water Supply Enhancement in Jordan." *Desalination* 124 (1): 163–74. doi:10.1016/S0011-9164(99)00101-0.

Mohsen, Mousa S., Jamal O. Jaber, and Maria Diná Afonso. 2003. "Desalination of Brackish Water by Nanofiltration and Reverse Osmosis." *Desalination* 157 (1–3). Elsevier: 167.

Mu, Xiaoyu, Jinliang Yang, Dengliang Zhao, Yanming Feng, Hanbing Xing, and Guibin Pang. 2018. "Effect of Brackish Water Irrigation on Soil Water-Salt Distribution." *Earth and Environmental Science* 208. IOP Publishing: 012062.

"Multiple Effect Distillation (MED)." 2021. *Veolia*. Accessed April 30 , 2021. https://www.veoliawatertechnologies.com/en/solutions/technologies/multiple-effect-distillation-med.

Ndlwana, Lwazi, Mxolisi M. Motsa, and Bhekie B. Mamba. 2020. "A New Method for a Polyethersulfone-Based Dopamine-Graphene (XGnP-DA/PES) Nanocomposite Membrane in Low/Ultra-Low Pressure Reverse Osmosis (L/ULPRO) Desalination." *Membranes* 10 (12). Multidisciplinary Digital Publishing Institute: 439. doi:10.3390/membranes10120439.

Nguyen, Hoang Huu, Paul Dargusch, and Patrick Moss. 2016. "A Review of the Drivers of 200 Years of Wetland Degradation in the Mekong Delta of Vietnam." *Regional Environmental Change* 16 (8). Springer: 2303–15.

Nthunya, Lebea N., Sebabatso Maifadi, Bhekie B. Mamba, Arne R. Verliefde, and Sabelo D. Mhlanga. 2018. "Spectroscopic Determination of Water Salinity in Brackish Surface Water in Nandoni Dam, at Vhembe District, Limpopo Province, South Africa." *Water* 10 (8). Multidisciplinary Digital Publishing Institute: 990.

Obotey Ezugbe, Elorm, and Sudesh Rathilal. 2020. "Membrane Technologies in Wastewater Treatment: A Review." *Membranes* 10 (5). Multidisciplinary Digital Publishing Institute: 89.

Pan, Shu-Yuan, Seth W. Snyder, Yupo J. Lin, and Pen-Chi Chiang. 2018. "Electrokinetic Desalination of Brackish Water and Associated Challenges in the Water and Energy Nexus." *Environmental Science: Water Research & Technology* 4 (5). The Royal Society of Chemistry: 613–38. doi:10.1039/C7EW00550D.

Pilon-Smits, Elizabeth. 2005. "Phytoremediation." *Annual Review of Plant Biology* 56: 15–39.

Rheinheimer, Gerhard. 2012. *Microbial Ecology of a Brackish Water Environment.* Vol. 25. Springer Science & Business Media, Berlin.

Scruggs, Caroline E., and Bruce M. Thomson. 2017. "Opportunities and Challenges for Direct Potable Water Reuse in Arid Inland Communities." *Journal of Water Resources Planning and Management* 143 (10). American Society of Civil Engineers: 04017064.

Sridhar, Sundergopal, and Siddhartha Moulik. 2018. *Membrane Processes: Pervaporation, Vapor Permeation and Membrane Distillation for Industrial Scale Separations.* John Wiley & Sons, Hoboken, NJ.

Stanton, Jennifer S., David W. Anning, Craig J. Brown, Richard B. Moore, Virginia L. McGuire, Sharon L. Qi, Alta C. Harris, Kevin F. Dennehy, Peter B. McMahon, and James R. Degnan. 2017. *Brackish Groundwater in the United States* (2330-7102). US Geological Survey, Washington, DC.

Stover, Richard L. 2013. "Industrial and Brackish Water Treatment with Closed Circuit Reverse Osmosis." *Desalination and Water Treatment* 51 (4–6). Taylor & Francis: 1124–30.

Subramani, Arun, James DeCarolis, William Pearce, and Joseph G. Jacangelo. 2012. "Vibratory Shear Enhanced Process (VSEP) for Treating Brackish Water Reverse Osmosis Concentrate with High Silica Content." *Desalination* 291. Elsevier: 15–22.

Sun, ZeQiang, XiaoXia Dong, XueJun Wang, DongFeng Zheng, and ZhaoHui Liu. 2011. "Review of the Effect of Sprinkler-Irrigation with Brackish Water on Crops." *Zhongguo Shengtai Nongye Xuebao/Chinese Journal of Eco-Agriculture* 19 (6). Science Press: 1475–79.

Suwaileh, Wafa, Daniel Johnson, Daniel Jones, and Nidal Hilal. 2019. "An Integrated Fertilizer Driven Forward Osmosis- Renewables Powered Membrane Distillation System for Brackish Water Desalination: A Combined Experimental and Theoretical Approach." *Desalination* 471 (December): 114126. doi:10.1016/j.desal.2019.114126.

Thomson, Bruce M., Sugam Tandukar, Ayush Shahi, Carson Odell Lee, and Kerry J. Howe. 2020. "Mineral Recovery Enhanced Desalination (MRED) Process: An Innovative Technology for Desalinating Hard Brackish Water." *Desalination* 496. Elsevier: 114761.

Toth, Andras Jozsef. 2020. "Modelling and Optimisation of Multi-Stage Flash Distillation and Reverse Osmosis for Desalination of Saline Process Wastewater Sources." *Membranes* 10 (10). Multidisciplinary Digital Publishing Institute: 265.

Tsiakis, Panagiotis, and Lazaros G. Papageorgiou. 2005. "Optimal Design of an Electrodialysis Brackish Water Desalination Plant." *Desalination* 173 (2). Elsevier: 173–86.

US EPA, OW. 2014. "What Are Water Quality Standards?" *Overviews and Factsheets.* US EPA. Accessed June 24, 2021. https://www.epa.gov/standards-water-body-health/what-are-water-quality-standards.

USGS. 2021. Government webpage. What Is Brackish? Accessed April 30, 2021. https://water.usgs.gov/ogw/gwrp/brackishgw/brackish.html.

Vane, Leland M. 2017. "Water Recovery from Brines and Salt-Saturated Solutions: Operability and Thermodynamic Efficiency Considerations for Desalination Technologies." *Journal of Chemical Technology & Biotechnology* 92 (10). Wiley Online Library: 2506–18.

Vliet, Michelle T. H. van, Edward R. Jones, Martina Flörke, Wietse H. P. Franssen, Naota Hanasaki, Yoshihide Wada, and John R. Yearsley. 2021. "Global Water Scarcity Including Surface Water Quality and Expansions of Clean Water Technologies." *Environmental Research Letters* 16 (2). IOP Publishing: 024020. doi:10.1088/1748-9326/abbfc3.

Wang, Jingbo, Dian Tanuwidjaja, Subir Bhattacharjee, Arian Edalat, David Jassby, and Eric Hoek. 2020. "Produced Water Desalination via Pervaporative Distillation." *Water* 12 (12). Multidisciplinary Digital Publishing Institute: 3560.

Wang, Wenke, Jiahui Zhao, and Lei Duan. 2021. "Simulation of Irrigation-Induced Groundwater Recharge in an Arid Area of China." *Hydrogeology Journal* 29 (2): 525–40. doi:10.1007/s10040-020-02270-3.

WHO. 2021. "Home." Accessed May 1, 2021. https://www.who.int.

World Energy Council. 2016. *World Energy Resources 2016*. World Energy Council, London.

Xiaobin, Li, Kang Yaohu, and Mohamed Abdelmoneim Shehata Wahba. 2019. "A Case Study of Achieving WFE-Nexus: Drip Irrigation with Brackish Water for Urbanization and Reclamation of Coastal Saline Soil of China." *3rd World Irrigation Forum*, 1–7 September 2019, Bali, Indonesia.

Xu, Pei, Jörg E. Drewes, Dean Heil, and Gary Wang. 2008. "Treatment of Brackish Produced Water Using Carbon Aerogel-Based Capacitive Deionization Technology." *Water Research* 42 (10–11). Elsevier: 2605–17.

Zhang, Bo, and Zhen He. 2013. "Improving Water Desalination by Hydraulically Coupling an Osmotic Microbial Fuel Cell with a Microbial Desalination Cell." *Journal of Membrane Science* 441. Elsevier: 18–24.

Zhang, Zemin, Zhanyu Zhang, Peirong Lu, Genxiang Feng, and Wei Qi. 2020. "Soil Water-Salt Dynamics and Maize Growth as Affected by Cutting Length of Topsoil Incorporation Straw under Brackish Water Irrigation." *Agronomy* 10 (2). Multidisciplinary Digital Publishing Institute: 246. doi:10.3390/agronomy10020246.

11 Treatment of Pollutants from Brackish Water
Technologies and Challenges

Faizan Ahmed and Shaik Feroz
Prince Mohammad Bin Fahd University

CONTENTS

11.1   INTRODUCTION

The supply of freshwater is becoming a challenging aspect throughout the world due to increase in the growth of population, climate change, and depleting of freshwater resources. The exploitation of water resources for domestic, industrial, and irrigation is leading to a shortage of freshwater supply. Desalination of saline water is considered as a significant option to provide the freshwater supply, especially in the arid areas, where there is much scarcity of water supply. The sources of saline water are seawater, brackish water, and industrial saline wastewater. The desalination industry makes a distinction between seawater and brackish water. If the total dissolved solids (TDS) less than 1,000 mg/L, then the waters are considered as potable water. If TDS is between 1,000 and 5,000 mg/L, the waters are categorized as mildly brackish waters; if TDS is between 5,000 and 15,000 mg/L, it is called moderately brackish waters. If TDS is between 15,000 and 35,000 mg/L, it is heavily brackish waters. Seawater typically has a salt concentration in the order of 35,000 mg/L. More than 70 elements are dissolved in seawater, but only two elements (Chloride and Sodium) make up greater than 85% by weight of all the dissolved water. Seawater is a saline solution of nearly constant composition.

Brackish water contains less TDS than seawater, but more than freshwater. The TDS concentrations in brackish water can range between 1,000 to 15,000 mg/L; most brackish water environments are dynamic, and TDS levels in these environments

DOI: 10.1201/9781003185437-13

TABLE 11.1

Constituents in Typical Brackish Groundwater [2]

1. Mean Dissolved Solid (mg/L)	Dominant Constituents
2. 1,800	$NaHCO_3$–SO_4^{2-} accounting for 1/3 of anion concentration
2,500	$CaSO_4$–Na^+, Mg_2^+ each accounting for 1/4 of cation concentration
8,400	NaCl
1,400	Mixture of cations and anions with low solubility–high silica content

fluctuate spatially and temporally. Table 11.1 shows the constituents of a typical brackish water sample. Brackish water aquifers around the world are getting depleted and being polluted due to multiple problems of saltwater intrusion, soil erosion, inadequate sanitation, contamination of ground/surface water by algal blooms, detergents, fertilizers, pesticides, chemicals, heavy metals, and so forth [1].

Various advanced sustainable, robust, and cost-effective technologies are implemented to treat the pollutants present in brackish waters. All these technologies depend on factors like quality and quantity of pollutants, salinity, the kind of matrix, pH, temperature, cost, permissible limit in discharge effluents, and volume to be treated. Some techniques that are employed for the removal of pollutants from brackish water are coagulation-flocculation, ozonization, ion exchange, biodegradation, adsorption, reverse osmosis, incineration, electrodialysis, chemical degradation, phytoremediation, ultrasonic, and photocatalytic degradation (advanced oxidation processes). The main factors that need to be considered in selecting a treatment technology are the quality of wastewater, treatment options and flexibility, efficiency, economics and life cycle assessment of the treatment technology, and more importantly the environmental compatibility and the use of treated water [3].

Some technologies are expensive and may result in toxic byproducts. Each technology has its own advantages and disadvantages in terms of efficiency, reliability, environmental impact, and specific operational requirements. Due to exhaustive usage of non-renewable resources, efforts are also being made to utilize low-cost, natural, renewable, and environmentally friendly materials during the application of these technologies. Single and hybrid treatment technologies were employed for the removal of targeted pollutants or groups of pollutants from brackish water. Mostly, hybrid treatment technologies provide better removal of pollutants along with salinity and also provide opportunities to develop innovative approaches for the treatment of brackish water. This chapter discusses the technologies used for the removal of pollutants present in brackish waters; this is described in the following sections.

11.2 TECHNOLOGIES FOR REMOVAL OF POLLUTANTS FROM BRACKISH WATER

11.2.1 ADSORPTION TECHNOLOGY

Adsorption technology is extensively used for the removal of pollutants from brackish water. An adsorption process is considered simple to design, easy to use, economical, and highly effective. The criteria for the selection of suitable sorbents are the cost of the medium and running costs, ease of operation, adsorption capacity, potential for reuse, a number of useful cycles, and the possibility of regeneration. Various adsorbents such as activated carbon adsorbents, nanomaterial adsorbents, bioadsorbents, natural materials, and agricultural and industrial waste materials are used for this purpose; among them, the usage of activated carbon comes first and it is considered as a universal adsorbent. Though the adsorption process using activated carbon is simple, the manufacturing process using highly activated carbon is expensive. The adsorbent capacity depends upon the number of active sites, surface area, porosity, and the attractive forces associated with the targeted pollutant present in brackish water. Pollutants can be removed from water either by adding adsorbents directly to the contaminated water or by employing columns assembled with adsorbents where the brackish water flows through the columns. Efforts are being made to use natural materials as adsorbents because of their low cost and easy availability. Adsorbents from agricultural wastes have also shown great potential for the removal of pollutants from brackish water. The disadvantage of adsorption process is that the pollutant cannot be mineralized and also dispose of the spent adsorbent to the environment is a major concern. To counteract the detrimental effects of brackish water upon human consumption, an appropriate water treatment technology is necessary. Adsorption technique is one such technique which is used to purify water. It is a more advantageous approach owing to its cheap cost and low-energy consumption requirements [4].

Silica scaling is a well-known issue in membranes, heat exchangers, and pipelines. In brackish water, silica concentrations vary between 12 and 60 mg/L. Monomeric silica forms a hard, durable scale while colloidal silica forms a softer gel-like scale initially. Chemical adsorption is the dominant mechanism for silica elimination. Because of its porous structure and large surface area, activated alumina has been reported to be an excellent silica adsorbent [5,6]. Adsorbed silica forms a soft, gel-like coating surrounding the alumina, which may aid in adsorbent renewal. Pure graphene consists of a single sheet of carbon atoms which are linked in some kind of sp^2-bonded system. Graphene, like the bulk of other nanomaterials, is a classic example of high-surface substances, with a potential surface area of around $2,630\,m^2/g$. This is due to the exposure of each atom in a single-layer graphene sheet to the surrounding environment. The corrosion-resistant nature, big surface area, existence of oxygen-based functional groups, customizable surface chemistry, and scalability of graphene have all contributed to its popularity as an adsorption material [7]. Figure 11.1 shows the various methods of application of graphene-based materials. There is a good potential to use graphene as an adsorbent for the treatment of pollutants present in brackish water. Selective pollutants such as oil contaminates can be removed totally from the brackish water using graphene adsorbents. Graphene can also be used in the manufacturing of membranes for filtration purposes and as a photocatalytic material in advanced oxidation processes to treat recalcitrant pollutants present in brackish water.

Zeolites are naturally occurring permeable minerals with pore sizes ranging from 3 to 8. A zeolite is composed essentially of three components: an aluminosilicate structure, exchangeable cations, and zeolitic water. Zeolites are made up of

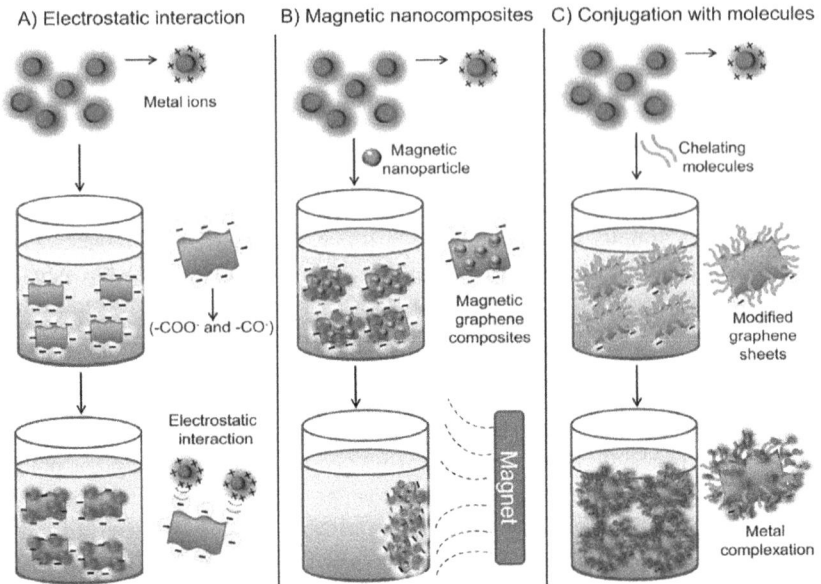

FIGURE 11.1   Methods of utilizing graphene-based materials as adsorbents (a) Electrostatic interaction, (b) Magnetic nanocomposites, (c) Conjugation with molecules. (See [7].)

TABLE 11.2
Brackish Water Treatment Using Zeolite

3. References	Feed type	4. Adsorbent	5. Sorbent	6. Performance
7. [10]	Brackish water	Zeolite	Clinoptilolite	Salinity reduction by 27%
[11]			Clinoptilolite	Salinity reduced by 99.5%
[12]			Clinoptilolite	TDS and chloride concentration reduced
[13]			Clinoptilolite	Salinity reduction by 9%
[14]			Modernite	Sodium ions reduced

tetrahedral silicate and aluminate elements that are connected together by shared oxygen atoms to form a three-dimensional network of cages and channels. The zeolite structure is negatively charged because of the valences of silicon and aluminium. The negative charge of the zeolite structure results in a strong affinity for cations and non-polar organic molecules, as well as the capacity to exchange ions. In an aqueous solution, zeolitic water may act as a bridge between framework and exchangeable cations [8,9]. Table 11.2 summarizes some selected studies on brackish water treatment using zeolite adsorbent.

Commercial activated carbon, low-cost adsorbents, and biosorbent from different agricultural wastes have the potential to treat the pollutants, especially trace heavy metals from brackish waters in many similar ways to other industrial saline wastewaters.

11.2.2 COAGULATION AND FLOCCULATION TECHNOLOGIES

Use of coagulation and flocculation is a process where an insoluble precipitant is formed after bonding between pollutants and coagulants. The removal of pollutants from brackish waters through coagulation and flocculation depends upon the nature of the coagulant, the pH of the water, chemical composition of brackish water, and concentrations of pollutants. This is a relatively simple process, although sometimes it can be expensive depending upon the quality and quantity of coagulants used. This process needs to be carefully controlled and monitored while adding sufficient coagulants required for the removal of pollutants. An excess quantity may lead to other pollution problems. The presence of various pollutants in brackish water can make the process difficult. Therefore, the selection of coagulant and optimizing the removal process of pollutants can be a tedious task. The biggest disadvantage of coagulation and flocculation processes is that they produce significant amounts of sludge which needs to be carefully disposed otherwise the re-entry of pollutants into water bodies will take place. The disposal process adds additional cost and also requires careful monitoring as per environmental regulations.

Electrocoagulation (EC) is an electrochemical-driven technology that yields coagulant species in situ from the electro-dissolution of sacrificial anodes, usually made of aluminum or iron. An EC reactor mainly consists of an electrolytic cell made of plexiglass, glass, an anode, and a cathode. The anode and cathode consist of metal plates with the anode recognized as a sacrificial electrode. The electrodes are commonly made of same or different materials. Overall, the EC process utilizes a simple tool and generates a relatively low amount of sludge and has been successfully employed to remove salts and other pollutants from brackish water. This method is characterized by its simplicity, safety, ease of control, selectivity, versatility, amenable to automation, and environmental compatibility. In one study, the EC method was employed for the removal of salts and other pollutants from brackish lake water using aluminum electrodes. The effects of current density, reaction time, pH, temperature of the brackish water, the distance between electrodes, and stirrer speed on the removal of pollutants were investigated. Using optimal conditions of process, above 90% removal efficiency was achieved [15]. Electrocoagulation creates hydrolysis products and hydrogen at the electrode, resulting in the development of contaminating flocs that are separated from the anodes by the created hydrogen. Electrocoagulation eliminates the need for coagulant chemicals. However, extracoagulant may be employed to aid in particle settling. Electrocoagulation increases the effectiveness of sludge removal and decreases the quantity of sludge formed as compared to conventional coagulation. Electrocoagulation has also been shown to be very effective in removing silica, humic acid (HA), cyanobacteria, natural organic matter (NOM), and arsenite from brackish water. Electrocoagulation's primary disadvantages are its high operating costs and the need for anode maintenance [16–20].

The coagulation-flocculation technique is often used in the treatment of brackish water because of its improved performance and cheap cost which makes the removal of pollutants more efficient. The coagulation process is linked to chemical products being added to brackish water to modify the physical condition of the dissolved and suspended substances and stimulate their removal via sedimentation [21,22]. In coagulation and flocculation processes, the choice of coagulant plays an important role in removing contaminants. There are several types of inorganic and organic coagulants. Inorganic coagulants like aluminum and iron salts were widely utilized for the treatment of wastewater. However, the inorganic coagulant-produced sludge is hazardous and created in enormous amounts and significantly alters the pH of treated water. Aluminium is also a neurotoxic substance that may contribute to Alzheimer's illness. The use of standard coagulants is thus uncertain and owing to biodegradability, not-toxicity, diversity, and availability; plant-derived polymer macromolecules known as organic coagulants are more interesting in brackish water treatment [23]. The utilization of natural coagulants obtained from plants was documented in recent literature concerning brackish water treatment. The natural coagulants investigated include tannins, *Jatropha curcas*, maize and *Moringa oleifera*. Many studies have also examined *Moringa oleifera* Lam and found it to be a promising natural coagulant. There is a high potential to use natural coagulants for the removal of pollutants from brackish water [24–30].

11.2.3 MEMBRANE, BIOLOGICAL, AND FILTRATION TECHNOLOGIES

Various microorganisms were tested for the treatment of pollutants present in brackish waters. The digestion of pollutants present in brackish water by microorganisms can be an aerobic or anaerobic process, depending upon the survival nature of the microorganisms. Usually carbon dioxide and water are the end products of digestion; however, it has also been observed to form some organic byproducts like methane. A membrane is an interface between two adjacent phases acting as a selective barrier, regulating the transport of substances between the two compartments. The main advantages of membrane technology as compared with other unit operations are related to this unique separation principle, i.e., the transport selectivity of the membrane. Separations with membranes do not require additives, and they can be performed isothermally at low temperatures and compared to other thermal separation processes at low energy consumption. Also, upscaling and downscaling of membrane processes as well as their integration into other separation or reaction processes are easy. In membrane processes, a membrane separates two phases. The membrane allows the transport of one or fewer components more readily than that of other components. The driving force for transport can be either a pressure gradient, a temperature gradient, a concentration gradient, or an electrical potential gradient. The selection of membrane having appropriate characteristics of pore size and molecular weight cutoff is very important for any membrane-based operation. Advancements in novel materials and pore structure are extensively researched for improving filter performance to achieve higher efficiency. Reverse osmosis (RO) is capable of rejecting smaller molecules, whereas nanofiltration (NF), ultrafiltration (UF), and microfiltration (MF) can remove large size molecules [31].

Figure 11.2 shows a schematic of RO process. RO is the most energy-efficient membrane method for desalination of brackish water with a greater salinity, while

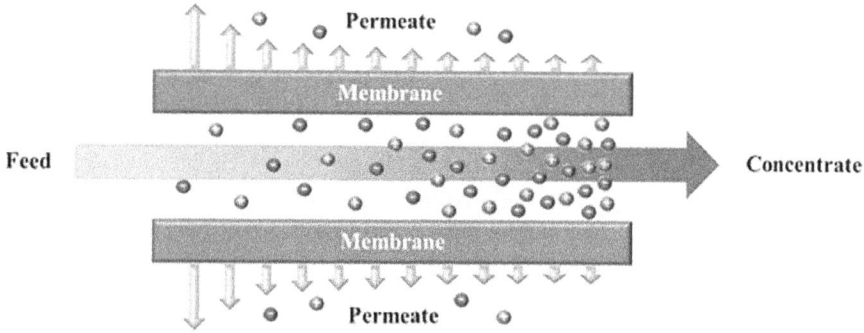

FIGURE 11.2   Reverse osmosis phenomena.

electrodialysis (ED) and membrane capacitive de-ionization (MCDI) are more energy-efficient for lower salinity ranged brackish water. NF is a suitable alternative for water softening due to its selective removal of multivalent components [32].

RO is an established technology that is well suited for large-scale treatment of brackish water with a high salinity (more than 5,000 mg/L). Water recovery is reported to be poor for current brackish water-RO systems, resulting in rather large concentrate waste discharges. Additionally, the effectiveness of energy recovery devices in RO is dependent on the salinity and volume of water. Energy recovery devices are less efficient with lower salinity input water and in small-scale desalination, reducing the amount of energy that can be recovered [33,34]. While RO has dominated present desalination procedures, alternate techniques may provide benefits for treating water with lower salinity ranges and selective ion removal. Hybrid membrane procedures are often used in conjunction with other treatment methods, including ion exchange, oxidation, coagulation, biological treatment, ion exchange, and adsorption to overcome the limits of individual methods. Membranes can be utilized as a pre-treatment, a post-treatment, or in conjunction with another unit. Following adsorption, it is difficult to separate micron-sized adsorbents from aqueous solution. Membrane filtration is a good method for overcoming this impediment. Membranes provide a number of benefits, including minimal chemical dose, good water quality, decreased sludge generation, and a minimal environmental imprint. The performance of a hybrid adsorption/membrane system is dependent on the capability and dosage of adsorbent, configuration of the reactor, quality of feed water, regulation of the quality of product water, and operating conditions such as flow rates, abrasion, flux, and membrane fouling. The primary disadvantages of hybrid membrane processes are the complexity of system design and the potential for high prices [35–39].

Figure 11.3 shows a schematic of nanofiltration process. NF has been commercialized and is generating about 3% of the global desalinated water. Due to lower operating pressure of NF compared to RO, it is a very energy-efficient approach for desalinating low-salinity waters completely or partially. Because NF is capable of removing multivalent ions, it may be used to selectively remove scale precipitating ions and soften water. This unique property makes NF a better desalination

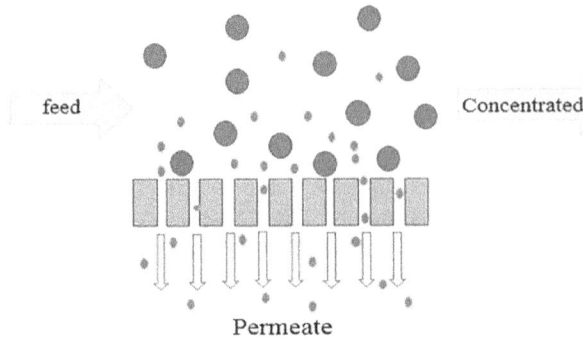

FIGURE 11.3   Nanofiltration phenomena.

solution for brackish water with a salinity of low to moderate that is dominated by multivalent ions. For brackish groundwater with salinity less than 6000 mg/L, NF is a successful method for producing freshwater with a salinity of less than 800 mg/L at a greater permeate flow than RO [40]. ED and ED-based technologies produce about 3% of the world's desalinated water volume. ED has been shown to be a cost-effective method for desalinating brackish water with salinity 1,000–10,000 mg/L. When water recovery exceeds 80%, electro dialysis reversal (EDR) requires less energy than continuous RO or semi-batch RO with the same input water salinity. The ability to operate at low pressures and with minimal pretreatment is one of the primary benefits of ED, since ion exchange membranes (IEM) are more resilient than RO membranes [41,42].

MCDI has been explored mostly at the laboratory scale, with a few reports of large-scale applications and pilot trials. However, laboratory investigations indicate that capacitive de-ionization (CDI) is energy efficient for desalination of brackish water with a salinity less than 3,000 mg/L and that MCDI uses even less energy than CDI under the same operating circumstances. These findings show that MCDI may be a viable method for desalinating brackish water with a low salinity. However, some investigations have showed that MCDI consumes more energy than ED or RO when treating brackish water under identical desalination conditions. Recent theoretical work showed that MCDI can be efficient when it operates with intermittent flow with substantial water recovery. MCDI was able to outperform RO when water recovery was set to 95%; however, fouling and scaling issues are more prevalent at this setting [43–47].

Trickling filter combined with ozonation was employed for the treatment of brackish water. The treated water was used for aquaculture ponds. After an initial conditioning period of the biofilter, biochemical oxygen demand (BOD) varied from 4.5 to 6.0 mg O_2/L, and ammonium levels were maintained at less than 1 mg/L and nitrite concentrations averaged 1 mg/L. The average efficiency (oxidation rate) of the biofilter for NH_4^+- and NO_2^--oxidation was 31% and 13.2%, respectively. The pH was stabilized slightly above 7.0 when a denitrification unit was connected to the system. Nitrate concentration of the system levelled of between 200 and 400 mg/L and was regulated by the addition of an electron donor (first glucose solution, then methanol) to the denitrification unit; the elimination rate averaged 50% with a maximum of 98%. High nitrite levels were avoided by ozone treatment of the recycled water. The accumulation of low-biodegradable substances was also successfully counteracted by ozonation [48]. If a high-quality filtration material, such as Filtralite expanded clay, is used for microbial attachment, it is possible to effectively remove both microbial energy and nutrient compounds even at low temperatures (6°C). Pilot and full-scale plant experiences from the Canadian prairies using biological filtration have advanced these treatment processes from experimental to proven technologies and potentially became "best available technology" in the treatment of extremely poor quality brackish groundwater. The first Integrated Biological and RO Treatment Plant was commissioned in December 2003, and after 2 years of full-scale testing, two more plants were commissioned in December 2005. At one of these plants, conventional manganese greensand treatment was followed by RO treatment resulting in frequent chemical RO cleanings as well as membrane replacements every 8 months. Removing the manganese greensand in the existing filters and replacing them with Filtralite material resulted in a rapid improvement of treated water quality and a literal stop to frequent RO cleanings. The biological filters need to be backwashed 36 times less than the manganese greensand filters (100 filter backwashes per year vs. 3,600). Backwash water use decreased to 0.4 million L from 23 million L and backwash labour decreased to 40 hours from 1,440 hours/year [49]. Biological membrane technologies are widely accepted technologies and commercially available for the treatment of pollutants present in brackish waters in addition to desalination.

11.2.4 ADVANCED OXIDATION TECHNOLOGIES

Advanced oxidation technologies (AOPs), both photochemical and non-photochemical such as homogenous and heterogeneous photocatalysis, photolysis, ozonation, sonication, and electrochemical oxidation are extensively employed for the degradation of various pollutants present in brackish waters. AOPs have the advantage of non-pollution of the environment with no sludge or toxic by-product formation and neither transferring the pollutants from one phase to another. Solar-driven photocatalysis is considered as greener, sustainable technology. In this technology, the photonic energy from natural sunlight is absorbed by the catalyst to produce hydroxyl radicals and superoxide. These powerful oxidative agents mineralize the organic compounds completely without any sludge formation. AOPs have attracted substantial focus in the past and are among the existing methods for bio-pollutant removal which are appropriate for post-treatment of brackish water effluents. AOPs are often suggested if the chemical oxygen demand (COD) is minimal. They are thus ideal for the treatment of effluents in wastewater. AOPs are now utilized to successfully segregate contaminants from drinking water plants and to remove bio-recalcitrant from wastewater treatment systems and disinfection techniques. Fenton-based processes are well investigated and are promising AOPs utilized in brackish water detoxification and purification. Photo-Fenton is recommended as one of the best accessible wastewater treatment systems [50–55]. Table 11.3 summarizes some selected studies on the photocatalytic removal of diclofenac pollutants.

11.3 CHALLENGES AND WAY FORWARD

Membrane-based brackish water desalination faces a number of obstacles. Membrane fouling and scaling are frequent problems with these processes, resulting in a decrease in water recovery and energy efficiency. To mitigate these harmful consequences, pretreatment processes must be designed. The surface of membranes and operating conditions must be modified. The efficiency of these treatments is contingent upon the feed water's fouling tendency. Apart from fouling and scaling, each process may confront additional difficulties based on its stage of development and maturity. The pollutants responsible for fouling and scaling should be removed before the brackish water is fed to desalination process. The effectiveness of brackish water treatment by electrodialysis (ED) is influenced by the performance of ion exchange membrane (IEM), fouling and scaling, and the cell's electrical resistance. Although membrane capacitive de-ionization (MCDI) has been examined for a variety of applications, a number of difficulties remain. Limited electrode capacitance, high electrical resistance of the cell components, fouling and scaling, and irreversible Faradaic reactions are only a few of the significant problems that must be overcome in order to progress MCDI technology. There are also certain fundamental limits in AOPs which are particularly critical in the treatment of brackish water streams containing radical species (chlorides, carbonate) and refractory pollutants [62]. As a result, AOPs themselves require reliable pretreatment. The utilization of ozone in brackish and marine systems is a difficulty, as ozone produces by-products. Moreover, ozone dose and management are hard since there are limited alternatives in ozone and total residual oxidants measurement technology [63]. It is of utmost significance to evaluate water toxicity during AOP treatment since, in most situations, more harmful than parental chemicals are generated. Maximal process efficiency with regard to COD and/or elimination of TOCs does not correlate with maximum detoxification.

Brackish water composition varies considerably, as do the quality requirements for diverse purposes. To make the most use of brackish water desalination technologies, it is necessary to define both the feed quality and the application quality criteria. Standards for drinkable water quality vary by area, with some allowing for greater salinities (500–1,000 mg/L) and others requiring stricter rules (250 mg/L). Greater treatment flexibility may be gained by combining high- and low-salinity waters to satisfy final water quality targets. Water quality requirements for industrial uses are very process-dependent and vary much more. While boilers normally need ultrapure water, water used for hydraulic fracturing in oil and gas production may have very high concentrations of non-precipitating salts as long as scale-producing elements are kept to a minimum. Brackish water needs for irrigation purposes are defined by the crops' tolerance for salt and the need to preserve soil quality. Each of the methods examined has distinct advantages and disadvantages, and no one methodology can be considered the final answer without taking

TABLE 11.3
Photocatalytic Removal of Diclofenac Contaminant in Brackish Water

8. Catalyst	Concentration (mg/L)	9. Time (minutes)	10. Catalyst Dosage (g/L)	11. pH	12. Degradation Efficiency (%)	13. References
14. ZnO-WO$_3$	10	260	0.8	6	82	[56]
Ce-Mn-TiO$_2$	10	240	0.05	6	94	[57]
Hydroxyapatite–TiO$_2$	5	1,440	4	7	99	[58]
TiO$_2$@ZnFe$_2$O$_4$/Pd	10	120	0.03	4	98	[59]
Co$_3$O$_4$/g-C$_3$N$_4$	10	120	0.5	6.7	100	[60]
Ag-Mn-TiO$_2$	10	120	0.05	7	0.05	[61]

into account the amount and quality of accessible water as well as the needed uses. Thus, in order to design the most effective fit-for-purpose treatment, the technologies for the treatment of brackish water should be chosen in accordance with the desired goals.

11.4 CONCLUDING REMARKS

There are a range of methods available for extracting freshwater from brackish inland waterways. The most suited technologies are those that can take advantage of brackish waters' low salinity and create water at a high rate of recovery to reduce concentrate production. Membrane desalination methods have been actively investigated for brackish water treatment due to their relatively low-energy needs. Desalination has become more affordable and viable over the last several decades as a result of process optimization and operation, the development of innovative membranes and electrodes, and the introduction of energy recovery devices. Membrane performance, fouling and scaling, and concentrate management, however, continue to be bottlenecks for brackish water desalination. Due to the diverse characteristics of brackish water and the disparate purification objectives, a fit-for-purpose treatment approach should be considered when selecting the appropriate treatment technique, taking into account both the volume and quality of feed and the criteria of the product water. All processes can be adapted and designed to achieve the desired level of purification. The major issues confronting processes used for brackish water treatment are addressed in this chapter, as well as current research and development attempts to enhance such technologies.

REFERENCES

1. Amin, M.T., Alazba, A.A., Manzoor, U. A review of removal of pollutants from water/wastewater using different types of nanomaterials. *Adv. Mater. Sci. Eng.* 2014, 1–25 (2014).
2. Stanton, J.S.; Anning, D.W., Brown, C.J., Moore, R.B., McGuire, V.L., Qi, S.L., Harris, A.C., Dennehy, K.F., McMahon, P.B., Degnan, J.R. *Brackish Groundwater in the United States; 2330–7102*; US Geological Survey: Reston, VA, 2017.
3. Oller, I., Malato, S., Sanchez-Perez, J. Combination of advanced oxidation processes and biological treatments for wastewater decontamination – A review. *Sci. Total Environ.* 409(20), 4141–4166 (2011).
4. Alaei Shahmirzadi, M.A., Hosseini, S.S., Luo, J., Ortiz, I. Significance, evolution and recent advances in adsorption technology, materials and processes for desalination, water softening and salt removal. *J. Environ. Manage.* 215, 324–344 (2018). doi: 10.1016/j.jenvman.2018.03.040.
5. Bouguerra, W., Ben Sik Ali, M., Hamrouni, B., Dhahbi, M. Equilibrium and kinetic studies of adsorption of silica onto activated alumina. *Desalination.* 206, 141–146 (2007).
6. Sasan, K., Brady, P.V., Krumhans. J.L., Nenoff, T.M. Exceptional selectivity for dissolved silicas in industrial waters using mixed oxides. *J. Water Pro. Eng.* 20, 187–192 (2017).
7. Perreault, F., Fonseca de Faria, A., Elimelech, M. Environmental applications of graphene-based nanomaterials. *Chem. Soc. Rev.* 44, 5861–5896 (2015).

8. Sivasankar, V., Ramachandramoorthy, T. Water softening behaviour of sand materials mimicking natural zeolites in some locations of Rameswaram Island, India. *Chem. Eng. J.* 171, 24–32 (2011).

9. Wajima, T.W., Imizu, T.S. Effect of HNO_3 and H_3PO_4 on ion exchange of natural zeolite for making agricultural cultivation solution from seawater. *Int. J. Soc. Mater. Eng. Resour.* 20, 109–112 (2014).

10. Darmawansa, M., Wahyuni, N., and Jati, D.R. Desalination of brackish water using zeolite adsorbent media in coastal areas as clean water. *J. Water Environ. Technol.* 2(1), 1–10 (2014).

11. Aziza, F.N., Kusumastuti, E. Utilization of natural zeolite activated ammonium nitrate (NH_4NO_3) to reduce salinity of brackish well water. *Indones. J. Chem. Sci.* 3(3) 233–238 (2014).

12. Kurniawan, A., Bambang, R., Susanawati, L.D. Study of the effect of HDTMA modified natural zeolite on reduction of brackish water salinity. *J. Nat. Resour. Environ.* 1, 38–46 (2014).

13. Wibowo, E., Rokhmat, M., Murniati, R., Abdullah, M. Utilization of natural zeolite as sorbent material for seawater. *Desal. Procedia Eng.* 170, 8–13 (2017).

14. Wajima, T. Ion exchange properties of Japanese natural zeolites in seawater. *Anal. Sci.* 29(1), 139–141 (2013).

15. Al-Raad, A.A., Hanafiah, M.M., Naje, A.S., Ajeel, M.A., O Basheer, A., Ali Aljayashi, T., Ekhwan Toriman, M. Treatment of saline water using electrocoagulation with combined electrical connection of electrodes. *Processes.* 7(242), 1–13 (2019).

16. Moosavirad, S.M. Treatment and operation cost analysis of greywater by electrocoagulation and comparison with coagulation process in mining areas. *Sep. Sci. Technol.* 52, 1742–1750 (2017).

17. Zhang, X., Lu, M., Idrus, M.A.M., Crombie, C., Jegatheesan, V. Performance of precipitation and electrocoagulation as pretreatment of silica removal in brackish water and seawater. *Process. Saf. Environ. Prot.* 126, 18–24 (2019).

18. De la Fuente, A., Muro-Pastor, A.M., Merchán, F., Madrid, F., Pérez-Martínez, J.I., Undabeytia, T. Electrocoagulation/flocculation of cyanobacteria from surface waters. *J. Clean. Prod.* 238, 117964 (2019).

19. Hakizimana, J.N., Gourich, B., Vial, C., Drogui, P., Oumani, A., Naja, J., Hilali, L. Assessment of hardness, microorganism and organic matter removal from seawater by electrocoagulation as a pretreatment of desalination by reverse osmosis. *Desal.* 393, 90–101 (2016).

20. Shruthi, M., Mahesh, S., Sahana, M., Srikantha, H. Arsenite removal mechanism from groundwater in a two-dimensional batch electrochemical treatment process. *Desal. Water Treat.* 146, 266–277 (2019).

21. Huang, X., Bo, X., Zhao, Y., Gao, Y., Wang, Y., Sun, S., Yue, Q., Li, Q. Effects of compound bioflocculant on coagulation performance and floc properties for dye removal. *Bioresour. Technol.* 165, 116–121 (2014).

22. Verma, A.K., Dash, R.R., Bhunia, P. A review on chemical coagulation/flocculation technologies for removal of colour from textile wastewaters. *J. Environ. Manag.* 93, 154–168 (2012).

23. Furlan, F.R., de Melo da Silva, L.G., Morgado, A.F., de Souza, A.A.U., Guelli Ulson de Souza, S.M.A. Removal of reactive dyes from aqueous solutions using combined coagulation/flocculation and adsorption on activated carbon. *Resour. Conserv. Recycl.* 54, 283–290 (2010).

24. Vijayaraghavan, G., Sivakumar, T., Kumar, A.V. Application of plant based coagulants for waste water treatment. *Int. J. Adv. Res. Technol.* 1, 88–92 (2011).

25. Hameed, Y.T., Idris, A., Hussain, S.A., Abdullah, N., Man, H.C., Suja, F. A tannin-based agent for coagulation and flocculation of municipal wastewater as a pretreatment for biofilm process. *J. Clean. Prod.* 182, 198–205 (2018).

26. Abidin, Z.Z., Mohd Shamsudin, N.S., Madehi, N., Sobri, S. Optimisation of a method to extract the active coagulant agent from Jatropha curcas seeds for use in turbidity removal. *Ind. Crop. Prod.* 41, 319–323 (2013).

27. De Paula, H.M., Ilha, M.S.O., Sarmento, A.P., Andrade, L.S. Dosage optimization of Moringa oleifera seed and traditional chemical coagulants solutions for concrete plant wastewater treatment. *J. Clean. Prod.* 174, 123–132 (2018).

28. Madrona, G., Serpelloni, G., Salcedo Vieira, A., Nishi, L., Cardoso, K., Bergamasco, R. Study of the effect of saline solution on the extraction of the Moringa oleifera seed's active component for water treatment. *Water Air Soil Pollut.* 211, 409–415 (2010).

29. Pritchard, M., Craven, T., Mkandawire, T., Edmondson, A.S., O'Neill, J.G. A comparison between Moringa oleifera and chemical coagulants in the purification of drinking water – An alternative sustainable solution for developing countries. *Phys. Chem. Earth.* 35, 798–805 (2010).

30. Mangale, M.S., Chonde, S.G., Jadhav, A.S., Raut, P.D. Study of Moringa oleifera (Drumstick) seed as natural absorbent and antimicrobial agent for River water treatment. *J. Nat. Prod. Plant Resour.* 2, 89–100 (2012).

31. Saha, P. A comprehensive review of saline effluents disposal and treatment: Conventional practices, emerging technologies and future potential. *Water Reuse.* 11(1), 33–65 (2021).

32. Honarparvar, S., Zhang, X., Chen, T., Alborzi, A., Afroz, K., Reible, D. Frontiers of membrane desalination processes for brackish water treatment: A review. *Membranes.* 11, 246 (2021). doi:10.3390/membranes11040246.

33. Qiu, T., Davies, P.A. Comparison of configurations for high-recovery inland desalination systems. *Water,* 4, 690–706 (2012).

34. Drak, A., Adato, M. Energy recovery consideration in brackish water desalination. *Desalination,* 339, 34–39 (2014).

35. Drioli, E., Stankiewicz, A.I., Macedonio, F. Membrane engineering in process intensification – An overview. *J. Membr. Sci.* 380, 1–8 (2011).

36. Hilbrandt, I., Shemer, H., Ruhl, A.R., Semiat, R., Jekel, M. Comparing fine particulate iron hydroxide adsorbents for the removal of phosphate in a hybrid adsorption/ ultrafiltration system. *Sep. Purif. Technol.* 221, 23–28 (2019).

37. Ang, W.L., Mohammad, A.W., Hilal, N., Leo, C.P. A review on the applicability of integrated/hybrid membrane processes in water treatment and desalination plants. *Desalination.* 363, 2–18 (2015).

38. Kalaruban, M., Loganathan, P., Kandasamy, J., Vigneswaran, S. Submerged membrane adsorption hybrid system using four adsorbents to remove nitrate from water. *Environ. Sci. Pollut. Res.* 25, 20328–20335 (2018).

39. Stoquart, C., Servais, P., Berube, P.R., Barbeau, B. Hybrid membrane processes using activated carbon treatment for drinking water: A review. *J. Membr. Sci.* 411–412, 1–12 (2012).

40. Yacubowicz, H., Yacubowicz, J. Nanofiltration: Properties and uses. *Filtr. Sep.* 42, 16–21 (2005).

41. Sata, T., Yamaguchi, T., Matsusaki, K. Anion exchange membranes for nitrate ion removal from groundwater by electrodialysis. *J. Chem. Soc. Chem. Commun.* 11, 1153–1154 (1995).

42. Hell, F., Lahnsteiner, J., Frischherz, H., Baumgartner, G. Experience with full-scale electrodialysis for nitrate and hardness removal. *Desalination.* 117, 173–180 (1998).

43. Volfkovich, Y.M. Capacitive deionization of water (a review). *Russ. J. Electrochem.* 56, 18–51 (2020).

44. Suss, M., Porada, S., Sun, X., Biesheuvel, P., Yoon, J., Presser, V. Water desalination via capacitive deionization: What is it and what can we expect from it? *Energy Environ. Sci.* 8, 2296–2319 (2015).

45. Zhao, Y., Wang, Y., Wang, R., Wu, Y., Xu, S., Wang, J. Performance comparison and energy consumption analysis of capacitive deionization and membrane capacitive deionization processes. *Desalination.* 324, 127–133 (2013).

46. Qin, M., Deshmukh, A., Epsztein, R., Patel, S.K., Owoseni, O.M., Walker, W.S., Elimelech, M. Comparison of energy consumption in desalination by capacitive deionization and reverse osmosis. *Desalination.* 455, 100–114 (2019).

47. Porada, S., Zhang, L., Dykstra, J. Energy consumption in membrane capacitive deionization and comparison with reverse osmosis. *Desalination.* 488, 114383 (2020).

48. Otte, G., Rosenthal, H. Management of a closed brackish water system for high density fish culture by biological and chemical water treatment. *Aquaculture.* 18(2), 169–181 (1979).

49. Peterson, H., Pratt, R., Neapetung, R., Sortehaug, O. Biological filtration of poor quality brackish water reducing reverse osmosis membrane fouling. *IDA World Congress*, Maspalomas, Gran Canaria, Spain October 21–26, 2007 REF: IDAWC/MP07-205IDA WC2007.

50. Andreozzi, R. Advanced oxidation processes (AOP) for water purification and recovery. *Catal. Today.* 53, 51–59 (1999).

51. Boczkaj, G., Fernandes, A. Wastewater treatment by means of advanced oxidation processes at basic pH conditions: A review. *Chem. Eng. J.* 320, 608–633 (2017). doi:10.1016/j.cej.2017.03.084.

52. Miklos, D.B., Remy, C., Jekel, M., Linden, K.G., Drewes, J.E., Hubner, U. Evaluation of advanced oxidation processes for water and wastewater treatment – A critical review. *Water Res.* 139, 118–131 (2018). doi:10.1016/j.watres.2018.03.042.

53. Pignatello, J.J., Oliveros, E., MacKay, A. Advanced oxidation processes for organic contaminant destruction based on the Fenton reaction and related chemistry. *Crit. Rev. Environ. Sci. Technol.* 36(1), 1–84 (2006). doi:10.1080/10643380500326564.

54. Mugunthan, E., Saidutta, M.B., Jagadeesh Babu, P.E. Photocatalytic activity of $ZnOWO_3$ for diclofenac degradation under visible light irradiation. *J. Photochem. Photobiol. A Chem.* 383, 111993 (2019).

55. Tbessi, J.I., Benito, M., Molins, E., Liorca, J., Touati, A., Sayadi, S., Najjar, W. Effect of Ce and Mn co-doping on photocatalytic performance of sol-gel TiO_2. *J. Solid State Sci.* 88, 20–28 (2019).

56. Sapia Murgolo, I.S., Moreira, C.P., Paula, M.L.C., Gianrocco, V., Claudio, M., Giuseppe, C. Photocatalytic degradation of diclofenac by hydroxyapatite–TiO_2 composite material: identification of transformation products and assessment of toxicity, *Materials.* 11, 1779 (2018).

57. Najmeh, A., Mohammad, H., Sara, S., Mahmood, H. Photocatalytic degradation of model pharmaceutical pollutant by novel magnetic $TiO_2@ZnFe_2O_4$/Pd nanocomposite with enhanced photocatalytic activity and stability under solar light irradiation. *J. Environ. Manag.* 271, 110964 (2020).

58. Shao, H., Zhao, X., Wang, Y., Mao, R., Wang, Y., Qiao, M., Zhao, S., Zhu, Y. Synergetic activation of peroxymonosulfate by CO_3O_4 modified g-C3N4 for enhanced degradation of diclofenac sodium under visible light irradiation. *Appl. Catal. B Environ.* 218, 810–818 (2017).

59. Tbessi, I., Benito, M., Llorca, J., Molins, E., Sayadi, S., Najjar, W. Silver and manganese co-doped titanium oxide aerogel for effective diclofenac degradation under UV-A light irradiation. *J. Alloys Compd.* 779, 314–325 (2019).

60. Andreozzi, R., Caprio, V., Insola, A., Marotta, R. Advanced oxidation processes (AOP) for water purification and recovery. *Catal. Today*, 53, 51–59 (1999). doi:10.1016/S0920-5861(99)00102-9.

61. Buchan, K.A., Martin-Robichaud, D.J., Benfey, T.J. Measurement of dissolved ozone in sea water: A comparison of methods. *Aquac. Eng.* 33, 225–231 (2005).

62. Beltran, J.M. Irrigation with saline water: Benefits and environmental impact. *Agric. Water Manag.* 1999, 40, 183–194.

63. Feizi, M., Hajabbasi, M.A., Mostafazadeh-Fard, B. Saline irrigation water management strategies for better yield of safflower ('Carthamus tinctorius' L.) in an Arid Region. *Aust. J. Crop Sci.* 4, 408 (2010).

Section III

Industrial Saline Wastewater

12 Application of Polymers in the Treatment of Industrial Saline Wastewater

Syed Murtuza Ali
National University of Science and Technology

Shaik Feroz
Prince Mohammad Bin Fahd University

CONTENTS

12.1 INTRODUCTION

Scarcity of clean water has become a global concern as the world is witnessing an increase in demand for fresh water due to the increase in population, urbanization and industrial processes [1]. The increasing pressure on limited freshwater resources such as groundwater and fresh surface water from lakes and rivers which would reduce the availability of potable water has led to focus on seawater desalination to meet the needs. Distillation technology was the main approach to desalination, and it was later replaced by reverse osmosis membranes. Membrane technologies emerged as a result of the breakthrough in the use of polymer films in the late 1950s to separate salts from water. In 1959, the possibility of using polymeric cellulose films for water desalination came into existence, thus creating the first polymeric RO membranes. Then, in 1963, the possibility of using asymmetric cellulose acetate membranes in desalination was explored. However, the membrane permeability was low, so the reverse osmosis technique wasn't considered as a solution to water desalination. During the past five decades, there has been another major advance where

DOI: 10.1201/9781003185437-15

233

interfacial polymerization has been relied upon to make reverse osmosis membranes. Whereas improvements in manufacturing processes and procedures have rapidly enhanced sustainability and increased the use of this type of membranes in desalination of saline water from various sources. With the rapid improvement of reverse osmosis systems and membranes and the decrease in energy use in their operations, more attention is being given to the use of solar or wind energy in desalination units. However, currently, these applications are limited only to small-scale capacities [2].

Treating oily wastewater from its various sources, such as storing, refining and transporting oil processes, has become very significant due to high usage of oil in industrial processes and petrochemical industries. Oily wastewater pollution has major negative effects, including the effect on groundwater and drinking water resources. It also has negative effects on human health, causing atmospheric pollution and affecting agricultural crops. Hence, scientific communities across the globe discuss appropriate methods for treating oil wastewater, efficient removal of oil, dissolved organic matter, suspended solids and other impurities. There are various methods of treating oily wastewater such as flotation method where the flotation is poured into the water in the form of fine bubbles. The small air bubbles that adhere to the oil particles then suspended in the water, followed by the formation of scum layer due to the difference in flotation densities of water and oil. Dissolved air flotation, impeller flotation and jet flotation are the most commonly used methods for a long time. However, there are some drawbacks of dissolved air flotation and impeller flotation methods like manufacturing, repairing problems, and high consumption of energy. In contrast, the jet flotation method solves major problems of the other two methods as well as providing safety features [3]. Flotation method can be used to remove particles of 25 mm in size and can also be used as pretreatment to remove other contaminants of 3 mm in size if coagulant is added; however, soluble oil contents cannot be removed from water. The effectiveness of flotation depends on the size of gas bubbles, when the gas bubble size is less than the size of oil droplet the effectiveness becomes more. Moreover, low temperature is preferable since it leads to dissolving gas into water stream [4].

Coagulation is the traditional technique that has been widely used in recent years to treat oily wastewater. This technology is characterized by its adaptability which makes it effective in removing emulsified and dissolved oils. Composite coagulant CAX has been developed by the oil industry to enhance oil and COD removal efficiency. One evidence is that before coagulation treatment, oil concentration was 207 mg/L and COD concentration was 600 mg/L; after using composite coagulant CAX, the percent of oil and COD decreased by 98% and 80%, respectively [3]. Membrane filtration is another oily wastewater treatment technology; membranes consist of microporous films with particular pore size in order to separate fluid components. Types of membrane treatment processes include microfiltration (MF), ultrafiltration (UF), reverse osmosis (RO), and nanofiltration (NF). UF has pore sizes between 0.01 and 0.1 mm while MF has pore sizes between 0.1 and 3 mm [4]. MF is usually used to remove suspended solids and for reducing the turbidity of oilfield-produced water. However, compared to other conventional methods, UF is the most effective way to remove oil, hydrocarbons, suspended solids and dissolved contaminants. Along with UF and MF, polymeric and ceramic membranes are used for water treatment.

Polyacrylonitrile and polyvinylidene are used to make polymeric MF/UF membranes and clays of nitrides, carbides and oxides of metals are the components of ceramic membranes. RO and NF membranes work by applying pressure to the feed solution at one side of the selective membrane which leads to diffuse through it. RO membrane has the ability to remove seawater contaminants of 0.0001 mm in size. The experimental studies showed an outstanding future of RO membrane as a pretreatment technology in oilfield water treatment.

Macro-porous polymer extraction (MPPE) is another technology for oil and gas-produced water management. Liquid– liquid extraction is the technique used where the extraction liquid freezed inside polymer particles and porosity of 60%– 70%. As the produced water enters to MPPE unit passing through a packed column with MPPE particles with a particular extraction liquid inside it, the extraction liquid starts recovering hydrocarbons. This process operates simultaneously in two columns for extraction and regeneration purposes. Macro-porous polymers were designed at first to absorb oil from water, and later it was used for water treatment. Pretreatment of oilfield-produced water through flotation methods like hydrocyclones is necessary before entering to MPPE unit. The essential purpose of this technology is to minimize the poisonous contents, and even for the treatment of produced water containing salts, methanol, glycol, dissolved heavy metals and other undesired contaminates. However, the high cost of installation, processing and energy consumption of MPPE unit are some of the challenges posed by this technology.

12.2 STUDIES INVOLVING POLYMER COMPOSITES, GRAFT CO-POLYMERIZATION, POLYMER BLENDS AND INTERPENETRATING POLYMER NETWORKS

Polymer composites can be defined as a mixture of polymers with organic and inorganic materials, and these polymers can be natural or synthetic polymers [5]. These polymer composites aim to create materials with new and desirable properties to be used for water treatment and desalination such as hardness, low density, chemical and mechanical stability and thermal behavior [6,7]. Polymer composites consist of two or more materials that are combined to form a single compound with properties higher than those of the individual components [8]. A case studies involving epoxy resin/nano TiO_2 composites and polyurethane green composites for the treatment of industrial saline wastewater (OPW) are discussed in the following sections.

12.2.1 Case Study Employing Epoxy Resin/Nano TiO_2 Composites

A case study was conducted in our research school for the treatment of industrial saline wastewaters by developing polymer-based composites. Nano photocatalyst (TiO_2) has been immobilized on a low-cost polymer (epoxy resin) to study the degradation of organics present in the oil-produced water (OPW) [9]. A series of batch experiments were conducted by varying the solution pH, stirring time, dosage, stirring speed and concentration of TiO_2. Dissolved oxygen (DO), chemical oxygen demand (COD), turbidity, conductivity and total dissolved solids (TDS) were evaluated to study the performance of the composites. The maximum percentage removal efficiency of organics was achieved between 77% and 88% at optimum operating conditions depending upon the intensity of natural sunlight. The surface morphology of composites was investigated using scanning electron microscopy. Thermal gravimetric analysis (TGA) data revealed that the incorporation of TiO_2 into the polymer matrix improved the thermal stability of the composite. An FTIR spectrum confirmed the presence of TiO_2 band peaks in the composite between wavenumbers 1,000 and 1,400 cm^{-1}, which indicates TiO_2 remains intact and responsible for photocatalytic degradation of organics. To examine the effect of pH on the pollutant degradation, the pH of the effluent varied from 5 to 8 using 0.1 N NaOH and 0.1 N HCl. The batch experiment was conducted by mixing 100 mL of the OPW sample with 8g/100 mL of 60 weight% TiO_2 composite. The stirring speed was maintained at 200 RPM for 1-hour duration. The resulting solution was analyzed for DO, COD, turbidity, conductivity and TDS. The maximum percentage reduction in COD was obtained at pH 6 and thereafter the COD removal efficiency showed a decreasing trend. The optimum percentage reductions in DO, COD, turbidity, conductivity and TDS at pH 6 were 41.3%, 74.0%, 42.1%, 36.6% and 50.2%, respectively (Figure 12.1). The charge surface of TiO_2 depends on the solution pH. The surface of the TiO_2 is positively charged under acidic conditions and negatively charged under alkaline conditions. TiO_2 is reported to have high oxidizing potential in acidic media and hence the percentage removal of the parameters was higher.

Contact time is an important factor in determining the efficiency of parameter reductions. Optimum stirring time or contact time for the effective removal of pollutants was obtained by varying the contact time from 20 to 60 minutes. Other parameters such as dosage of the photocatalyst, the pH of the solution, agitation or stirring speed and concentration of polymer composite were kept constant. Experimental results indicate that the percentage removal efficiency, increased with increase in contact time (Figure 12.2). With increase in contact time, the availability of hydroxyl radicals for the oxidation of pollutants present in OPW increases. COD reduction increased from 71% to 81% when contact time was increased from 20 to 60 minutes.

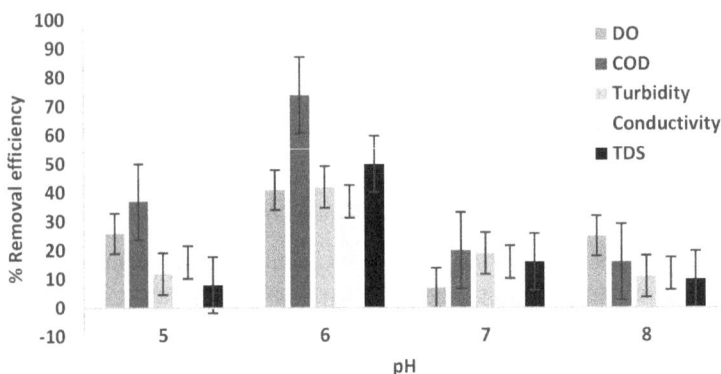

FIGURE 12.1   Effect of pH on the percentage removal efficiency (Contact time 60 minutes, Dosage 8 g/100 mL, 60 wt% composite, 200 RPM).

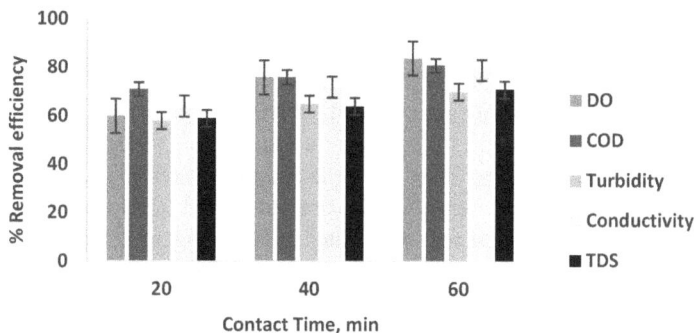

FIGURE 12.2 Effect of contact time on the percentage removal efficiency (pH 6, Dosage 8 g/100 mL, 60 wt% composite, 200 RPM).

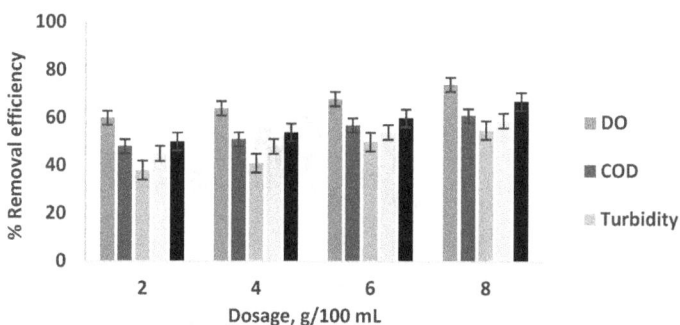

FIGURE 12.3 Effect of dosage on the percentage removal efficiency (Contact time 60 minutes, pH 6, 60 wt% composite, 200 RPM).

Loading of the photocatalyst in the matrix is a critical factor, which determines the economics of the treatment system. The efficiency in degradation of organics is generally improved with increasing the loading of photocatalyst in the matrix. This is expected because the higher the dosage of photocatalyst in the polymer matrix, the greater the availability of reaction sites on the surface of the composite. The dependence of parameter reduction in dosage was studied by varying the amount of dosage from 2 to 8 g/100 mL, and all other parameters such as stirring speed, stirring time, concentration of polymer composite and pH were kept constant. The maximum reduction in COD occurred at a dosage of 8 g/100 mL. The percentage removal of COD increased from 48% to 60%. Hence, it appears an optimum dosage of 8 g/100 mL is required for the effective degradation and removal of organic pollutants present in wastewater samples (Figure 12.3). The dosage plays an important role in the reduction of COD, DO, turbidity, TDS and conductivity. The TiO_2/epoxy showed good activity in reducing contaminants. It can be observed that as the TiO_2 content increases in the TiO_2/epoxy composites the photocatalytic activity also increases. With increase in the concentration of photocatalyst, the number of active sites will increase and leads to the enhancement of positive holes. Increase in positive holes produces a large number of hydroxyl radicals, which oxidize the pollutants present in OPW.

The effect of variation of stirring speed with percentage reduction in COD, TDS, DO, turbidity and conductivity was studied. Stirring speed was varied from 100 to 200 RPM, while keeping other parameters constant. The percentage reduction in COD increased from 80.4% to 88%, when the agitation speed increased from 100 to 200 RPM. Maximum reduction in COD occurred at 200 RPM indicating that this is the optimum stirring speed for wastewater treatment using a TiO_2 photocatalyst (Figure 12.4). With the increase in stirring speed, the mass transfer rate increases, which, in turn, increases the contact between pollutants and the TiO_2 surface. The percentage reduction of all the parameters increased, indicating a higher degradation of organic matter in OPW at 200 RPM.

The amount of TiO_2 content in the polymer composite was varied from 40 to 60 wt%, while maintaining the pH at 6, 8 g/100 mL of composite dosage, a stirring time of 1 hour and a stirring speed of 200 RPM. The maximum reduction in COD removal occurs at a concentration of 60% TiO_2 polymer composite as shown in Figure 12.5.

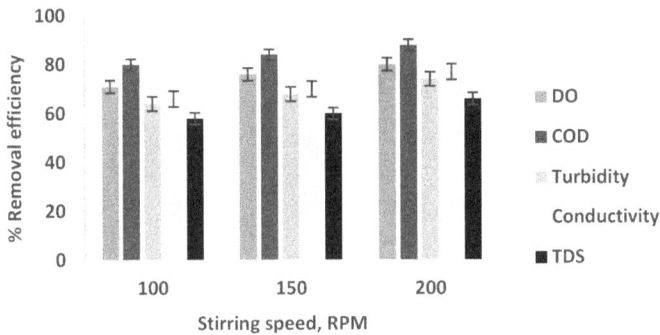

FIGURE 12.4 Effect of stirring speed on the percentage removal efficiency (Contact time 60 minutes, pH 6, Dosage 8 g/100 mL, 60 wt% composite).

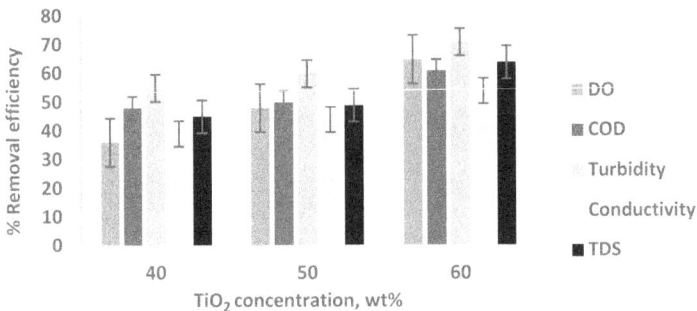

FIGURE 12.5 Effect of TiO_2 concentration on the percentage removal efficiency (Contact time 60 min, pH 6, Dosage 8 g/100 mL, 200 RPM).

The COD value increased from 48 to 61.1% with an increase in the percent of TiO_2 content in the polymer composite. With increase in photocatalyst content and corresponding to photonic efficiency, the available hydroxyl radicals increase on the surface of the composite, thereby enhancing the percentage removal efficiency. However, due to marginal variation in photonic efficiency from natural sunlight, the percentage removal efficiency was varying for the same set of experimental conditions.

The interaction between TiO_2 particles with the polymer matrix and dispersion/aggregation of TiO_2 particles was studied using SEM. SEM images for the composite (Figure 12.6a) suggest the structure of TiO_2 particles (white arrows) on and in the polymer matrix surface. For the composites, TiO_2 particles agglomeration in higher amounts is observed (Figure 12.6b). These particles are strongly attached to the surface of the polymer, and even a vigorous stirring in the OPW sample for 1 hour could not remove them from the polymer surface.

The TiO_2/epoxy composites were prepared with different amounts of TiO_2 (Degussa) and epoxy resin. Thermal Gravimetric Analyses (Figure 12.7) of the composites showed that epoxy composites without TiO_2 decompose at 450°C. The thermal degradation of the composite can be described in two stages. The first stage at a temperature range of 23°–274°C, about 10% weight loss is observed (Figure 12.7a and b), which may be attributed to water loss. The second stage at a temperature range of 274°–538°C, about 30% weight loss is observed (Figure 12.7b), which may be attributed to TiO_2 loss. This result suggests that all the TiO_2 added in the preparation is incorporated in the polymer matrix.

FTIR spectrum analyzed for neat epoxy and 60 wt% TiO_2 composites are shown in Figure 12.8a and b, respectively. Ti-O modes are clearly observed from the stretch

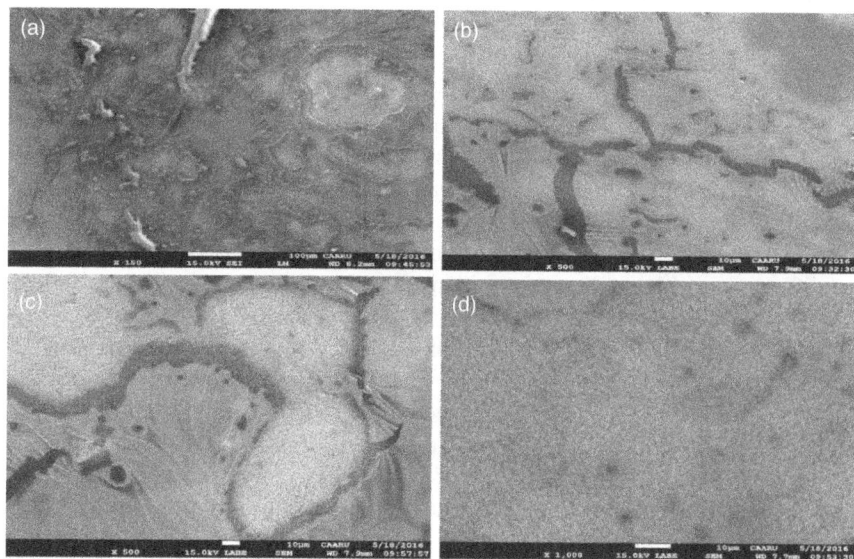

FIGURE 12.6 SEM images of TiO_2/epoxy composites, (a, b) Particles attached to the surface of polymer, (c) TiO_2 and matrix interface, (d) aggregated TiO_2 particles.

(a)   (b)

FIGURE 12.7   TGA theromograms. (a) 0% TiO_2/epoxy. (b) 60% TiO_2/epoxy composite.

of peaks between wave numbers 1,000 and 1400 cm^{-1} in Figure 12.8b. This confirms the presence of TiO_2 photocatalyst in the polymer composite.

12.2.2   CASE STUDY EMPLOYING POLYURETHANE GREEN COMPOSITES

In another case study, boron removal from OPW was carried out using polyurethane (PU)-based algae biocomposite [10]. The percentage of algae is varied in the bio-composite, and its physical and chemical properties are evaluated. The surface morphology, crystalline structure, and thermal stability are characterized using scanning electron microscope (SEM), Fourier transform infrared spectroscopy (FITR) and thermal gravimetric analysis (TGA). The density of synthesized PU/algae composites is in the range of 1.12 and 1.20 g/mL based on the content of filler in the PU matrix. Weight losses of the tested specimens in various chemical solutions are less than 10%. The boron removal efficiency is in the range of 84%–85%. Three specimens are taken from each sample to determine the average density of PU/Algae compositions. The changes in density in a single polymer sheet are due to localized differences in crystalline structure, loss of plasticizer, absorption of methyl isobutyl ketone solvent and differences in algae filler compositions. The densities of neat polymer and algae filler were estimated as 1.124 and 1.27 g/mL, respectively. The densities estimated are in the range of 1.12 and 1.20 g/mL based on the content of the filler in the PU matrix. The density increased with an increase of the filler in green composite. This may be due to the restriction of soft PU to pass through the tiny space between the rigid filler particles. The experimental results are slightly higher than theoretical results. This is due to the aggregation of filler at its higher content and maybe the filler not wetted completely in the PU matrix. The surface area of the uniform mass of specimens is decreased with an increase in the filler content. This may be due to the passage of the high density filler into the gaps of the PU matrix. Hence, the higher filler content polymer has less surface area compared to the analogous mass of less filler content polymer.

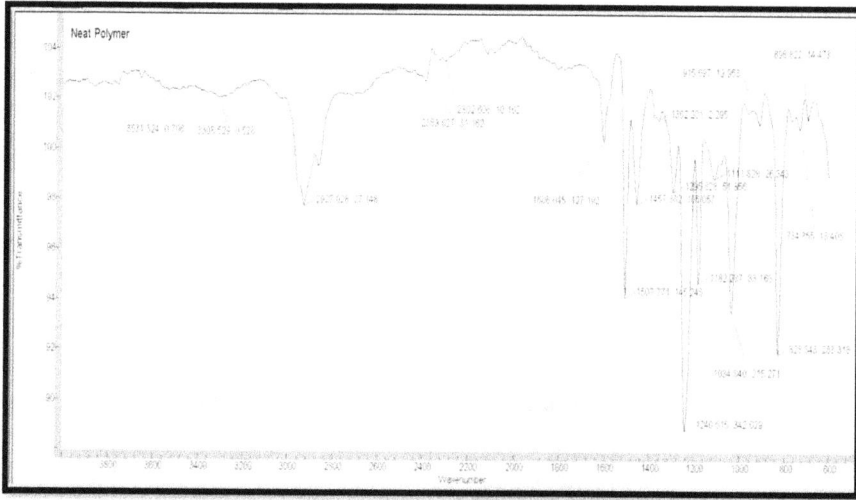

(a)

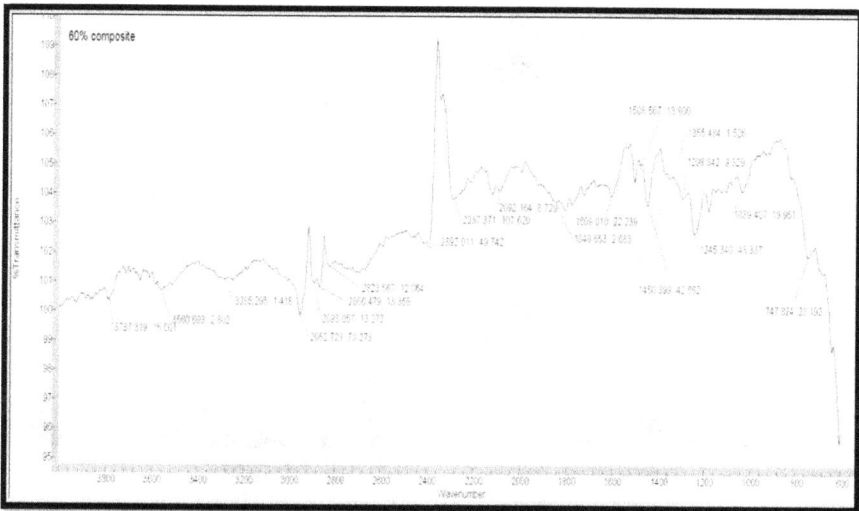

(b)

FIGURE 12.8 FTIR spectrum for (a) Neat epoxy and (b) 60 wt% TiO$_2$ polymer composites.

The chemical resistivity of synthesized PU/algae polymer composites is studied in the various chemical reagents. The chemical resistivity study helps to understand the applicable uses of the PU/algae. It shows to what extent the algae powder can stay in touch with the PU specimen under various chemical environments. The weight losses for different PU/Algae specimens are obtained after 7 days. The percentage weight losses of the tested samples in various chemical agents are within acceptable limit, which is less than 10%. Furthermore, it is observed that PU/algae composites are dissolved rapidly in the acidic and basic solution than other chemical solutions. There is no significant change in surface color and thickness of the PU/algae specimens in the various tested chemical solutions. The SEM images of different PU/algae composites are shown in Figures 12.9a–e. SEM images indicate that algae powder has been distributed and integrated properly in the PU matrix. The 5% of filler weight in the PU matrix (Figure 12.9c) is distributed uniformly compared to the other filler composites. All the SEM images showed good adhesion of the filler into PU matrix. It is observed that the roughness and deformation of the polymers increased with an increase of the algae filler content. It may be due to the restriction of soft PU to pass through the tiny space between the rigid filler particles.

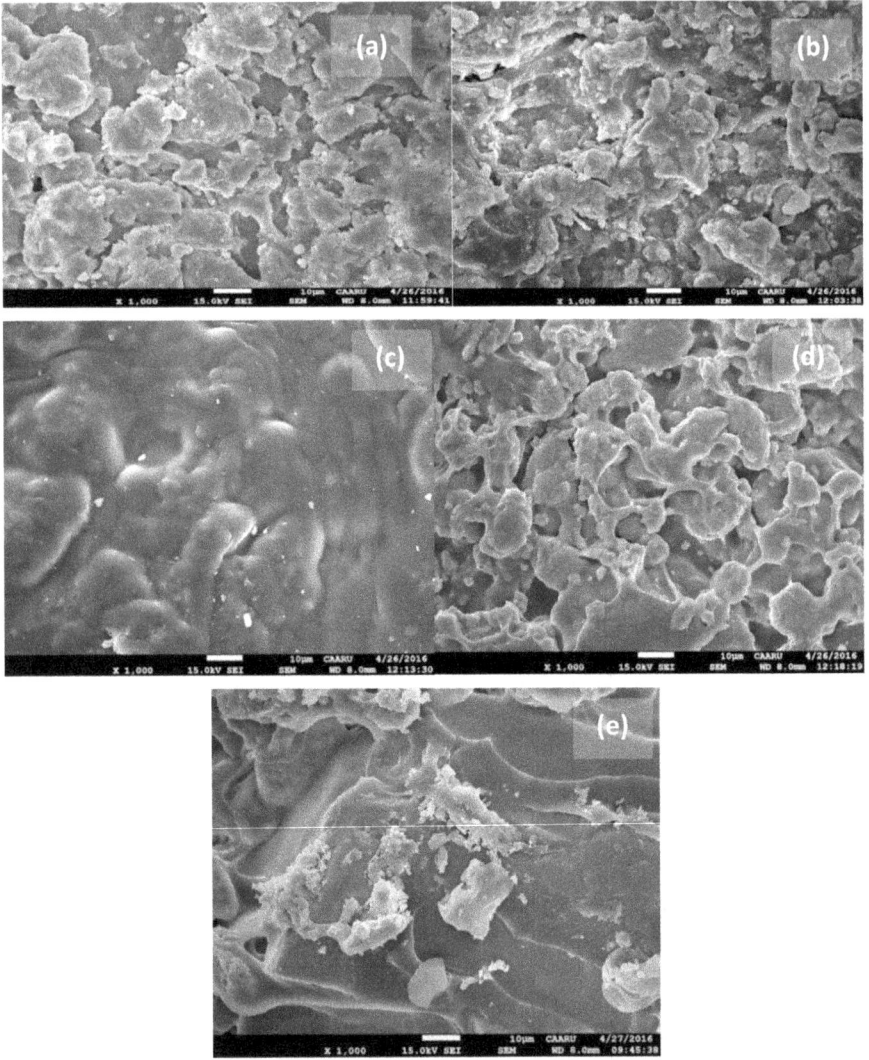

FIGURE 12.9   SEM images of PU green composites with (a) 0%, (b) 2.5%, (c) 5%, (d) 7.5% and (e) 10% of algae filler.

FTIR spectra of PU/algae composites are shown in Figure 12.10. Most of the synthesized biocomposite polymers shown the absence of the peak between 2,270 and 2,280 cm^{-1} indicating that no free NCO groups. All the isocyanate groups have reacted during the polymerization process. Furthermore, NCO group peak areas are increased slightly with an increase of algae mass ratio. This indicates the reactivity of urethane NCO group decreased with an increase in the algae green composition. OH group peaks are not observed for all the developed polymer composites as shown in Figure 12.10. Complete reaction of OH group with isocyanates groups. It is also observed in the absence of Urethane–C=O vibration in the range of 1,720–1,730 cm^{-1} for the 7.5% algae content and appearance of C=C aromatic vibration in all the synthesized PU/algae composites.

TGA is used to estimate the thermal stability of the PU/algae composites. The obtained thermograms are shown in Figure 12.11. The TGA thermograms indicated that PU matrix with different mass content of algae is stable up to 70°C–75°C. The PU/algae composite degradation can be described in three stages. First stage

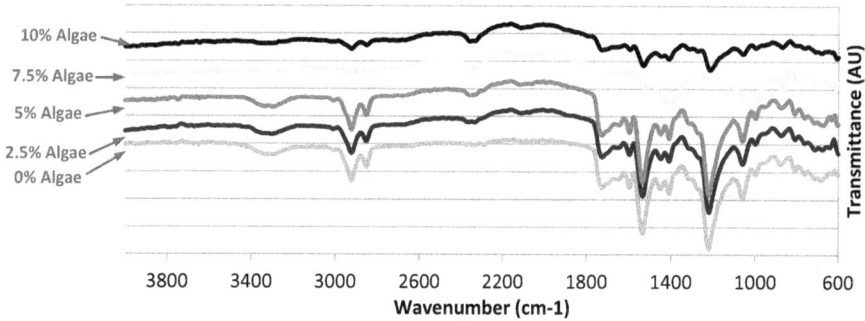

FIGURE 12.10 FTIR spectra of PU/algae composites.

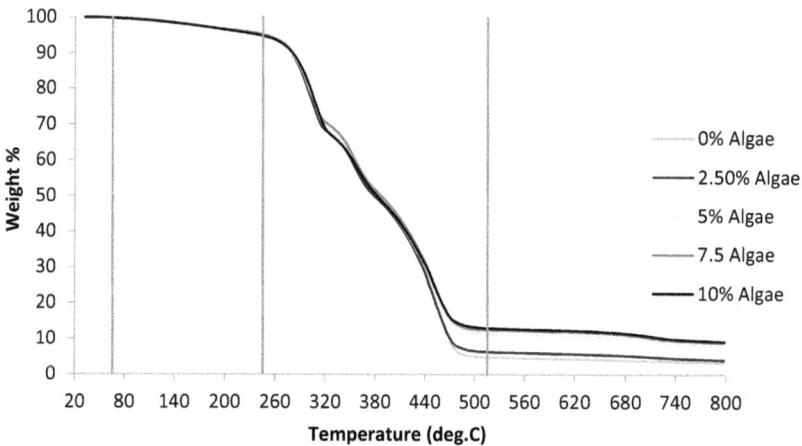

FIGURE 12.11 Thermograms of PU/algae composites.

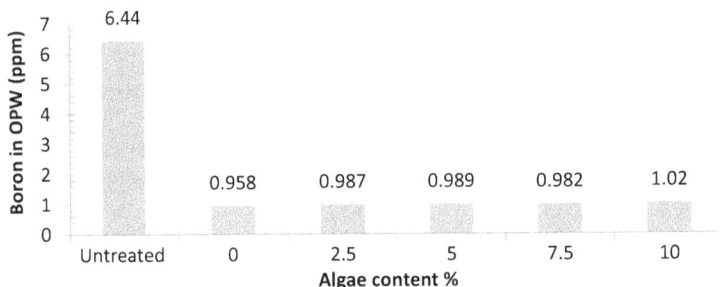

FIGURE 12.12   Boron content in OPW after adsorption on PU/algae composites.

at temperature range of 75°C–251.80°C, the sample loses about 5.29% of its initial weight due to the gradual evaporation of the methyl isobutyl ketone solvent and the slight degradation of the soft segment in the PU/algae composite. Second stage at temperature range of 251.80°C–520°C, the sample losses about 88.46% of its weight attributed to thermal degradation of the hard segment and partially degradation of the algae component. Finally, the third stage starting at temperature 520°C up to upper limit temperature considered for this study (800°C), the sample loses about 2.14% of total weight due to decomposition of the algae composite and degradation of the remains hard segment and other associated compartments. More algae content has less weight loss, which indicates that algae can act as a good filler for thermal degradation stability of PU polymer.

The relative thermal stability of the PU matrix with different algae content is estimated by comparing the decomposition temperatures at different weight loss percentages. The temperature of degradation of PU/algae composites is increased with an increase of algae content at different weight loss percentages. This lead to conclude that filler content is improving the relative thermal stability of the PU polymer. It can act as a thermal insulator due to the ability to block the volatile component during the thermal degradation analysis process.

The removal efficiency of boron is in the range of 84%–85% as shown in Figure 12.12. The maximum boron removal efficiency is observed in the case of neat PU polymer as adsorbent due to absence of filler and availability of space between PU matrix molecules. However, there is not much difference between the different PU/algae composites for boron removal efficiency. It was observed that the adsorption process is very slow and took 72 hours to reach the optimum values.

12.3   POLYMERIC MEMBRANES

Polymeric membranes have been widely used for treating oily wastewater due to their efficiency, low costs and energy consumption. Researchers in both the academic and industrial sectors are seeking to manufacture new polymers for use in the synthesis of polymeric membranes that solve previous problems. Researchers at Tehran's water refinery studied the effectiveness of polyacrylonitrile ultrafiltration membrane with a molecular weight cutoff of about 20 kDa for treating wastewater, and the end results were 99%, 30% and 100% of the oil, TDS, and TSS contents respectively were removed with a decrease in water turbidity by 98% [11]. In another study, the effectiveness of mixed cellulose esters MF membranes with a pore size of 0.1μm in treating offshore oil wastewater of a Brazilian plant, where the results showed the ability of the used membrane in reducing the amount of oil and grease content by 92% as well reducing COD and TOC content by 35% and 25%, respectively. However, it was noted that this membrane needs to be cleaned every 20 hours to avoid fouling [12]. A study was conducted to investigate the effectiveness of oily water filtration using highly hydrophilic polyethersulfone-based nanofiltration (NF) for pretreatment process and RO membranes as a final treatment, where RO membrane shown to be more effective in reducing Cl^- and SO_4^{2-} content by 77% and 81% respectively, and by 99% of F^- and S^{2-} content [13].

Graft copolymerization is a common method used to improve polymeric properties and modify their structure. Grafting is a technique used to modify the surface and performance of membranes by adding hydrophilic molecules. Radical, ionic and ring-opening polymerizations are the types that have been used in the graft polymerization of monomers onto cellulose [14]. A study was conducted by Elkony and team to create new types of grafted and cross-linked cellulose acetate (CA) reverse osmosis (RO) membranes using phase inversion mechanism. In this study, salt rejection was investigated in the treated CA-RO membranes and in the grafted/cross-linked CA-RO membranes. The results of salt rejection at 12 bar in the CA-RO membranes that grafted to 0.05%, 0.1%, 0.2% and 0.3% were 97.5%, 98.9%, 97.9% and 96.7%, respectively; while at 18 bar the salt rejections were 78.6%, 89%, 88% and 81.84%, respectively [15]. Polymer blends have been used in various fields including the development of ultrafiltration membranes for wastewater treatment. Cellulose acetate (CA) is a polymeric material that is widely used in the manufacture of ultrafiltration membranes. Despite the positives of the CA membranes, there are some defects that affect the membrane performance such as biofouling. Therefore, it has been addressed to modify CA membranes by blending with other polymers to enhance their properties and performance. Blending suitable polymers would provide the desired requirements in practical applications. For example, blending CA with the appropriate polymer may provide a higher flux and enhance membrane selectivity. Polyethersulfone (PES) is another polymer that gained wide attention in the manufacture of ultrafiltration membranes for gas separation and oil-water separation, whereas the sulfone unit gave PES distinct properties such as mechanical durability, chemical resistance and thermal stability. Moreover, the ether group provides flexible linkage in order to provide good treatment. However, PES faces a challenge when used in the aqueous phase due to its hydrophobicity in addition to its low gas permeability and low rejection rate [16].

Interpenetrating polymer networks (IPNs) can be defined as a group of cross-linked polymers that are manufactured or linked one network with another existing network [17]. This process usually aims to give the main attributes of one component while preserving the other sensitive features of the other component to obtain new and unique properties [18]. There are two types of IPNs: the type of which one network is inflated and polymerized in the presence of the other network, and this type is called a sequential IPN. As for the second type, both networks are manufactured by independent non-overlapping paths simultaneously, and this is called simultaneous

IPN [19]. Semi-IPN is produced when not all the polymers that make up the network are linear, while a homo-IPN is formed when two polymers are identical [18]. The incompatibility between the hydrophilic ion exchange resin particles and the matrix material of the inert membrane led to the leakage of resin particles from the membrane and formed the void during the electro-driven process. This leads to a rise in the electrical resistance of the membrane and a decrease in the selectivity of the ions. This, in turn, leads to decrease in the membrane's life due to their inadequacy to work under highly acidic or alkaline conditions, hence increase the process cost. To improve compatibility of heterogeneous cation exchange membranes, it has been working on the preparation of the membranes from new components such as poly(-ether sulfone) (PES), sulfonated poly(phenylene sulfide) (SPPS), sulfonated poly(ether ketone ketone) (SPEKK) and sulfonated poly(styrene-ethylene/butylene-styrene) (SSEBS), and also via restoring to physical modification techniques such as ultrasonic dispersion, the addition of inorganic fillers and coating the surface. However, these improvements will increase the cost of manufacturing the membranes. A study conducted by Lei and his colleagues for preparation semi-interpenetrating polymer network (semi-IPN) cation exchange membrane by merging the classic method of sulfonation polystyrene-based cation exchange resins with the thermoforming process of heterogeneous cation exchange membranes [20]. This technique has been proposed to reduce the cost of manufacturing heterogeneous ion-exchange membranes.

12.4 SUMMARY

Oil-produced water contains a lot of hazardous inorganic and organic contaminants which cannot be discharged directly into the environment. Pollutants include dissolved and dispersed oil components, grease, heavy metals, radionuclides, hydrofluoric chemicals, dissolved formation minerals, salts, dissolved gases, scale products, waxes, microorganisms and dissolved oxygen. The composition will vary widely as a function of geologic formation, lifetime of the reservoir and the type of hydrocarbon produced. In the oil fields, part of the produced water is reinjected for reservoir pressure maintenance. However, large amounts remain and cause a disposal problem. If the produced water is disposed of in shallow aquifers, a problem of contaminating potable water reservoirs is likely to happen. Therefore, the reuse/reclaim/recycling/recharging of treated OPW in arid areas could be a promising development. Polymer-based composites were used in the case studies for the treatment of pollutants present in the oil- produced water. Based on the outcomes of these case studies, the process for the utilization of polymer based composites especially green composites for the treatment of pollutants from oil produced water at commercial level can be explored.

REFERENCES

1. Odhiambo, G. (2016). Water scarcity in the Arabian Peninsula and socio-economic implications. *Applied Water Science, 7*(5), 2479–2492. doi:10.1007/s13201-016-0440-1
2. Kumar, M., Culp, T., & Shen, Y. (2017). Water desalination: History, advances, and challenges. *Frontiers of Engineering: Reports on Leading-Edge Engineering from the 2016 Symposium*, The National Academies Press, Washington, DC.

3. Yu, L., Han, M., & He, F. (2013). A review of treating oily wastewater. *Arabian Journal of Chemistry, 10*. doi:10.1016/j.arabjc.2013.07.020.

4. Igunnu, E. T., & Chen, G. Z. (2012). Produced water treatment technologies. *International Journal of Low-Carbon Technologies, 9*(3), 157–177. doi:10.1093/ijlct/cts049.

5. Syed, M. A., & Akheel, A. S. (2012). Development of a new inexpensive green thermoplastic composite and evaluation of its physicomechanical and wear properties. *Materials and Design, 36*, 421–427 (Elsevier).

6. Syed, M. A., Siddaramaiah, Syed, R. T., & Syed, A. A. (2010) Investigation on physico-mechanical properties, water, thermal and chemical ageing of unsaturated polyester/turmeric spent composites. *Polymer-Plastics Technology and Engineering, 49*(6), 555–559.

7. Syed, M. A., & Syed, A. A. (2016a). Development of green thermoplastic composites from centella spent and study of its physico-mechanical, tribological and morphological characteristics. *Journal of Thermoplastic Composite Materials, 29*(9), 1297–1311.

8. Syed, M. A., & Syed, A. A. (2016b). Investigation on physico-mechanical and wear properties of new green thermoplastic composites. *Polymer Composites, 37*(8), 2306–2312.

9. Syed, M. A., Mauriya, A. K., & Shaik, F. (2020a). Investigation of epoxy resin/nano-TiO_2 composites in photocatalytic degradation of organics present in oil-produced water. *International Journal of Environmental Analytical Chemistry*. doi:10.1080/03067319.2020.1784889.

10. Syed, M. A., Al Sawafi, M., Shaik, F., & Nayeemuddin, M. (2020b). Polyurethane green composites: Synthesize, characterization and treatment of boron present in the oil produced water. *International Journal of Engineering Research and Technology, 13*(8), 1866–1873. ISSN 0974-3154.

11. Salahi, A., Abbasi, M., & Mohammadi, T. (2010). Permeate flux decline during UF of oily wastewater: Experimental and modeling. *Desalination, 251*(251), 153–160.

12. Campos, J. C., Borges, R. M. H., Oliveira Filho, A. M. D., Nobrega, R., Sant'Anna, G. L. (2002). Oilfield wastewater treatment by combined microfiltration and biological processes. *Water Research, 36*(31), 95–104.

13. Alzahrani, S., Mohammad, A. W., Hilal, N., Abdullah, P., & Jaafar, O. (2013). Comparative study of NF and RO membranes in the treatment of produced water—Part I: Assessing water quality. *Desalination, 315*, 18–26.

14. Hizam, M., & Ngadi, N. (2019). Preparation of polyacrylamide grafted onto magnetic cellulose as flocculant in wastewater pre-treatment application. *IEOM Society International, International Conference on Industrial Engineering and Operations Management*, Bangkok, Thailand, 3401–3408.

15. Elkony, Y., Mansour, E., Elhusseiny, A., Hassan, H., & Ebrahim, S. (2020). Novel grafted/crosslinked cellulose acetate membrane with N-isopropylacrylamide/N, N-methylenebisacrylamide for water desalination. *Scientific Reports, 10*(1). doi:10.1038/s41598-020-67008-3.

16. Ibrahim, G. P. S., Isloor, A. M., Inamuddin, Asiri, A. M., Ismail, A. F., Kumar, R., & Ahamed, M. I. (2018). Performance intensification of the polysulfone ultrafiltration membrane by blending with copolymer encompassing novel derivative of poly(styrene-co-maleic anhydride) for heavy metal removal from wastewater. *Chemical Engineering Journal, 353*, 425–435. doi:10.1016/j.cej.2018.07.098.

17. Murtuza, A. S., Al-Shukaili, Z. S., Shaik, F. et al. (2021). Development and characterization of algae based semi-interpenetrating polymer network composite. *Arabian Journal of Science and Engineering*. doi:10.1007/s13369-021-05567-x.

18. Myung, D., Wiseman, M., Duhamel, P., & Waters, D. (2008). Progress in the development of interpenetrating polymer network hydrogels. *Polymers for Advanced Technologies, 19*(6), 647–657.

19. Murtuza, A. S., Siddaramaiah, Suresha, B., & Syed, A. A. (2009). Mechanical and abrasive wear behaviour of coleus spent filled unsaturated polyester/polymethyl methacrylate semi interpenetrating polymer network composites. *Journal of Composite Materials, 43*(21), 2387–2400.
20. Lei, Y., Luo, Y., Chen, F., & Mei, L. (2014). Sulfonation process and desalination effect of polystyrene/PVDF semi-interpenetrating polymer network cation exchange membrane. *Polymers, 6*(7), 1914–1928. doi:10.3390/polym6071914.

13 Membrane Process Technologies for Industrial Saline Wastewater Treatment and Its Applications

B. Sudeep and K. Senthil Kumar
Kongu Engineering College

V.P. Kamalakannan
Anna University

M. Venkata Ratnam
Mettu University

B. Bharathiraja
Vel Tech High Tech Engg College

CONTENTS

DOI: 10.1201/9781003185437-16

13.1 INTRODUCTION

Agricultural and industrial developments, globalization, climate change, urban development, and endless water consumption have all contributed to the exhaustion of available water supplies over the last few decades. This has necessitated the quest for new water sources all over the world [1]. When wastewater from agricultural and industrial operations, as well as domestic use of freshwater, is released into the aquatic environment without even being treated appropriately, it contains multiple micropollutants and poses severe environmental and human health threats [2,3]. Due to the population growth, escalating water requirement, and the weakening of water reservoir quality and quantity, water is one of the majority of valuable resources on the globe. The problem of water scarcity is not only a difficulty of suitable methods but also a communal and enlightening problem. Besides, effluent generation is unavoidable as it is an essential component of the value chain in most industrial sectors. Due to the natural and anthropogenic activities, different kinds of effluents and toxic substances are discharged into the environment. Out of which release of saline water is placing an important position when comparing with the industrial effluents [4]. Saline wastewater is an effluent containing a mixture of suspended salts, hardness-producing ions, organic matters, and other metals. These highly saline streams are produced mainly from different industries like tanneries and chlor-alkali, acid mine drainage, and desalination plants. For instance, in the food industry, saline waste streams are generated due to the use of brine solutions and sodium chloride salt for obtaining the finished product. Saline water generates from different sources such as ocean water, ground water, discharge from decantation industries, wastewater from STPs, salts effluent from salt lake or storage basins, and brine from salt yielding proceedings. Wastewater from agricultural fields, effluent from different salt using industries like food, pulp and paper, fertilizers, chemicals, paint, pharmaceuticals, oil and gas industries, mineral processing industries, and dyeing are also different sources for saline water generation. Micro-contaminant is a complex group of organic compounds contained in very low levels in the atmosphere (1 ngL–1 gL). Synthetic organic compounds, such as fertilizer, care products, inorganic wastes, dietary supplements, and detergents, as well as certain naturally occurring substances, such as estrogens, are among these substances [5]. Drugs and individual consideration items and steroid chemicals are not yet enrolled in the overall poisons. Pharmaceutical products and skincare products as well as steroid receptors are yet to be added to the list of general contaminants. Because of their persistent existence, these contaminants are considered to be toxic and bioaccumulative [6]. After all, in order to establish a legal limit for these substances, it is imperative to actually investigate their biological effects. Besides that, in order to recognize the true impact of these contaminants on human health and the climate, environmental conservation agencies and scientific research communities must consider their synergistic or antagonistic effects. Additionally, reasonable wastewater treatment and reusing by suitable strategies can restrict their delivery into the climate [7].

The conventional methods used to treat saline wastewater are infective in complete removal of salts due to the reason that it is only designed for the degradation of simple pollutants [8]. Besides that, WWTPs are ill-equipped to detect pollutant levels in wastewater [9]. Numerous attempts have been carried out over the years to begin different effluent treatment technologies such as filtration, coagulation, flocculation, and biological treatment systems among others. There is also the development of already accessible technologies to meet existing discharge or reuse standards. One of the wastewater treatment technologies that have seen a most important boost over this era is membrane technology. In wastewater management, membrane technology, which refers to separate chemical substances using synthetic membranes, has been accepted as the key technology for the separation of contaminants from polluted sources thus purifying salt waters. This chapter describes elaborately about various pollutants, membrane process for saline industrial wastewater treatment, and various integrated membrane processes that are being used for performing the desalination process and its applications in the treatment of saline industrial wastewater.

13.2 SOURCES OF DIFFERENT POLLUTANTS

Pollutants enter the ecosystem via several ways, but wastewater through industrial operations is one of the primary sources. Saline effluent from seafood processing units, oil, and gas production are generated offshore and then released into the sea directly without any effluent treatment. A large amount of water used at fish-processing plants eventually discharged as waste effluent, containing elevated levels of macrobiotic matter, phosphates, and nitrates. The effluent contains COD (1–32 gpL), ammonium (0.039–1.94 gpL), and salinity (1.3%–3.9% (w/v) sodium chloride). In fish farming, COD of the farm sludge is reported as higher (193 gpL). The BOD from textile effluent is a major part of the total BOD effluents in East Asian nations and varies from 15.47% in China, 24.71% in Cambodia, 35.4% in Thailand, and 64.18% in Bangladesh. Since around 50% of dyes are not fixed in fabrics, wastewaters from the dyeing process contain concentrated saline pollutants and are discharged in the effluent. When diluted with washing water, the usual salt concentration in wastewater is 2–3 gpL. Dyes effluents has a high salinity and alkalinity which may demolish the ecosystem, where the wastewater is discharged. The saltwater discharged from desalination plants includes extremely concentrated salts (>5 weight%), pretreatment chemicals, and heavy metals resulting from corrosion of the tubes and equipment. Over 100,000 commercially registered chemicals already exist in Europe and Asia, and there is an insufficient disposal location to block the residues from entering the water cycle. Primary pollutants are found in a wide range of items and processes used in modern life, including pharmaceuticals, textile clothes, pesticides, and fertilizers, as shown in Table 13.1. When these compounds are carcinogenic, pervasive, and bioaccumulative, they become a hazard. They are released in water from a wide range of production facilities. Community treatment facilities serve as unified collecting sites for industrial saline wastewater, rainwater, and household sewage, and are a target for regulators looking to reduce harmful compounds' impact on the environment.

TABLE 13.1
Various Pollutants' Sources

Category	Containment	Utilization	Surface water	Effluent
Disinfectants,	Atenolol	β-Blocker	205	843
pharmaceuticals	Azithromycin	Antibiotic	12	175
(prescriptions, over-the-	Diatrizoate (amidotrizoic	Contrast medium	206	598
counter drugs, veterinary	acid)			
drugs) [10,11]	Diclofenac	Analgesic	65	647
	Ethinylestradiol	Synthetic estrogen	5	2
	Sotalol	β-Blocker	63	435
	Sulfamethoxazole	Antibiotic	26	238
	Trimethoprim	Antibiotic	13	100
	Penicillin V	Personal care product	–	28.7
	Irbesartan	Antihypertensives	–	479.5
	Tramadol	Analgesics	–	255.8
	Risperidone	Neuroleptics	–	6.9
	Venlafaxine	Antidepressant	–	118.9
	Codeine	Morphine derivatives	–	70.6
	Diphenhydramine	Antihistamine	–	11.7
	Repaglinide	Antidiabetic medications	–	3.1
	Flecainide	Antiarrhythmic agents	–	45.5
	Bisoprolol	β-Blockers	–	41.6
	Alfuzosin	α-Blockers	–	2.8
	Bupropion	Antidepressant	–	1
	Ciprofloxacin	Antibiotics	–	96.3
	Oxazepam	Anxiolytics	–	161.7
	Carbamazepine	Antiepileptic drugs	–	832.3
	Haloperidol	Psychiatric medication	–	32.2
	Citalopram	Antidepressant	–	33.8
	Sulfamethoxazole (JRC)	Antibiotics	–	142.3
Detergents, dishwashing	Methylbenzotriazole	Personal care product	–	2900
liquids, personal care	Gadolinium	Personal care product	–	115
products (fragrances,	Loperamide	Personal care product	–	29.3
cosmetics, sunscreens), and	Miconazole	Personal care product	–	0.2
food products [11]	Chlorpromazine	Personal care product	–	0.1
	Acesulfame	Food additive	4010	22500
	Sucralose	Food additive	540	4600
Pesticides [10]	Diazinon	Insecticide	15	173
	Diethyltoluamide (DEET)	Insecticide	135	593
	Dimethoate	Insecticide	22	–
	MCPA	Insecticides	–	149.9
	Carbaryl	Insecticide	–	1.6

(Continued)

TABLE 13.1 (*Continued*)
Various Pollutants' Sources

Category	Containment	Utilization	Concentration (ng/L) [10,11]	
			Surface water	Effluent
Biocides [10]	2,4-D	Herbicide	67	13
	Carbendazim	Fungicide	16	81
	Diuron	Herbicide	54	201
	Glyphosate	Herbicide	373	–
	Irgarol (cybutryne)	Herbicide	3	30
	Isoproturon	Herbicide	315	12
	MCPA	Herbicide	40	25
	Mecoprop-P	Herbicide	45	424
	Triclosan	Microbiocide	20	116
	Terbutylazine-desethyl	Herbicide	–	68.8
	Simazine	Herbicide	–	26.3
	Atrazine-desethyl	Herbicide	–	13.8
	Chlortoluron	Herbicide	–	3.2
	Hexazinone	Herbicide	–	0.8
	Linuron	Herbicide	–	40.1
	2,4,5-T	Herbicide	–	0.3
Hormone active substances (effect on the hormone balance) [10]	Bisphenol A (BPA)	Additive	840	331
	Estradiol	Natural estrogens	2	3
	Estrone	Natural estrogens	2	15
	Nonylphenol	Additive	441	267
	Perfluorooctanesulfonate (PFOS)	Tenside	–	–

13.3 MEMBRANE PROCESSES FOR INDUSTRIAL SALINE WASTEWATER TREATMENT

Water treatment and purification membrane filtration systems include reverse osmosis (RO), nanofiltration (NF), ultrafiltration (UF), microfiltration (MF), and electrodialysis (ED), which are being used for saline treatment. A membrane is a semipermeable barrier that prevents two homogeneous fluid phases. The semipermeable barrier selectively allows the passage of certain components, but not others. The permeate side is rich with the permissible compound (water in general) while the least permissible compound is enriched with reject (salt) side. The efficiency of the membrane is dependent on a number of factors, including membrane selectiveness, influent flux, tensile properties, and chemical and thermal strength of the membrane and its module. The membrane must be very permeable and highly selective in order to be effective in a membrane process. For liquid separation, both hydrophilic and hydrophobic characteristics should be incorporated into the membrane. Not all membrane is suitable for the treatment of effluent as it depends on the constituent and the separation mechanism of the membrane. For instance, the permeate flow rate would be 70%–80% lower if UF is used as wastewater compared to filtered water, due to the accumulation (i.e. polarized concentration (PC)) of solutes on the membrane surface rejections. The inherent membrane properties and actual membrane performance so clearly differentiate. The intrinsic membrane properties and actual membrane performance are therefore clearly differentiated. The Loebe Sourirajan (L-S) membrane [12,13] is now used as the phase inversion method. This phase inversion method would give an asymmetric module (anisotropic).

In 1964, Riley investigated the membranes using an electron microscope, which was then shown to be skinned [14]. The next big advance was the creation of a new type of membrane material by Cadotte and Rozelle [14] in the early 1970s, known as the Thin-Film Composition (TFC) RO membrane, which was fabricated by depositing a thin polymer over a finely porous substrate using an inotropic gelation. The driving force for the microfiltration is the pore size of the particulate, whereas for ultrafiltration and reverse osmosis the transmembrane pressure is the pith for separation. The general particle rejects in micro, ultra, nano, and RO is >1 μm, >0.1 μm, >0.01 μm, and >0.001 μm respectively. Cadotte also created high-flux, high rejecting TFC membranes with a high permeability of divalent ions to watery chloride solutions, in 1976. In 1984, Petersen named these loose RO membranes the membrane 'Nanofiltration.' Among all methods available today, RO is obtaining universal recognition in both water treatment and desalination applications. RO membranes can be used to eliminate salinity and dissolved organic matter, while reducing total organic carbon (TOC), chemical oxygen demand (COD), and biological oxygen demand (BOD). The mass transfer in RO is maybe due to the solution-diffusion method, size elimination, charge elimination, and physio-chemical relations between solute, solvent, and the membrane [15].

13.4 MEMBRANE FOULING

Membrane systems can be configured to function in either constant permeate flux with transmembrane pressure (TMP) or constant TMP with variable permeate flux. The first is the typical mode. Membranes are classified as low-pressure membranes (LPMs) or high-pressure membranes (HPMs) based on the operating TMP. LPMs include microfiltration (MF) as well as weak ultrafiltration and tight UF, nanofiltration (NF), and reverse osmosis (RO) membranes. Membrane fouling happens while an ascend of the TMP is to keep the specific flux or when the system is operated at invariant pressure during the reduction of the flux. Membrane fouling can be divided into two categories: reversible fouling and irreversible fouling, with the distinction depending largely on the environment in which membranes are used and washed.

Fouling (both back washable and non-back washable) happens at the membrane rejection surface due to the cake deposition or concentration polarization of materials. Backwashing or hydrodynamic scrubbing (surface washing) can restore a membrane with back washable reversible fouling, whereas chemical cleaning can only remove non-back washable reversible fouling. Chemisorption and porosity blocking processes cause irreversible fouling. The decrease in transmembrane flow cannot be recovered hydrodynamically or chemically in the case of irreversible fouling. This necessitates substantial chemical cleaning or replacement of the membranes. Membrane fouling is induced by complicated mechanical and chemical interactions between the numerous fouling ingredients in feed and the membrane surface. Earlier research has shown that the feed water constitution, major constituent concentrations, water chemistry (i.e., pH, ionic strength, and divalent ions concentration), membrane characteristics (i.e., surface morphology, hydrophobicity, charge and molecular weight cut-off), temperature, operational type, and hydrodynamic conditions all influence membrane fouling and foulant characteristics (i.e., initial permeate flux and crossflow velocity) [16]. As a result, any parameters that alter the hydrodynamic properties of membrane modules or the chemical properties of input liquid will have an impact on the total membrane effectiveness [17]. The combined physical and chemical effect will therefore influence the degree of attachment and decide the severity and the techniques for managing it [18].

Foulants are classified into the following four categories: (i) Particles: inorganic or organic particles/colloids can behave as foulants, which blind the membrane surface mechanically, restrict porosity, or delay the transfer of the cake layers to the surface; (ii) Organic: Soluble ions and colloids that would attach to the membrane by a sorption, (e.g., humic and fulvic agents, hydrophilic and hydrophobic agents, and nutrients) [19]; (iii) Synthetic components (e.g., iron, manganese, silica) often have a tendency to fall on the surface of the membrane due to a change in pH (i.e., scaling) or oxidation; and (iv) Micro-biological organisms: Vegetative materials, including algae, microorganisms, and bacteria, that can attach to the membranes and create biofouling (biofilm formation) are included in the microbiological category [20].

13.5 INTEGRATED MEMBRANE PROCESS

13.5.1 ELECTRODIALYSIS AND ELECTRODIALYSIS REVERSE FLOW

In electrodialysis (ED) and electrodialysis reverse flow (EDR), electrical gradient and permeable membrane are configured to separate high concentration salt solutions. These processes utilize electrical gradient for ion permeability [21]. Two different types of ion-exchange membranes are utilized in electrodialysis. The first is permeable to anions but rejects cations, while the second is permeable to cations but rejects anions. When an electronic charge passes into the system, dilute ions migrate through the recharged membranes into the concentrate (cations to cathode whiles anions to anode). During the transfer, anions will be barriered by the cation exchange membrane and cations by the anion exchange membrane [21,22]. EDR requires a regular reversal of membrane electrodes and, hence, reversal of ion transport. This results in dilution and concentration of concentrated streams and hence reduction of the membrane fouling [23]. ED and EDR have a very rapid recuperation rate. They need low water pretreatment due to process inversion. Membrane fouling is also low , and renewable energy may be incorporated [24]. Korngold et al. [25] studied the treatment of brine released by an RO plant with ED with a split gypsum precipitator in a pilot-scale trial in order to reduce the scaling issue in ED. A cost analysis and environmental assessment of the RO feed treatment process employing ED were carried out at Zhang et al. [26] in other research. For this reason, both laboratory and a small-scale industrial ED technology were evaluated, and the current density and feed rate influenced the salt recovery rate among the many parameters tested.

The production and effectiveness of ED processes diminish at high input salinities as a result of the diluent and concentrate chemical characteristic impedance of salt. In order to make the procedure possible for the desalination of high salinity water, a reduced ratio of power to its limiting value was advised [27]. There have been a few documented EDR systems for RO saline treatment and water recovery volume reduction. These processes have shown strong recuperation (85%) with considerable reduction in volume (diminishes six and a half times). Salt works Technologies Inc. developed EDR devices that can handle saltwater supplies up to 8 weight% on average [28]. These encouraging results foster the growth of massive EDR technologies.

13.5.2 PERVAPORATION

This separation process integrates the infiltration and evaporation of the membrane into distinct, preferably liquid mixes. On the one hand, the liquid feed is given to the membrane and the permeate on the other side which evaporates [29]. The permeate in the upstream is absorbed during this procedure. This causes the liquid mixture's more permeable component to be sorbed over membrane (anisotropic membrane or porous inorganic membrane). The gas (vapour) is condensed and then regenerated as fluid. The solution diffusion model is recognized as this mechanism of mass transfer throughout the membrane [29,30]. Edgar et al. [31] employed micro-irrigation pervaporation of wastewater plants. A heavily hydrophilic pervaporative membrane was arranged in the ground during this experiment. Synthesized waste water was cycled in the feed tank via the membranes to check the permeate flow and wastewater concentration (contaminant refusal). The studies have shown that this method offers promise for micro-irrigation of brackish water. In an experiment, 100 organophilic membranes have been utilized to extract organic solvents (benzene, amines, esters, ethers, ketones, etc.) from weak water streams [32]. It was found that organic compounds were concentrated at least 50–100 times, and a clean wastewater stream was made accessible for reuse or disposal. A similar study [33] employed a membrane of the polyetherblock amide to extract phenol from waste water drained from a modified epoxy procedure. Roughly 10% phenol and other pollutants were present in waste water. The quantities of phenol found were below 300 mg/L after testing. The pervaporation characteristics allow it to be appropriate for certain targeted pollutants treatment.

13.5.3 FORWARD OSMOSIS

Forward osmosis (FO) takes place through conventional osmosis when water is pulled across a semipermeable membrane from one side towards the other. In this scenario, a highly concentrated draw solution is employed to create a concentration difference that draws water from the source. The osmotic pressure difference required to move water molecules from the source to the draw is provided by this differential. This transport proceeds until a chemical potential balance is reached from the source to the draw. This flow continues until a chemical potential balance is attained [34]. Applied FO in the concentration of anaerobic digester utilized to recuperate and replenish the draw solution [35]. In a series of FO trials, Haupt et al. [36] investigated FO applications in the automotive production plant and dairy business. The freshwater recovery method following FO was highly dependent on the type of drainage solution utilized [37]. For higher saline feed solution, additional energy would be necessary by RO to conquer the osmotic pressure; for this reason, the better option is FO. In the case of monovalent ions like salt and chloride, RO requires mostly regenerative process whereas for bivalent ions, hydrophilic nanoparticles, micelles, and polyelectrolytes need membranes with high diameters of pore, such as ultrafiltration and nanofiltration. FO has a number of benefits [37–40]:

a. The technique requires no outside pressure which reduces energy use in comparison with pressurized procedures.
b. The use of osmotic pressure for separation and water purification is also easier
c. High flexibility due to option for choosing of draw solution.
d. The regeneration and reuse of draw solution are advantageous in saving cost.

FO is also paying attention to the concentration of saline among other novel procedures. The concentrations in generated water from natural gas extraction and production were attained in some of the rare academic research of 73–180 g/L −1 using NH_3/CO_2 FO membrane saline concentrate [41]. This procedure has resulted in higher concentration and 42% less energy than the mechanical vapour compression evaporator [42]. As a result, 90% water recovery was accomplished by FO, greater than MD for identical streams from brine [42].

With land mining businesses continually facing growing challenges with respect to high-level ore depletion, critical water and energy consumption, and environmental problems, marine and marine water with enormous amounts of precious minerals have developed into an appealing alternative to mineral harvest [43]. Van Wyk et al. [43] have examined and experimented with lithium and boron rejection and related processes in order to better analyse the mechanism of rejection. The RO membrane, which can feature typically 90% boron rejection [44], studies conducted in FO reported lower and very variable rejection (from 10% to 100%). Coday et al. [45] evaluated lithium reject from the solution and reported TFC-FO membrane effectiveness of 78%–88%. Van Wyk et al. [43] validated the value by seeing increased reject of lithium while working in the membrane direction of an active-lagging solution (AL-DS) and concluded that the rejection of lithium was mediated by steric and electrostatic interactions. The membrane modules and devices used for FO include plate and frame, spiral-wound, tubular, and hydration bags. The applications of FO in various domestic and industrial effluent treatments as mentioned in Table 13.2 [46].

13.5.4 MEMBRANE DISTILLATION

This process is integration of the distillation and membrane permeation. The feed is separated based on the differential in vapour pressure across the stream with selective permeation of particles across the membrane, as shown in Figure 13.1.

TABLE 13.2

Applications of Forward Osmosis in Saline Water Treatment

S. No	Industrial Effluent	Solute Present	Outcome	Applications
1	Coal mine wastewater	High concentration of saline mine waster	Greater than 80% mine water recovered	To remove salt from coal mine salt water
2	Coak-oven effluent	Sodium, magnesium, and calcium salts	Upto 98% toxic substance removal	Treatment of coak-oven wastewater
3	Gas field produced water	Sodium chloride salts	Upto 50% volume could be minimized	Diminish in volume of gas field generated water
4	Municipal effluent	Sodium chloride and magnesium chloride salts	Upto 70% recovery of water	Treatment of municipal effluent
5	Primary sewage effluent	Sodium chloride and magnesium chloride salts	Fouling causes minimum water recovery	Treatment of sewage effluent
6	Domestic effluent	Sodium chloride salts	More than 90% salt removal	Treatment of domestic effluent

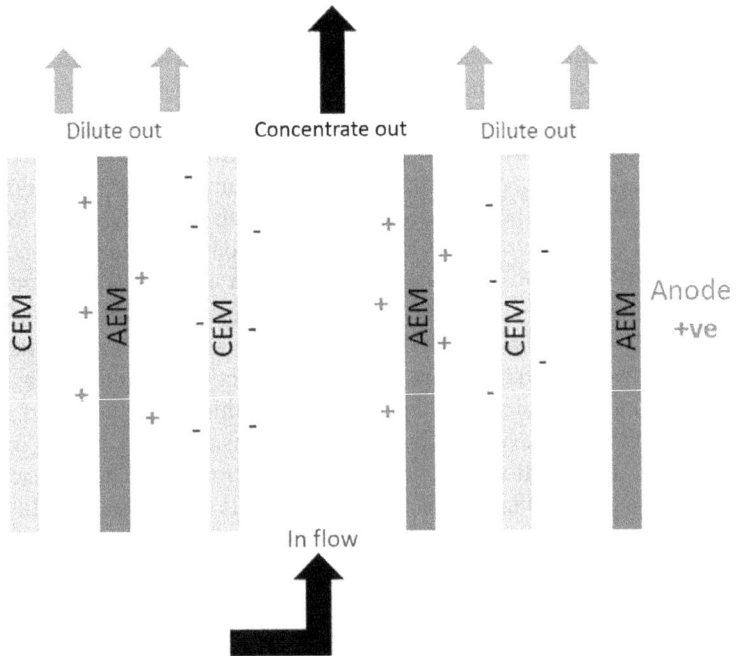

FIGURE 13.1   Schematic for MD.

This process of distillation is very much useful for the separation of azeotropic mixture with very narrow boiling point difference [47]. The fundamental advantage of this process is the separation of feed solutions with high water concentration. MD is developed to utilize low-level (<100°C) heating energy to ensure the necessary difference in vapour pressure between the feed and the product side [39,48,49]. For example, applying MD in treating manufactured water from the Arab Gulf, Alkhudhiri et al. [50] observed that MD was extremely potential to achieve high permeate flow and power use. Kiai et al. [51] in order to test impacts on permeate quality and phenol content, membranes of varied pore sizes were utilized. Product quality with phenolic content less than 16 mg TYE/L (tyrosol equivalent per litre) and conductivity less than 193 μS/cm indicated good product quality. Table 13.3 represents configurations of membrane distillation. Singh et al. [52] have investigated the power consumption of an MD system for distillation flux, purity of the distillate, and degree polarization in the industrial wastewater treatment process. And the study showed that MD is a potential process with high permeate and with lower salt concentration.

DCMD was used by Abdelkader et al. [52] for saline milk effluent treatment. The pretreatment utilized for membrane clogging in MD was macrofiltration and ultrafiltration procedures. Notable factors include enhanced pre-treatment of effluent before DCMD with steady fluxing and higher temperature performance. Osman carried out a computer modelling and experimental investigation on the desalination of industrial petrochemical wastewater using DCMD [60]. Compared to the RO process, FO and MD processes are able to handle high TDS effluent (up to 100,000 mg/L). MD

TABLE 13.3
Configurations of MD

Configuration	Remarks	References
Direct contact membrane distillation (DCMD)	• Widely applied • Vapour pressure difference is generated through the difference ΔT on membrane sides • Significant loss in heat of conduction	[52,53]
Air gap membrane distillation (AGMD)	• Air gap introduced between membrane and condensation plane • Vapour pressure difference due to natural convection • Less loss in heat of conduction	[54]
Vacuum membrane distillation (VMD	• Constant vacuum applied on membrane side • No temperature difference	[55]
Sweeping gas membrane distillation (SGMD)	• Gas to scrub vaporized permeate • Complex external condensing module	[53,56]
Thermostatic sweeping gas membrane distillation (TSGMD)	• Addition of cold wall to AGMD • Cold wall reduces sweeping gas temperature	[43,54,57]
Liquid gap membrane distillation (LGMD)	• Non-permeable layer introduced for separation of permeate • Addition of coolant in permeate side	[58,59]

might be a good alternative for treating saline fluids throughout the development phase, along with low-cost pretreatment. It is important to examine further the feasibility of utilizing waste heat in MD treatment.

13.6   MEMBRANE BIOPROCESS

A membrane bioreactor (MBR), along with a membrane separation technology, may be regarded as a unit built for a biochemical conversion. The membrane can be utilized for several purposes, such as adding a reactant or selectively removing one of the reactive species within the bioreactor [61]. In addition, membranes could be deployed to preserve or maintain the biocatalyst or to separate the enzymes by excluding the size. The first reported study for the MBR was in the late 1970s for the applications in beverage industries. Later on with the advancements in the technology and the requirements in cleaner water, this technology was inbuilt to replace the conventional activated sludge process. This MBR has a better effluent grade and minimal impact on nature. This MBR also gives higher hydraulic retention time as well as solid retention time compared to conventional activated sludge process. Though this method has better performance than CAS, it has many disadvantages such as extensive fouling and high-power consumption [62].

MBRs are viewed as complicated systems in many applications, combining biological waste degradation with membrane filtering. There are therefore two pieces in an MBR: a biological component and a membrane component [63]. Membrane based on geography is categorized as external and internal module. In internal module, the membrane is inside the bio-chamber and permeate would recirculate internally. The membrane in an external MBR is positioned outside the bioreactor, and the treated wastewater is recycled back through the membrane module's main side. The pressure from the axial flow volumetric flux along with the membrane surface provides the driving gradient. MBRs considered as an excellent technology in the removal of organic and inorganic pollutants from waste water, have a proper control over bioactivity – the effluents are usually devoid of any bacteria and pathogens [61,64].

MBR is now extensively used in the bio-gasification process very similar to anaerobic digestion by incorporating a micro-organic consortium in their biological unit. Methane, CO_2, and H_2S are frequently the primary components. Membranes can be used to separate these chemicals; however, these membranes have selectiveness in the gaseous composition over one component. The driving force is very similar to the transmembrane pressure in UF and MF. Because of the negative effects of high concentration inorganic salts on carbonaceous oxidation, nitrification, denitrification, and phosphorous removal, the standalone activated sludge method has not proven successful in the treatment of saline wastewater.

An innovative use of electrobiochemical reactors to remove salts and organic contaminants from wastewater has recently been investigated. Anode and cathode are separated by an ion-exchange membrane in bioelectrochemical systems, which are capable of transforming chemical energy inherent in biodegradable materials (wastewater, sludge, and sediments) into electrical energy. With activated sludge from a household waste management facility in Sydney, Australia, the MBR was

TABLE 13.4
Advantages of Different Methods Used in Treatment of Industrial Saline Wastewater

S. No	Method	Advantages
1	Reverse osmosis (RO)	• Low investment • High treatment efficiency, unproblematic scale-up • Easy to control
2	Electrodialysis	• High water recovery and • More salt removal • More metals separation • Low-energy requirement
3	Membrane bioreactor (MBR) treatment	• Low footprint • Able to treat more suspended solids • Able to treat more organic pollutants and good-quality wastewater
4	Membrane distillation (MD)	• Require low electrical energy • More salt rejection • Able to treat high salinity (up to 20 wt%)
5	Forward osmosis (FO)	• Low energy-intensive compared to RO

utilized to treat saline (0.35 g/L NaCl) wastewater. With salinity rising from 0 to 35 g/L NaCl, organic carbon and ammonia removals dropped from 77% to 93% to 10% and 0%, respectively. At high salt concentrations, the transmembrane pressures (TMP) formed in the MBR were greater. Nadjafic et al. [46] proved that integral MBR-UF and MBR-RO systems are feasible for refining wastewater treatment in another instance. The system MBR–RO hybrid has shown improved effectiveness of salt and organic waste treatment (turning to 98%) in comparison with the MBR–UF system. After treatment, product quality was determined to fulfil the minimal criteria to be re-used or discharged in heaters and coolers. The FO-RO system has a massive benefit of small exterior energy necessities for the applications of salt removal. A comparison of remarkable methods used for the treatment of saline and industrial effluents is presented in Table 13.4, which shows the advantages, limitations, technology status, potential applications, and future prospects of each method used in saline effluent treatment [65].

13.7 CONCLUDING REMARKS AND FUTURE PROSPECTIVE

Both traditional and emerging methods of saline wastewater treatment are significantly reviewed. Membrane process has changed the orientation of the tertiary treatment methodology. In this chapter, the overview of membrane technology and applications of integrated membrane process with respect to saline water treatment are discussed. The conventional membrane process has many drawbacks which are overcome by the integration with other separation processes such as distillation, dialysis, osmosis, centrifugation, and activated sludge process. Recent research in membrane bioprocess has led to the integration of many conventional bio-process with membrane technology and has been reported with very high permeate fluxes, solid retention, and less fouling in saline water treatment. Thus, in saline water treatment, membrane technology has the potential of bridging the cost-effective and sustainability gap, among possibilities of low/no chemical usage, ecological easiness, and easy accessibility to many. That is, membrane technology has established to be a more encouraging alternative in saline water treatment processes in recent times. Though membrane technologies are hopeful at different stages of saline wastewater treatment, further research and developmental attempts towards membrane material and design are necessary for improving the efficiency and economy of the entire treatment process.

REFERENCES

1. A.T. Besha, A.Y. Gebreyohannes, R.A. Tufa, D.N. Bekele, E. Curcio, L. Giorno, Removal of emerging micro pollutants by activated sludge process and membrane bioreactors and the effects of micro pollutants on membrane fouling: a review, *J. Environ. Chem. Eng.* 5 (2017) 2395–2414.
2. C. Grandclément, I. Seyssiecq, A. Piram, P. Wong-Wah-Chung, G. Vanot, N. Tiliacos, P. Doumenq, From the conventional biological wastewater treatment to hybrid processes, the evaluation of organic micro pollutant removal: a review, *Water Res.* 111 (2017) 297–317.
3. R.P. Schwarzenbach, P.M. Gschwend, D.M. Imboden, *Environmental Organic Chemistry*, John Wiley & Sons, Hoboken, NJ (2005).
4. A.R. Ribeiro, O.C. Nunes, M.F. Pereira, A.M. Silva, An overview on the advanced oxidation processes applied for the treatment of water pollutants defined in the recently launched Directive 2013/39/EU, *Environ. Int.* 75 (2015) 33–51.
5. M.O. Barbosa, A.R. Ribeiro, M.F.R. Pereira, A.M.T. Silva, Eco-friendly LC–MS/MS method for analysis of multi-class micro pollutants in tap, fountain, and well water from northern Portugal, *Anal. Bioanal. Chem.* 408 (2016) 8355–8367.
6. Canadian Environmental Protection Act, Priority substances list assessment report, "nonylphenol and its ethoxylates" (1999).
7. L. Goswami, N.A. Manikandan, K. Pakshirajan, G. Pugazhenthi, Simultaneous heavy metal removal and anthracene biodegradation by the oleaginous bacteria Rhodococcusopacus, *3 Biotech.* 7 (2017) 37.
8. S. Arriaga, N. de Jonge, M.L. Nielsen, H.R. Andersen, V. Borregaard, K. Jewel, T.A. Ternes, J.L. Nielsen, Evaluation of a membrane bioreactor system as post treatment in waste water treatment for better removal of micro pollutants, *Water Res.* 107 (2016) 37–46.
9. N. Bolong, A.F. Ismail, M.R. Salim, T. Matsuura, A review of the effects of emerging contaminants in wastewater and options for their removal, Desalination 239 (2009) 229–246.
10. R. Kase, R. Eggen, M. Junghans, C. Götz, J. Hollender, *Assessment of Micro Pollutants from Municipal Wastewater-Combination of Exposure and Ecotoxicological Effect Data for Switzerland*, InTech-Open Access Publisher, Lausanne (2011).
11. R. Loos, R. Carvalho, D.C. António, S. Comero, G. Locoro, S. Tavazzi, B. Paracchini, M. Ghiani, T. Lettieri, L. Blaha, B. Jarosova, EU-wide monitoring survey on emerging polar organic contaminants in wastewater treatment plant effluents, *Water Res.* 47(17) (2013) 6475–6487.

12. S. Loeb, S. Sourirajan, Sea water demineralisation by means of an osmotic membrane, *Adv. Chem. Ser.* 38 (1963) 117e132.

13. S. Loeb, The Loeb-Sourirajan membrane: how it came about, *ACS Symp. Ser.* 153 (1981) 1e9.

14. H.K. Lonsdale, The growth of membrane technology, *J. Membr. Sci.* 10 (1982) 81e181.

15. R.J. Petersen, Composite reverse osmosis and nanofiltration membranes, *J. Membr. Sci.* 83 (1993) 81e150.

16. Q. Li, M. Elimelech, Organic fouling and chemical cleaning of nanofiltration membranes: measurements and mechanisms, *Environ. Sci. Technol.* 38(17) (2004) 4683–4693.

17. H. Zengin, W. Zhou, J. Jin, R. Czerw, D.W. Smith Jr., L. Echegoyen, D.L. Carroll, S.H. Foulger, J. Ballato, Carbon nanotube doped polyaniline, *Adv. Mater.* 14(20) (2002) 1480–1483.

18. G. Pearce, Introduction to membranes: fouling control, *Filtr. Sep.* 44(6) (2007) 30–32.

19. E. Tummons, Q. Han, H.J. Tanudjaja, C.A. Hejase, J.W. Chew, V.V. Tarabara, Membrane fouling by emulsified oil: a review, *Sep. Purif. Technol.* 248 (2020) 116919.

20. K.R. Zodrow, E. Bar-Zeev, M.J. Giannetto, M. Elimelech, Biofouling and microbial communities in membrane distillation and reverse osmosis, *Environ. Sci. Technol.* 48(22) (2014) 13155–13164.

21. A. Campione, L. Gurreri, M. Ciofalo, G. Micale, A. Tamburini, A. Cipollina, Electrodialysis for water desalination: a critical assessment of recent developments on process fundamentals, models and applications. *Desalination.* 434 (2018) 121–160.

22. H. Strathmann, Assessment of electrodialysis water desalination process costs. In *Proceedings of the International Conference on Desalination Costing*, Limassol, Cyprus, 6 December 2004; pp. 32–54.

23. Y.-M. Chao, T.M. Liang, A feasibility study of industrial wastewater recovery using electrodialysis reversal. *Desalination.* 221 (2008) 433–439.

24. A.W.W. Association, *Electrodialysis and Electrodialysis Reversal: M38*, American Water Works Association, Denver, CO (1995); Volume 38.

25. E. Korngold, L. Aronov, N. Daltrophe, Electrodialysis of brine solutions discharged from an RO plant. *Desalination.* 242 (2009) 215–227.

26. Y. Zhang, K. Ghyselbrecht, R. Vanherpe, B. Meesschaert, L. Pinoy, B. Van der Bruggen, RO concentrate minimization by electrodialysis: techno-economic analysis and environmental concerns, *J. Environ. Manage.* 107 (2012) 28–36.

27. R.K. McGovern, A.M. Weiner, L. Sun, C.G. Chambers, S.M. Zubair, V.J.H. Lienhard, On the cost of electrodialysis for the desalination of high salinity feeds, *Appl. Energy.* 136 (2014) 649–661.

28. A. Panagopoulos, K.-J. Haralambous, M. Loizidou, Desalination brine disposal methods and treatment technologies – a review, *Sci. Total Environ.* 693 (2019) 133545.

29. K. Nagai, Fundamentals and perspectives for pervaporation. In E. Drioli, L. Giorno (eds.). *Comprehensive Membrane Science and Engineering*, Elsevier Inc., Amsterdam (2010); pp. 135–163.

30. S. Zhang, E. Drioli, Pervaporation membranes, *Sep. Sci. Technol.* 30(1995) 1–31.

31. Q.-B. Edgar, Z. Hongde, P. Gary, Membrane pervaporation for wastewater reuse in micro irrigation, *J. Environ. Eng.* 131 (2005) 1633–1643.

32. J.G. Wijmans, J. Kaschemekat, J.E. Davidson, R.W. Baker, Treatment of organic-contaminated wastewater streams by pervaporation, *Environ. Prog.* 9 (1990) 262–268.

33. M. Kondo, H. Sato, Treatment of wastewater from phenolic resin process by pervaporation, *Desalination.* 98 (1994) 147–154.

34. W.A. Suwaileh, D.J. Johnson, S. Sarp, N. Hilal, Advances in forward osmosis membranes: altering the sub-layer structure via recent fabrication and chemical modification approaches, *Desalination.* 436 (2018) 176–201.

35. R.J. York, R.S. Thiel, E.G. Beaudry, Full-scale experience of direct osmosis concentration applied to Leachate management. In *Proceedings of the Seventh International Waste and Landfill Symposium (Sardinia '99)*, Cagliari, Sardinia, Italy, 1999; pp. 4–8.

36. A. Haupt, A. Lerch, Forward osmosis treatment of effluents from dairy and automobile industry – results from short-term experiments to show general applicability, *Water Sci. Technol.* 78 (2018) 467–475.

37. L.-C. Shen, N.P. Hankins, *Forward Osmosis for Sustainable Water Treatment*, Elsevier, Amsterdam (2016); pp. 55–76.

38. D.L. Shaer, J.R. Werber, H. Jaramillo, S. Lin, M. Elimelech, Forward osmosis: where are we now?, *Desalination.* 356 (2015) 271–284.

39. Y.-N. Wang, K. Goh, X. Li, L. Setiawan, R. Wang, Membranes and processes for forward osmosis-based desalination: recent advances and future prospects, *Desalination.* 434 (2018) 81–99.

40. P. Wang, T.-S. Chung, Recent advances in membrane distillation processes: membrane development, configuration design and application exploring, *J. Membr. Sci.* 474 (2015) 39–56.

41. R.L. McGinnis, N.T. Hancock, M.S. Nowosielski-Slepowron, G.D. McGurgan, Pilot demonstration of the NH_3/CO_2 forward osmosis desalination process on high salinity brines, *Desalination.* 312 (2013) 67–74.

42. C.R. Martinetti, A.E. Childress, T.Y. Cath, High recovery of concentrated RO brines using forward osmosis and membrane distillation, *J. Membr. Sci.* 331 (2009) 31–39.

43. P. Loganathan, G. Naidu, S. Vigneswaran, Mining valuable minerals from seawater: a critical review, *Environ. Sci. Water Res. Technol.* 3 (2017) 37–53.

44. L.A. VanWyk, Solute transport in a submerged forward osmosis membrane system. PhD diss., Stellenbosch University, Stellenbosch (2019).

45. A. Farhat, F. Ahmad, N. Hilal, H.A. Arafat, Boron removal in new generation reverse osmosis (RO) membranes using two-pass RO without pH adjustment, *Desalination.* 310 (2013) 50–59.

46. E. Obotey Ezugbe, S. Rathilal, Membrane technologies in wastewater treatment: a review. *Membranes.* 10 (2020) 89.

47. L. Salonen (retired) et al., *Environmental Liquid Scintillation Analysis, Elsevier Handbook of Radioactivity Analysis* (Third Edition), Elsevier, Amsterdam (2012).

48. D. González, J. Amigo, F. Suárez, Membrane distillation: perspectives for sustainable and improved desalination. *Renew. Sustain. Energy Rev.* 80 (2017) 238–259.

49. J.A. Sanmartino, M. Khayet, M.C. García-Payo, N.P. Hankins, R. Singh, *Desalination by Membrane Distillation*, Elsevier, Madrid (2016); pp. 77–109.

50. A. Alkhudhiri, N. Darwish, N. Hilal, Produced water treatment: application of air gap membrane distillation, *Desalination.* 309 (2013) 46–51.

51. H. Kiai, M.C. García-Payo, A. Hafidi, M. Khayet, Application of membrane distillation technology in the treatment of table olive wastewaters for phenolic compounds concentration and high quality water production. *Chem. Eng. Process. Process Intensif.* 86 (2014) 153–161.

52. R. Singh, N. Hankins, *Emerging Membrane Technology for Sustainable Water Treatment*, Elsevier, Amsterdam (2016).

53. M. Khayet, T. Matsuura, *Membrane Distillation: Principles and Applications*, Elsevier, Amsterdam (2011).

54. A. Alkhudhiri, N. Hilal, 3-Membrane distillation – principles, applications, configurations, design, and implementation. In *Emerging Technologies for Sustainable Desalination Handbook*, Gude, V.G., Ed., Butterworth- Heinemann, Oxford (2018); pp. 55–106.

55. D. González, J. Amigo, F. Suárez, Membrane distillation: perspectives for sustainable and improved desalination, *Renew. Sustain. Energy Rev.* 80 (2017) 238–259.

56. A. Alkhudhiri, N. Darwish, N. Hilal, Membrane distillation: a comprehensive review, *Desalination.* 287 (2012) 2–18.
57. C.A. Rivier, M.C. García-Payo, I.W. Marison, U. von Stockar, Separation of binary mixtures by thermostatic sweeping gas membrane distillation: I. Theory and simulations, *J. Membr. Sci.* 201 (2002) 1–16.
58. D. Winter, Permeate gap membrane distillation (PGMD). In *Encyclopedia of Membranes*, Drioli, E., Giorno, L., Eds., Springer, Berlin (2016); pp. 1–2.
59. D. Winter, J. Koschikowski, M. Wieghaus, Desalination using membrane distillation: experimental studies on full scale spiral wound modules, *J. Membr. Sci.* 375 (2011) 104–112.
60. M.S. Osman, V. Masindi, A.M. Abu-Mahfouz, Computational and experimental study for the desalination of petrochemical industrial effluents using direct contact membrane distillation, *Appl. Water Sci.* 9 (2019) 1–13.
61. V. Calabro, Engineering aspects of membrane bioreactors. In *Membrane Reactor Engineering: Applications for a Greener Process Industry*, John Wiley & Sons, Hoboken, NJ, (2013); Volume 2. ISBN 9780857097347.
62. K. Xiao, S. Liang, X. Wang, C. Chen, X. Huang, Current state and challenges of full-scale membrane bioreactor applications: a critical review, *Bioresour. Technol.* 271 (2019) 473–481.
63. N. Cicek, A review on membrane bioreactors and their potential application in the treatment of agricultural wastewater, *Can. Biosyst. Eng.* 45 (2003) 637–649.
64. S.H. Yoon, *Membrane Bioreactor Processes*, CRC Press, Boca Raton, FL (2016).
65. P. Sahu, A comprehensive review of saline effluent disposal and treatment: conventional practices, emerging technologies, and future potential, *J. Water Reuse Desal.* 11 (1) (2021) 33–65.

14 Industrial Saline Wastewater Treatment Technologies Using Microalgae Biomass

Manoj Kumar Enamala and A. Sai Kumar
Bioserve Biotechnologies (India) Private Limited

P. Divya Sruthi
Andhra Pradesh State Warehousing Corporation

Murthy Chavali
NTRC-MCETRC and Aarshanano
Composite Technologies Pvt. Ltd.

Meenakshi Singh
The M.S. University of Baroda

Chandrasekhar Kuppam
Yonsei University

CONTENTS

DOI: 10.1201/9781003185437-17

14.1   INTRODUCTION

Human activities have made water to be unusable and also causing damage to the environment. The discharge of wastewater into the water bodies is posing a serious threat to human health and environmental concerns, and main reasons for this huge amount of discharge are due to rapid industrialization and urbanization which are major concerns that need to be addressed. The wastewater usually consists of 99.9% of water and the remaining 0.1% which is seen is present in various forms like nutrients, fats, oils, cooking oils and various harmful pathogens. Various other contaminants include oils, dissolved heavy metals, suspended solids and various other organic compounds. There should be a defined set of limits that determines the quality of discharge of wastes, and these should be legally met or before discharging the wastes should be pre-treated to remove the majority of the contaminants. The water is always polluted, and there is no clear distinction between clean water and polluted water. The only difference which defines the type of water is dependent upon the type and concentration of the impurities which are found in water, and it also depends upon the kind of the usage of water, for example, the type of its use like drinking, swimming and fishing. Therefore, water pollution is primarily caused by the mixing of contaminated wastewater with the groundwater/surface water.

Several originating sources for water pollutants are known as point sources or dispersed sources. Point sources are defined as the source which originates from one particular channel such as a single pipeline or channel, whereas dispersed sources are defined as any unconfined areas from which pollutants enter the body of the water. Point sources are easier to control when compared with dispersed source pollutants. There are usually two different types of discharged sewage wastewater. Black water consists of human waste and harmful pathogens which are discharged from bathrooms and cannot be used, whereas the other type of wastewater is known as greywater which is discharged from domestic discharge like sinks, washing machines, showers and does not contain harmful pathogens.

14.1.1   Sources of Wastewater

Wastewater is defined as any water which has been contaminated by human use which originates from any combination of domestic, industrial, commercial or agricultural activities. In other words, it is defined as a by-product of domestic, industrial, commercial or agricultural activities. The wastewater which is found in sewage is an ideal source for the growth of a wide variety of microorganisms like bacteria, fungi, protozoa and viruses. It also contains several pathogenic microorganisms, and these microorganisms cause various dreadful diseases like cholera, typhoid, dysentery and hepatitis (Abdel-raouf, 2012). The various sources of wastewater are shown in Table 14.1.

A representation of the various sources of wastewater is depicted in Figure 14.1.

TABLE 14.1
Sources of Wastewater Generation

Source of Wastewater	Various Activities/Pollutants Released into the Environment
Domestic/ Household wastewater	Human excreta (mixing of faeces, urine and other bodily fluids) Washing water (personal hygiene, clothes, dishes, floors, etc.)
Industrial wastewater	Acids, bases, reactive wastes, waste inks, heavy metals, paint wastes, nuclear wastes, industrial site drainage, cooling waters, solids, emulsions from paper mills, etc.
Other sources	Urban run-off from highways, railway tracks, food wastes, herbicides and pesticides run off from garden, agricultural pollution, etc.

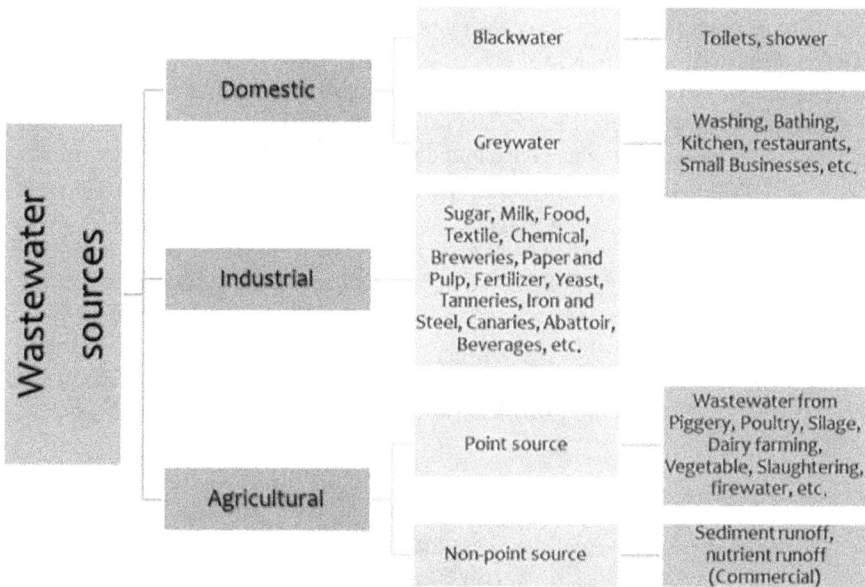

FIGURE 14.1 Wastewater origination sources.

14.1.2 SALINE WATER SOURCES AND CHARACTERISTICS

Saline water usually termed saltwater is defined as the water which contains a high concentration of dissolved salts like sodium chloride (NaCl), and the unit is expressed in terms of parts per million (ppm). As per the universal theory, the salinity is classified into three different categories: slight saline water (1,000–3,000 ppm), moderately saline water (3,000–10,000 ppm) and highly saline water (10,000–35,000 ppm). The seawater on average has a salinity range of up to 35 g of salt in one 1 L/kg of water. Saline water is classified into four types, based on salinity (concentration of dissolved salts) and based on electrical conductivity:

   i. Freshwater
   ii. Brackish water
   iii. Saline water
   iv. Brine water

A representation is given in Figure 14.2 for a clear picture of the salinities of the water.

14.1.3 MICROALGAE BIOMASS IN WASTEWATER TREATMENT

Water pollution has become one of the major concerns which are leading to the insufficiency of water due to the contamination of freshwater bodies on a larger scale. This has become a serious concern in protecting the health of human individuals in most developing countries. The freshwater bodies are getting polluted with the dumping of heavy metals, eutrophication, various organic pollutants, sewage, acidifications and human errors, which are also causing pollution of water bodies on a larger scale. Many researchers have found a suitable solution for the treatment of wastewater using the microalgae which survive on the Earth's ecosystem in various water bodies like freshwater, brackish water, seawater and salinity water. Among the various microbes available around us, microalgae have found to be of very high importance in treating the wastewater as they have the capability of absorbing organic and various heavy metal pollutants from the wastewater.

    There are several other species of microalgae available (about 60 genera and 80 species) which is listed by Palmer. There are still many more species which have been reported in various research works, and these species of microalgae are being used in a wide range of study in their respective works. The wastewater also serves as a nutrient source in form of nitrates, phosphates, ammonia, urea, trace metals, etc., which is useful for the growth of microalgae in the treatment of wastewater

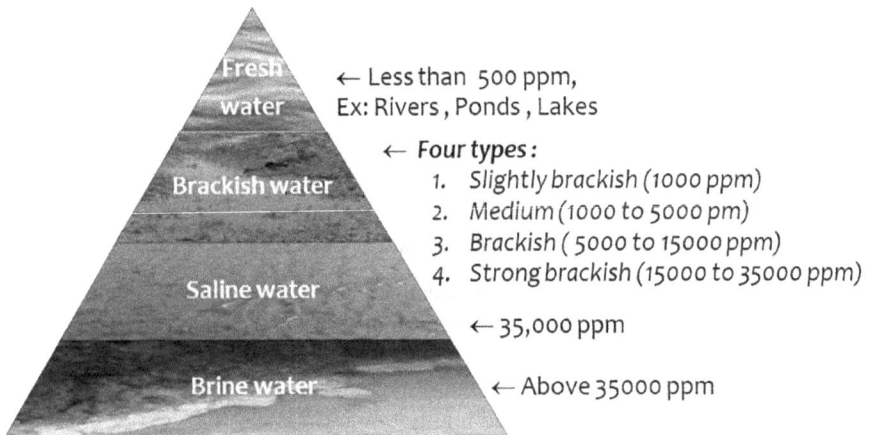

FIGURE 14.2 Salinities of the water.

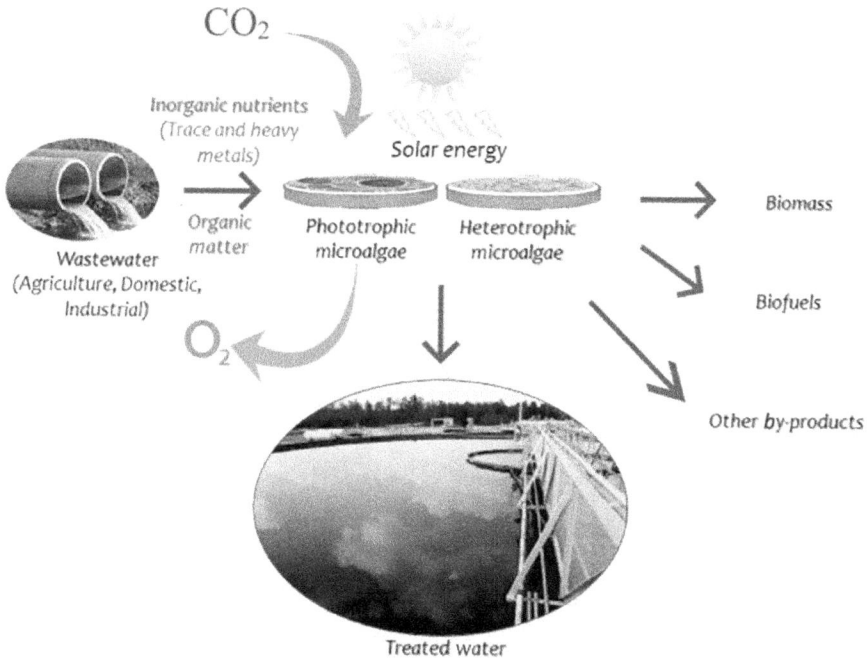

FIGURE 14.3 Representation of wastewater treatment using microalgae.

(Salama et al., 2017). Microalgae have been found to be more effective in treating wastewater (see Figure 14.3) through conventional methods which make the microalgae an efficient source to degrade the wastes through the metabolic pathways which help these microalgae to treat the wide variety of sources of wastewaters. Currently, among the various species of microalgae available, only a few species like *Chlorella* sp., *Arthospira* sp., *Scendesmus* sp. and *Nanochloropsis* sp. have been widely studied in the treated wastewater, and the reason behind this is that they have the potential of accumulating starch and high levels of lipids (Wollmann et al., 2019). Microalgae are autotrophic unicellular organism which can perform photosynthesis through which they convert the water and CO_2 into O_2 and carbohydrates which provide energy for their growth along with various nutrients are consumed in form of nitrogen and phosphorous which helps in algae cell reproduction. Moreover, since the waste water is considered to be a growth medium for the cultivation of various kinds of algae in various growth conditions, these nutrients are taken up and are transformed within the algal cell, and hence algae metabolism is considered to be a key factor behind the breaking of various elements present in the wastewater (Abdel-raouf, 2012).

14.1.4 IMPACT OF WASTEWATER ON LIVING ECOSYSTEMS

The discharge of the wastewater into the water bodies poses a serious threat to the aquatic animals as they contain some serious chemicals which cause serious health problems to the freshwater ecosystem and in turn also affect the health of human beings who are consuming them. Most of the water bodies are generally at risk of the harmful effects of the wastewater. When an industrial effluent is discharged into the water bodies, there are several toxic compounds/chemicals which are also released into the aquatic environment. The microbes present in the water start to break these substances, and in turn, they use a lot of dissolved oxygen and when this dissolved oxygen is being used up by the microbes it becomes life-threatening for the animals thriving in the water bodies. Also when humans consume freshwater animals like prawns and fish, they can suffer from serious health issues as they have already ingested various heavy metals. Various studies reported that there are severe behavioural problems, the disappearance of species from the freshwater bodies which leads to a lot of disturbance to the ecosystem of the freshwater bodies (Balamurugan, Sathishkumar, and Li, 2020).

14.2 MICROALGAE IN TREATMENT OF INDUSTRIAL WASTEWATER EFFLUENTS

The water demand in this current scenario has increased the consumption of water, and various reports indicate by 2030 there will be a 40% of water deficit due to rapid industrialization, urbanization and groundwater contamination. Due to these reasons, there are severe shortages of water. Apart from this, there is also a rapid discharge of industrial effluents into the water bodies causing eutrophication of the water bodies which is leading to the death of several aquatic species dwelling in the particular area and further making the deficit of water increase. Hence beneficially reusing the wastewater would boost the circular economy by recycling resources. Eutrophication is one of the major problems in the current scenario, and this has led to serious issues to the ecosystem of the water bodies present all around the globe.

International Lake committee reported from their survey that 48% in North America, 54% in Asia and the Pacific, 53% in Europe, 43% in South America and 28% in Africa the lakes and water bodies have become eutrophic (Cai, Park, and Li, 2013). In this aspect, many researchers are working towards green photosynthetic microorganisms the microalgae are ubiquitous microscopic organisms that flourish in water bodies and utilize necessary nutrients like nitrogen, phosphorous and various other essential nutrients which are required for their growth (Al-jabri et al., 2021). Microalgae are considered to be the most diverse group of eukaryotes, and there are about 3,50,000 species discovered. This microalga is an alternative approach for the traditional feedstocks, and these species of microalgae contain as high as 60% carbohydrates, 70% lipids and 65% of proteins and amino acids (Shahid et al., 2019). A general representation in treating industrial wastewater is depicted in Figure 14.4. The microalgae species can treat a wide variety of wastewater originating from various industries like food, agricultural, thermal power plants, food processing and pharmaceutical, and these microalgae species can be grown on a variety of different wastewaters which makes microalgae be considered as an efficient source in treating a wide variety of industrial wastewaters. Many microalgae species (around 3,000 strains) have been found/adapted to grow efficiently in wastewater (Molazadeh et al.,

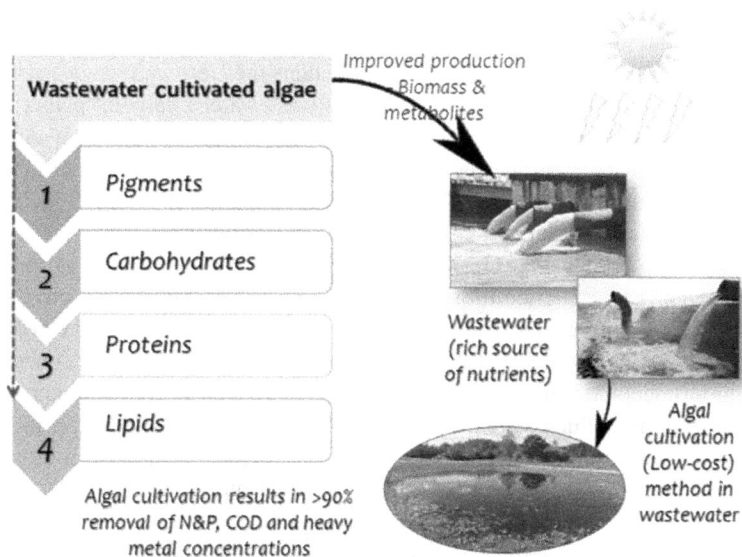

FIGURE 14.4 Representation of industrial wastewater treatment using microalgae.

2019). The most favourable species available in treating wastewaters are considered to be the microalgae because they take up nutrients from the surroundings and convert them into biomass, and these nutrients from the surroundings are incorporated for their growth, thus reducing the nutrients from the wastewater and reducing the COD and BOD from the wastewater. The major advantage to use microalgae in wastewater treatment is its small size and has a large surface area to absorb all the essential nutrients like CO_2, NO_3^- and PO_4^{3-} required for its growth which makes it stand at the top position for usage in wastewater treatment (Farhan et al., 2017).

Numerous species of microalgae available are adapted to their growth in various wastewaters, and by doing this, there is a dramatic change/decrease in the investment invested in the wastewater treatment plants, more economic, cost-effective and even it is considered eco-friendly. The main aim of wastewater treatment is to remove the excess amount of micropollutants in terms of nitrogen, sulphur, copper, phosphorous and heavy metals like cadmium, copper, zinc, nickel, cadmium, lead and various organic pollutants like phenolic compounds, aromatic hydrocarbons, biocides and surfactants. There are several conventional treatments available for wastewaters but when the biological approach is preferred it depends upon the type of wastewater which is being treated (Shahid et al., 2019).

14.2.1 Algae in Industrial Saline Effluent Treatment

Saline wastewater is defined as the type of water which has various levels of salinity in the range of 3%–15% and the units which it is measured is Artificial Sea Water (ASW), Practical Salinity Unit (PSU), g/L, % and ppt. Saline wastewater is a type of pollution that consists of various levels of types of salts and is considered to be one of the most serious environmental problems which relate to water reservoirs. Inorganic salts from saline wastewater can cause damage to crops, inhibit the growth of crops and can even lead to eutrophication which threatens the freshwater and marine ecosystems. The investment invested in the saline wastewater treatment plants is at an alarming rate as their investment cost is soaring. Due to these reasons, an alternative source of microalgae is being used for the treatment of saline wastewater. The various types of saline wastewaters and their salinities are given in Table 14.2. Saline wastewater is a recalcitrant source that consists of various pollutants and inorganic salts like NaCl, Na_2SO_4, $MgSO_4$, KNO_3 and $NaHCO_3$. The sources of saline wastewater include various agricultural run-offs, coastal area industrial activities and various effluents sources, and the concentration of these effluents is very high in the range of 3%–15% and varies in a broad range, because of this highly concentrated water this has become a major serious concern towards the environmental protection. When these saline effluents are being let off into ecological systems, it is leading to major health issues for various aquatic animals as well as to the health of humans. There has been a lot of attention in treating this wastewater, and various governments all around the globe are spending a lump sum amount for the welfare of the people. Researchers have seen that the microalgae are being extensively used for treating the saline wastewater as it can remove the various micropollutants, heavy metals and various organic pollutants extensively. Salinity is a necessary factor for the growth of microalgae as it is required for adjusting the osmotic environment around its environment. The minimum required amount of salinity is termed as euryhaline (little or no need of NaCl) whereas in the halotolerant strains the salinity is much greater and ranges around (1,000 ASW). The level of salinity depends upon the type of strain selected, and it is always species independent. The adjustments of salinity by microalgae cells are shown in Figure 14.5.

TABLE 14.2
Salinity % of Various Industrial Effluents

Origin of Saline Wastewater Effluent	Salinity (%)
Pharmaceutical effluent	0–6
Tannery effluent	5.4
	2.1–5.7
Petrochemical wastewater	1.9–4.2
Coal mine	9.2
Vegetable picked production	15
Marine culture	3.45
Oily saline wastewater (contains heavy metals)	3.3
Saline antibiotic wastewater	2.6
Food processing	3
Textile	0.1–1
Fishing cannery	0.2–0.5
	1.3–3.1
Agricultural run-off	0.12–1.4

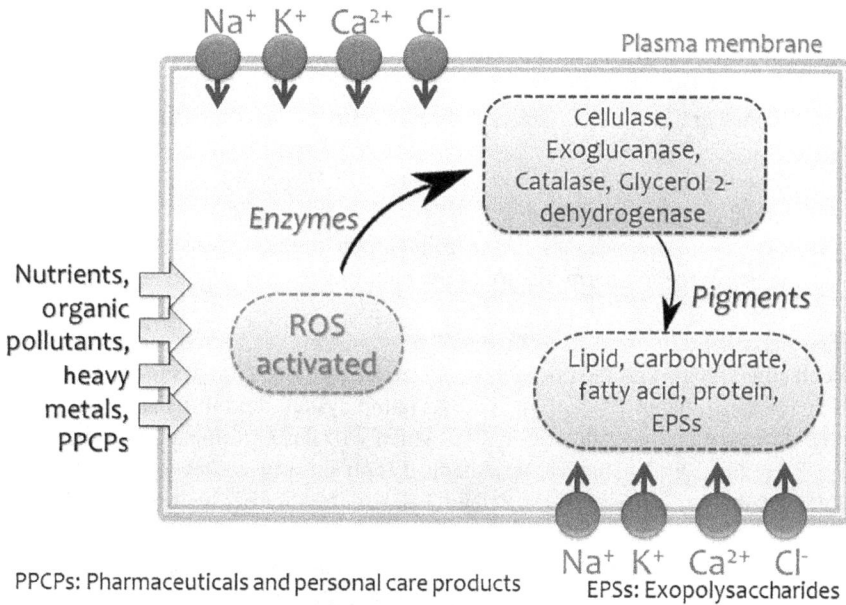

FIGURE 14.5   Biochemical adjustments due to high salinity and stress in its environment by microalgal cells.

Even though the algae is being extensively used in saline wastewater, certain limitations need to be addressed:

i. The performance of freshwater and marine strains has not been studied extensively, and the comparison among various strains in saline wastewater treatment is to be explored which will increase the ability of microalgae towards saline treatment.

ii. The mechanism of assimilation of various pollutants with salinity is not studied completely which again will have insufficient data for comparison for microalgae (Vo et al., 2019).

14.2.2   Effect of Salt Concentration on the Growth of Microalgae

Salinization is significantly increasing globally and is one of the serious problems which is present during the last few decades. The treatment of the saline wastewaters presents a complicated task for the wastewater treating professionals due to a large amount of salt content present, and these wastewaters cannot be introduced as such directly into surface waters or groundwater without proper pre-treatment. Physical and chemical treatment of saline wastewaters like distillation, vacuum evaporation, electrodialysis and reverse osmosis are extremely expensive. So, to replace these costly procedures, researchers have employed the biological methods of treating the saline wastewaters. As discussed in the above sections that microalgae is an efficient treatment used for treating various industrial waters, some species of microalgae can tolerate high salinity and some of the species cannot even tolerate the salinity (Figler et al., 2019). The changes in the salt content affect the growth of algae in various ways,

  i. When there is osmotic stress, it directly affects the water potential of the cell,
 ii. When any unavoidable uptake or release of ions occurs, ionic salt stress is created among the cells, and
iii. When there are changes in the cellular ion ratios, there will be a slight change in the ion permeability of the membrane.

When the algae are exposed to high salt content, ion concentration is increased in all cell constituents and the micro/macroalgae adapt to the changes in the salinity by biochemical strategies through the Na^+/K^+ pump system, and this pump contributes to the development of osmotic potential which will have an impact on the cell volume. The study shows that the treatment of high salinity wastewater is difficult to treat when the salt concentrations exceed 2. A very few studies have reported biological nutrient removal with salt concentrations >3, and removal efficiencies were significantly lower (Church et al., 2017). Sui and Vlaeminck (2019) have experimented by varying the salt concentrations (NaCl) to 1M, 2M and 3M on *Dunaliella salina* (a type of halophile green microalgae found especially in sea salt fields, which are known for their antioxidant activity because of their ability to create a large number of carotenoids). These salt concentrations did not have an impact on their overall growth, and the pattern followed was an increase and decrease pattern during the exponential phase, whereas there was a sharp rise in the protein content and a sharp decrease for about 50–60 in the stationary phase and the biomass content at the highest early exponential phase with a fast decline at around 40–54.

Another theory that is supporting during the hypersaline treatment is that there is a vacuole formation inside the cells of microalgae to store ions like Na^+ and Cl^- when there are moderate changes in the salinity which is being characterized by the excess of water inflow/outflow with the osmotic gradient. The compositions of saline wastewaters could be variable, and due to the presence of heavy metals, non-metallic compounds exhibit complicated interactions with each other. A few have conducted experiments on freshwater species *Chlorella* and *Chlorococcum* species (Figler et al., 2019) for the higher salt tolerance. They have found that long-term salinity was confirmed by these species, and most of the experiments reported by showing that the microalgae species had tolerance towards salinity/NaCl. Even though there were salinity changes in terrestrial habitat, it could not be able to explain for a few of the species as it could be modelled exclusively using NaCl. The microalgae cultures adaptation (see Figure 14.6) towards the relatively fast adaptation was able to see in certain green microalgae, and it was also proved that seawater with the same salinity caused a much lower inhibition than artificial medium with the nutrient content.

14.2.3 Photobioreactors in Saline Wastewater Treatment

Microalgae are commonly cultivated in the various kinds of photobioreactors, and they are commonly used in various applications like biofuel production, various

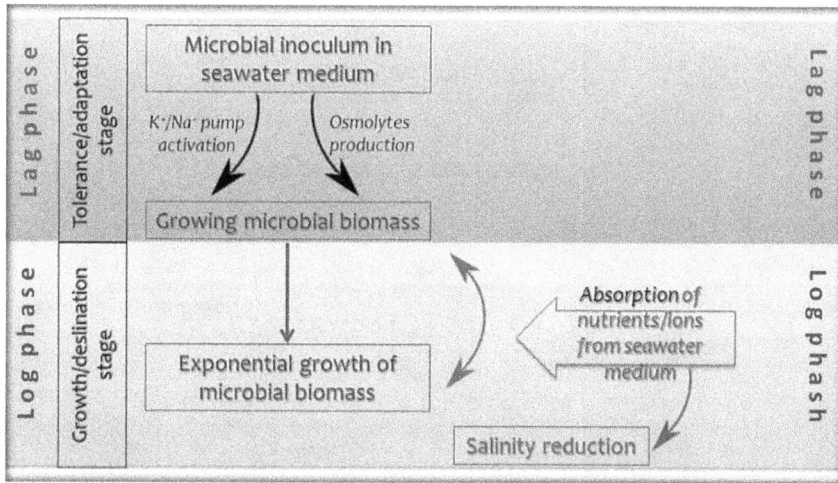

FIGURE 14.6 Mechanism of direct desalination of seawater by microalgae.

range of food products, removal of unwanted wastewater wastes and wastewater treatment. There are various types of photobioreactors available depending upon the application/type of microalgae growth required for their growth. When the advent of photobioreactors, the growth of microalgae was carried out in the open-pond cultivation systems/ natural systems where all the factors were controlled based upon the environmental conditions. In an open-pond cultivation technique, there is a lesser control of the physical parameters (uneven illumination, evaporation of media, temperature losses, exchange of gases, selection of species, etc.). However, all these drawbacks were eliminated in the closed photobioreactors. In the closed photobioreactors, the control for the growth of microalgae is being controlled by the instrument at regular intervals, and there is no deviation from it. These photobioreactors have a greater advantage over natural systems. Several types of photobioreactors available like tank/box type, cylindrical/column type, tubular type, helical type, flat plate/flat panel and several other designs are also being available in the market/research areas (Ashok, Sanjay, and Amritanshu 2019). Various parameters which affect the operation of the photobioreactor are shown in Figure 14.7.

In saline wastewater treatment, there are several interactions between microalgae and other microbes that have also been studied in the wastewater treatment, and the most studied common interaction is the microalgae-bacteria consortia as their complex interactions provide a solution for the treatment of wastewater. The microalgae and bacteria interactions work quite opposite in treating the wastewaters. The microalgae release oxygen to heterotrophic bacteria which then convert this O_2 into CO_2 and other essential mineral nutrients through respiration and organic matter degradation which are efficiently converted into biomass by microalgae. The exchange of these nutrients in this way helps microalgae to increase the biomass of microalgae during wastewater treatment (Guzzon, Di Pippo, and Congestri, 2019).

Wastewater treatment
Metal removal
Sequestration - Flue gases
Bio-fertilizer, biofuel and
bio-H₂ production
Desulphurization
Extraction of Chemicals
and proteins

Nutrient load, temp. & pH
Light – Intensity & duration
Hydraulic retention time
Aeration rate & Biomass
recycling rate
Gas removal
Mixing rates (High for
shear stress and low for
biomass settling)
Regular cleaning

Type of reactor
(cleaning/output)
Wastewater – Collection &
distribution
Climatic conditions
Species selection
Target population
Economics
Heating & cooling option
Harvesting Tech
Min. power an land coverage

Output

Design Parameters

Operating Conditions

Measurable
Parameters

Physiological
Parameters

Cell count – Algae, bacteria, biomass
Temp. pH, BOD, COD, DO, alkalinity
Carbon – Organic & inorganic
Others – Metals, nutrients, lipid,
polysaccharide concentrations

Density, odour & colour of the culture
Coagulation
Flocculation

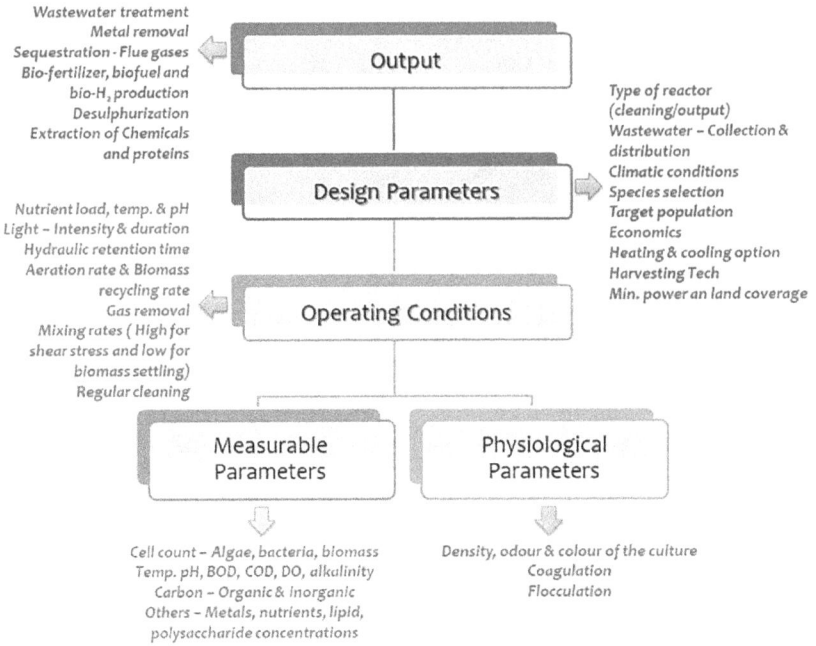

FIGURE 14.7   Parameters that affect the operation of photobioreactors.

14.3   SALINE WASTEWATER TREATMENT TECHNOLOGIES

Saline wastewater contains numerous pollutants such as nutrients, heavy metals, micropollutants and organic pollutants. These wastewaters can be treated by conventional physico-chemical methods but result in extremely high cost, large volumes of water, turbid conditions and high costs for treating the organic matter in salty conditions. Regular wastewater treatments may prove to be highly ineffective in treating saline wastewaters. So the alternative is to use the bacterial processes or in combinations with alternative technologies to eliminate organic matter and other nutrients that are inhibited under highly saline environments.

Microalgae can be a sustainable alternative resource for wastewater treatment either in combination or independently. When compared with other available technologies, treating saline wastewaters with microalgae is greener and environmentally friendly while generating significant with no secondary pollutants and creating profit via by-product production during the processes. Importantly, during the microalgae process, the saline wastewater transformed into a source for biofuel and at times pigment production. Here in this section let us discuss the saline wastewater pollutants and the industries that generate turbid saline wastewaters in high concentrations.

14.3.1   SALINE WASTEWATER POLLUTANTS

Many industrial sectors are likely to generate highly saline wastewaters. Primarily such saline wastewater contains high levels of salinity, turbidity, dissolved solids, suspended solids and high organic content. Pollutants such as nutrients, heavy metals, micropollutants, biological oxygen demand (BOD) and organic pollutants are largely present in saline wastewater. Even carbonaceous, nitrogenous and phosphorous pollution is seen in saline wastewaters. Organic matter comprises chemical oxygen demand (COD), azo dyes, aromatic hydrocarbons, phenols, N-aliphatic and N-aromatic compounds and other hydrocarbons.

14.3.2 Industries Generating Saline Water

Almost all industrial segments are likely to produce saline wastewater at one point or the other but get diluted during their course to effluent treatment plant (ETP). Several major industries generate highly concentrated saline wastewater, and these include agriculture, food, beverage, metal/electroplating, petroleum, meat processing, textile and leather industries. These discharges contain high turbidity, high salinity and high organic content which can cause a drastic effect on aquatic life, groundwater and agriculture. Food and beverage, meat processing industry, textile and leather industries that are devoted to all kinds of brined products, devoted to the production of canned fish, meat processing (like pickled vegetables, tannery products, fish product preserving, aquaculture) generate high saline wastewater along with high concentrations of biodegradable organic matter, suspended solids, nutrients (sodium, potassium, calcium, nitrogen and phosphorus) and salt concentrations ($\leq 15\%$). The wastewater from these sectors is becoming enormously expensive and difficult to treat. The untreated signifies a major threat to the environment.

Many microalgae species are efficient in removing toxicants from wastewater and are highly promising, eco-friendly and sustainable option for tertiary wastewater treatment with a possible advantage of biofuel production. An algal treatment especially using microalgae (halophilic microorganisms) in combination with altered anaerobic digestion and desalination technologies became a cost-effective and novel solution treating saline wastewater while significant production of valuable algae-based products. The salinity % of various industrial effluents is shown in Table 14.2.

The textile industry generates large volumes of wastewater, say an average of $\sim 7 \times 10^5$ tons of organic dyes/year that are highly saline, treating such water with microalgae started way back in the 1960s in Japan by a company called *Nihon Chlorella*. Since then large-scale microalgae cultivation in industries started commercially for diversified applications (Cai, Park, and Li, 2013; Robinson et al., 2001). The wastewater generated during the wet-blue tanning process is treated with a dense population of *Scenedesmus* sp., making it a potential growth medium for biomass production of the test alga for phycoremediation of toxicants in tannery wastewaters (Ajayan et al., 2015). *Scenedesmus* sp. were also used in treating wastewater from the chicken slaughterhouse (WWCSH). *Scenedesmus* sp. biomass production with high contents of carbohydrates, proteins and lipids were determined at 30°C and after 24 hours.

The ability of microalgae, say, *Oscillatoria acuminate* and *Phormidium irrigum*, was validated against the heavy metals from tannery effluent of Ranipet industrial area, India. Microalgae species were cultured in media containing tannery effluent in two different volumes, and parameters like specific growth rate, protein content and antioxidant enzyme activities were studied. Antioxidant enzymes namely superoxide dismutase (SOD), catalase (CAT) and glutathione (GSH) contents were increased in microalgae species indicating the free radical scavenging mechanism under heavy metal stress. Results revealed *Oscillatoria acuminate* possess high antioxidant activity and biosorption efficiency than *Phormidium irrigum* and hence well-thought-out to be very useful in treating heavy metals contaminated effluents (Balaji et al., 2016).

The food and beverage industry is one of the major contributors to the nation's economy, nevertheless associated with severe environmental issues like depleting groundwater levels by consuming high levels of water consumption and huge generation of wastewater. Holding and treating by conventional ways is highly expensive (Valta et al., 2015). Algae have full potential in treating food industry wastewater and also produce biodiesel from the spent wash. Microalgal strains of *Cladophora* sp. and *Spyrogyra* sp. with different dilutions (10%, 20%, 40%, 80% and 100%) of wastewater were made with distilled water and microalgal strains were grown. It is cheaper and presents an environment-friendly way to handle food industry wastewater. With 100% growth, *Cladophora* sp. produced higher biomass than *Spyrogyra* sp. while growing on food industry wastewater (Mureed et al., 2018). Using microalgae for brewery wastewater treatment has its potential benefits (Amenorfenyo et al., 2019). Brewery wastewater holds enormous amounts of organic compounds that can cause severe environmental pollution, utilizing these microalgae grew well in nutrient-rich wastewater by absorbing organic nutrients and converting them into useful biomass. Microalgae-based wastewater treatment method thus emerged as an environmentally friendly biotechnological process.

Hydraulic fracturing technology (HFT) is widely used in modern times for recovering natural gas and oil from tight oil and gas reserves. During the process, large volumes of wastewater and flow-back water are produced. In the oil and gas industry during hydraulic fracturing technology, the wastewater generated was treated with microalgae. Thirteen known microalgae strains consisting of cyanobacteria and green algae were investigated. Over 65% total dissolved solids (TDS), 100% nitrates and >95% boron reduction were observed and noted that microalgae treatment of flow-back water can significantly reduce the impact of hydraulic fracturing technology on the environment while converting the biomass useful to bioproducts (Lutzu and Dunford, 2019).

14.4   TREATMENT TECHNOLOGIES APPLIED TO SALINE EFFLUENTS

The adequate treatment of saline waste/ effluents can minimize the safety of discharge of these effluents into the water bodies. Over the years, several techniques are being adopted in saline industrial wastewater treatment. Using such methods has proven that the water can be reused for potable or non-potable purposes. Various types of treatment methods used for the treatment of saline wastewater are given in Table 14.3 (Sahu, 2021).

TABLE 14.3
Types of Treatment Methods and Technologies Used for the Treatment of Saline Wastewater

Treatment Methods	Technologies
Membrane based methods	Filtration, reverse osmosis, electrodialysis, membrane bioreactor, membrane distillation, forward osmosis
Thermal methods	Multi-effect distillation, spay evaporation, freeze crystallization, electric freeze crystallization
Biological methods	Aerobic treatment, anaerobic treatment, biological reduction, nitrification and denitrification
Chemical methods	Coagulation, ion exchange, biooxidation, advanced oxidation process

14.4.1 MEMBRANE TECHNOLOGY APPLIED TO SALINE EFFLUENTS

The membrane technology evolved in the 18th century, and since then, there has been a lot of developments being taken place and there have been used in a wide variety of applications. The membrane is a barrier that is defined as a medium that separates two phases from each other by restricting the movement of components through them. Several types of membranes make up their structures using isotropic, organic and inorganic membranes. Isotropic membranes are non-porous, microporous membranes which are made up of different layers with different structures and compositions whereas organic membranes are made up of synthetic organic polymers like polyethylene (PE), polytetrafluoroethylene (PTFE) and cellulose acetate, and these materials are chemically and thermally stable and can be used in a wide variety of industrial applications (see Figure 14.8) like hydrogen separation, ultrafiltration and microfiltration (Elorm Obotey Ezugbe et al., 2020).

14.4.2 FILTRATION TECHNIQUES TECHNOLOGY APPLIED TO SALINE EFFLUENTS

The reverse osmosis method has been commercially exploited for wastewater treatment, and the production of drinking water worldwide is being done using this method. This method uses the pressure-driven membrane for purifying the impurities present in the contaminated waters. This method is carried out by application of a positive hydrostatic pressure to force water to leave and which then goes into the desalted zone. During this process, small monovalent ions like Na^+, Cl^- are rejected/ removed, and the process nanofiltration (NF), ultrafiltration (UF) and microfiltration (MF) can remove the molecules of increased sizes. Though these processes are not suitable for treating high salinity water, they are becoming attractive pre-treatment techniques for reverse osmosis (RO) which helps in the removal of bacteria and reduces turbidity. The MF and UF can be used efficiently in treating low-grade industrial water and can efficiently eliminate organic substances like micropollutants and multivalent ions present in the saline effluents. The types of filtration methods and the types of contaminants removed are given in Table 14.4.

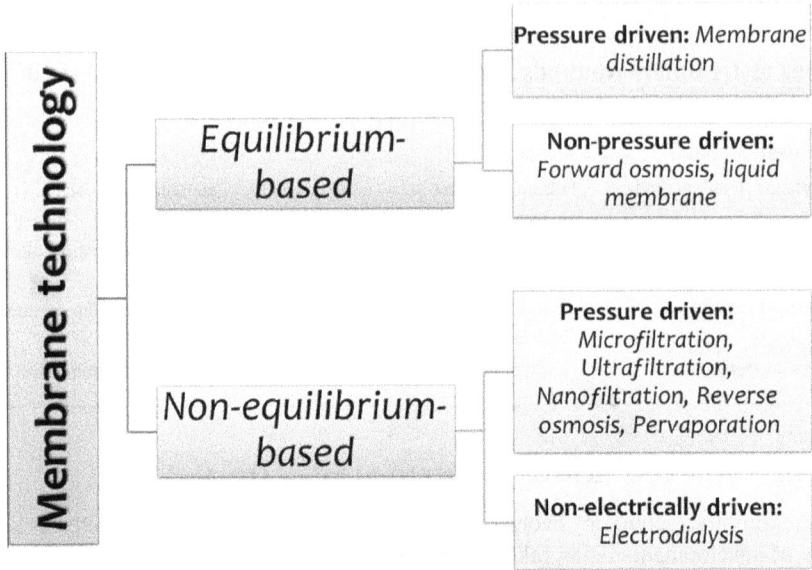

FIGURE 14.8   Representation of membrane technology process.

TABLE 14.4
Types of Filtration Methods and the Removable Contaminants

Membrane Process	Size of Particles Removed (μm)	Type of Contaminants Removed
Microfiltration	0.1–10	Suspended solids, bacteria and protozoa,
Ultrafiltration	0.01–0.1	Colloids, proteins, polysaccharides, most bacteria, viruses, etc.
Nanofiltration	0.001	Viruses, natural organic matter, multivalent ions
Reverse osmosis	0.001	Almost all impurities including multivalent ions

The nanofiltration method is a recent membrane filtration process (see Figure 14.9) that plays an important role in the wastewater treatment field. A recent development in membrane technology can be used in either the aqueous phase or non-aqueous phase. This technique is used for treatment in a wide variety of wastewaters and is used in different applications, and the main job is to remove all of the ions and organic substances which is being used in specified saline water application. These nanofiltration membranes are classified according to the membrane structure, pore shape and based on electrical charge and microporous and non-porous. A lot of research done by various researchers also found that the nanofiltration techniques have gained a special attraction in various applications like water reuse, industrial wastewater treatment and even in the treatment of drinking water (Abdel-fatah, 2018).

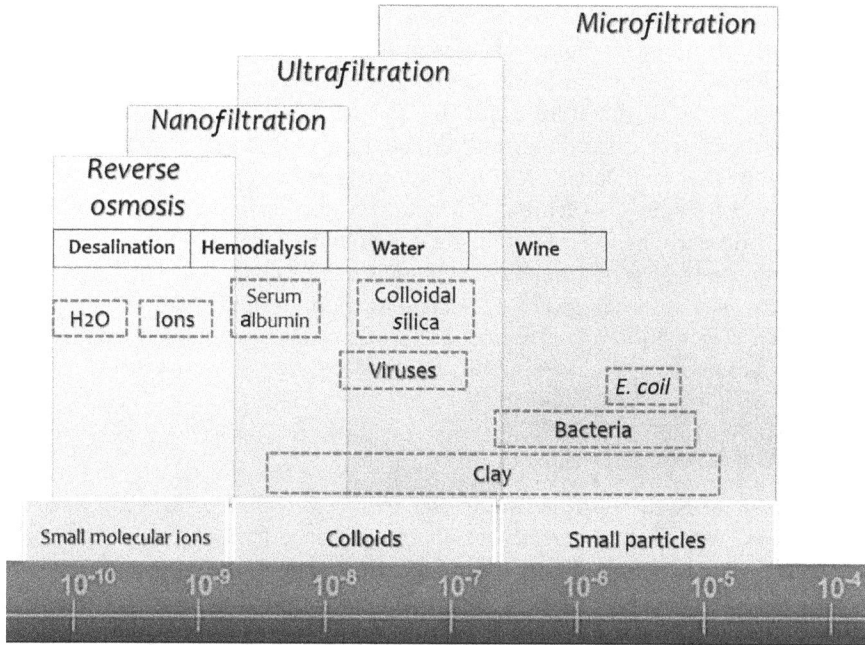

FIGURE 14.9 Spectrum of filtration methods used in wastewater treatment.

14.4.3 BIODEGRADATION OF POLLUTANTS IN SALINE WASTEWATER

Many industries like agro, food, petroleum and textile industries generate saline wastewater from their effluent bodies, and they have a high content of various organic pollutants such as aromatic hydrocarbons, phenols, nitroaromatics and various other dyes. The halophilic microbes are gaining increasing interest in industrial waste treatment and yet need to be explored further. These halophilic organisms have the potential to degrade hazardous substances under high salt conditions. The various pollutants emerging from saline wastewater include various aromatic compounds, hydrocarbons, phenols, amines, azo dyes and nitro-substituted compounds. The halophilic microbes can grow in hypersaline conditions. The non-halophilic microbes show optimal growth below 2% NaCl; whereas halotolerant and halo-dependent microbes can grow upto 30% NaCl and these halophilic microbes are classified according to the concentrations of the salt content (NaCl). Due to various industrial activities, these saline environments are frequently contaminated with various organic compounds, and this affects aquatic life severely, water portability, etc. These halophilic organisms have excellent efficiency in removing hydrocarbon pollutants. For halophilic microalgae or any type of microalgae, organic carbon is an important source of its cell for cell buildup.

Many reports gave us a clear picture that microalgae have the potential to degrade petroleum hydrocarbons. *Chlorella vulgaris* could remediate 88% of the crude oil hydrocarbons, the microalgae species *S. obliquus* and *C. vulgaris* can degrade the crude oil hydrocarbons, *Nostoc punctiforme* and *Spirulina plantesis* can degrade the aliphatic compounds present in the crude oil and various other species have also been reported in degradation of various hydrocarbons as per the various works of literature available. The experiments carried out by various researchers were carried out in the standard algal media for maintenance of their growth (Das and Deka, 2019). The microalgae *Prototheca zopci* was found to degrade petroleum hydrocarbons. Microalgae can degrade 60% of saturated aliphatic hydrocarbons, 41% of aromatic compounds, 23% of saturated aliphatic hydrocarbons and 26% of aromatic compounds can be degraded. The microalgae *Chlamydomonas* sp. when grown on acetate medium in light and dark conditions it produced a metabolite identified as 2-hydroxymuconic semialdehyde (Semple, Cain, and Schmidt 1999).

14.5   SUMMARY

This chapter focussed on saline wastewater treatment technologies using microalgae biomass. Water salinity creates unsuitable existence for many important water bodies. Saline wastewater discharged out of industries must be subjected to highly expensive treatment methods. Microalgae especially halophilic species provide an alternative solution for treating saline wastewater; in addition, it also generates many by-products. In this chapter, we have discussed various sources of wastewater, saline wastewater sources and their characteristics, microalgae biomass and their impact of wastewater on living ecosystems. Several microalgae are used in the treatment of saline industrial wastewater effluents, salinity and use of photobioreactors and other saline wastewater treatment technologies were mentioned. Additionally, saline wastewater pollutants, industries generating saline water and several treatment technologies applied to saline effluents were also discussed. Nonetheless, we conclude that algae have the potential to be involved in the biological segment of saline wastewater treatment and, therefore, demand further useful research and development in this direction. The successful development of sustainable saline wastewater treatment technology not only a solution to all types of industries but also create a considerable employment opportunity for algae growers, to save a lot of revenue and producing alternative by-products.

REFERENCES

Abdel-fatah, M. A. 2018. Nanofiltration Systems and Applications in Wastewater Treatment : Review Article. *Ain Shams Engineering Journal* 9(4): 3077–92. doi:10.1016/j. asej.2018.08.001.

Abdel-raouf, N. 2012. Microalgae and Wastewater Treatment. *Saudi Journal of Biological Sciences* 19 (3): 257–75. doi:10.1016/j.sjbs.2012.04.005.

Ajayan, K. V., Selvaraju, M., Unnikannan, P., Sruthi, P. 2015. Phycoremediation of Tannery Wastewater Using Microalgae Scenedesmus Species. *International Journal of Phytoremediation* 17(10): 907–16. doi: 10.1080/15226514.2014.989313. PMID: 25580934.

Al-jabri, H., Das, P., Khan, S., Thaher, M., Abdulquadir, M. 2021. Treatment of Wastewaters by Microalgae and the Potential Applications of the Produced Biomass — A Review. *Water* 13(1): 27. doi:10.3390/w13010027.

Amenorfenyo, D. K., Huang, X., Zhang, Y., Zeng, Q., Zhang, N., Ren, J., Huang, Q. 2019. Microalgae Brewery Wastewater Treatment: Potentials, Benefits and the Challenges. *International Journal of Environmental Research and Public Health* 16(11): 1910. doi: 10.3390/ijerph16111910. PMID: 31151156; PMCID: PMC6603649.

Ashok, V., Sanjay, K. G., Amritanshu, S. 2019. Photobioreactors for Wastewater Treatment Photobioreactors for Wastewater. In: Sanjay Kumar Gupta and Faizal Bux, (eds), *Application of Microalgae in Wastewater Treatment*, Vol 1: Domestic and Industrial Wastewater Treatment. doi: 10.1007/978-3-030-13913-1.

Balaji, S., Kalaivani, T., Sushma, B., Pillai, C. V., Shalini, M., Rajasekaran, C. 2016. Characterization of sorption sites and differential stress response of microalgae isolates against tannery effluents from Ranipet industrial area-An application towards phycoremediation. *International Journal of Phytoremediation* 18(8): 747–53. doi: 10.1080/15226514.2015.1115960. PMID: 26587690.

Balamurugan, S., Sathishkumar, R., Li, H. Y. 2020. Biotechnological Perspectives to Augment the Synthesis of Valuable Biomolecules from Microalgae by Employing Wastewater. *Journal of Water Process Engineering*: 101713. doi: 10.1016/j.jwpe.2020.101713.

Cai, T., Park, S. Y., Li, Y. 2013. Nutrient Recovery from Wastewater Streams by Microalgae : Status and Prospects. *Renewable and Sustainable Energy Reviews* 19: 360–9. doi:10.1016/j.rser.2012.11.030.

Church, J., Hwang, J. H., Kim, K. T., Mclean, R., Oh, Y. K., Nam, B., Chul, J., Hyoung, W. 2017. Effect of Salt Type and Concentration on the Growth and Lipid Content of Chlorella Vulgaris in Synthetic Saline Wastewater for Biofuel Production. *Bioresource Technology* 243: 147–53. doi: 10.1016/j.biortech.2017.06.081.

Das, B., Deka, S. 2019. A Cost-Effective and Environmentally Sustainable Process for Phycoremediation of Oil Field Formation Water for its Safe Disposal and Reuse. *Scientific Reports*: 1–15. doi: 10.1038/s41598-019-51806-5.

Farhan, A., Udaiyappan, M., Abu, H., Sobri, M. 2017. A Review of the Potentials, Challenges and Current Status of Microalgae Biomass Applications in Industrial Wastewater Treatment. *Journal of Water Processing Engineering* 20: 8–21. doi: 10.1016/j.jwpe.2017.09.006.

Figler, A., et al. 2019. Salt Tolerance and Desalination Abilities of Nine Common Green Microalgae Isolates. *Water* 11(12): 2527. doi:10.3390/w11122527.

Guzzon, A., Di Pippo, F., Congestri, R. 2019. Wastewater Biofilm Photosynthesis in Photobioreactors. *Microorganisms* 7(8): 252. doi: 10.3390/microorganisms7080252.

Lutzu, G. A., Dunford, N. T. 2019. Algal Treatment of Wastewater Generated during Oil and Gas Production Using Hydraulic Fracturing Technology. *Environmental Technology* 40(8): 1027–34. doi: 10.1080/09593330.2017.1415983. Epub 2017 Dec 19. PMID: 29226768.

Molazadeh, M., Ahmadzadeh, H., Pourianfar, H. R., Lyon, S. 2019. The Use of Microalgae for Coupling Wastewater Treatment with CO_2 Biofixation. *Frontiers in Bioengineering and Biotechnology*. doi: 10.3389/fbioe.2019.00042.

Mureed, K., Kanwal, S., Hussain, A., Noureen, S., Hussain, S., Ahmad, S., Ahmad, M., Waqas, R. 2018. Biodiesel Production from Algae Grown on Food Industry Wastewater. *Environmental Monitoring and Assessment* 190(5): 271. doi: 10.1007/s10661-018-6641-3. PMID: 29633020.

Obotey Ezugbe, E., Rathilal, S. 2020. Membrane Technologies in Wastewater Treatment. *Membranes* 10(5): 89. doi: 10.3390/membranes10050089.

Robinson, T., Mcmullan, G., Marchant, R., et al. 2001. Remediation of Dyes in Textile Effluent: A Critical Review on Current Treatment Technologies with a Proposed Alternative. *Bioresource Technology* 77(3): 247–55.

Sahu, P. 2021. A Comprehensive Review of Saline Effluent Disposal and Treatment: Conventional Practices, Emerging Technologies, and Future Potential. *Journal of Water Reuse and Desalination* 11(1): 33–65. doi:10.2166/wrd.2020.065.

Salama, E. S., Kurade, M. B., Abou-shanab, R. A., El-dalatony, M. M., et al. 2017. Recent Progress in Microalgal Biomass Production Coupled with Wastewater Treatment for Biofuel Generation. *Renewable and Sustainable Energy Reviews* 79 (March): 1189–211.

Semple, K. T, Cain, R. B., Schmidt, S. 1999. FEMS Microbiology Letters Biodegradation of Aromatic Compounds by Microalgae. *FEMS Microbiology Letters* 170(2): 291–300.

Shahid, A., Malik, S., Zhu, H., Xu, J., Nawaz, M. Z., Nawaz, S., Alam, A., Mehmood, M. A. 2019. Cultivating Microalgae in Wastewater for Biomass Production, Pollutant Removal, and Atmospheric Carbon Mitigation: a Review. *Science of the Total Environment*: 135303. doi: 10.1016/j.scitotenv.2019.135303.

Sui, Y., Vlaeminck, S. E. 2019. *Journal of Chemical Technology and Biotechnology* 94(4): 1032–40. doi: 10.1002/jctb.5850.

Valta, K., Kosanovic, T., Malamis, D., Moustakas, K., Loizidou, M. 2015. Overview of Water Usage and Wastewater Management in the Food and Beverage Industry. *Desalination Water Treatment* 53: 3335–47. doi: 10.1080/19443994.2014.934100.

Vo, H. N. P., Ngo, H. H., Guo, W., Liu, Y., Chang, S. W., Nguyen, D. D., Nguyen, P. D., Bui, X. T., Ren, J.. 2019. Identification of the Pollutants' Removal and Mechanism by Microalgae in Saline Wastewater. *Bioresource Technology* 275: 44–52 doi: 10.1016/j.biortech.2018.12.026.

Wollmann, F., Walther, T., Dietze, S., Ackermann, J. U., Krujatz, F., Bley, T., Steingroewer, J. 2019. Microalgae Wastewater Treatment: Biological and Technological Approaches. *Engineering in Life Sciences*: 860–71. doi: 10.1002/elsc.201900071.

15 Advanced Treatment Technologies of Industrial Saline Wastewater Using Microalgal Culture
Challenges and Future Prospective

K. Senthil Kumar, M. Naveen Kumar, and K. Jeevith
Kongu Engineering College

CONTENTS

DOI: 10.1201/9781003185437-18

15.1 INTRODUCTION

The Earth is covered with 71% of the water. In this, the usable form is merely 0.3%. Water resources play a vital role in a country's economy. Since most industries require water for every process, it decides the nation's growth. Perhaps, the most hazardous environmental issue existing on the planet is the absence of treatment of saline wastewater generated as a result of rapid industrialization. It consists of a large number of dissolved salts such as sodium, magnesium, calcium, potassium, sulfate, and chloride salts (Li et al., 2014). The units used for the measurement of salinity are the Practical Salinity Unit (PSU), percentage (%), g/kg, and parts per million. These waste salts are generated from process industries like food processing, textile, petrochemical, coal mine, pharmaceutical, fishing, tannery, mining, pulp and paper industry, and agricultural runoff (Song et al., 2018). The expanding interest in energy has prompted more interest in biofuels. Crudely biofuels are generated from agricultural wastes like oil crops, wheat, sugar beet, maize, and corn (Noraini et al., 2014). Compared to the traditional crops, the use of microalgae as biofuel offers many benefits like high development rate, high-effectiveness CO$_2$ moderation, and non-rivalry for farmland. An ideal answer for the synthesis of biofuel, as well as lessening the pressure factor put on freshwater utilization, is the use of saline water to develop microalgae, and it will likewise diminish the rate of production; in this way, enabling the biofuel to be financially attainable (Brennan and Owende, 2010).

In recent times, microalgae culture for treating saline wastewater has attracted more interest since it can adjust to an assortment of salinities. Microalgae have the ability to adjust saline wastewater conditions by adjusting pigment formation, yield of biomass, biochemical identity, and removal of contaminants efficiently. The treatment of saline wastewater utilizing microalgae and examinations with different advancements are not surveyed systematically and extensively till now (Cai et al., 2013). Hydrocarbon has played a dominating in building up the country's economy. Hydrocarbon has played a predominant in developing the nation's economy. The decline in fossil fuels has directed attention toward biofuel concepts. Numerous researches are being done in biofuel areas, and some percentage of biofuel is blended with hydrocarbon-based fuels. The ability of algae to sustain under a heavily polluted environment makes it effective to treat saline wastewater as well as for the production of biofuels making it a double advantage. In this chapter, the importance and

TABLE 15.1
Percentage Salinity in Industrial Wastewater

S.N.	Industry	% Salinity	References
1.	Textile	0.1–1	Mirbolooki et al. (2017)
2.	Tannery	5.4–5.7	Cataldo et al. (2016)
3.	Pharmaceutical	0–6	Guo et al. (2018)
4.	Coal mine	9.3	Ericsson and Hallmans (1994)
5.	Food processing	3	Shi et al. (2018)
6.	Agriculture run-off	0.12–1.4	Faria & Cohen (2013)
7.	Marine Culture	3–4.5	Song et al. (2018)
8.	Petrochemical	1.9–4.2	Jorfi et al. (2017)

necessity of treating saline wastewater with microalgae that also serves as a precursor for third-generation biofuels have been investigated.

15.2 SOURCES OF SALINE WASTEWATER

Industrial saline wastewater mainly consists of sodium chloride, sodium sulfate, magnesium sulfate, potassium nitrate, sodium carbonate, and similar inorganics. It also comprises heavy metals, organic pollutants, and pharmaceuticals and personal care products (PPCPs). The main sources of saline wastewater are process industries like pharmaceuticals, tanneries, textiles, pickling wastewater, and petroleum industries (Kubo et al., 2001). Another source is the farmland which is most prone to inorganic fertilizers. Due to excess rainwater, the nutrients are eroded and runoff occurs (Jiang et al., 2006). The drainage water is enriched in salts like nitrates, borates, sulfates, carbonates, chlorides, nitrogen, and phosphorous (Table 15.1).

15.3 EFFECTS DUE TO SALINE WASTEWATER

15.3.1 ECOLOGICAL EFFECTS

Because of modernization and fast industrialization, the significant contribution of contamination is coming from processing and manufacturing plants. A scope of toxins is released into the environment including the landfills, water bodies, and barren lands. All modern impacts are liable for establishing a risky climate for living beings due to the release of contaminants having harmful and profoundly focused poisoning elements. Likewise, it makes issues for living and nonliving things like habitats, infection, death toll, leading to environmental imbalance. Because of saline wastewater, there are some contamination impacts alongside their hazardous effects.

15.3.1.1 Wildlife Impacts

The untreated effluent released from the factories and production lines massively affects the water environment. For the most part, these wastewaters are released into the stream and influence the hydroponics, animal life cycles, and so forth. The use of unacceptable polluted water and the release of contaminated industrial effluents into water bodies indirectly affect the life of people depending on that water. Quite possibly, the most major issue in nations like India is the inappropriate administration of huge wastes created by different process industries. The great challenge involved in the perilous removal of these scraps to the encompassing circumstances (Kanu and Achi, 2011).

15.3.1.2 Marine Impacts

Exhaustion of dissolved oxygen (DO) level in water is the most dangerous impact of discharging waste into water bodies since oxygen is extremely fundamental for marine life. The main environmental issues related to the process industry are the utilization of water, treatment of water, and removal of effluent. Water shortage is the main manageability problem confronting the process industry. Consequently, reducing the release of effluent into the diverse ecological pathways can lessen the ecological danger.

Saline wastewaters produced from various phases of manufacturing and process industries contain large amounts of contaminations that are extremely dangerous to the ecological balance if discharged without appropriate discharge methods. The degree of natural contamination by the discharge of saline wastewater is extremely high.

15.3.2 HEALTH EFFECTS

Untreated industrial wastewater has high levels of toxins and dangerous chemical matters that are discharged to waterways, lakes, ponds, and the sea. It may end with sickness, diseases, human cell rupture, and so forth; for instance, inorganic arsenic causes the development of tumors in a person's well-being. Thus, untreated process effluents might be liable for a large number of sicknesses and early passing across the world. Aside from that, it may cause serious medical illness for the individuals who are consuming such harvests, since soil and yields are defiled by poisonous toxic substances and dangerous chemicals that are debased in water.

The commotion level coming due to the machines equipped manufacturing industry may abuse the cut-off permitted by the law and mess hearing up. Health impacts might be applied straightforwardly at the place of use (influencing the specialists) and in the existence cycle (influencing the user). Since clothes come in full-time contact with skin, toxins are stuck within the skin, particularly when the victim is warm and skin pores are opened to percolate sweat. When consumed by people, heavy metals affect all body parts. The impacts on physical and mental well-being of human will be very serious when significant degrees of metals are present in the body. The impact is especially genuine in youngsters because the harmful color and additionally weighty metal aggregation may contrarily influence their development and might be their life too.

15.4 STAGES IN INDUSTRIAL WASTEWATER TREATMENT

Industrial wastewater treatment is a tedious process that involves series of processing stages. In most cases, industrial wastewater treatment has three stages. In some special cases, another stage is added resulting in quaternary stages. The first is the preliminary step which involves the elimination of large suspended solids through a series of screens. Second, the wastewater is allowed for the physical process where the organic and inorganic fines are settled and removal of biochemical oxygen demand (BOD) content, dissolving all the organic loads by supplying oxygen to aerobic bacteria. The third stage is the use of chemicals to remove the remaining contaminants. The final stage comprises modern membrane processes for the effective recirculation of water.

15.4.1 PRIMARY OR PRELIMINARY TREATMENT

The wastewater collected from all the possible vents is gathered in single drainage. Initial screening is done with the help of screens and grit-like structures and are allowed to settle in large tanks. Solids that are heavier settle down faster, while the lighter particles tend to float. Once the sedimentation process is over, the remaining liquid is moved to the next stage. Normally, sedimentation is carried out series of tanks to get maximum removal of sludge. The sludge retained is fed to the digester for further processing like gasification and composting. Nearly 50% of the suspended solids are removed in the primary treatment.

15.4.1.1 Grit Removal

Grit or screen-like structures made with steel are used to remove the suspended solids in wastewater. Wastewater collected from the sources is sent through a series of screens where the solids are retained based on the size of screens used. The suspended solids retained must be removed regularly from the consecutive grits for the better efficiency of separation.

15.4.1.2 Sedimentation

Primary sedimentation is done sequentially after grit removal. Sedimentation is the removal of organic solids that can settle down due to external force, especially gravitational force. The wastewater is allowed to settle in large tanks for optimal time so that maximum solids settle down and are recovered as sludge. The sludge can be used as manure, fuel, source of carbon, etc., and much more based on the type of wastewater.

15.4.2 SECONDARY TREATMENT

The secondary stage of wastewater treatment deals with the next level to the primary stage and is designed purposefully to degrade and reduce the organic contaminant through aerobic and anaerobic biological processes. It is done in one of the three ways: biofiltration, aeration, and oxidation ponds. Oxidation of wastewater further reduces the contaminant level in the water (Figure 15.1).

15.4.2.1 Biofiltration

Traditional biofiltration alludes to filtration measures where the channel contains permeable material that is colonized by indigenous microbial networks. These

Primary treatment (Suspended solids removal)	Physical process involving screening, sedimentation and grids.
Secondary treatment (dissolved organics removal)	Biological oxidation in aeration tanks like activated sludge process.
Tertiary treatment (nutrients removal)	Microalgal culture in ponds, photobioreactor, etc. for efficient removal of minerals like nitrogen, phosphorous, etc.

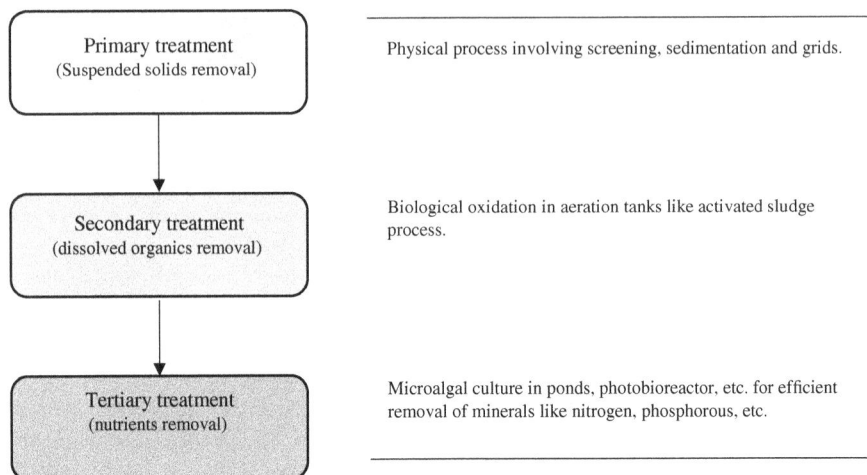

FIGURE 15.1   Overview of saline wastewater treatment.

microorganisms perform in any event as one of the fundamental elements of the filtration cycle. Biofilters like granular dynamic carbon (GAC) channels, rapid sand channels (RSFs), and slow sand channels (SSFs) are key cycles in the biotreatment of wastewater in both industrialized and developing nations. The principle motivation behind biofiltration is the evacuation of biodegradable organic carbon, yet a wide assortment of conceivable filter setups, along with an expansive scope of various water types and pretreatment techniques. This strategy of secondary treatment of wastewater utilizes sand beds, contacting beds, or trickling channels to guarantee that extra residue is eliminated from wastewater. Among the three types, trickling filters are normally the best for batch process wastewater treatment.

15.4.2.2   Aeration

Aeration is a time-consuming step that builds oxygen immersion by introducing air with wastewater. However, it is lengthy yet powerful step that involves mixing effluent in a pool of microorganisms. The high-impact measure produces CO_2, H_2O, nitrates, sulfates, and heat with stable material usually utilized as manure for crops. The aerobic microscopic organisms that survive in oxygen-rich conditions digest the waste. A few unique strategies are commonly used to ensure the oxygen-consuming deterioration of organic matter including windrow treating the soil, aerated static heap, and in-vessel fertilizing the soil. The aeration step acts to improve porosity and oxygen accessibility to enhance the disintegration rate. Significantly, the exothermic behavior of aerobic decomposition is used to keep the fertilizer heap at the essential raised temperatures for the microbes to flourish. In aerated heaps, the material is laid out in heaps as opposed to lines and doesn't go through a similar turning methodology as windrow composting. Also, air blowers might be added to the framework to additional aid the air circulation of the compost stack. In conclusion, in-vessel treating the soil gives the best control of conditions and is acted in a shut drum or silo where conditions like dampness, aeration, and temperature can be all the more decisively controlled.

15.4.2.3 Oxidation Ponds

Normally used in warmer climates, this technique characterized water bodies such as lagoons, permitting wastewater to go through for a determined time before being retained for half a month. Biodegradable contaminants can be reduced down to safe levels with the help of secondary water treatment and released to surrounding environments with the lower potential of risks associated. Remarkably, microalgae have a hugely influential part in the treatment of indigenous wastewater with the help of oxidation pools. Microalgae consume elemental N and P for producing biomass. They additionally can supply O_2 to heterotrophic oxygen-consuming microorganisms to mineralize organic pollutants. Photosynthetic oxidation is fascinating to decrease operational expenses and to restrict the dangers of contamination volatilization under mechanical oxidation.

15.4.3 TERTIARY TREATMENT

Tertiary treatment is the final stage in treatment process where all the remaining organic contaminants like nitrogen and phosphorous, bacteria, viruses, and parasites are removed. Processes like filtration, ion exchange, and activated carbon adsorption are commonly used techniques for purifying wastewater.

15.4.3.1 Filtration

The process of removing the particulate matter in wastewater by passing through a porous media is called filtration. It generally consists of different levels of media like sand, gravel, and charcoal. Treated water is being used in agricultural fields and aquaculture and recycled for industrial uses, thus increasing water sustainability in an ecofriendly manner.

15.4.3.2 Ion Exchange

A chemical reaction in which the number of free mobile ions of a solid is exchanged in a solution is termed an ion exchange process. Ion exchange is a process that employs demineralization or water softening, resins like zeolite, sodium polystyrene sulfonate, clay, and soil humus. mainly used for this process. It is classified into cationic and anionic exchangers. Ion exchange is broadly utilized in the sugar refineries, breweries, distilleries, casting, petrochemical, drug innovation, food industry, drinking water treatment, cement, paper industry, mining industries, and many other process plants.

15.4.3.3 Activated Carbon Adsorption

Activated carbon adsorption is a surface phenomenon process commonly used technology for the adsorption of contaminants in wastewater. The efficiency of adsorption relies on the formation of activated carbon deployed water characteristics and operating parameters. Desorption is done at particular intervals for the regeneration of active sites of the adsorbent. Various types of activated carbon adsorption filters are available for household and industry requirements. The advantages of using activated carbon filters are simple in process, maximum removal of VOCs, easy maintenance, etc.

15.4.3.4 Treatment of Saline Wastewater Using Microalgal Culture

Existing treatment of saline wastewater includes biological processes and physicochemical processes. Among these, biological processes are preferred only when the salinity ranges from 4% to 12% (low salinity levels). Microalgae represent a bioremediation technique in saline effluent treatment since they could deal with a huge load of contaminants, such as (Huang et al., 2015) organic substances, heavy metals, and micropollutants (Bai and Acharya, 2017). The fundamental contaminations eliminated in the tertiary stage are nitrogen and phosphorous. Common strategies used to eliminate these poisons are anaerobic digestion and post-treatment with nitrification and denitrification. However, several cycles of anaerobic, nitrification, and denitrification are needed to meet the criteria of contaminant levels permitted by the legislative norms. Also, disadvantages of these methods are the requirement of more tanks and complexity involved in recycling activated sludge, which brings about a rise in cycle expenses, intricacy, and power consumption levels. Likewise, nitrogen and phosphorus evacuation can be accomplished by chemical strategies, utilizing precipitants like aluminum and Fe salts. Consequently, these methods require further treatment of solid sludge contaminated with the huge amount of salts produced as a result of precipitation (Figure 15.2).

To overthrow the shortcomings related to the ordinarily utilized tertiary treatment techniques, biological treatment utilizing microalgae (a word generally used to allude to photosynthetic microorganisms) was widely concentrated in the previous many years. Since microalgae need enormous amounts of elemental nitrogen (N) and elemental phosphorus (P) for their turn of events, these microalgae can adequately allow elemental N and elemental P from effluents. In fact, high nitrogen and phosphorous expulsion effectiveness (85%–95%) from effluents of various sources (e.g., industrial, rural, and corporation) were recognized in ongoing explore for microalgae.

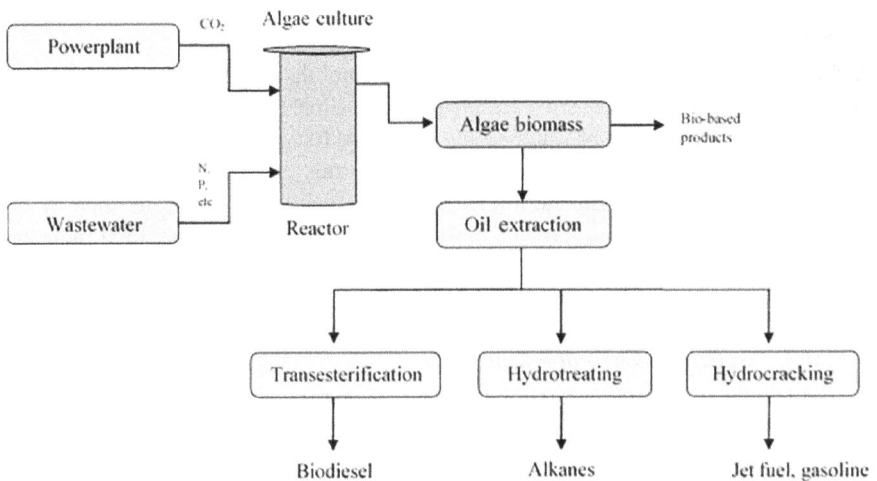

FIGURE 15.2 Schematic depiction of microalgal culture for wastewater treatment.

15.5 FACTORS CONCERNING MICROALGAL CULTURE

Microalgae growth depends on conditions like light, pH, nutrient amount, temperature, level of salinity, quality, and aeration that impact photosynthesis motion and conduct of the microalgae development rate. The major factors concerning the microalgal culture are discussed in the following sections.

15.5.1 GAS TRANSFER RATE

Microalgal cultivation involves the consumption of carbon dioxide joined by oxygen evolution because of photosynthesis. The DO content accumulates in the growth medium and the exchange of the oxygen to the atmosphere evolved during photosynthesis is minimal. This accumulation causes photooxidation. Removal of oxygen is observed to be a pivotal aspect in the open photobioreactors and can be enhanced by stirring (Fernández et al., 2013). Since the atmospheric CO_2 is not sufficiently consumed by microalgae, CO_2 is fed to the ponds to circumvent carbon limitation. The CO_2 fed in open ponds relies on the ideal culture medium, stirring, pH, mass transfer coefficient, time of contact for gas-liquid, and the sparger type (Kumar et al., 2015). Some ponds are installed with micropore air diffusers at the base to feed gas as tiny air bubbles. Another arrangement is utilizing, CO_2 from exhaust gas through ethylene propylene diene monomer (EPDM) rubber permeable film tubes which are put at the lower part of the tanks.

15.5.2 AVAILABILITY OF LIGHT

The open photobioreactor ought to be planned in a manner so that light is effectively accessible for the microalgae to have ideal development. In open lakes, the accessibility of light is influenced by pond depth and cell concentration (Chisti, 2016). The arrangement of the reactor respective with the sun is additionally a significant factor to consider so that microalgae get daylight for the entire day. In contrast, high intensities of light can also lead to a phenomenon called photoinhibition. This phenomenon also relies upon different parameters like temperature, oxygen fixation, and algae species. Minimal light intensities also become a restricting variable for microalgal growth. Photo limitation can also cause shading of cells on the deeper side. These limitations can be overcome by rising the recurrence of the L/D cycles (Chen et al., 2016). During higher illumination, microalgal cells at the surface side are photoinhibition, whereas the inner cells in microalgal cells are photo limited. Thus it is imperative to look after L/D cycles in photobioreactor by mixing the ponds effectively so that cells are equally exposed to light.

15.5.3 TEMPERATURE

Temperature is the crucial factor that influences the metabolism order of microorganisms and differs for different species. The optimal temperature range lies between 25°C and 35°C for most of the species (Pires et al., 2012). Marine organisms are more susceptible to temperature changes, and temperature aside mentioned can cause the destruction of cells. Temperature in open pond reactors depends on sunlight, ambient air temperature, and evaporation rate. Culture temperatures may change due to seasonal changes and region of reactors. Hence it is essential to select the right microalgal species considering the environmental factors. Uniform temperature throughout the year yields the maximum amount of biomass (Slegers et al., 2013). Research shows that changes in environmental conditions (Eustance et al., 2016), like increase in light and temperature, ended with an elevated rate of fatty acids content for the application in biodiesel generation.

15.5.4 pH

Most of the algae develop under the alkaline pH of 7.0–9.0. *Spirulina platensis* can develop in conditions with a pH running somewhere in the range of 8 and 11, whereas the microalgae *Chlorococcum littorale* in acidic conditions, with pH, somewhere in the range of 4.0 and 7.0 (Ota et al., 2009). Since pH is a crucial factor in physiological changes of microalgae, it is important to maintain the culture medium to have maximum growth. The supply of CO_2 can also alter the culture medium. The increase of CO_2 concentration in the feed can reduce the pH of the growth medium. Subsequently, the CO_2 supplied to microalgal cells ought to be cautiously observed, so that reducing pH does not influence the growth of microalgal cells (Pires et al., 2013). Also, during the development period, it is recognizable to notice a rise in pH, because of carbon dioxide consumption. Subsequently, regular checking of the pH in the medium is vital to inspect the microalgal development and to follow SOP to keep up the medium in the necessary pH range.

15.6 UTILIZATION OF MICROALGAE GROWN IN WASTEWATER

15.6.1 Bio-based Products

The capability of microalgae to grow in the uninhabitable environment and the capacity for metabolic transformation below pressure conditions make microalgal biomass an alluring crude substance ensuring scale-up possibilities and for application in industries. The improvement of novel bio-based items is a creative dimension in the stream of microalgae research. It is feasible to show their utilization in the drug, dietnourishment, corrective, and biofuel productions (Yu et al., 2015). It is possible to extract various bioproducts from microalgae that are grown in a different culture with various parameters. In addition to biofuels, there are opportunities for utilizing crude microalgae for applications in anti-inflammation treatment, malignant growth control, diabetes control, and antibacterial activities. The ability to cure various medical issues helps in enabling microalgae commercialization.

15.6.2 In Food Industry

A large assortment of microalgae can produce high protein content. The insufficiency of protein supplies in human food supply prompted malnutrition due to the increasing population and shortage of food production. Hence there emerges a necessity for an alternative source for protein which is highly fortified with protein extracted from microalgae. Microorganisms like fungi, algae, bacteria, and yeast are often utilized as the root of protein. Since microalgae consist of higher levels of amino acids than bacteria and fungi, it is recommended for the consumption of humans. The amino acids example of practically all green growth are similar to that of the Food and Agriculture Organization (FAO) requirements with some traces of sulfur. Chlorophyll extracted from microalgae is permitted as a food coloring agent for application in food industries. Microalgae stand as a source of vitamins that are essential for the development of all human beings. Nutritional supplements are also being synthesized from microalgae. Spirulina is an outstanding source of vitamin B complex for vegetarian consumers and a good source of vitamin B_{12}.

15.6.3 Conversion into Biofuels

Biofuels from algae are considered to be third-generation biofuels that overwhelmed the disadvantage of first- and second-generation biofuels. It is feasible to convert algal biomass into biofuel economically using different technologies. Since the oil yield is very high (Nascimento et al., 2013), algae have much more capacity to generate biomass and oil suitable for conversion into biofuels. They have higher growth per unit area compared with oil crops (Griffiths and Harrison, 2009). A lot of researches are being performed in thrust areas of microalgae culture to identify optimum conditions for higher growth and extraction of maximum oil lipids from it (Table 15.2).

15.6.4 CO₂ Sequestration

In the photobioreactor, conversion of biomass is carried out to mitigate the CO_2 levels in the environment. Both marine and freshwater microalgae are minuscule photosynthetic entity. CO_2 emissions from sources like power plants and process industries can be captured by microalgae like Cyanophyceae and eukaryotic microalgae, Chlorophyta (Fulke et al., 2010). Photosynthesis helps the microalgae to capture CO_2 from the atmosphere and thereby reducing CO_2 from the environment. Exhaust gas from power plants can be redirected to pond systems and utilized for the growth of microalgae. Nonetheless, only a minimal number of microalgae can endure the

TABLE 15.2
Comparison of Oil Yield from Different Sources

S.No.	Oil Sources	Yield (per hectare)	References
1.	Microalgae	1,00,000	Dermirbas et al. (2011)
2.	Castor	1,413	Dermirbas et al. (2011)
3.	Groundnut	2,689	Dermirbas et al. (2011)
4.	Palm	5,366	Ahmad et al. (2011)
5.	Sunflower	952	Sing et al. (2013)
6.	Corn/maize	172	Mubaraka et al. (2015)
7.	Soyabean	636	Ahmad et al. (2011)

significant levels of NO_x and SO_x present in vent gases. Some microalgae traits can assimilate CO_2 from saline wastewater which has Na_2CO_3 and $NaHCO_3$ and other minerals. The high cost involved in processing and price affordability of bio-oil extraction from microalgae could be repaid with bio-alleviation of CO_2 discharges which might be used to lessen cost. Hence, microalgal culture could be joined with CO_2 sequestration and biofuel production.

15.7 CHALLENGES

Microalgal development is by all accounts straightforward since it is feasible to check the characteristic presence of these microorganisms in several conditions; nonetheless, the change of biomass to bioproducts on a huge scale is energizing. Notably, microalgae need organic sources for heterotrophic growth and light energy and CO_2 sources for autotrophic development. Microalgae have revealed a great prospective for the seizure of CO_2 and the production of bioenergy. In any case, their way of life is as yet not financially workable for these uses. The incorporation of one of the two applications with effluent management ought to be tried at a business scale to wipe out the requirement for fresh water and supplements, decreasing the way of culture expenses and its current circumstance. Investigation ought to be concentrated around the advancement of microalgal culture and collecting, reaping, and discharging with boundaries beneath as far as possible. Concerning culture, studies ought to be accomplished to assess the microalgal development in an extensive reach and outrageous ecological conditions such as light, pH, and toxin focus. Utilizing effluent as a culture medium, the investigation of pollutants is critical for the determination of microalgal species.

15.8 FUTURE PERSPECTIVES

Study effort concerning the linkage amid microalgae and other microbial strains ought to be made. The research reveals that microalgae can accomplish higher development rates. Besides, this kind of microbial affiliation can be gainful for biomass production. The micro-sized and low density of microalgal cells are the principal viewpoints that intensify the trouble and budget of collecting and utilizing traditional strategies (filtration, centrifugation, gravity sedimentation). Contrastingly, bio flocculation and natural conglomeration promote basic gravity settling, improving the detachment of cells from the medium. For certain species, the conglomeration might be accomplished by limiting elemental N and CO_2 supply. The physiological attributes of the microalgae and the collection instrument ought to be concentrated added to discover culture conditions that endorse the favored culturing process. The challenge existing is to discover minimal cost-effective, straightforward, easy-to-use strategies, without compromising our significant water resources and save guarding water relied occupations. For the production of biofuel on large-scale, microalgae biorefineries can be applied jointly with economically feasible strategies. The fundamental benefits of the open lake are moderately low implementation cost and lower power consumption costs, joined with biorefinery ideas can be viewed as probably the most ideal alternative for microalgal biofuel generation. Studies about the ecological and novel techniques for microalgal production are increasing day by day. Even though microalgal development for saline effluent treatment drives has been broadly depicted in the previous works, the utilization of microalgae in the water treatment field is yet in a beginning phase of information, which is mainly due to the wide assortment of substitute mixes that might be possible. Comparing the efficiency of phosphorus and nitrogen evacuation got with microalgae, it is suggested to utilize the microalgae in the treatment of saline wastewater, earlier to salt removal by physicochemical treatment.

REFERENCES

Ahmad, A. L., Mat Yasin, N. H., Derek, C. J. C., & Lim, J. K. (2011). A review–microalgae as a sustainable energy source for biodiesel production. *Renewable and Sustainable Energy Reviews*, 15, 584–593.

Bai, X., & Acharya, K. (2017). Algae-mediated removal of selected pharmaceutical and personal care products (PPCPs) from Lake Mead water. *Science of the Total Environment*, 581–582, 734–740. doi:10.1016/j.scitotenv.2016.12.192.

Brennan, L., & Owende, P. (2010). Biofuels from microalgae—A review of technologies for production, processing, and extractions of biofuels and co-products. *Renewable and Sustainable Energy Reviews*, 14, 557–77.

Cai, T., Park, S. Y., & Li, Y. (2013). Nutrient recovery from wastewater streams by microalgae—Status and prospects. *Renewable and Sustainable Energy Reviews*, 19, 360–369. doi:10.1016/j.rser.2012.11.030.

Cataldo, S., Loddo, V., Mirenda, E., Palmisano, L., Parrino, F., & Piazzese, D. (2016). Combination of advanced oxidation processes and active carbons adsorption for the treatment of simulated saline wastewater. *Separation and Purification Technology*, 171, 101–111. doi:10.1016/j.seppur.2016.07.026.

Chen, Z., Zhang, X., Jiang, Z., Chen, X., He, H., & Zhang, X. (2016). Light/dark cycle of microalgae cells in raceway ponds—effects of paddlewheel rotational speeds and baffles installation, *Bioresource Technology*, 219, 387–391. doi:10.1016/j.biortech.2016.07.108.

Chisti, Y. Large-scale production of algal biomass—raceway ponds, in: F. Bux, Y. Chisti (Eds.), *Algae Biotechnology—Products and Processes*, Springer, Cham, 2016, pp. 21–40.

Ericsson, B., & Hallmans, B. (1994). Treatment and disposal of saline wastewater from coal mines in Poland. *Desalination*, 98(1–3), 239–248. doi:10.1016/0011-9164(94)00148-0.

Eustance, E., Wray, J. T., Badvipour, S., & Sommerfeld, M. R. (2016). The effects of cultivation depth, areal density, and nutrient level on lipid accumulation of Scenedesmusacutus in outdoor raceway ponds. *Journal of Applied Phycology*, 28, 1459–1469. doi:10.1007/s10811-015-0709-z.

Fernández, F. G. A., Sevilla, J. M. F., & Grima, E. M. (2013). Photobioreactors for the production of microalgae. *Reviews in Environmental Science and Biotechnology*, 12, 131–151. doi:10.1007/s11157-012-9307-6.

Fulke, A. B., Mudliar, S. N., Yadav, R., Shekh, A., Srinivasan, N., et al. (2010). Bio-mitigation of CO_2, calcite formation and simultaneous biodiesel precursors production using Chlorella sp. *Bioresource Technology*, 101, 8473–8476.

Griffiths, M. J., & Harrison, S. T. L. (2009). Lipid productivity as a key characteristic for choosing algal species for biodiesel production. *Journal of Applied Phycology*, 21, 493–507.

Guo, N., Wang, Y., Tong, T., & Wang, S. (2018). The fate of antibiotic resistance genes and their potential hosts during bio-electrochemical treatment of high-salinity pharmaceutical wastewater. *Water Research*, 133, 79–86. doi:10.1016/j.watres.2018.01.020.

Huang, W., Li, B., Zhang, C., Zhang, Z., Lei, Z., Lu, B., & Zhou, B. (2015). Effect of algae growth on aerobic granulation and nutrients removal from synthetic wastewater by using sequencing batch reactors. *Bioresource Technology*, 179, 187–192. doi:10.1016/j.biortech.2014.12.024.

Jiang, D. M., Wang, X. Y., Liu, M. H., & Lu, G. F. (2006). Study on agricultural structure and non-point source pollution—a case in Dapu Town of Yixing City. *Ecological Economy*, 2, 270–280.

Jorfi, S., Pourfadakari, S., & Ahmadi, M. (2017). Electrokinetic treatment of high saline petrochemical wastewater—Evaluation and scale-up. *Journal of Environmental Management*, 204, 221–229. doi:10.1016/j.jenvman.2017.08.058.

Kanu, I, & Achi, O. K. (2011). Industrial effluents and their impact on water quality of receiving rivers in Nigeria. *Journal of Applied Technology in Environmental Sanitation*, 1(1), 75–86.

Kubo, M., Hiroe, J., Murakami, M., Fukami, H., & Tachiki, T. (2001). Treatment of hypersaline-containing wastewater with salt-tolerant microorganisms. *Journal of Bioscience and Bioengineering*, 91(2), 222–224. doi:10.1263/jbb.91.222.

Kumar, K., Mishra, S.K., Shrivastav, A., Park, M.S., & Yang, J.W. (2015). Recent trends in the mass cultivation of algae in raceway ponds. *Renewable and Sustainable Energy Reviews*, 51, 875–885. doi:10.1016/j.rser.2015.06.033.

Li, J., Pu, L., Han, M., Zhu, M., Zhang, R., & Xiang, Y. (2014). Soil salinization research in China—Advances and prospects. *Journal of Geographical Sciences*, 24(5), 943–960. doi:10.1007/s11442-014-1130-2.

Mirbolooki, H., Amirnezhad, R., & Pendashteh, A. R. (2017). Treatment of high saline textile wastewater by activated sludge microorganisms. *Journal of Applied Research and Technology*, 15(2), 167–172. doi:10.1016/j.jart.2017.01.012.

Mubaraka, M., Shaijaa, A., & Suchithrab, T. V. (2015). A review–the extraction of lipid from microalgae for biodiesel production. *Algal Research*, 7, 117–123.

Nascimento, I. A., Marques, S. S. I., Cabanelas, I. T. D., Pereira, S. A., Druzian, J. I., de Souza, C. O., et al. (2013). Screening microalgae strains for biodiesel production—lipid productivity and estimation of fuel quality based on fatty acids profiles as selective criteria. *Bioenergy Research*, 6, 1–13. doi:10.1007/s12155-012-9222-2.

Noraini, M. Y., Ong, H. C., Badrul, M. J., & Chong, W. T. (2014). A review on potential enzymatic reaction for biofuel production from algae. *Renewable and Sustainable Energy Reviews*, 39, 24–34.

Ota, M., Kato, Y., Watanabe, H., Watanabe, M., Sato, Y., Smith, R. L., & Inomata, H. (2009). Effect of inorganic carbon on photoautotrophic growth of microalga Chlorococcumlittorale. *Biotechnology Progress*, 25, 492–498.

Pires, J. C. M., Alvim-Ferraz, M. C. M., Martins, F. G., & Simões, M. (2012). Carbon dioxide capture from flue gases using microalgae—engineering aspects and biorefinery concept. *Renewable and Sustainable Energy Reviews*, 16, 3043–3053. doi:10.1016/j.rser.2012.02.055.

Pires, J. C. M., Alvim-Ferraz, M. C. M., Martins, F. G., & Simões, M. (2013). Wastewater treatment to enhance the economic viability of microalgae culture. *Environmental Science and Pollution Research*, 20, 5096–5105.

Shi, X., Yeap, T. S., Huang, S., Chen, J., & Ng, H. Y. (2018). Pretreatment of saline antibiotic wastewater using marine microalga. *Bioresource Technology*, 258, 240–246. doi:10.1016/j.biortech.2018.02.110.

Sing, S. F., Isdepsky, A., Borowitzka, M. A., & Moheimani, N. R. (2013). Production of biofuels from microalgae. *Mitigation and Adaptation Strategies for Global Change*, 18, 47–72.

Slegers, P. M., Lösing, M. B., Wijffels, R. H., Straten, G., & Boxtel, A. J. B. (2013). Scenario evaluation of open pond microalgae production. *Algal Research*, 2, 358–368. doi:10.1016/j.algal.2013.05.001.

Song, W., Li, Z., Ding, Y., Liu, F., You, H., Qi, P., & Jin, C. (2018). Performance of a novel hybrid membrane bioreactor for treating saline wastewater from mariculture—Assessment of pollutants removal and membrane filtration performance. *Chemical Engineering Journal*, 331, 695–703. doi:10.1016/j.cej.2017.09.032.

Yu, X., Chen, L., & Zhang, W. (2015). Chemicals to enhance microalgal growth and accumulation of high-value bioproducts. *Frontiers in Microbiology*, 56(6), 1–10.

16 Treatment of Industrial Saline Wastewater Using Phytoremediation

M. Naveen Kumar and K. Senthil Kumar
Kongu Engineering College

M. Venkata Ratnam
Mettu University, Ethiopia

S. Samraj
MVJ Engineering College

M. Neeraja and M. Chithra
Kongu Engineering College

CONTENTS

DOI: 10.1201/9781003185437-19

16.1 INTRODUCTION

Tanneries are of tremendous importance worldwide. It involves the processing of raw animal hides to produce a resilient and durable product which is leather. As the process involves many kinds of treatments to make skins and hides to give out a product for our daily use, various chemicals are used for their production. Tanneries are also huge consumers of freshwater. Thus, it is obvious that wastes generated from these industries are released along with used water. This saline wastewater generated is potentially toxic with chemicals containing mineral salts, chromium, fats, acids, cyanides, amines and other organic compounds. These chemicals pose serious threats to the environment through the discharge of toxins into natural resources like water, air, soil and eventually pollute them. Also, human beings are also affected by those chemicals which would lead to serious health problems like cancer, diabetes, cardiac arrest and feto-maternal death. Conventional technologies like coagulation and flocculation methods are found to be less effective and require huge installation cost. Therefore, a better remediation method is of high importance. Biological treatment like phytoremediation has found to be an efficient and eco-friendly method in accumulating elements and compounds in recent years (Table 16.1).

This technique involves the use of green living plants along with microorganisms that have made their home in the roots of those plants with proper farming techniques to remove toxic contaminants from the environment. This method has also been approved as an economical one in enhancing the environmental clean-up. It adds up to the ability of green plants to suck, accumulate and pile up elements and various organic compounds directly from the natural water bodies

TABLE 16.1
Availability of Hides and Skins in India in 2018

Category	No. of Pieces (in millions)
Cattle hides	30.770
Buffalo hides	35.875
Goat skins	96.900
Sheep skins	40.660

as well as wetlands, thus rendering them harmless. The greatest advantage of this novel technique is that plants used for the remediation process are efficient in taking up heavy metals like cadmium, arsenic, antimony, lead, aluminium and zinc (Adetutu et al., 2012).

In this chapter, information relevant to the phytoremediation process of tannery soak saline effluents with selected plant species like *Sesuvium portulacastrum* and *Salicornia brachiata* has been discussed and analysed. These two plants have been proved to be efficient in accumulating chromium and salt from the tannery effluent and reducing their concentration in the effluent. This novel technique shows the possibility of efficient treatment and an alternate effective method.

16.2 PROCESS OF TANNERY INDUSTRIES

16.2.1 Pre-tanning (Beam House Operation): Soaking

In this process, raw animal hides which are preserved using chemicals like sodium hydroxide and bactericides regain their inherent water content. Other substances like dirt, blood, manure and preservatives are removed and cleaned.

16.2.2 Fleshing and Trimming

In this operation, the extraneous tissue from the animal hide is removed which consists of hair and the epidermis layer. Chemical dissolution of the hide removes the hair, and the epidermis layer is removed by treating the hide in an alkaline medium of sulphide and lime. These have to be removed because excessive meat content remains stuck to the hide during skinning at the slaughterhouse. This will be removed by fleshing process proceeded by unhairing and liming. Liming and unhairing produce the effluent stream with the highest chemical oxygen demand (COD) value.

16.2.3 De-liming and Bating

The un-haired, fleshed and alkaline hides are neutralized with ammonium salt and treated with enzymes similar to removing hair remnants and degrading proteins. During this process, hair roots and pigment are removed. This results in the major part of the ammonium load in the effluent.

16.2.4 Pickling

Pickling increases the pH of the hide to the value of 3 by adding acid liquor and chromium salts to enter into the hide. Swelling of the hide can be prevented by the addition of salts. Fungicides and bactericides of about 0.03%–2% by weight are usually applied to preserve the hides.

16.2.5 TANNING (TAN YARD OPERATIONS)

Principally, there are two possible processes: (i) Chrome tanning and (ii) Vegetable tanning.

16.2.5.1 Chrome Tanning

After the pickling operations pH value reduces. During this pH reduction, chromium (Ш) salts are added. The process of chromium tanning is based on the cross-linkage of chromium ions with a free carboxyl group in the collagen. This process makes the hide to become resistant to bacteria and high temperatures. The chromium-tanned hide contains about 2%–3% by dry weight of Cr (Ш) or Wet blue. The rawhide obtained after the chrome tanning process will contain about 40% of dry matter.

16.2.5.2 Vegetable Tanning

The process of vegetable tanning is usually accomplished in a series of vats. A vat is a large tank where hides are dipped in the liquor of tan to steep. Initially, hides are taken to the rocker-section vat where the liquor is agitated and then they are taken to the lay-away vats where the liquor is not agitated with increasing concentrations of tanning liquor.

16.2.5.3 Wet Finishing (Post-tanning)

The process of wet finishing is sometimes performed in one single float. Often re-tanning process is done to chromium-tanned hides or wet blue; therefore, the desirable properties of more than one tanning agent can be combined or mixed together, and when treated with dyes and fats, hides will obtain proper filling, smoothness and colour. Then to make hides suitable for splitting and shaving processes, large quantities of water are removed before the actual drying process is allowed to take place. Splitting and shaving processes are done to obtain the desired thickness of the hide. Due to the presence of different kinds of dyes, fats and combined tanning agents, the effluent produced during wet finishing makes it highly complex.

16.2.6 FINISHING

Finishing operation is done on the crust material obtained after re-tanning and drying processes. This crust is then treated with an organic solvent or water-based dye and then varnish. This is done to make the hide softer and curb all the other small mistakes. The finished end-product will contain dry matter of about 66%–85% by weight. Those chemicals which are used in the finishing operations also have the capacity to pollute the water bodies. Besides this, tanning is the fundamental stage which gives leather its essential characters and stability. Raw hides after several stages of pre and post-treatments are converted into a final product with specific properties such as stability, appearance, water resistance, temperature resistance, elasticity and perspiration permeability. In India, as per 2008 statistics, around 2 billion sq. ft. of finished leather per annum was produced, and the industry has set a target to double this figure by the year 2011–2012. On the other hand, the industry has been facing serious challenges on account of pollution-related problems. The presence of eco-sensitive chemicals in leathers is an area of concern (An et al., 2010).

The malodorous atmosphere around traditional tannery clusters creates the impression of a highly polluting industry. Tanneries fall under the "red" category of industries in India based on the pollution potential. Tanning operations imply a tremendous impact on nature, and activities taking place inside tanneries give out enormous amounts of pollutants which are of varied kinds, like the chemicals applied, effluents produced from the treatment of raw hides and the generation of flue gas from the factories. Since the early 1980s, tanning industry began to tackle environmental pollution problems of liquid and solid wastes generated during tanning processes.

16.3   ENUMERATION OF TANNERIES IN INDIA

Primarily, the concentration of tanning industries is directly related to places which are abundantly available with superior quality water and raw materials. In India, tanning clusters are largely located in the states of Tamil Nadu, Andhra Pradesh, Punjab, West Bengal and Uttar Pradesh, and a few isolated tanneries are located in the states of Bihar, Madhya Pradesh and Kerala.

16.3.1   WATER USAGE AND POLLUTION POTENTIAL

The quantity of chemicals used in tanning processes is almost in the same range in all regions in India. Depending on the availability of water, the wash water volume usage varies, and wash water contains chemicals in various concentrations in the sectional and composite wastewater. The concentration of the toxins that are present in the wastewater discharge also depends on the cleaner technology adopted in tanneries like dusting of salt, recovery and reuse of chromium and quality of the chemicals used. In Tamil Nadu, it is mandatory for the tanneries processing raw to semi-finish (chrome tanning process) to have a chrome recovery system. Due to scarcity of available natural waterbodies or water supply, washing done using water is comparatively less in the state of Tamil Nadu than other parts of the country. Due to this, the concentrations of the COD and biochemical oxygen demand (BOD) are comparatively high in wastewater. But in the case of tanneries in Uttar Pradesh, Punjab and West Bengal, the volume of water used for washing is high and this results in dilution of pollutants. In Andhra Pradesh and Tamil Nadu, soak and pickle liquor are segregated and conveyed to solar evaporation pans. But in other parts of the country, no segregation of soak and pickle is done, and they are combined with the remaining sectional streams and hence saline in nature.

16.3.2   POLLUTION AND THREATS POSED BY TANNERY EFFLUENT

Tanning process undergoes three major steps: (i) beam house operations – where unhairing and liming processes take place; (ii) tanning operations – where shaving, washing, and tanning (chrome and vegetable tanning) operations take place; (iii) finishing operations – where re-tanning, dying and fat liquoring processes are handled. Due to these repeated soaking, washing and tanning operations, a large amount of wastewater is generated which is contaminated with different chemicals used in those operations. Some of the chemicals which may be present in the wastewater produced are chromium, sulphate, amines, cyanides, acids, sodium chloride and sodium chlorate. Chromium contamination in the wastewater produced leads to increase in the COD and total dissolved solids (TDS) which would eventually lead to the contamination of the natural water body if it is not disposed properly. Agricultural crops cannot be grown in the tannery dumpsites as chemicals (mainly chromium) contained in the wastewater are carcinogenic and teratogenic. Most of the chemicals are non-biodegradable. Therefore, disposal tannery effluents require a novel method, and phytoremediation has proved to be an eco-friendly and efficient technique. The quantity of wastewater produced and the pollution created during each major tanning operation in a typical tanning industry are presented.

16.4   MECHANISM

16.4.1   UPTAKE AND TRANSLOCATION OF HEAVY METAL IN PLANTS

There are numerous processes involved in the accumulation and concentration of heavy metals in plants, including heavy metal mobilization, root uptake, cellular compartmentation and sequestration (Bahadar et al., 2014). However, heavy metals are not directly available to plants as they possess insoluble form in the soil. By releasing several root exudates, plants can increase their bioavailability which can change the rhizosphere pH and increase heavy metal solubility. The metal ions available in the subsoil are sucked at the root level, and the movement of these ions into the root cells takes place through the cellular membrane. Transportation of heavy metal ions can take place in two ways: apoplastic pathway and symplastic pathway. Apoplastic pathways area passive diffusion whereas the symplastic pathway is an active transport against electrochemical potential gradients and concentration across the plasma membrane (Bento et al., 2005). The uptake of heavy metals most commonly occurs through the symplastic pathway which is an energy-dependent process that is mediated by metal ion carriers or complexing agents.

16.4.2   BIOREMEDIATION

The importance of bioremediation process for the accumulation of substances and various other uses of this process are mentioned. These mechanisms are dealt with in an orderly fashion as the sequence of how contaminants come into contact with the plant system, rhizosphere and transportation processes (Kim et al., 2004). Several mechanisms are involved in bioremediation of soil, water and air. These mechanisms are interrelated and dependent upon plant physiological processes driven by solar energy, rhizospheric processes and other available precursors. Therefore, in bioremediation applications, multiple mechanisms are involved depending on the designed application.

16.4.3 PHYTOREMEDIATION

It is a process of using living green plant species to eliminate, transport and destroy various pollutants in natural resources like soil and water. Using the technique of phytoremediation, stabilization of the quantity of certain heavy metals and other substances can be done (Sun et al., 2013). There are numerous numbers of phytoremediation mechanisms that can be involved in the restoration of natural resources. Phytoremediation comprises a group of novel biological remediation technologies that use green living plants to eliminate contaminants from the environment and render them harmless (Salt et al., 1998). Plants can be used to partially or completely remediate miscellaneous inorganic and organic pollutants. Various approaches of phytoremediation are shown in Figure 16.1. Phytoremediation removal technologies imply the cleaning-up of the contaminated sites, while phytoremediation containment technologies entail toxicity of the pollutant in the environment (Gobran et al., 2000). Phytoremediation is based on natural physiological processes of green plants that include water and nutrient uptake, translocation of ions, accumulation of ions, transpiration and gas exchange leading to different types of phytoremediation mechanisms that conduct contaminants (Tsao, 2003). These main phytoremediation technologies are namely phytostabilization, phytoextraction, phytodegradation, rhizodegradation,

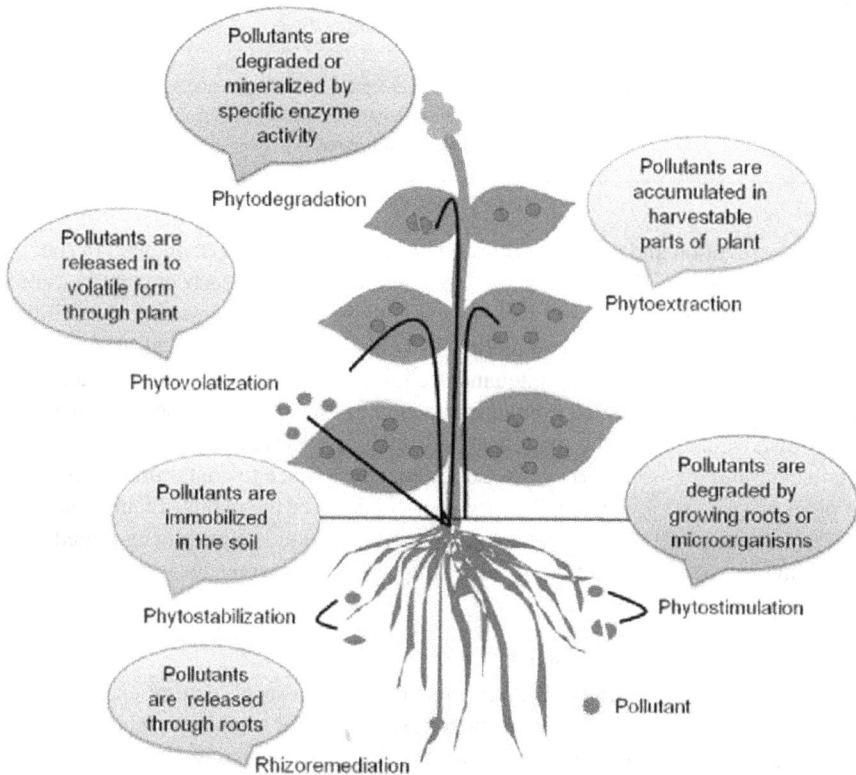

FIGURE 16.1   Represent the various approaches of phytoremediation.

TABLE 16.2

Characteristics of Each Phytoremediation Technology

Phytotechnology	Clean-up Goal	Mechanism of Remediation	Type of Contaminants
Phytostabilization	Containment	Here, contaminants are immobilized at the root level through the processes of absorption, adsorption and precipitation.	Inorganic and organic
Phytoextraction	Remediation	Here, contaminants are extracted by plant roots and are translocated to the above ground tissue.	Inorganic
Phytodegradation	Remediation	Plant enzymes accumulate and transform the contaminants. Then they are metabolized by rhizosphere microorganisms.	Organic
Rhizodegradation	Remediation	Here, contaminants are metabolized by rhizosphere microorganisms whose growth and activity are supported by the release of plant root exudates.	Organic (moderately hydrophobic compounds)
Phytovolatilization	Remediation	Here, contaminants are transformed into more volatile and less polluting substances by plants and are then released to the atmosphere by transpiration.	Inorganic and organic (moderately hydrophobic compounds)
Evapotranspiration	Containment	Rain water interception, evaporation and plant transpiration that reduces contaminant infiltration.	Organic and inorganic water

phytovolatilization and evapotranspiration, each of which is experimented in specific design applications. Table 16.2 shows the characteristics of each phytoremediation technology. Major advantages of phyto-technologies, when compared to traditional chemical and physical treatment methods like soil washing, chemical oxidation and air venting are cheap, low maintenance requirement, applicable to simultaneously remediated sites with mixed contaminants and eco-friendly. On the other hand, its major disadvantages can be the longer refurbishment period that must be needed to attain clear out goals (Susarla et al., 2002), plant tolerance to contaminants, disposal of plant wastes and low bioavailability of pollutants for the plants (Evangelou et al., 2007). In spite of all these drawbacks, phytoremediation is a promising remediation technology, whose development is increasing since its emergence.

16.4.3.1   *Sesuvium portulacastrum*

Sesuvium portulacastrum is also rich in salt, and pH is acid, neutral and alkalinity compounds. It is a pioneer plant species, mostly used for sand-dune fixation, desalination and phytoremediation along coastal region. This plant tolerates stringent conditions like salinity, drought and toxic metals. *S. portulacastrum* is also used as a fodder for domestic animals and as a beautiful ornamental plant. It further grows at severe saline conditions of 1,000 mM NaCl without any poisonous symptoms on leaves.

This perennial herb can grow up to 30 cm (12 inch) high, with thick, smooth stems upto 1 metre (3.3 feet) long and spreads out to a wide range of area. The plant leaves appear to be smooth and glossy. The fleshy leaves of the plant are linear and lanceolate having a length of about 10–70 mm (0.39–2.76 inch) and are 2–15 mm (0.079–0.591 inch) wide. Flowers commonly appear to be in pink or purple colour.

Durai Ayyappan and Ravindran (2014) stated that the soil electrical conductivity in the tannery effluent treated soil was considerably decreased to about 4 times its initial value by *S. portulacastrum*. Also the pH level was considerably decreased in both tannery effluent and salt-treated soils. There was also a massive decline in the sodium and chloride level in both tannery effluent and salt-treated soils by 60% and 55%, respectively. *Sesuvium portulacastrum* is a game changer when it comes to heavy metals like chromium, cadmium, copper and zinc from tannery effluent treated soil. *Sesuvium portulacastrum* has nearly reduced 51% of heavy metal contaminants and reduced about 45% from salt-treated soil.

16.4.3.2   *Salicornia brachiata*

Salicornia brachiata (a common name of saltwort and pickleweed) is salt tolerant, suitable at neutral and alkaline pH is naturally occurring and allows growth of plants. This salt-tolerant halophyte is a drought-resistant plant and commonly grows in salt marshes and mangroves. *S. brachiata* may be also used as a raw material in paper and board industries. Its seeds also yield high-quality edible oil similar to sunflower oil and are observed to be a beneficial feedstock for biodiesel production. *Salicornia brachiata* is an excellent candidate for salt succulence and TDS removal. Since osmotic pressure is crucially involved in the growth of *Salicornia brachiata*, it was found that an increase in the succulence of salt molecules decreases plant growth. Removal of TDS was achieved by about 40% from tannery wastewater. The evidence prove that phytoremediation using *Salicornia brachiata* is an efficient way to reduce TDS levels in tannery soak effluent.

16.5   POTENTIAL OF PLANTS FOR THE EXTRACTION OF METAL AND MINERAL SALTS FROM TANNERY EFFLUENT

16.5.1   Water Hyacinth

Water hyacinth is one of the rapidly growing and very reproductive free-floating aquatic plants. It has high affinity and accumulation capability for several metals and mineral salts and has shown a high potential to grow in heavily saline water. A heavy metal phytoremediation has been demonstrated by using water hyacinth, which has a greater succulent capacity for metals like Pb, Cu and Zn. We can see the seriousness of Cr pollution so that some alternative methods are developed for the process of remediation of Cr contaminated tannery wastewater and found that water hyacinth has increased affinity for nutrients uptake and metal ion accumulation (Sas-Nowosielska et al., 2004). The potential of water hyacinth in remediation of Cr-contaminated tannery effluent has been investigated to show that the growth of water hyacinth was inhibited at high concentrations due to Cr toxicity. Therefore, the application of water hyacinth will be sustainable only in low concentration of Cr in tannery effluent.

16.5.2 VETIVER GRASS

Vetiver grass (*Chrysopogon zinanioides*) is a perennial grass providing bacterial and fungal growth and multiplication which was required to absorb and break down contaminants. It has high affinity, salinity and heavy metal. Phytoremediation of tannery effluent uses plant species as vetiver grass. Thus, it was allowed to grow in tannery effluent. The post-test analysis study clearly shows that the concentration of COD and BOD reduced remarkably. The vetiver grass had better removable efficiency for COD removal (80.8%). Thus, the phytoremediation experiments demonstrate that removal of selected pollutants due to vegetation provided remediation pollutants through different mechanisms including rhizodegradation, rhizofiltration, photo degradation and stabilization (Clemens, 2002). The chromium metal was largely accumulated in root tissues of vetiver grass. It concludes that vetiver grass is going to be a candidate as an alternative solution through post-treatment method for small-scale tannery Industry.

16.6 CONCLUDING REMARKS

The biological treatment has been found to be an efficient and eco-friendly method in accumulating various elements and compounds. Since it has been a reassuring remediation technique for the elimination of heavy metal polluted soils which has a reliable public acceptance, it shows various advantages compared with other remediation techniques. By using phytoextraction study, it is concluded that both halophytes *S. portulacastrum* and *S. brachiata* are more efficient for this process. Since *S. brachiata* which is considered as a high salt-tolerant halophyte has the excess ability to grow under high salinity conditions. And hence removal of sodium and chromium from the tannery effluent waste was effective by using *S. brachiate*. The harvested waste obtained from the removal process can be used for preserving skins, and this preservation doesn't affect the quality of leathers and skins. This process of harvesting will also overcomes the environmental constraints of using a high concentration of salt in temporary preservations. This effective alternate method results almost zero percentage of salt discharge.

REFERENCES

Adetutu, E.M., Ball, A.S., Weber, J., Aleer, S., Dandie, C.E., Juhasz, A.L., 2012. Impact of bacterial and fungal processes on[14]C-hexadecane mineralisation in weathered hydrocarbon contaminated soil. *Science of the Total Environment* 414, 585–591.
An, C., Huang, G., Yu, H., Chen, W., Li, G., 2010. Effect of short-chain organic acids and pH on the behaviors of pyrene. *Chemosphere* 81, 1423–1429.

Ayyappan, D., Ravindran, K.C., 2014. Potentiality of Suaeda monoica Forsk. A salt marsh halophyte on bioaccumulation of heavy metals from tannery effluent. *International Journal of Modern Research and Reviews* 2, 267–274.

Bahadar, H., Mostafalou, S., Abdollahi, M., 2014. Current understanding and perspectives on non-cancer health effects of benzene: a global concern. *Toxicology and Applied Pharmacology* 276, 83–94.

Bento, F.M., Camargo, F.A.O., Okeke, B.C., Frankenberger, W.T., 2005. Comparative bioremediation of soils contaminated with diesel oil by natural attenuation, biostimulation and bioaugmentation. *Bioresource Technology* 96, 1049–1055.

Clemens, S., Palmgren, M.G., Krämer, U., 2002. A long way ahead: understanding and engineering plant metal accumulation. *Trends in Plant Science* 7, 309–315.

Evangelou, M.W.H., Ebel, M., Schaeffer, A., 2007. Chelate assisted phytoextraction of heavy metals from soil. Effect, mechanism, toxicity and fate of chelating agents. *Chemosphere* 68, 989–1003.

Gobran, G.R., Wenzel, W.W., Lombi, E., 2000. *Trace Elements in the Rhizosphere*. CRC Press Inc., Boca Raton, FL.

Kim, Y.B., Park, K.Y., Chung, Y., Oh, K.C., Buchanan, B.B., 2004. Phytoremediation of anthracene contaminated soils by different plant species. *Journal of Plant Biology* 47, 174–178.

Salt, D.E., Smith, R.D., Raskin, I., 1998. Phytoremediation. *Annual Review of Plant Physiology and Plant Molecular Biology* 49, 643–668.

Sas-Nowosielska, A., Kucharski, R., Malkowski, E., Pogrzeba, M., Kuperberg, J.M., Krynski, K., 2004. Phytoextraction crop disposal: an unsolved problem. *Environmental Pollution* 128, 373–379.

Sun, Y., Xu, Y., Zhou, Q., Wang, L., Lin, D., Liang, X., 2013. The potential of gibberellic acid 3 (GA$_3$) and Tween-80 induced phytoremediation of co- contamination of Cd and Benzo[a]pyrene (B[a]P) using *Tagetes patula*. *Journal of Environmental Management* 114, 202–208.

Susarla, S., Medina, V.F., McCutcheon, S.C., 2002. Phytoremediation: An ecological solution to organic chemical contamination. *Ecological Engineering* 18, 647–658.

Tsao, D., 2003. Overview of Phytotechnologies. in: Tsao, D. (Ed.). *Phytotechnologies*. Springer-Verlag, Heidelberg, pp. 1–50.

17 Bio-coagulation-Flocculation of Land-Based Saline Aquaculture Effluent Using *Parkia biglobosa* Seeds
Turbidity, Salinity, and Colour Reduction

Chinenye Adaobi Igwegbe
Nnamdi Azikiwe University

Joshua O. Ighalo
Nnamdi Azikiwe University
University of Ilorin

Okechukwu Dominic Onukwuli
Nnamdi Azikiwe University

Shahin Ahmadi
Zabol University of Medical Sciences

CONTENTS

DOI: 10.1201/9781003185437-20

17.1   INTRODUCTION

Due to high fish demand, aquaculture is a common practice in Nigeria and generates a large volume of liquid effluent [1]. The aquaculture saline effluents contain a variety of pollutants that need to be removed/reduced before release into the environment [2]. These pollutants are the constituents of uneaten feed and fish faeces [3]. It is important to treat the saline effluents so that environmental protection can be achieved [4]. Aquaculture saline effluent can be treated by bioremediation [5, 6], biomass packed-bed filtration [7], foam fractionation [8], electrochemical reduction [9–11], and a host of others. Coagulation-flocculation can be used for treatment because of the dilute nature of the effluent, and the efficiency of the process is catching the suspended and dissolved solids into flocs [1,12].

Several investigations have explored the coagulation-flocculation process for different aspects of the treatment of the aquaculture saline effluent. Different researchers have examined the reduction of phosphorus [13,14], organic matter [12], suspended solids [13,14], dissolved solids [15], turbidity [10], colour [16,17], and biochemical and chemical oxygen demand [3,18]. The reduction of turbidity, salinity, and color of aquacultural effluent via the coagulation-flocculation process using *Parkia biglobosa* seeds has not been reported. Herein lies the novelty of the current research reported in this chapter. The material of interest in the current investigation for the production of extracts for the coagulation-flocculation process is Locust bean (*Parkia biglobosa*) seeds. The African locust bean (*Parkia biglobosa*) is a deciduous and vascular perennial legume belonging to the family *leguminosae* [19,20]. *Parkia biglobosa* is utilized in the preparation of numerous delicacies in West Africa [21–23]. The pods of the plant have previously been explored as a coagulant [19] and an adsorbent [24] for water treatment. Natural coagulants such as extracts of plant seeds are more environmentally friendly and cheap when compared to chemical coagulants [25–28]. The reduction of turbidity (TUR), salinity (SAL), and colour (COL) from land-based saline aquaculture effluent (LSAE) using *Parkia biglobosa* seeds coagulant (PBC) was investigated and presented as a case in this chapter. The physicochemical characteristics of the seeds were determined. The coagulant obtained from seeds by the salt extraction technique was also analyzed for its morphological and spectral characteristics through the scanning electron microscopy (SEM) and Fourier transform infrared spectroscopy (FTIR). The coagulation-flocculation kinetics of the process was studied. Also, the cost of employing PBC for the treatment of 1 L of LSAE was estimated. The following sections in this chapter represent the steps involved during this case study.

17.2 MATERIALS AND METHODS EMPLOYED IN THIS CASE STUDY

17.2.1 LAND-BASED SALINE AQUACULTURE EFFLUENT COLLECTION

The land-based saline aquaculture effluent collection (LSAE) was obtained from a local business aquaculture pond (Figure 17.1) in Agu-Awka, Nigeria (Latitude: 6°12′45.68″ N, Longitude: 7°04′19.16″ E). The effluent samples were kept at 40°C to prevent modifications in their original properties. Table 17.1 shows the characteristics of the LSAE used in the analysis using conventional techniques [29–31].

17.2.2 CHEMICALS AND EQUIPMENT USED IN THE CASE STUDY

The reagents, sodium hydroxide pellet (NaOH, 98%), sodium chloride (NaCl, 99.999%), calcium chloride (CaCl, 95%), hydrochloric acid (HCl, 99.0%), magnesium chloride (MgCl, 98%), potassium chloride (KCl, 99.5%), and n-Hexane (C_6H_{14}, 99.9%) as well as distilled water and deionized water were of analytical grade. All chemicals were used without additional refinement. The TUR was obtained through the EPA method 180.1 [31]. To determine the maximum wavelength (λ_{max}) for effective measurement of the absorbance (or COL) of the untreated and treated LSAE, the absorbance of the LSAE was measured at different wavelengths with the range of 200–600 nm using a UV-visible spectrophotometer (APEL PD-3000UV Spectrophotometer). For the absorbance calculations, the wavelength with the highest

FIGURE 17.1 Place of collection of the LSAE.

TABLE 17.1
Features of the LSAE

Parameter	Unit	Value
pH		7.9
Temperature	°C	30.2
Appearance		yellowish green
Absorbance at 275 nm	–	0.842
Biochemical oxygen demand (BOD$_5$)	mg/L	317
Chemical oxygen demand (COD)	mg/L	758
Biodegrability index (BI)	–	0.42
Electrical conductivity (EC)	μS/cm	1,963
Salinity (SAL)	g/L	0.972
Turbidity (TUR)	NTU	404
Total solids (TS)	mg/L	695
Total suspended solids (TSS)	mg/L	45
Total dissolved solids (TDS)	mg/L	650
Nitrogen	mg/L	0.829
Total phosphorus	mg/L	0.91
Ammoniacal nitrogen	mg/L	0.829
Chloride	mg/L	2.44
Calcium	mg/L	5.40
Potassium	mg/L	5.45
Iron	mg/L	0.425

absorbance was selected. SAL is often obtained from conductivity measurements rather than being measured directly. The SAL values were obtained by first taking the electrical conductivities (EC) and converting them to salinity using the relationship [32]:

$$SAL \left(\frac{g}{L} \right) = EC^{1.0878} \times 0.4665 \qquad (17.1)$$

The TUR meter was first calibrated before use by using 400 NTU standard solution prepared with $H_6N_2O_4S$ and $C_6H_{12}N_4$. The pH meter (Hanna instruments) was calibrated using buffer solutions (pH 4 and 7). The calibration of the pH meter, turbidity meter, and conductivity/TDS meter was carried out daily at room temperature (27°C ± 2°C).

17.2.3 COLLECTION AND PROCESSING OF COAGULANT PRECURSOR

The coagulant precursor, *Parkia biglobosa* (locust bean) seeds, was obtained at a local market in Nsukka, Enugu state, Nigeria. To eliminate debris and impurities, the seeds (Figure 17.2) were washed in de-ionized water. 500 g of the washed seeds were dried at room temperature. The dried samples were ground with a local grinder and sieved using 75 μm sieve. The ground samples were then put in an airtight container and refrigerated until required.

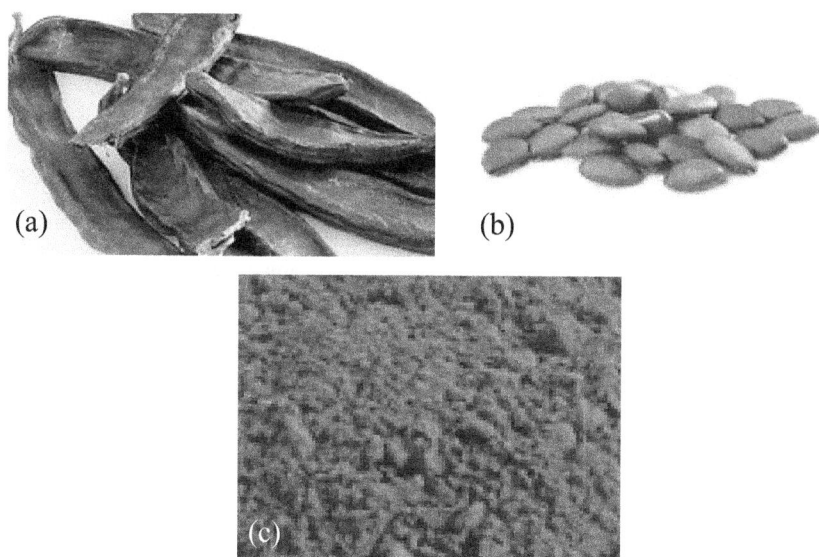

FIGURE 17.2   *Parkia biglobosa* (locust bean) (a) pods, (b) seeds and (c) powder.

17.2.4   Preparation of the Active Coagulant

300g of the ground seeds were defatted to remove the oil components via the Soxhlet extraction method described by Krzyczkowska and Kozłowska [33]; n-Hexane (99% purity) was used as the solvent. The oil was extracted from 30 g of each sample with 250 mL n-hexane using a Soxhlet extractor at 70°C for 6 hours. Complex salts (1 g of $CaCl_2$, 25 g of NaCl, 0.75 g of KCl, and 4 g of $MgCl_2$) were dissolved in 1,000 mL of distilled water to make a salt solution. 10 g of the ground seeds was dissolved in 250 mL of the salt solution and shook constantly using a hotplate/magnetic stirrer (78HW-1) for 1 hour at 50°C. The samples were then passed through a filter paper (Whatmann no. 90 mm) to collect the filtrate (stock solution), which was then allowed to solidify at room temperature as the coagulant.

17.2.5   Characterization of the PBC Coagulant

The functional groups present in the PBC were determined to ascertain the functional groups present and taking part in the coagulation experiment. FTIR analysis was carried out using the potassium bromide (KBr) pellet method at room temperature. The FTIR analysis was carried out by mixing the PBC with dry finely powdered potassium bromide (KBr) in a ratio of 1:100, and the spectra were collected from 4,000 to 400 cm^{-1} through an FTIR transmission machine (Buck M520 Infrared spectrophotometer, USA). SEM was used to observe the surface morphology of the PBC using a Carl Zeiss Analytical SEM Series.MA 10.EVO-10-09-49, Germany. The 3D reconstruction was done using ImageJ v1.53 [34] image processing and analysis software at a grid size of 1,024 and smoothening of 10.0.

17.2.6   Bio-coagulation-Flocculation Experiments

The BCF experiments were carried out using a jar test apparatus. The LSAE was stirred before taking the samples for treatment to make its concentration consistent. The effects of PBC dosage, pH, temperature, and settling time on the TUR, SAL, and COL reductions using PBC were investigated. A measurement cylinder was used to measure 500 mL of LSAE, which was then poured into various beakers. Then, their pHs were adjusted to a specified pH using HCl or NaOH 1M solutions. A known PBC dosage was introduced into each of the beakers which were stirred using a magnetic stirrer at a constant speed of 120 rpm for 5 minutes and followed by 20 minutes of shaking at a lower speed of 30 rpm at a specified temperature. Then, the mixture was allowed to settle for different settling time. A syringe was used to extract 20 mL of the samples from each of the beakers at a depth of 2 cm for testing for the coagulation efficiency (%R) in terms of TUR, SAL, and COL. The %R was evaluated using Equation 17.2.

$$\%R = \frac{R_i - R_f}{R_i} \times 100 \tag{17.2}$$

where R_i and R_f are the initial and final values of the TUR, SAL, and COL.

17.2.7   Coagulation-Flocculation Kinetics

The kinetic data can be fitted into Equation 17.3 to inspect the line of best fitting when $\alpha = 2$ (second-order reaction kinetics) [9,35], where K_{MK} (the reaction rate constant) can be estimated using Equation 17.3 [36]:

$$\frac{1}{C} - \frac{1}{C_0} = K_{MK}t \tag{17.3}$$

where C is the particle's concentration at any period, t and C_0 is the initial particle concentration.

To see if the coagulation-flocculation treatment of LSAE with PBC follows the perikinetics concept, the regression coefficient (R^2) was used as a basis. The kinetics of aggregation was studied by making a regression plot of $1/C_t$ versus time (t) using Equation 17.3. The particle concentration for turbidity was determined by converting the turbidity (in NTU) to turbidity in mg/L.

β_{BR} is a function of flocculation transport for the mechanisms of shear, Brownian, and differential sedimentation which is described as [37]:

$$\beta_{BR} = \frac{8}{3} \frac{\varepsilon_c K_{BZ}T}{\eta} \tag{17.4}$$

where K_{BZ} is $1.38064852 \times 10^{-23}$ J/K (the Boltzmann's constant), η is 2.6 m Pa s (the viscosity of the LSAE), ε_c is the efficiency of collision, and $T =$ absolute temperature.

The rate of decline in concentration of LSAE particles ($-r_p$) at early Brownian kinetic coagulation (i.e., 30 minutes) is determined through Equation 17.5 [1,38]:

$$-r_p = -\frac{dC_t}{dt} = K_{MK}C_t^{\alpha} \tag{17.5}$$

where α is the coagulation reaction order, K_{MK} is the Menkonu constant rate of coagulation, and C_t is particles concentration (the total suspended and dissolved particles (TDSP) at t).

K_{MK} can be obtained through Equation 17.6 [39,40]:

$$K_{MK} = \frac{1}{2}\beta_{BR} = \varepsilon_c\,K_R \qquad (17.6)$$

$$K_R = \frac{4K_{BZ}T}{3\eta} \qquad (17.7)$$

where K_R is the Smoluchowski rate constant for swift coagulation.

The diffusivity (D^1) can be evaluated through Equation 17.8 [41,42]:

$$D^1 = \frac{K_R}{8\pi r} \qquad (17.8)$$

where r is the particle's radius.

The r and B_f (the friction factor) can be evaluated using Equations 17.9 and 17.10:

$$r = \frac{\beta_{BR}}{6\eta\pi} \qquad (17.9)$$

$$B_f = \frac{D^1}{T}K_{MK} \qquad (17.10)$$

The fast coagulation time (τ_F) and fast coagulation half-life $(\tau_{F1/2})$ were evaluated using Equations 17.11 and 17.12 [42–44]:

$$\tau_F = \frac{1}{C_0 K_{MK}} \qquad (17.11)$$

$$\tau_{F1/2} = \frac{1}{0.5C_0 K_{MK}} \qquad (17.12)$$

17.2.8 Cost Estimation

Total cost for the treatment of 1 L of the LSAE was evaluated using the expression shown as Equation 17.13:

$$\text{Total cost} = \text{Cost of coagulant production} + \text{Cost of labour} + \text{Cost of energy} \qquad (17.13)$$

The energy consumption (E_C) was evaluated as 1.3€ using Equation 17.14 [45]:

$$E_C = P_C\,(b \times t \times C) \qquad (17.14)$$

where P_C is the power consumed by device (kW), b is a load factor (in full mode so $b=1$), t is the time of usage of the device (0.25 hour), and C is the energy cost (0.13 €/KWh).

17.3   CASE STUDY FINDINGS

17.3.1   Physio-chemical Characteristics of PBC

Table 17.2 shows the proximate composition of the *Parkia biglobosa* seeds. The relatively high protein content of 37.9% obtained for *Parkia biglobosa* seeds is an indication of the likelihood of the seeds as a good coagulant precursor for the treatment of LSAE [46]. Protein is needed for the neutralization of colloidal particles, which activates the coagulation mechanism by bridging, which promotes the forming of flocs [47]. Also, these seeds can serve as coagulants for wastewater treatment due to their high carbohydrate content (which consist of cellulose and polysaccharides) [15]. This natural polymer- based coagulant will encourage the formation of particle aggregates, which can help to reduce pollution levels. Bridging is a coagulation process that occurs when coagulants with threads or fibres are added to multiple colloids, trapping and linking them together [48].

17.3.2   Characterization of the Coagulant (PBC)

The SEM image (at 500× and 1,000×) of the PBC is shown in Figure 17.3. The PBC had a high porosity, indicating that there were a number of active adsorption sites on the material for capturing contaminants [56,57]. The PBC has irregular and uneven granular arrangements which are necessary characteristics of coagulants for adsorption and bridging of colloidal particles, hence, encouraging the sedimentation of particles. Furthermore, the PBC seemed coarse and fibrous, as shown by its fibre content as determined by proximate examination (Table 17.4). The 3D reconstructed surface revealed that the irregular and uneven granular arrangements are jagged and appear at an elevation above the smoother areas of the adsorbent surface. The general heterogeneous outlook of the surface reveals that the material will have a good potential for pollutant uptake by adsorption [58].

Table 17.3 lists the functional groups present in the PBC that are involved in the coagulation-flocculation mechanism, along with their bond origins and explanations. The presence of N-H stretching in the PBC spectrum suggests the presence of amino compounds (protein), which is consistent with Table 17.2's proximate analysis.

TABLE 17.2
Proximate Analysis of the *Parkia biglobosa* Seeds

Parameters	Unit	Procedure	Values in %
Bulk density	g/mL	AOAC method PA 105 [49]	0.28
Ash content	%	AOAC method 942.05 [50]	4.24
Moisture content	%	AACC method number: 44-15A [51]	6.21
Crude fibre	%	AOAC method 978.10 [52]	3.93
Fat content	%	AOAC method 920.39 [53]	15.48
Crude protein	%	AOAC method 945.18-B [54]	37.9
Carbohydrate content	%	FAO [55]	32.24
Yield of the coagulant	%	Menkiti and Ezemagu [46]	71.39

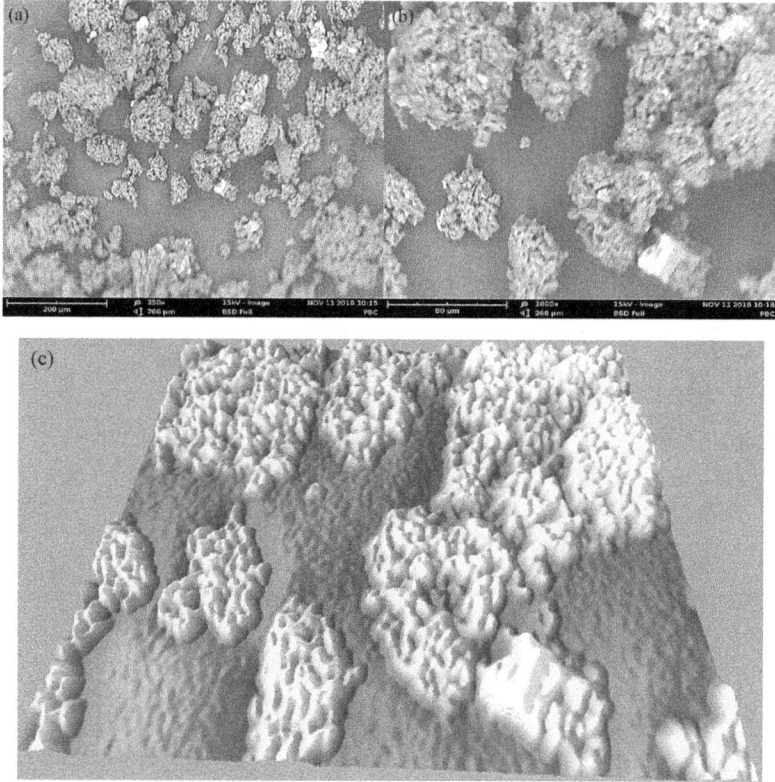

FIGURE 17.3   SEM image of PBC at (a) 350×, (b) 1,000× and (c) 3D reconstruction of adsorbent surface (at magnification ×1,000).

The hydroxyl (O–H), carboxyl (C=O), and amino or amide (N-H) groups, as well as hydrogen bonding, were found on the PBC, which are the favoured groups for the coagulation-flocculation mechanism [59]. The C=O groups can serve as an ion bridge at the particle's surface, causing divalent metal cations like Mg^{2+} and Ca^{2+} to cause coagulation behaviour [60]. The medium alkane group (C–H stretch) was present on the PBC. A powerful and sharp band called the O–H stretch, or free hydroxyl of alcohols and phenols, was observed (Figure 17.4).

17.3.3   EFFECT OF PROCESS FACTORS

17.3.3.1   Effect of PBC Dosage on TUR, SAL, and COL Reduction

The better understanding of the scientific principles following coagulation-flocculation and the process had allowed the optimization of dosages of coagulants. This ensures that enough coagulant is added to attain charge neutralization. An optimal biocoagulant dosage is detected before which the coagulant is not enough to form enough bridging among particles and beyond which the biocoagulant saturates the colloidal surfaces leaving no sites for the development of polymer bridges [61]. The effect of PBC dosage on the reduction of pollutants (TUR, SAL, and COL) on LSAE was studied.

TABLE 17.3

Functional Groups Present in PBC with Their Peaks and Description

Wave Number (cm⁻¹)	Bond Source	Functional Group	Peak Description
722.2474	=C–H bend	Alkynes	Medium-strong broad
944.6713	O–H bend	Carboxylic acids	Medium
1,037.028	C–O stretch	Alcohols, carboxylic acids, esters, ethers	Strong
1,354.643	N–O symmetric stretch	Nitro compounds	Medium
1,612.148	N–H bend	1° amines	Medium
2,097.93, 2,184.903	–C≡C– stretch	Alkynes	Weak
2,501.196, 2,590.966	O–H stretch	Carboxylic acids	Strong and broad
2,705.018	H–C=O: C–H stretch	Aldehydes	Medium
2,978.842, 2,843.809	C–H stretch	Alkanes	Medium
3,192.811	O–H stretch	Carboxylic acids	Medium
3,506.122	O–H stretch, H–bonded	Alcohols, phenols	Strong and broad
3,808.635	O–H stretch, free hydroxyl	Alcohols, phenols	Strong and sharp

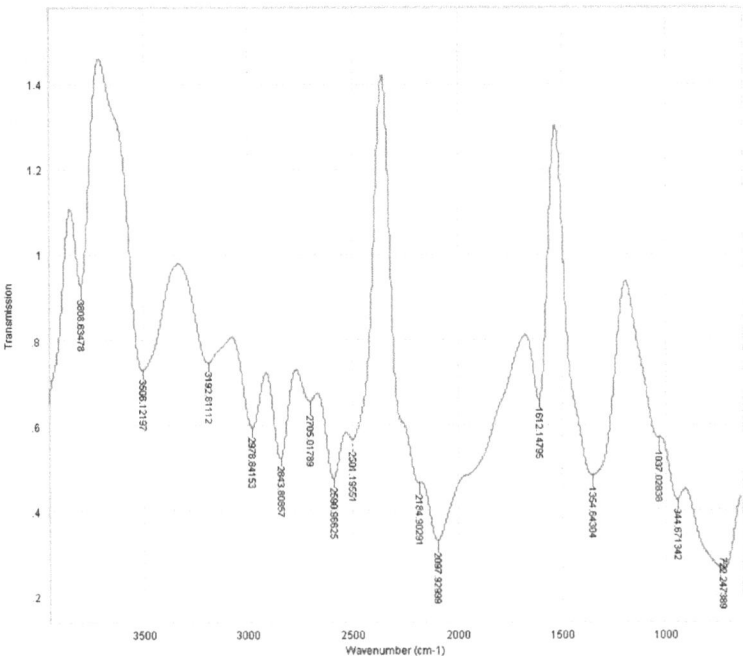

FIGURE 17.4 FTIR spectra on the PBC.

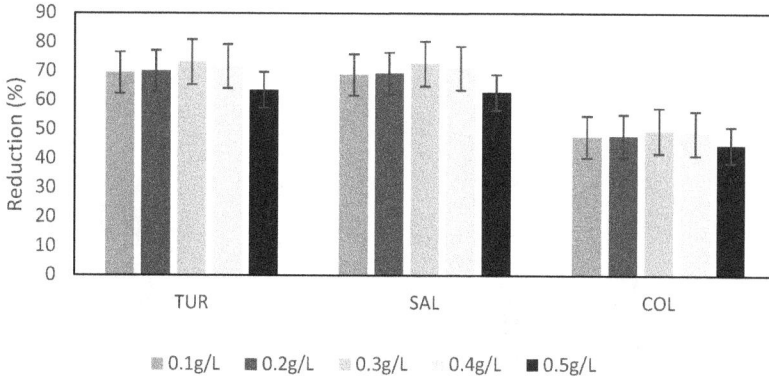

FIGURE 17.5 Effect of PBC dosage on TUR, SAL, and COL reduction (at effluent pH = 7.9, and settling time of 60 minutes).

The effect of coagulant dosage on TUR, SAL, and COL removals was studied by varying the PBC dosages from 0.1 to 0.5 g/L at settling time of 60 minutes and effluent pH 7.9. The effect of coagulant dosage on TUR, SAL, and COL removals is shown in Figure 17.5. Maximum TUR, SAL, and COL reductions of 73.27%, 72.67%, and 49.64%, respectively, on LSAE was achieved at PBC dosage of 0.3 g/L with poorest performance at the highest studied dosage, 0.5 g/L. The addition of excessive coagulant above the optimum dosage reversed the surface charges of the pollutants and reduced its removal efficiency [62,63]. The correlation between the optimal dosage and percentage removal gives uncertainty on the involvement of sweep flocculation mechanism [64,65]. Also, the low optimum dose obtained using PBC indicates that the neutralization and adsorption mechanism is the chief mechanism in the BCF process [65,66].

17.3.3.2 Effect of Initial pH on TUR, SAL, and COL Removal

The pH is established to play a significant role in coagulation-flocculation processes [67]. In most cases, pH is labelled as the 'master variable' [68]. The effectiveness of using PBC for TUR, SAL, and COL reduction in LSAE were tested by varying the solution pH from 2 to 10 at settling time of 60 minutes, and PBC dosage of 0.3 g/L (Figure 17.6). Maximum TUR, SAL, and COL removals of 80.44%, 79.98%, and 54.47%, respectively. The removal was greatest in acidic environments. Also, pH of 6 gave the worst performance; this is due to a poor impact on particle electrophoretic mobility, resulting in low surface charge neutralization [62]. Increasing the pH above the optimum resulted in an increase in electrostatic repulsion between the LSAE particles and PBC due to similar charge, leading to low reduction [25]. Hence, lower pH was favourable for the process using the PBC. In addition, agglomeration of the particles was increased at low pH due to reduced electrostatic repulsion [69,70]. Charge neutralization mechanism was responsible for the TUR, SAL, and COL reduction including bridging of the polymeric material, adsorption, and entrapment of flocs played important roles in pollutants removal. The pH effect on pollutants reduction is related to the pH of the solution and functional groups present in the coagulants which affect its surface charge.

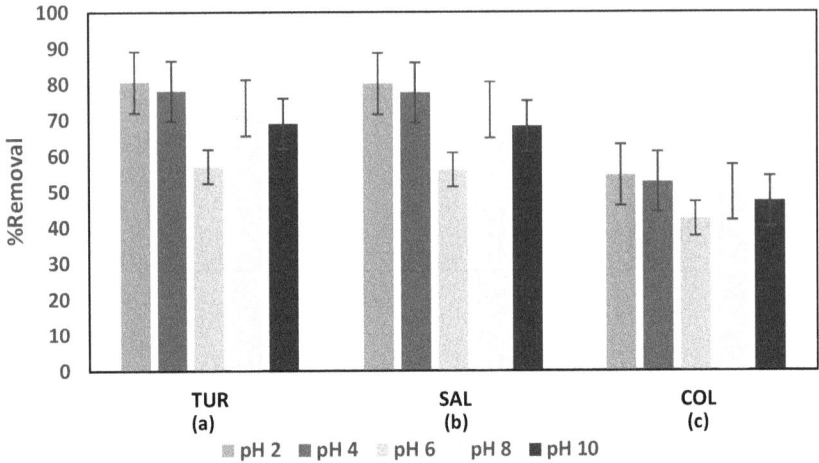

FIGURE 17.6   Effect of initial pH on (a) TUR, (b) SAL, and (c) COL reduction on LSAE using PBC at settling time of 60 minutes and optimum coagulant dosage.

17.3.3.3   Effect of Temperature on TUR, SAL, and COL Removal

During coagulation-flocculation processes, temperature affects the removal efficiency of pollutants [71]. The effect of temperature on TUR, SAL, and COL reductions on LSAE using PBC was investigated at 303, 313, and 323 K (that is, 30°C, 40°C and 50°C) and at the optimum dosage and pH in order to determine the optimum temperature for the coagulation-flocculation process. Maximum TUR, SAL, and COL removals of 80.45%, 79.98%, and 54.47% were achieved at a temperature of 30°C (Figure 17.7). Also, a temperature of 50^{0}C gave the poorest performance on TUR, SAL, and COL reductions. The TUR, SAL, and COL reductions were declined with increase in temperature; therefore, the optimum temperature was obtained as 303 K (30°C) for further experiments. The reduction in TUR, SAL, and COL with increasing temperature may be due to random motion of the colloidal particles caused by the increase in kinetic energy which may interfere with the attachment of particles onto the PBC to form flocs and reduction in flocs sizes [72]. Hence, making the particles spread further apart instead of agglomerating together to form larger flocs. The floc strength will eventually become weaker and easily broken [25]. In addition, lesser large flocs were formed. The viscosity of the effluents could change with increased temperature and affect the treatment efficiency [73]. Also, increase in temperature may destroy the coagulating property and structure (functional groups) of the coagulant. Furthermore, the sedimentation of the flocs was not effective since the flocs remained dispersed in the solution due to the convection currents which prevented the particles from settling faster [69,74].

17.3.3.4   Effect of Settling Time

The effect of settling time on the process of coagulation-flocculation on the pollutants (TUR, SAL, and COL) was studied by varying the settling time from 0 to 60 minutes. This study was performed at the optimum dosage and temperature at different pHs.

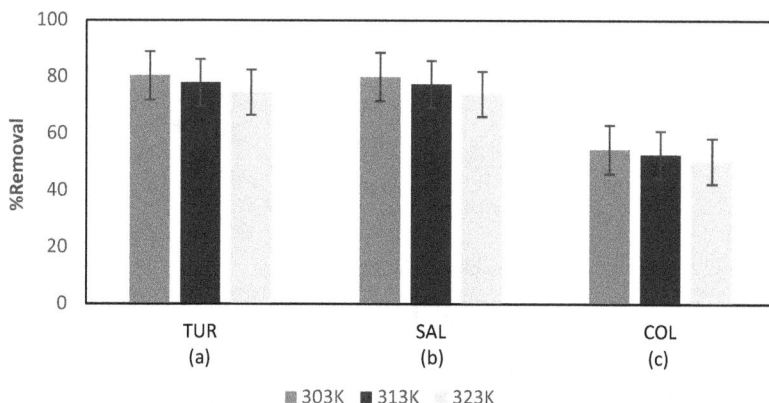

FIGURE 17.7 Effect of temperature on (a) TUR, (b) SAL, and (c) COL reduction on LSAE using PBC at settling time of 60 minutes, PBC dosage of 0.3 g/L and pH 2.

Figure 17.8 shows the effect of settling time on TUR, SAL, and COL reduction on LSAE using PBC. The removal of pollutants was improved with increasing settling time and remained constant after some time with little or negligible increment. This implies that no further increase in time of settling is needed after equilibrium has been reached. The high settling speed was promoted due to the formation of larger and denser flocs owing to the attachment of the biopolymer chain onto the particles [25]. Equilibrium was reached at 40 minutes for all pollutants removal. Maximum TUR, SAL, and COL removals of 80.44%, 79.98%, and 54.47% (at 60 minutes), respectively, were obtained at the optimum dosage (0.3 g/L) and equilibrium settling time. A similar observation was made by Menkiti and Ezemagu [46] using *Tympanotonos fuscatus* coagulant (TFC) for the treatment of produced water. In the present study, slow coagulation (optimum settling time > 30 minutes) was reached using PBC. In addition, as the TUR was decreased, the SAL and COL were reduced simultaneously as observed in Figure 17.8a–c.

17.3.3.5 Brownian Coagulation-Flocculation Kinetics on the BCF Process

The TDSP plots of TUR (NTU) versus TUR (mg/L) which is shown in Figure 17.9a were used. The regression equations $y = 1.0912x$ were used for the conversions, where y is the TUR in mg/L, x is the TUR in NTU, and 1.0912 is the conversion factor determined. The coagulation rate, k_m and α, were obtained from the intercept and slope of the regression plots (Figure 17.9b). The kinetic parameters at the optimum PBC dosage of 0.3 g/L and temperature of 303 K at different pHs are presented in Table 17.4. The rate of coagulation $(-r_p)$ was determined using Equation 17.5. The kinetics parameters β_{BR}, ε_p, K_R, D^1, τ_F, and $\tau_{F1/2}$ were evaluated using Equations 17.4, 17.6–17.8, 17.11, and 17.12, respectively. The R^2 values (Table 17.4) indicate that the kinetic data obeyed the second-order kinetic model (perikinetics flocculation theory). The K_R values remained unchanged at all pHs due to negligible changes in temperature and viscosity of the coag-flocculation medium [75]. The rate $(-r_p)$, which

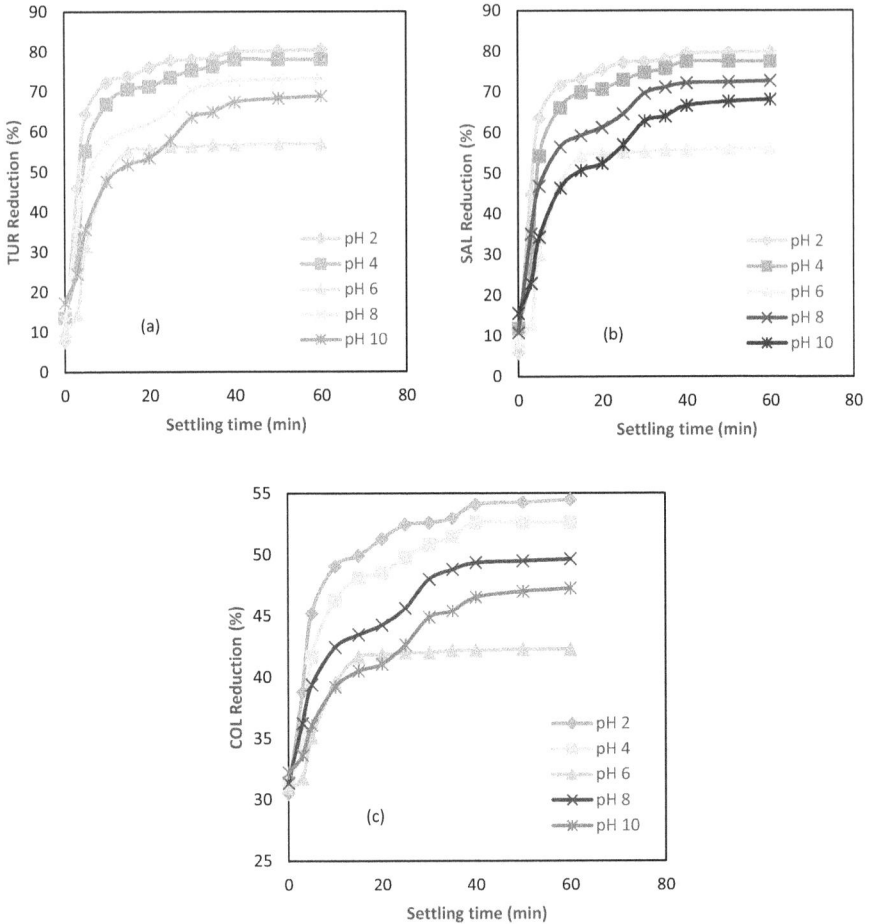

FIGURE 17.8   Effect of settling time on (a) TUR, (b) SAL, and (c) COL reduction on LSAE (at PBC dosage of 0.3 g/L and temperature of 30°C).

accounts for the rate of particle concentration depletion, is derived from Equation 17.5. The values of ε_p and $\tau_{F1/2}$ are used to determine the degree of coagulation. Higher ε_p (collision efficiency) value was observed for pH 2 (the optimum pH) and pH 4 which is in agreement with the statement by Hunter [76] and Menkiti and Ezemagu [46] that high ε_p value results in a high kinetic energy, which tends to lower the zeta potential, leading to double-layer collapse or colloidal destabilization at low $\tau_{1/2}$ to produce optimum coagulation efficiency [77]. Finally, the values of τ_F and $\tau_{F1/2}$ are presented in Table 17.4. In a coagulation process, the value $\tau_{F1/2}$ is an important parameter since it is linked to ion or particle aggregation [78].

17.3.3.6   Cost Analysis on the Process

Cost analysis is commonly used in decision making. The cost of biocoagulation-flocculation process is dependent on the cost of the coagulant used for pollutants removal. However,

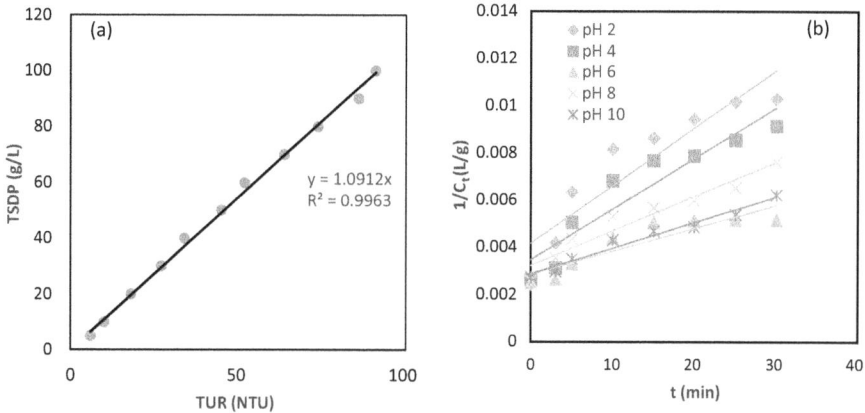

FIGURE 17.9 TDSP plots for LSAE (a) and particle concentration versus time graph for TUR reduction on LSAE using second-order kinetics at 303 K and PBC dosage of 0.3 g/L (b).

TABLE 17.4
Brownian Coagulation-Flocculation Kinetics at 303 K and the Optimum Dosage

Parameter	pH 2	pH 4	pH 6	pH 8	pH 10
R^2	0.8436	0.8818	0.8201	0.9365	0.9688
K_{MZ} (L/mg min)	0.0002	0.0002	0.0004	0.0001	0.0001
C_0 (mg/L)	238.10	285.71	344.83	312.50	344.83
$-r_p$ (mg/min)	$0.0002C_t^2$	$0.0002C_t^2$	$0.0004C_t^2$	$0.0001C_t^2$	$0.0001C_t^2$
$\tau_{F1/2}$ (min)	42	35	14.5	64	58
τ_F (min)	84	70	29	128	116
β_{BR} (L/mg min)	0.0004	0.0004	0.0008	0.0002	0.0002
K_R (L/min)	2.14532E−18	2.14532E−18	2.14532E−18	2.14532E−18	2.14532E−18
ε_p (L/mg)	9.32264E+13	9.32264E+13	1.86453E+14	4.66132E+13	4.66132E+13
D^1	0.008160734	0.008160734	0.016321468	0.004080367	0.004080367

cheap materials are required for water and wastewater treatment. The cost of treatment of 1 L of LSAE was evaluated considering the cost of preparation of 0.3 g/L (optimum dose) of the coagulant dosage, cost of energy, and labour cost. The cost of preparation of 0.3 g/L of PBC was 0.27€. The cost of labour was estimated as 2.16€. The total cost was calculated as 3.73€ by summing up the cost of preparation o0.3 g/L PBC, labour cost, and energy cost.

17.4 CONCLUDING REMARKS

Turbidity (TUR), salinity (SAL), and colour (COL) removals from land-based saline aquaculture effluent (LSAE) using *Parkia biglobosa* seeds coagulant (PBC) were investigated. Maximum removals of TUR = 80.44%, SAL = 79.98%, and COL = 54.47% (at 60 minutes) were obtained at PBC dosage:0.3 g/L, pH 2, and temperature: 303 K. Charge neutralization, adsorption, and intraparticle bridging mechanisms were responsible for TUR, SAL, and COD removals using PBC at the optimum conditions. The coagulation-flocculation kinetics parameters were successfully determined. The data fitted into the perikinetics theory (with $R^2 > 0.8$) at order of the coagulation reaction (α): 2. The reaction rate (K_m) and coagulation-flocculation half-life ($\tau_{F1/2}$) were evaluated as 0.0002 L/g.min and 42 minutes, respectively, at the optimum conditions. Also, the cost of using PBC for the treatment of 1 L of LSAE was estimated as 3.73€ including material, energy, and labour costs. It is possible to assume that these seeds can be used for LSAE pretreatment.

ACKNOWLEDGEMENTS

The first author wish to acknowledge her husband Engr. George C. Igwegbe, and children: Divine Chinenye Igwegbe, Marvel Ifechukwu Igwegbe, Victory Chukwudubem Igwegbe, and Great Ekene Igwegbe for their spiritual and moral support.

REFERENCES

1. Igwegbe, C.A. and O.D. Onukwuli, Removal of total dissolved solids (TDS) from aquaculture wastewater by coagulation-flocculation process using *Sesamum indicum* extract: Effect of operating parameters and coagulation-flocculation kinetics. *The Pharmaceutical and Chemical Journal*, 2019. **6**(4): pp. 32–45.
2. Ebeling, J.M., C.F. Welsh, and K.L. Rishel, Performance evaluation of an inclined belt filter using coagulation/flocculation aids for the removal of suspended solids and phosphorus from microscreen backwash effluent. *Aquacultural Engineering*, 2006. **35**(1): pp. 61–77.
3. Sharrer, M.J., K. Rishel, and S. Summerfelt, Evaluation of geotextile filtration applying coagulant and flocculant amendments for aquaculture biosolids dewatering and phosphorus removal. *Aquacultural Engineering*, 2009. **40**(1): pp. 1–10.
4. Martins, C., E.H. Eding, M.C. Verdegem, L.T. Heinsbroek, O. Schneider, J.-P. Blancheton, E.R. d'Orbcastel, and J. Verreth, New developments in recirculating aquaculture systems in Europe: A perspective on environmental sustainability. *Aquacultural Engineering*, 2010. **43**(3): pp. 83–93.
5. Dong, D., H. Sun, Z. Qi, and X. Liu, Improving microbial bioremediation efficiency of intensive aquacultural wastewater based on bacterial pollutant metabolism kinetics analysis. *Chemosphere*, 2021. **265**: p. 129151.
6. Cao, L., R. Naylor, P. Henriksson, D. Leadbitter, M. Metian, M. Troell, and W. Zhang, China's aquaculture and the world's wild fisheries. *Science*, 2015. **347**(6218): p. 133–135.
7. Ighalo, J.O., A.G. Adeniyi, A.A.O. Eletta, N.I. Ojetimi, and O.J. Ajala, Evaluation of luffa cylindrica fibers in a biomass packed bed for the treatment of fish pond effluent before environmental release. *Sustainable Water Resources Management*, 2020. **6**(6): pp. 1–11.
8. Weeks, N.C., M.B. Timmons, and S. Chen, Feasibility of using foam fractionation for the removal of dissolved and suspended solids from fish culture water. *Aquacultural Engineering*, 1992. **11**(4): pp. 251–265.
9. Igwegbe, C.A., Evaluation of bio- and electro- coagulants' activities on Fish pond wastewater and Solid waste leachate. 2019, Ph.D. dissertation submitted to the Department of Chemical Engineering.

10. Igwegbe, C.A., O.D. Onukwuli, and P.C. Onyechi, Optimal route for turbidity removal from aquaculture wastewater by electrocoagulation-flocculation process. *Journal of Engineering and Applied Sciences*, 2019. **15**(1): pp. 99–108.

11. Igwegbe, C.A., O.D. Onukwuli, J.O. Ighalo, and C.J. Umembamalu, Electrocoagulation-flocculation of aquaculture effluent using hybrid iron and aluminium electrodes: A comparative study. *Chemical Engineering Journal Advances*, 2021. **6**: p. 100107.

12. Letelier-Gordo, C.O. and P.M. Fernandes, Coagulation of phosphorous and organic matter from marine, land-based recirculating aquaculture system effluents. *Aquacultural Engineering*, 2020. **92**: p. 102144.

13. Ebeling, J.M., P.L. Sibrell, S.R. Ogden, and S.T. Summerfelt, Evaluation of chemical coagulation–flocculation aids for the removal of suspended solids and phosphorus from intensive recirculating aquaculture effluent discharge. *Aquacultural Engineering*, 2003. **29**(1–2): pp. 23–42.

14. Ebeling, J.M., S.R. Ogden, P.L. Sibrell, and K.L. Rishel, Application of chemical coagulation aids for the removal of suspended solids (TSS) and phosphorus from the microscreen effluent discharge of an intensive recirculating aquaculture system. *North American Journal of Aquaculture*, 2004. **66**(3): pp. 198–207.

15. Igwegbe, C.A. and O.D. Onukwuli, Removal of Total Dissolved Solids (TDS) from Aquaculture Wastewater by Coagulation-flocculation Process using Sesamum indicum extract: Effect of operating parameters and coagulation-flocculation kinetics. *The Pharmaceutical and Chemical Journal*, 2019.**6**(4): pp. 32–45.

16. Ohale, P.E., C.E. Onu, N.J. Ohale, and S.N. Oba, Adsorptive kinetics, isotherm and thermodynamic analysis of fishpond effluent coagulation using chitin derived coagulant from waste Brachyura shell. *Chemical Engineering Journal Advances*, 2020. **4**: p. 100036.

17. Obiora-Okafo, I., O. Onukwuli, and C. Ezugwu, Application of kinetics and mathematical modelling for the study of colour removal from aqueous solution using natural organic polymer. *Desalination Water Treatment*, 2019. **165**: pp. 362–373.

18. Rishel, K.L. and J.M. Ebeling, Screening and evaluation of alum and polymer combinations as coagulation/flocculation aids to treat effluents from intensive aquaculture systems. *Journal of the World Aquaculture Society*, 2006. **37**(2): pp. 191–199.

19. Akpenpuun, T.D., B.A. Akinyemi, K.A. Adeniran, M. Sulaiman, and B. Adelodun, Activated locust bean pod (Parliament Biglobosa) as a potential coagulant for domestic sewage treatment. *Environmental Research, Engineering and Management*, 2017. **73**(4): pp. 70–81.

20. Wafar, R., B. Lalabe, Z. Adi, C. Obun, P. Shinggu, H. Yusuf, M. Caleb, U. Doofan, and G. Obogo, Impact of toasted African locust bean (*Parkia Biglobosa*) meal (Tabm) on the growth performance, nutrient digestibility, blood profile and carcass characteristics of grower rabbits. *Nigerian Journal of Animal Science and Technology (NJAST)*, 2019. **2**(1): pp. 85–96.

21. Akinoso, R. and N.E. El-alawa, Some engineering and chemical properties of cooked locust bean seed (*Parkia biglobosa*). *West Indian Journal of Engineering*, 2013. **35**(2): pp. 51–57.

22. Oluwaniyi, O. and I. Bazambo, Nutritional and amino acid analysis of raw, partially fermented and completely fermented locust bean (*Parkia biglobosa*) seeds. *African Journal of Food, Agriculture, Nutrition and Development*, 2016. **16**(2): pp. 10866–10883.

23. Ahodegnon, D.K., M. Gnansounou, R.G. Bogninou, E.R. Kanfon, B. Chabi, P.C.A. Dossa, E.A. Anago, E. Ahoussi, V. Wotto, and D.C. Sohounhloue, Biochemical profile and antioxidant activity of *Parkia biglobosa* and *Tamarindus indica* fruits acclimated in Benin. *International Journal of Advanced Research*, 2018. **6**(11): pp. 702–711.

24. Bello, O.S., K.A. Adegoke, O.O. Sarumi, and O.S. Lameed, Functionalized locust bean pod (*Parkia biglobosa*) activated carbon for Rhodamine B dye removal. *Heliyon*, 2019. **5**(8): p. e02323.

25. Teh, C.Y., T.Y. Wu, and J.C. Juan, Potential use of rice starch in coagulation–flocculation process of agro-industrial wastewater: Treatment performance and flocs characterization. *Ecological Engineering*, 2014. **71**: pp. 509–519.

26. Teh, C.Y., T.Y. Wu, and J.C. Juan, Optimization of agro-industrial wastewater treatment using unmodified rice starch as a natural coagulant. *Industrial Crops and Products*, 2014. **56**: pp. 17–26.

27. Teh, C.Y., P.M. Budiman, K.P.Y. Shak, and T.Y. Wu, Recent advancement of coagulation–flocculation and its application in wastewater treatment. *Industrial & Engineering Chemistry Research*, 2016. **55**(16): pp. 4363–4389.

28. Othmani, B., M.G. Rasteiro, and M. Khadhraoui, Toward green technology: A review on some efficient model plant-based coagulants/flocculants for freshwater and wastewater remediation. *Clean Technologies and Environmental Policy*, 2020. **22**: pp. 1025–1040.

29. APHA, *Standard Methods for the Examination of Water and Wastewater.* 1995: American Public Health Association, American Water Works Association, and Water Environment Federation, Washington, DC. 9th ed.

30. APHA, *Standard Methods for the Examination of Water and Wastewater.* 1998: American Public Health Association, American Water Works Association, and Water Environment Federation, Washington, DC, pp. 3–37.

31. EPA U, Method 180.1: Determination of Turbidity by Nephelometry. Rev. 2.0. Methods for Chemical Analysis of Water and Wastes, 1993.

32. Guiry, M.D., G.M. Guiry, L. Morrison, F. Rindi, S.V. Miranda, A.C. Mathieson, B.C. Parker, A. Langangen, D.M. John, and I. Bárbara, AlgaeBase: An on-line resource for algae. *Cryptogamie, Algologie*, 2014. **35**(2): pp. 105–115.

33. Krzyczkowska, J. and M. Kozłowska, Effect of oils extracted from plant seeds on the growth and lipolytic activity of Yarrowia lipolytica yeast. *Journal of the American Oil Chemists' Society*, 2017. **94**(5): pp. 661–671.

34. Collins, T.J., Image J for microscopy. *Biotechniques*, 2007. **43**(S1): pp. S25–S30.

35. Menkiti, M., Studies on coagulation and flocculation of Coal Washery Effluent: Turbid metric approach. 2007, M.Sc thesis, Nnamdi Azikiwe University Awka, Nigeria.

36. Nwabanne, J.T. and C.C. Obi, Coagulation-flocculation performance of snail shell biomass in abattoir wastewater treatment. *Journal of Chemical Technology & Metallurgy*, 2019. **54**(6).

37. Smoluchowski, M., Versuch einer mathematischen Theorie der Koagulationskinetik kolloider Lösungen. 1917.

38. Menkiti, M., P. Igbokwe, F. Ugodulunwa, and O. Onukwuli, Rapid coagulation/flocculation kinetics of coal effluent with high organic content using blended and unblended chitin derived coagulant (CSC). *Research Journal of Applied Sciences*, 2008. **3**(4): pp. 317–323.

39. Menkiti, M.C., C.I. Nwoye, C.A. Onyechi, and O.D. Onukwuli, Factorial optimization and kinetics of coal washery effluent coag-flocculation by Moringa oleifera seed biomass. *Advances in Chemical Engineering and science*, 2011. **1**(03): p. 125.

40. Emembolu, L.N., C.A. Igwegbe, and V.I. Ugonabo, Effect of natural biomass treatment on vegetable oil industry effluent via coag-flocculation. *Saudi Journal of Engineering and Technology*, 2016. **1**(4): pp. 172–179.

41. Fridrikhsberg, D.A., *A Course in Colloid Chemistry.* 1986: Imported Publication, Mascow, Russia.

42. Ugonabo, V.I., L.N. Emembolu, and C.A. Igwegbe, Bio-coag-flocculation of refined petroleum wastewater using plant extract: A turbidimeric approach. *International Journal*, 2016. **19**.

43. Menkiti, M.C. and M.I. Ejimofor, Experimental and artificial neural network application on the optimization of paint effluent (PE) coagulation using novel Achatinoidea shell extract (ASE). *Journal of Water Process Engineering*, 2016. **10**: pp. 172–187.
44. Okey-Onyesolu, C., O. Onukwuli, M. Ejimofor, and C. Okoye, Kinetics and mechanistic analysis of particles decontamination from abattoir wastewater (ABW) using novel Fish Bone Chito-protein (FBC). *Heliyon*, 2020. **6**(8): p. e04468.
45. Ahmadi, S., L. Mohammadi, A. Rahdar, S. Rahdar, R. Dehghani, C.A. Igwegbe, and G.Z. Kyzas, Acid dye removal from aqueous solution by using Neodymium (III) Oxide Nanoadsorbents. *Nanomaterials*, 2020. **10**(3): p. 556.
46. Menkiti, M.C. and I.G. Ezemagu, Sludge characterization and treatment of produced water (PW) using Tympanotonus fuscatus coagulant (TFC). *Petroleum*, 2015. **1**: pp. 51–62.
47. Choy, S., K. Prasad, T. Wu, and R. Ramanan, A review on common vegetables and legumes as promising plant-based natural coagulants in water clarification. *International Journal of Environmental Science and Technology*, 2015. **12**(1): pp. 367–390.
48. Coruh, H.A., Use of calcium alginate as a coagulant in water treatment. 2005.
49. AOAC, *Official Methods of Analysis*. 1984: Association of Official Analytical Chemists, Arlington, VA, 14th ed.
50. AOAC, *Official Methods of Analysis, Method 935.14, 942.05 and 992.24*. 2005: Association of Official Analytical Chemists, Washington DC. 18th ed.
51. AACC, *AACC Method 44-15A. Approved Methods of the American Association for Clinical Chemistry*. 2000: American Association of Cereal Chemists, St. Paul. 10th ed.
52. AOAC, Total, soluble, and insoluble dietary fiber in foods: Enzymatic gravimetric method, MES-TRIS buffer, in *Official Methods of Analysis of AOAC International, Method 991.43*. 2000: Association of Official Analytical Chemists, Arlington, VA. 17th ed.
53. AOAC, Oil in cereal adjuncts: Petroleum ether extraction method, in *Official Methods of Analysis of AOAC International, Method 945.16*. 2000: Association of Official Analytical Chemists, Arlington, VA, 17th ed.
54. AOAC, Determination of Protein Content in Food, Method 945.18-B, in *Official Methods of Analysis*. 2005: AOAC International Publisher, Gaithersburg.
55. FAO, *Rice is Life*. 2004: Food and Agricultural Organization of the United Nations, Rome.
56. Ahmadi, S., C.A. Igwegbe, S. Rahdar, and Z. Asadi, The survey of application of the linear and nonlinear kinetic models for the adsorption of nickel (II) by modified multi-walled carbon nanotubes. *Applied Water Science*, 2019. **9**(4): p. 98.
57. Banerjee, S., S. Dubey, R.K. Gautam, M. Chattopadhyaya, and Y.C. Sharma, Adsorption characteristics of alumina nanoparticles for the removal of hazardous dye, Orange G from aqueous solutions. *Arabian Journal of Chemistry*, 2019, **12**(8): pp. 5339–5354.
58. Ighalo, J.O. and A.G. Adeniyi, A mini-review of the morphological properties of biosorbents derived from plant leaves. *SN Applied Sciences*, 2020. **2**(3): p. 509.
59. Zhang, Z., S. Xia, J. Zhao, and J. Zhang, Characterization and flocculation mechanism of high efficiency microbial flocculant TJ-F1 from Proteus mirabilis. *Colloids and Surfaces B: Biointerfaces*, 2010. **75**(1): pp. 247–251.
60. Awang, N.A. and H.A. Aziz, Hibiscus rosa-sinensis leaf extract as coagulant aid in leachate treatment. *Applied Water Science*, 2012. **2**(4): pp. 293–298.
61. Nharingo, T., M. Zivurawa, and U. Guyo, Exploring the use of cactus Opuntia ficus indica in the biocoagulation–flocculation of Pb (II) ions from wastewaters. *International Journal of Environmental Science and Technology*, 2015. **12**(12): pp. 3791–3802.
62. Ejimofor, M.I., M.C. Menkiti, and I.G. Ezemagu, Comparative studies on removal of turbid-metric particles (TDSP) using animal based chito-protein and aluminium sulfate on paint wastewater (PWW). *Sigma: Journal of Engineering & Natural Sciences/ Mühendislik ve Fen Bilimleri Dergisi*, 2020. **38**(3).

63. Bhandari, V.M. and V.V. Ranade, Advanced physico-chemical methods of treatment for industrial wastewaters. *Industrial Wastewater Treatment, Recycling and Reuse*, 2014: pp. 81–140.

64. Fedala, N., H. Lounici, N. Drouiche, N. Mameri, and M. Drouiche, *RETRACTED: Physical Parameters Affecting Coagulation of Turbid Water with Opuntia ficus-indica Cactus*. 2015: Elsevier.

65. Cruz, D., M. Pimentel, A. Russo, and W. Cabral, Charge neutralization mechanism efficiency in water with high color turbidity ratio using aluminium sulfate and flocculation index. *Water*, 2020. **12**(2): p. 572.

66. Obiora-Okafo, I., O. Onukwuli, and N. Eli-Chukwu, Evaluation of bio-coagulants for colour removal from dye synthetic wastewater: Characterization, adsorption kinetics, and modelling approach. *Water SA*, 2020. **46**(2): pp. 300–312.

67. Beltrán-Heredia, J., J. Sánchez-Martín, and C. Gómez-Muñoz, Performance and characterization of a new tannin-based coagulant. *Applied Water Science*, 2012. **2**(3): pp. 199–208.

68. Banerjee, S. and M. Chattopadhyaya, Adsorption characteristics for the removal of a toxic dye, tartrazine from aqueous solutions by a low cost agricultural by-product. *Arabian Journal of Chemistry*, 2017. **10**: pp. S1629–S1638.

69. Shak, K.P.Y. and T.Y. Wu, Coagulation–flocculation treatment of high-strength agro-industrial wastewater using natural Cassia obtusifolia seed gum: Treatment efficiencies and flocs characterization. *Chemical Engineering Journal*, 2014. **256**: pp. 293–305.

70. Zuki, N.M., N. Ismail, and F.M. Omar, Evaluation of zeta potential and particle size measurements of multiple coagulants in semiconductor wastewater. in *AIP Conference Proceedings*. AIP Publishing LLC, 2019.

71. Zhou, Z., Y. Yang, X. Li, W. Wang, Y. Wu, C. Wang, and J. Luo, Coagulation performance and flocs characteristics of recycling pre-sonicated condensate sludge for low-turbidity surface water treatment. *Separation and Purification Technology*, 2014. **123**: pp. 1–8.

72. Phalakornkule, C., J. Mangmeemak, K. Intrachod, and B. Nuntakumjorn, Pretreatment of palm oil mill effluent by electrocoagulation and coagulation. *Science Asia*, 2010. **36**(2): pp. 142–149.

73. Bhatia, S., Z. Othman, and A.L. Ahmad, Pretreatment of palm oil mill effluent (POME) using Moringa oleifera seeds as natural coagulant. *Journal of Hazardous Materials*, 2007. **145**(1–2): pp. 120–126.

74. Marriott, N.G. and G. Robertson, *Essentials of Food Sanitation*. 1997: Springer Science & Business Media.

75. Okolo, B.I., P.C. Nnaji, M.C. Menkiti, V.I. Ugonabo, and O.D. Onukwuli, Application of single angle turbidimetry on coag-flocculation effect of detarium microcarpum seed in brewery effluent. *Materials Sciences and Applications*, 2014. **2014**.

76. Hunter, R., *Introduction to Modern Colloid Science*. 1993: Oxford University Press, Oxford, New York.

77. de Oliveira Reis, G., *Study of the Mechanism of Acid Coagulation of Hevea latex and of the Rheological Properties of Resulting Gels*. 2015: Université de Montpellier.

78. Ifeanyi, U., M.M. Chukwudi, and O.D. Okechukwu, Effect of coag-flocculation kinetics on telfairia occidentalis seed coagulant (TOC) in pharmaceutical wastewater. *International Journal of Multidisciplinary Sciences and Engineering*, 2012. 3(9): pp. 22–33.

18 Low-Cost Adsorbents for the Treatment of Saline Oil Produced Water

Lakkimsetty Nageswara Rao and Varghese Manappallil Joy
National University of Science and Technology

Shaik Feroz
Prince Mohammad Bin Fahd University

CONTENTS

18.1 INTRODUCTION

Oil-produced water (OPW) is the major waste stream generated from oil and gas industries. It is a mixture of organic and inorganic compounds and is a byproduct of oil and gas operation. OPW is a mixture of formation water and injected water that is produced to surface facilities with oil and gas, not including drilling or well treatment fluids. The water varies greatly in quality and quantity. OPW handling methodology depends on the composition of produced water, location, quantity, and the availability of resources. OPW management and cost control can be done by choosing appropriate water disposal options or by finding an appropriate beneficial use for the water. Dissolved and dispersed oil components, dissolved minerals, production chemicals, production solids, and dissolved gases are the main components in produced water. Due to the growing volume of waste all over the world, the effect of discharging OPW on the environment has lately become an important issue of

DOI: 10.1201/9781003185437-21

environmental concern. OPW is treated with physical, chemical, and biological methods. Because of space constraints in offshore platforms, compact physical and chemical systems were used for operations. However, present technologies cannot eliminate few suspended oil particles as well as dissolved elements. Many chemical treatments, initial and operating costs are high and produce hazardous sludge. In onshore facilities, biological pretreatment of oily wastewater can be a cost-effective and environmentally friendly method. As high salt concentration and differences of influent characteristics have a straight effect on the turbidity of the effluent, it is suitable to incorporate a physical treatment, e.g., membrane to refine the final effluent. For these reasons, major research efforts in the future could focus on the optimization of existing technologies and the use of combined physicochemical and biological treatment of produced water in order to comply with reuse and discharge limits.

The significance of oil and natural gas in modern civilization is well known. Nevertheless, like most production activities, oil and gas production processes generate large volumes of liquid waste. Discharging produced water can pollute surface and underground water and soil. In order to meet environmental regulations as well as to reuse and recycle produced water, many researchers have focused on treating oily saline-produced water. The treatment of produced water is a significantly growing contest for the oil and gas industry that requires serious attention. It is basically a combination of the water originally exiting in the wells, which is the formation water, and any additional water injected for enhanced oil and gas recovery. The pollutants constituting OPW can vary severely depending on the geo-location and the petroleum product obtained. OPW contaminants could have detrimental effects on the aquatic environment, as this contaminated water is often discharged to surface water, especially for offshore operations. Depending on the geolocation of the discharge, a specific regulatory limit tends for the amount of contaminant present in OPW. OPW flows upward in large volume in the case of oil wells and in lesser volume in the case of gas wells [1]. Large quantities of OPW are generated globally, and it was estimated to be in the range of 250 million barrels per day. OPW was injurious to the environment if it is discharged without treatment. It contains high concentrations of dissolved solids, metals, radioactive materials, inorganic and organic toxic compounds, biological content, etc. Most of the OPW has been injected back to the oil wells to enhance oil production as a part of enhanced oil recovery (EOR) operations, and some of it was injected for the purpose of disposal. The composition of OPW can vary from one well to another based on the nature of the geological formation [2]. The typical composition of OPW is shown in Table 18.1.

18.2 MANAGEMENT OF SALINE OIL-PRODUCED WATER

Some of the options available for managing oil produced water:

- **Avoid production of water onto the surface:** Using polymer gels that block water contributing fissures or fractures or downhole water separators which separate water from oil or gas streams. This option eliminates wastewater and is one of the more elegant solutions, but is not always possible.
- **Inject produced water:** Inject the produced water into the same formation or another suitable formation; involves transportation of produced water

TABLE 18.1

Typical Composition of Oil-Produced Water [3]

Parameter	Minimum Value	Maximum Value	Heavy Metal	Minimum Value (mg/L)	Maximum Value (mg/L)
Density (kg/m³)	1,014	1,140	Calcium	13	25,800
Conductivity (µS/cm)	4,200	58,600	Sodium	132	97,000
Surface tension (dyn/cm)	43	78	Potassium	24	4,300
pH	4.3	10	Magnesium	8	6,000
TOC (mg/L)	0	1,500	Iron	<0.1	100
TSS (mg/L)	1.2	1,000	Aluminum	310	410
Total oil (IR; mg/L)	2	565	Boron	5	95
Volatile (BTEX; mg/L)	0.39	35	Barium	1.3	650
Base/neutrals (mg/L)	-	<140	Cadmium	<0.005	0.2
Chloride (mg/L)	80	200,000	Copper	<0.02	1.5
Bicarbonate (mg/L)	77	3,990	Chromium	0.02	1.1
Sulphate (mg/L)	<2	1,650	Lithium	3	50
Ammonium nitrogen (mg/L)	10	300	Manganese	<0.004	175
Sulphite (mg/L)	-	10	Lead	0.002	8.8
Total polar (mg/L)	9.7	600	Strontium	0.02	1,000
Higher acids (mg/L)	<1	63	Titanium	<0.01	0.7
Phenol (mg/L)	0.009	23	Zinc	0.01	35
Volatile fatty acids (mg/L)	2	4,900	Arsenic	<0.005	0.3
			Mercury	<0.005	0.3
			Silver	<0.001	0.15
			Beryllium	<0.001	0.004

from the producing site to the injection site. Treatment of the injectate to reduce fouling and scaling agents and bacteria might be necessary.

- **Discharge produced water:** Treat the produced water to meet onshore or offshore discharge regulations.
- **Reuse in oil and gas operations:** Treat the produced water to meet the quality required to use it for drilling, stimulation, and work over operations.
- **Consume in beneficial use:** In some cases, significant treatment of produced water is required to meet the quality required for beneficial uses such as irrigation, rangeland restoration, cattle and animal consumption, and drinking water for private use or in public water systems.

18.3 TREATMENT OF SALINE OIL-PRODUCED WATER

Various technologies are used for treating oil-produced water, but the selection of a particular technology depends on the essential quality of treated water. Actually, there were environmental and federal regulations controlling oil-produced water

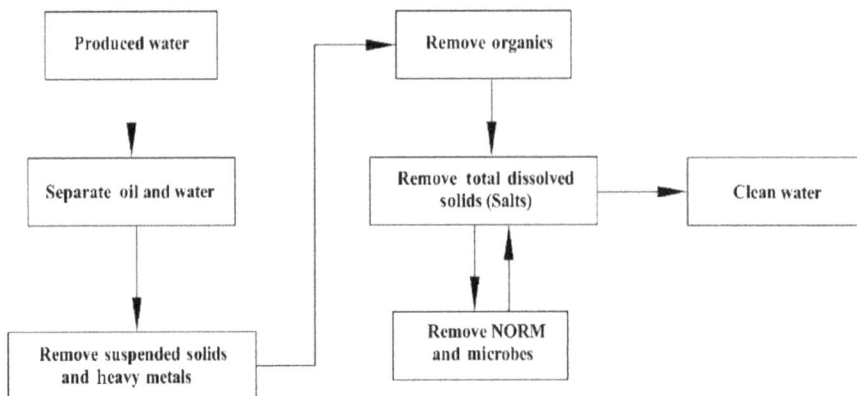

FIGURE 18.1   Treatments steps for OPW.

treatment and discharge processes. If oil-produced water has to be discharged into the ocean, then toxic chemicals, oil, and grease contents have to be removed. While if it was to be discharged onshore, then along with other impurities, salt must also be removed and if it was to be injected then oil content, solids, and bacteria have to be eliminated [4]. Treatment steps of OPW (Figure 18.1) include various processing steps such as a de-oiling process to remove the free and dispersed oil and grease present including polyaromatic hydrocarbons and alkyl phenol. Removal process of suspended solid particles, sand, turbidity, etc. Treatment of dissolved organics including benzene, toluene, ethylbenzene, xylene compounds (BTEX), aliphatic hydrocarbons, phenols, carboxylic acids, and light molecular weight (LMW) aromatic compounds. Disinfection step involves the removal of bacteria, microorganisms, algae, etc. Removal process of dissolved gas including light hydrocarbon gases, carbon dioxide, hydrogen sulfide, etc. Desalination or demineralization process step removes dissolved salts (1,000–300,000 ppm), sulfates, nitrates, contaminants, scaling agents, etc. Softening step removes excess water hardness. If the treated water is used for irrigation, sodium adsorption ratio (SAR) adjustment will be done by adding a required quantity of calcium or magnesium ions. In some cases, the treatment step also involves the removal of naturally occurring radioactive materials (NORM) like radium.

Pretreatment and treatment processes of oil-produced water can be classified based on the targeted removal of pollutants. The gravity separation, physical separation, and adsorption techniques were used to eliminate oil and grease. Membrane filtration, ion exchange, and electrocoagulation were used for heavy metal and inorganic removal. Chemical biocides, ultraviolet light, and ozone techniques were used for microbial treatment [5]. The criteria to choose technology to achieve a desired specification depend upon: off or onshore location; produced water characterization (species of ions, contaminants present); volume of water expected (over the field life cycle); any special needs/initiatives of operator company; scope for profitable use by reinjecting into the reservoir; and economic feasibility thereof. Local discharge specifications: eventual fate of water in the localized environment, scope for reuse of water for local population, playoff between equipment cost, available footprint, operability, ease of maintenance, energy consumption, associated chemical consumption, and byproducts generated. Most of the technologies employed for the treatment of OPW are costly and energy intensive. Many researchers are working to provide cost-effective technologies to treat OPW or at least to employ as a pretreatment step. Amorphous carbon thin film was synthesized from palm oil tree leaves. A series of batch adsorption experiments were conducted for the treatment of the OPW and evaluated the pollutant removal efficiency [6]. TiO_2 photocatalyst was employed in suspension and immobilization on a glass substrate to treat the organics present in OPW. The experimental results showed over 80% of organics removal from OPW [7,8]. Commercial reverse osmosis (RO) membrane technology was widely employed for the treatment of OPW due to its high salinity. Fouling of membranes is still considered as a major functional issue with this technology. The use of low-cost adsorbent technology as a pretreatment step to RO was considered an effective option. Low-cost adsorbents can be prepared from various organic waste materials such walnut shells, date seeds, coconut shells, etc. This also minimizes the waste disposal to the environment and can be used for the pretreatment of OPW. The expected increase of OPW is mainly related to two factors: increased production of oil/gas and age of the oil/gas field. With these two factors, the increase of OPW is projected to be more than the current levels. A treatment method that can handle large volumes of OPW is required, and this is while economically covering a range of contaminants. With these requirements and based on the literature available, the adsorption approach is the best treatment option that can deliver these requirements for the OPW.

The adsorption treatment approach is reported to operate at low concentrations with less operation time and much lesser cost than other treatment methods. Lowering the cost of treatment gives potential to OPW for reuse in several practices. Such practices can range from injection into petroleum wells to industrial and domestic usages. Managing OPW by reusing it gives an incentive for industries, as treatment assigns a value to OPW in the economy. Factors affecting production volume of oil-produced water include: method of well drilling, single zone and commingled, water injection or water flooding for enhanced oil recovery (EOR), the type of water separation technologies, location of the well within homogeneous/heterogeneous reservoirs, poor mechanical integrity, and underground communications.

18.4 ADSORPTION TECHNOLOGY AND LOW-COST ADSORBENTS

18.4.1 ADSORPTION TECHNOLOGY

The adsorption takes place through four staged process. The first stage involves the movement of ions from the bulk solution to the surface of the adsorbent followed by the second stage where adsorption takes place on the adsorbent surface. Then the third stage involves the movement of ions inside the adsorbent surface. The last stage is desorption where the adsorbed ions are released from the adsorbent surface for reuse. Adsorption can be divided into two types: physisorption and chemisorption. In physisorption, the interaction between the adsorbent and adsorbate is by weak forces, whereas in chemisorption the interaction is by means of a chemical bond. Adsorbents vary among themselves based on factors such as chemical composition, density, particle size distribution, and porosity. A good absorbent must possess high selectivity, high adsorption capacity, good mechanical properties, stability, and reusability [9]. The adsorption isotherm describes the relationship between the concentration of an adsorbate on a solid surface and its concentration in an aqueous solution for specified temperatures. An appropriate isotherm model gives essential physicochemical information that may be used to determine the suitability of sorption processes and to build successful adsorption systems. Typically, two kinds of studies are used to assess the adsorption process in an aqueous phase, depending on whether the adsorption system is in batch mode or continuous mode. They are (i) equilibrium batch adsorption and (ii) dynamic continuous-flow adsorption. There are two types of equilibrium isotherms: empirical and mechanistic. Mechanistic models provide critical information about the process of metal ion adsorption, while empirical models may be used to predict experimental outcomes [10].

18.4.2 Low-Cost Adsorbents

In recent years, scientists and researchers have focused their efforts on developing low-cost and readily accessible adsorbents for the separation of ionic species. This is shown by many review papers evaluating the usage of low-cost adsorbents in water and wastewater treatment. In several studies, several different kinds of inexpensive adsorbents have been evaluated for the extraction of pollutants from industrial saline wastewater. Figure 18.2 illustrates a generic categorization of low-cost adsorbents.

Various types of adsorbent materials have been used for the treatment of OPW. These materials are often considered natural adsorbents or synthetic adsorbents. The most common adsorbents that have been used for the treatment of contaminated water, especially for produced water are the absorbents derived from the Earth's crust

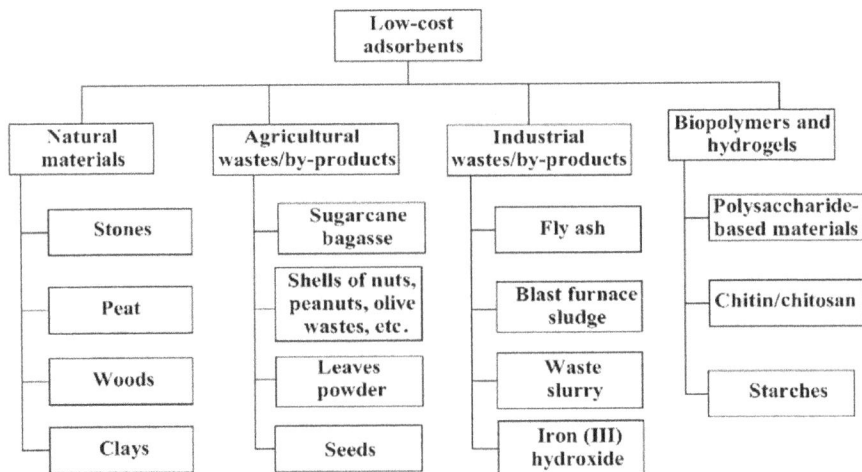

FIGURE 18.2 Generic categorization of low-cost adsorbents.

and biological sources. On the other hand, non-natural adsorbents included adsorbents synthesized in the lab or obtained by buying them from commercial entities. It is important to note that biological adsorbents are those found from agricultural trees and their byproducts and the food chain waste. Adsorbents obtained through natural means were explored by a number of studies for the treatment of OPW. Most of these studies reported better economics when compared to commercial adsorbents, making them more favorable. This natural adsorbents can be obtained from the Earth's crust or biological sources; agriculture trees and their derivatives or food wastes. Several researchers have studied the effectiveness of adsorbents that originated from the Earth's crust and mantle for OPW treatment.

18.5 CASE STUDY USING LOW-COST ADSORBENT FOR TREATMENT OF OPW

18.5.1 PROCESS

Batch experiments were conducted using prepared low-cost adsorbents for the treatment of OPW. OPW samples were collected from the Oman oil field and stored at 4°C to avoid any contamination. Raw date seeds and coconut shells were collected from domestic uses and farms in the Muscat area, Sultanate of Oman. Analytical grade reagents were procured and used as pure as supplied. Dissolved oxygen (DO), total dissolved solids (TDS), chemical oxygen demand (COD), total organic carbon (TOC), and oil and grease are the parameters measured to evaluate the performance of low-cost adsorbents in the treatment of OPW. 100 mL of OPW was taken in a glass beaker for each experimental run. A calculated amount of prepared activated carbon was added. The glass beaker was placed on a magnetic stirrer and allowed constant stirring at 500 RPM. Samples were collected at each time interval and analyzed for different parameters using standard procedures. The % removal efficiency was estimated using Equation 18.1.

$$\% \text{ removal efficiency} = \frac{(C_0 - C_f)}{C_0} \times 100 \qquad (18.1)$$

where C_0 and C_f are the initial and final concentrations in mg/L, respectively. The characteristics of OPW before the treatment was shown in Table 18.2.

The collected raw date seeds materials were washed to remove impurities and then soaked in distilled water. After soaking for sufficient time, the material was dried at 105°C in a drying oven for 24 hours. The dried material was then washed with 60% phosphoric acid (H_3PO_4) for chemical activation, thereafter washed multiple times using distilled water till pH becomes neutral. The drying procedure was repeated for the complete removal of moisture from the materials. Dried materials were crushed in a ball mill and sieved to 2 mm size. The sieved material (2 mm) was stored in a desiccator. Again, the material was chemically activated with 60% H_3PO_4 acid. The material was filled in stainless steel perforated cylinders. These cylinders were kept in a muffle furnace, and partial oxidation (control carbonization) was carried out at

TABLE 18.2
Characteristics of OPW
before Treatment

Parameter	Value
DO, mg/L	3.9
COD, mg/L	1136
TDS, mg/L	8890
TSS, mg/L	55
BOD, mg/L	355
TOC, mg/L	265
Oil and grease, mg/L	22.5

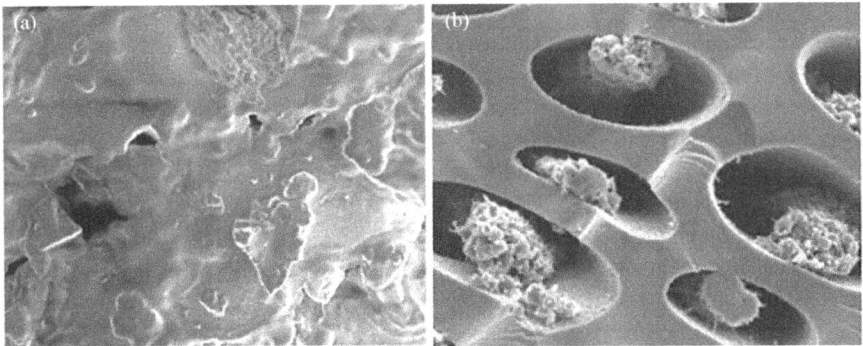

FIGURE 18.3   (a) DSA before activation. (b) DSA after activation.

300°C for 3 hours. The activated carbon from date seeds (DSA) was then taken out, cooled, and stored in air-tight polyethylene covers. The same procedure was followed to prepare activated carbon from coconut shells (CSA).

18.5.2 OUTCOMES

Date seeds were characterized using SEM and XRD before and after activation. Coconut shell characterized using SEM before and after activation. OPW was characterized before and after treatment using FT-IR and GC-MS. Figure 18.3a and b showed the SEM images of DSA before and after activation. Figure 18.4a and b showed the SEM images of CSA before and after activation. It was observed that calcination breaks the lignocellulose bonds in the raw adsorbent, which resulted in a dense spherical hole as an active site for the adsorption process. Figure 18.5a and b showed the XRD spectra of DSA before and after activation. The peaks at angles 23.4°, 44°, and 27° in Figure 18.5b indicated the presence of carbon and graphite in DSA.

The FT-IR spectra of OPW before treatment was shown in Figure 18.6a. Stretching peaks were observed at wave numbers 1,540, 1,640, 2,870, 2,926, and

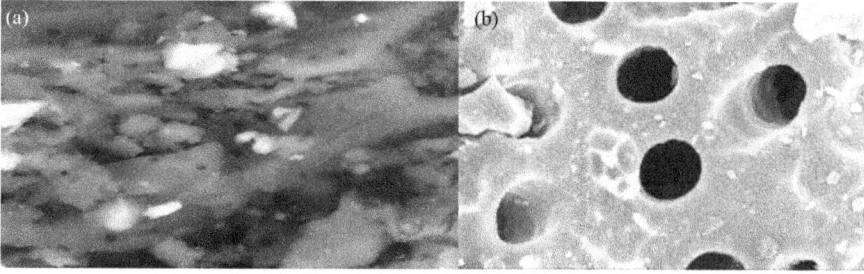

FIGURE 18.4 (a) CSA before activation. (b) CSA after activation.

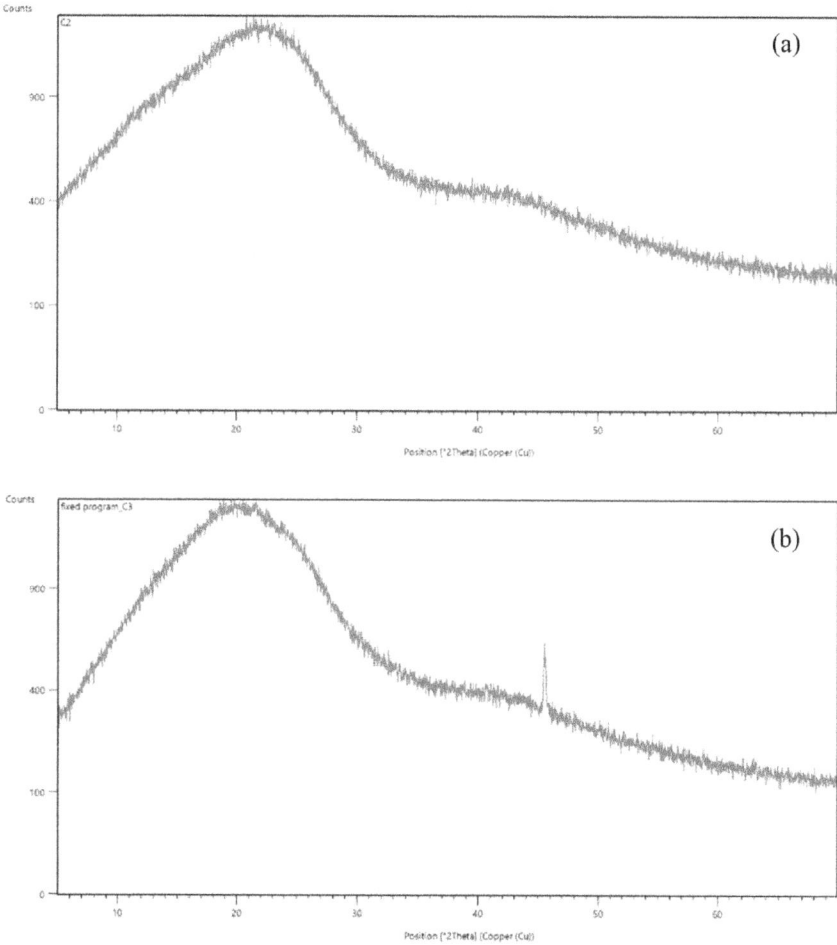

FIGURE 18.5 (a) DSA before activation. (b) DSA after activation.

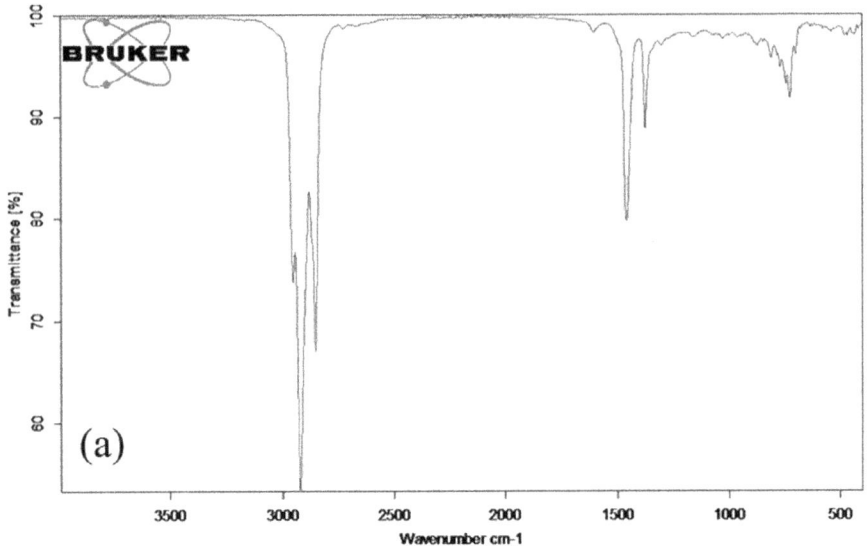

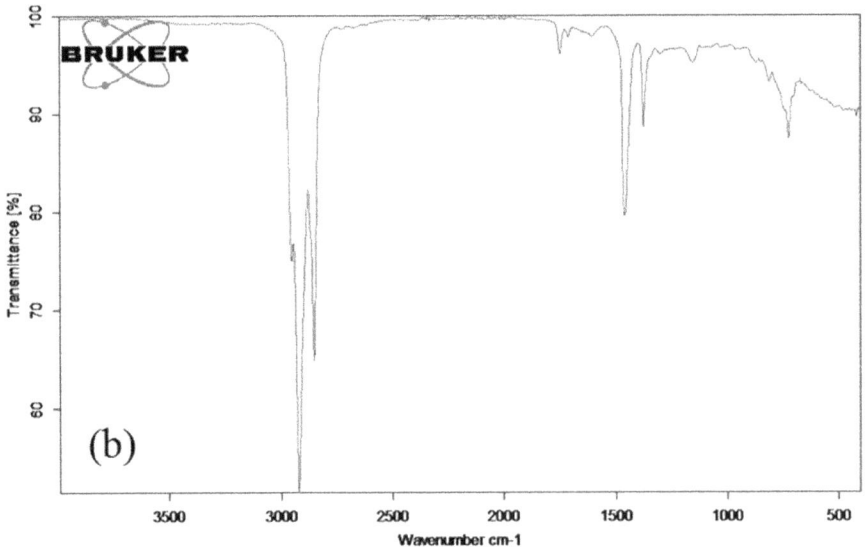

FIGURE 18.6   (a) FTIR spectra before treatment. (b) FTIR spectra after treatment.

3,400 cm^{-1}. These peaks indicated the presence of high aliphatic, hydroxyl, and C=C bonds group compounds. However, after treatment using DSA, the FITR spectra as shown in Figure 18.6b exhibited less intense peaks at wave numbers 3,200 and 3,500 cm^{-1}. This indicates the adsorption of hydroxyl group compounds. The peaks between 1,540 and 1,640 cm^{-1} almost disappeared, which indicated that carbonyl group (C=O) compounds were completely adsorbed by DSA. The peak between 1,200 and 1,000 cm^{-1} indicates the presence of C-O bond group compounds, and at 830 cm^{-1}, the peak represents the presence of Si-O bond compounds.

Figure 18.7a and b showed the GC–MS spectra of OPW before and after treatment with DSA, respectively. The intense peaks that appeared before treatment almost disappeared after treatment, which indicated that DSA effectively removed pollutants from OPW.

Two sets of experiments were run with a dosage of 10 g DSA and CSA each in 100 mL of OPW. A constant stirring was maintained, and samples were collected at each time interval of 15, 30, 45 and 60 minutes. Samples were analyzed for various parameters. The pH of the OPW was maintained at neutral. It was observed that the adsorption process was very slow till 30 minutes for both DSA and CSA and then took place rapidly between 30 and 60 minutes as showed in Figures 18.8 and 18.9. The maximum percentage removal efficiency of parameters was at 60 minutes contact time.

Another set of experiments were run by varying the adsorbent dosages from 2 to 10 g at a constant contact time of 60 minutes. Constant stirring was maintained, and samples were collected at the end of 60 minutes and analyzed. The percentage reduction of parameters was linear with increase in adsorbent dosage. Figures 18.10 and 18.11 showed the percentage reduction of parameters with a variation of adsorbent dosage. All the experiments were conducted maintaining constant pH of 7. Neutral pH was maintained to minimize the interference of either H$^+$ or OH$^-$ ions during the adsorption process with high or low pH, respectively,

Table 18.3 showed the compared performance of DSA and CSA. The percentage reduction of parameters was higher in DSA than CSA. They may be due to structural changes during activation and availability of more active sites on DSA.

18.6 CONCLUDING REMARKS

The discharge of OPW is irrefutably foreseen to increase in the near future. Hence, there is a need for an economic path of treatment. This was reasoned by multiple studies to be based on their inexpensive economic potential. However, to decisively make this conclusion, it is recommended to explore each adsorbent in terms of essential factors related to its usage. Such factors should encompass the adsorbent's potential by reporting their availability, processing, environmental, and economic aspects in more detail. Knowing the fate of a treatment system is of a great value, as it gives a better image of the adsorbent's potential. Activated carbons were prepared from low-cost adsorbents waste date seeds (DSA) and coconut shells (CSA) and used in the case study for the treatment of saline oil-produced water. The adsorbents were characterized before and after activation using SEM and XRD. The performance of DSA was marginally higher than CSA for the treatment of pollutants

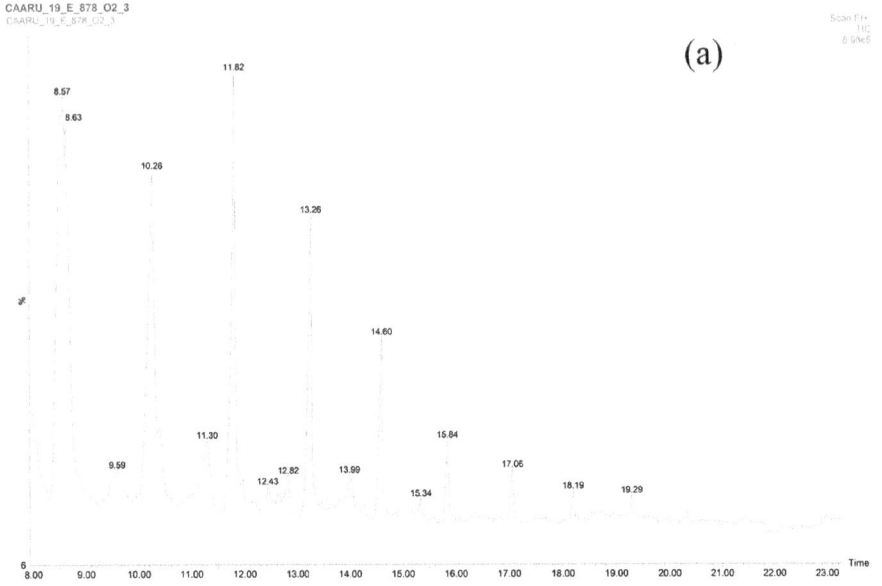

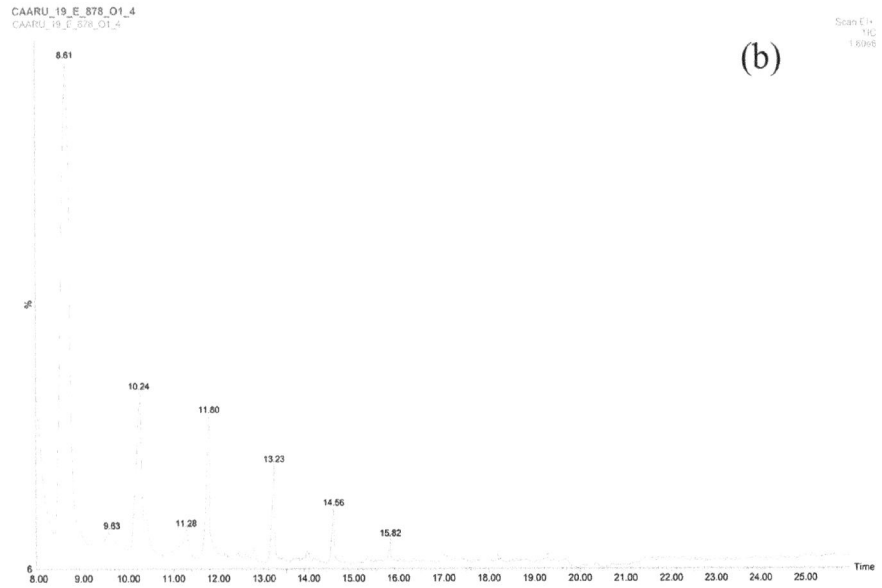

FIGURE 18.7 (a) GS-MS spectra before treatment. (b) GS-MS spectra after treatment.

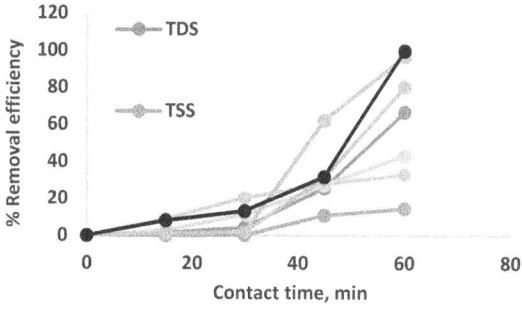

FIGURE 18.8   Effect of contact time for DSA.

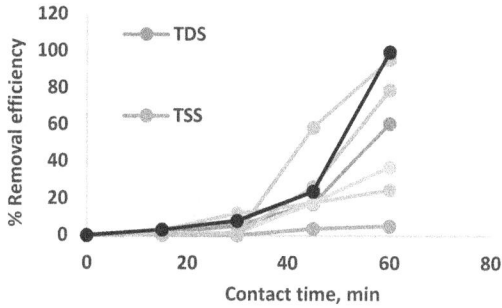

FIGURE 18.9   Effect of contact time for CSA.

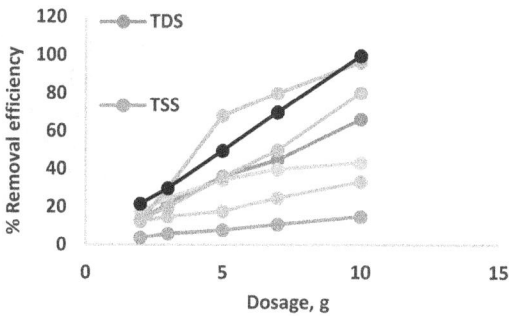

FIGURE 18.10   Effect of DSA dosage.

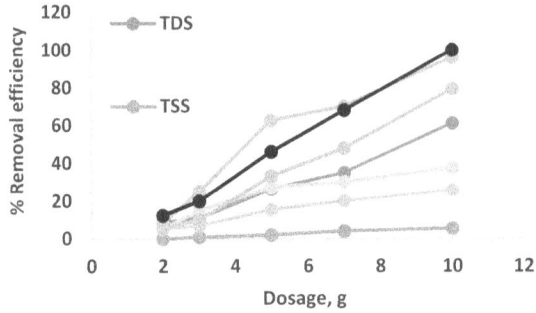

FIGURE 18.11   Effect of CSA dosage.

TABLE 18.3
Performance Comparison at
Optimum Conditions

Parameter	DSA (%)	CSA (%)
DO	33.3	25.6
COD	43.4	37.4
TDS	66.4	61.1
TSS	15.0	5.3
BOD	96.6	95.8
TOC	80.4	79.2
Oil & Grease	99.9	99.8

present in oil-produced water (OPW). OPW was characterized before and after treatment using FT-IR and GC-MS. The maximum percentage removal efficiency of DO = 33.3%, COD = 43.4%, TDS = 66.4%, TSS = 15%, BOD = 96.6%, TOC = 80.4%, and oil and grease = 99.9% was achieved with DSA and DO = 25.6%, COD = 37.4%, TDS = 61.1%, TSS = 5.3%, BOD = 96.6%, TOC = 79.2%, and oil and grease = 99.8% was achieved with CSA, respectively. Based on the outcome of this case study, the application of low-cost adsorbents at the commercial level for the treatment or pretreatment of OPW can be explored.

REFERENCES

1. Huishu, L. 2013. Oil produced water quality characterization and prediction for Wattenberg field. *Oil Produced Water*, 3(2), 36–75.
2. Al Anezi, K., Belkharchouche, M., Alali, S., Abuhaimed, W. 2012. Oil produced water characterization in Kuwait and its impact on environment. *Desalination and Water Treatment*, 56(5), 35–66.
3. Fakhru'l-Razi, A., Pendashteh, A., Abdullah, L. C., Biak, D. R. A., Madaeni, S. S., Abidin, Z. Z. 2009. Review of technologies for oil and gas produced water treatment. *Journal of Hazardous Materials*, 170, 530–551.

4. Nasiri, M., Jafari, I. 2017. Oil produced water from oil-gas plants: A short review on challenges and opportunities. *Oil Produced Water from Oil and Gas Plants*, 61(2), 73–81.
5. Igunnu, E., Chen, G. 2012. Oil produced water treatment technologies. *International Journal of Low-Carbon Technologies*, 9(3), 157–177.
6. Fathy, M., El-Sayed, M., Ramzi, M., Abdelraheem, O. 2018. Adsorption separation of condensate oil from oil produced water using ACTF prepared of oil palm leaves by batch and fixed bed techniques. *Egyptian Journal of Petroleum*, 27(3), 319–326.
7. Feroz, S., Al Saadi, K. H., Nageswara Rao, L., Varghese, M. 2017. Performance studies on solar photocatalysis and Photo Fenton process in treatment of polymer-based oil produced water. *International Journal of Photochemistry*, 3, 1–6.
8. Al Mashari., M. A., Varghese, M., Feroz, S., Nageswara Rao, L. 2017. Characterization and photocatalytic treatment of oil produced water using TiO_2. *International Journal of Applied Nanotechnology*. 3, 1–10.
9. Do, D. D. 1998. *Adsorption Analysis: Equilibria and Kinetics*. Published by Imperial College Press and Distributed by World Scientific Publishing CO., Singapore.
10. Smith, K. A. 1976. *Fundamentals of Momentum, Heat and Mass*; Welty, J.R., Wicks, C.E., Wilson, R.E., Eds.; Wiley: Hoboken, NJ, 789, pp. 27–55.

19 Sulphate Radical-Based Advanced Oxidation Process for Treatment of Organic Contaminants from Industrial Saline Wastewater

M. Venkata Ratnam
Mettu University

K. Nagamalleswara Rao
Vellore Institute of Technology

K. Senthil Kumar
Kongu Engineering College

Shaik Feroz
Prince Mohammad Bin Fahd University

CONTENTS

DOI: 10.1201/9781003185437-22

19.1 INTRODUCTION

Water is a key component of the earth's ecological structure and a fundamental part of life. The use of water is growing gradually due to population growth and daily comfort expectations (Senthil Kumar et al. 2020; Aljubourya et al. 2016a; Myneni, Kanidarapu, and Vangalapati 2020). The quality of water resources is decreasing day by day because of the constant adding of problematic synthetic chemicals into our water assets (Venkata Ratnam et al. 2019). Salt amounts above 1% are high salinity wastewaters (Linaric, Markic, and Sipos 2013; Z. Li et al. 2020; Mirbolooki, Amirnezhad, and Pendashteh 2017). Salt wastewater usually produced in various industrial activities, such as the leather industry, marine products, pharmaceuticals, and industries relating to crude oil extraction and gas refining, is rich in organic compounds and has at least 1–3.5 g of total dissolved solids (TDS) (Alipour, Moein, and Rezaei 2016). The salinity of wastewaters produced by oil field activities is higher than that of normal seawaters (nearly eight times) (Al-Raad et al. 2019; Brüninghoff et al. 2019; Aljuboury et al. 2015). The rejection from the reverse osmosis (RO) method used for desalination of seawater and industrial wastewater for reuse contains a considerable amount of salts as well as chemical pollutants that must be detoxified (Yang et al. 2014). Untreated wastewater discharges have significant impacts on marine life and cause migration, mortality, extinction of species, and ecosystem imbalance (Alipour, Moein, and Rezaei 2016; Cai et al. 2020; Laskar 2014; Saravanakumar et al. 2020; Aljuboury et al. 2016b). Design of highly efficient treatment procedures for salt wastes at high levels of toxic and inhibitor compounds is more difficult (Asgari et al. 2016). Many methods for treating the high salinity wastewater have been applied in industrial (Al-Raad et al. 2019; Asgari et al. 2016; Borges, Roux-De Balmann, and Guardani 2009, 2010; Brüninghoff et al. 2019; Cai et al. 2020; Darvishmotevalli et al. 2019). However, no single technology has been considered as a good method, as each has significant drawbacks. Biological wastewater treatment solution is the most cost-efficient approach compared to other methods (Zhuo et al. 2017; Tan et al. 2019). Based on bacteria resistance to sodium chloride, non-halophiles expand optimally at sodium chloride concentration less than 2%, so the high salinity wastewater must be diluted with a lot of fresh water before biological treatment. Not only does dilution absorb a significant quantity of freshwater, but also boosts the drainage volumes. On the other hand, evaporation and crystallization are popular methods for salt and water separation, which is used for pre-processing of highly salinized organic wastewater by multifaceted evaporation and mechanical vapour recompression (Stasinakis 2008). However, the evaporation process crystallized salt contains a great deal of organic pollutants, which according to regulations should be disposed of as dangerous waste. The cost of disposing of dangerous waste, though, is very costly. Hence, there is a need for the creation of an alternative method for highly saline wastewater treatment.

Advanced oxidation processes (AOPs) were thought to have the ability to remove organic pollutants (Giannakis, Lin, and Ghanbari 2021; Kakavandi and Ahmadi 2019; Z. Li et al. 2020; Lu, Fan, and Roddick 2013; Lutze, Kerlin, and Schmidt 2015). AOPs are methods used for oxidizing pollutants that are toxic, carcinogenic, and non-biodegradable to produce potent reactive oxidants such as $OH^{\bullet}$ and $SO_4^{\bullet-}$. In recent years, AOPs have gained attention in the management of sewage and waste-water. A variety of organic radicals are used to remove organic pollution in fresh waters, as hydroxyl radicals react with a number of organic compounds at almost disseminating levels. These processes include O_3/H_2O_2, H_2O_2/Fe^{2+}, O_3/UV, and H_2O_2/UV processes that produce hydroxyl radicals by decomposition of added O_3 or H_2O_2 (Yang et al. 2014; Zhang et al. 2020). Sulfate radical AOPs (SR-AOP) out-perform conventional hydroxyl radical AOPs due to their higher reduction potential (2.5–3.1 V), longer half-life, and applicability over a wider pH range (L. Yang et al. 2019; Q. Yang et al. 2019; Y. Yang et al. 2018). AOPs based on sulfate radicals have been shown to effectively strip aromatic components from induced wastewaters. In general, $SO_4^{•-}$ is more susceptible to one electron oxidation reactions than $OH^{•}$, thus broadening the spectrum of contaminant transition pathways (S. Yang et al. 2018; Yang et al. 2017; Y. Yang et al. 2018). Persulfate (PS, $S_2O_8^{-}$) and peroxymonosulfate are the most widely used reagents for sulfate radical generation (PMS, HSO_5^{-}). Since the O–O bond energy of PS and PMS is smaller than that of H_2O_2, they are easier to activate (Hu et al. 2017; Hu and Long 2016). Recent advances in the activation path-ways of precursors (PS/PMS), such as sophisticated catalytic materials and the use of innovative or expensive activator technologies, have a once-in-a-lifetime opportunity for integrated wastewater treatment (Giannakis, Lin, and Ghanbari 2021). In this chapter, various PS and PMS activation methods and the application of SR-AOPs for contaminant degradation are presented. Efforts were made to resolve knowledge gaps and needs for further research in this field, particularly in treating industrial wastewaters under saline conditions. Finally, the potential research directions were discussed.

19.2 ACTIVATION MECHANISM OF SULFATE RADICALS

By cleaving the peroxide bonds from reagents such as persulfate (PS, $S_2O_8^{-}$) and peroxymonosulfate (PMS, HSO_5^{-}), persulfate oxidation generates $SO_4^{•-}$ and $OH^{•}$ radi-cals. PS has a few distinct benefits over the other reagents, including a lower cost, reduced energy consumption in the generation of sulfate radicals, and a longer reten-tion time (Wacławek et al. 2017; Wang and Wang 2018). Without activation, the deg-radation rates of persulfates were low; therefore, the focus is on activation methods (Wacławek et al. 2017). An energy supply in the form of heat or UV radiation may be used to activate peroxide bonds. Ultrasonic energy, transitional metals, and recently carbonaceous materials were also employed in persulfate activation. Figure 19.1 summarizes the activation mechanisms employed.

19.2.1 ACTIVATION BY HEAT

While thermal activation is an efficient method of SR activation, it is not suitable for large-scale pollutant remediation due to the higher energy demands (Wang and Wang 2018). A few experiments used lower temperatures/room temperatures for SR activation, but the procedure resulted in lower removal efficiencies. Ultrasonic acti-vation, like heat activation, was thought to be energy intensive due to the higher

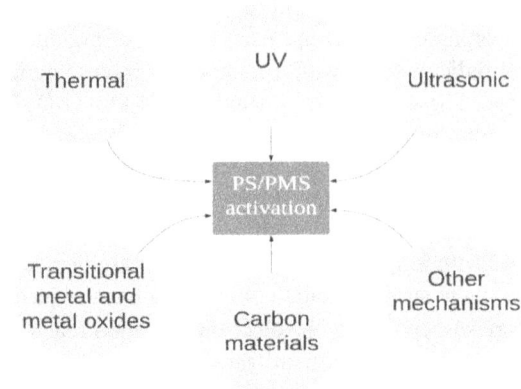

FIGURE 19.1   PS/PMS activation mechanisms.

energy requirements needed in cavitation. Thermal activation of the sulfate radical is activated by the fission of O–O bonds at temperatures greater than 50°C (Liu et al. 2018; Wang et al. 2017) as shown in Figure 19.2. Hydroxyl radicals were predominant in the thermal activation process as the sulfate radicals formed will transform into hydroxyl radicals as shown in Equation 19.3.

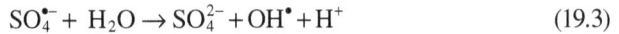

$$S_2O_8^{2-} \rightarrow 2SO_4^{\bullet-} \qquad (19.1)$$

$$HSO_5^- \rightarrow SO_4^{\bullet-} + OH^{\bullet} \qquad (19.2)$$

$$SO_4^{\bullet-} + H_2O \rightarrow SO_4^{2-} + OH^{\bullet} + H^+ \qquad (19.3)$$

The rather slow reaction rates at lower temperatures and the increase in reaction rate at higher temperatures at Equation 19.3 hold the secret to the treatment process's effectiveness. Table 19.1 summarizes the literature findings on thermal amplification of sulfate radicals.

In general, rising temperatures promote sulfate radical activation, which accelerates pollutant degradation (Tan et al. 2013; Pan et al. 2018; Norzaee et al. 2018). However, for the degradation tests, there should be certain optimum temperature conditions that can be checked. The pH of the solution is also essential. Both $SO_4^{\bullet-}$ and $OH^{\bullet}$ were significant at pH 9, $SO_4^{\bullet-}$ is the prevalent radical at pH < 7, and $OH^{\bullet}$ is predominant for pH above 12 (Milh et al. 2020; Hu et al. 2020; Lebik-Elhadi et al. 2020). According to the literature, most organic pollutants are effectively degraded at pH values about 7–8, implying that the coexistence of sulfate radicals and hydroxyl radicals can work better in eliminating organic pollutants than either hydroxyl radicals or sulfate radicals alone. The effect of coexisting inorganic anions, i.e., carbonate/bicarbonate $\left(CO_3^{2-}/HCO^{3-}\right)$, nitrate ions $\left(NO_3^-\right)$, and chloride ions (Cl⁻) on BisphenolS (BPS), was investigated by (Wang et al. 2017), and it was discovered that the presence of NO_3^- anions had no effect BPS degradation. The Cl⁻ anions play a dual role in the oxidation of BPS and HCO³⁻, on the other hand has the potential to greatly delay the degradation rate of BPS.

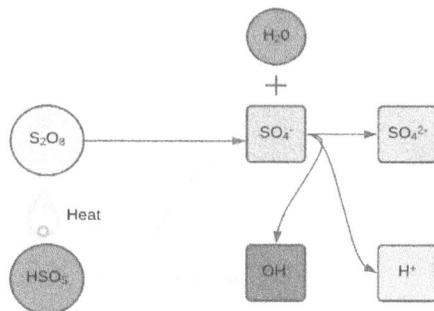

FIGURE 19.2   Heat-activated PS/PMS.

19.2.2   UV Irradiation

The sulfate radical-advanced oxidation method (SR-AOP) under light irradiation is gaining popularity in the elimination of different organic pollutants (Z. Zhou et al. 2019; Zhang et al. 2020; Yang et al. 2017). As persulfate is exposed to UV light, the peroxide bond is cleaved, resulting in the formation of two sulfate anions (Resources et al. 2018). In general, the radiation wavelength for persulfate is 254 nm, which is sufficient to break the $-O-O-$ bond in PS/PMS (Q. Yang et al. 2019). Sulfate radicals can be activated via two mechanisms: one involving the breakage of O-O bonds and the other involving the activation of PS/PMS via the breakage of water molecules with UV radiation. Some organic chemicals can be directly degraded by UV radiation alone; however, a small number of organic compounds are not UV sensitive. In persulfate activation, both UV-A and UV-C lamps are used. UV-C radiation is preferable because it is more efficient at 254 nm wavelength. However, it has disadvantages such as poor light exposure, intrusion from other water components, and the presence of mercury owing to its poisonous nature. Only a few researchers investigated the activation mechanism using UV-A lamps. Despite its poor effectiveness as compared to UV-C lamps, it has promise due to its low-energy intake and lack of radioactive heavy metals (Ioannidi, Frontistis, and Mantzavinos 2018). However, the UV activation mechanism is hampered by the fact that direct UV radiation alone is in several cases insufficient to cause PS/PMS, as seen in many studies outlined in Table 19.2 (Izadifard, Achari, and Langford 2017). As a result, for successful activation, researchers are focusing on catalytic UV-irradiated activation processes summarized in Figure 19.3. Homogeneous and heterogeneous metal and metal oxide catalysts, as well as carbon-based materials, are used; a thorough understanding of their functionality is given below.

19.2.3   Ultrasonic Activation

Ultrasonic activation of persulfate is more effective due to the higher temperatures and pressures caused by cavitation (L. Yang et al. 2019). Several papers have been written on the US-assisted SR-AOP process for organic pollutant degradation (Gągol, Przyjazny, and Boczkaj 2018; Gao et al. 2018; Chen and

TABLE 19.1

Heat Activated Per Sulfate for Pollutant Degradation

S. No	Pollutant	Conditions	Kinetics	Degradation Efficiency	Observations	References
1	Antipyrine	[PS]=0.1 MT=60°C Time=0–120 minutes pH 4.5–11	Pseudo-first-order kinetics	80%	At chloride to per sulfate ratio 10:1 Chloride has enhanced the degradation	Tan et al. (2013)
2	Bisphenol S (BPS)	[BPS]=0.25–2.0 mM [PS]=0.00625–0.05 mM T=55°C–70°C pH 7	Pseudo-zero-order Activation energy(E_a)=112.9 ± 6.8 kJ/mol	84%	The Cl⁻ anions has dual role on BPS oxidization	Wang et al. (2017)
3	Methylparaben (MeP) and Ethylparaben (EtP)	[PS]=14.8 mM T=20°C–60°C Time=240 min pH 5–9	Pseudo-first-order E_a=95.2 and 98.9 kJ/mol	94.8% and 95.3%	The presence of chloride, bicarbonate, or humic acid was found to inhibit the degradation of the two parabens to some extent	Chen et al. (2017)
4	Phenol	Phenol 1 mM [PS]=20–320 mM T=30°C–70°C pH 1.3–13.9	Pseudo-first-order E_a=139.7 ± 1.3 kJ/mol	99%	Increases in the solution pH from 1.3 to 13.9 significantly reduced the E_a (~50%)	Ma et al. (2017)
5	Sulfachloropyri-dazine	[PS]=0–280 μM T=30°C–40°C pH 3.0–10.0	Pseudo-first-order E_a=155 ± 15 kJ/mol	85%	The presence of chloride (Cl⁻) showed inhibitory effects	L. Liu et al. (2018)
6	Benzophenone-3 (BP-3)	[BP-3]:[PS]=1:500 T=40°C pH 7.0 Time=3 hours		100		Pan et al. (2018)

(Continued)

TABLE 19.1 (*Continued*)
Heat Activated Per Sulfate for Pollutant Degradation

S. No	Pollutant	Conditions	Kinetics	Degradation Efficiency	Observations	References
7	Penicillin G	[PS] = 0.05–0.5 μM $T = 40°C–80°C$ pH 3.0–10.0	Pseudo-first-order $E_a = 94.8$ kJ/mol	97.25		Norzaee et al. (2018)
8	Sulfamethoxazole (SMZ)	[SMZ] = 25–100 ppm [PS] = 0–35 mM $T = 40°C–80°C$ pH 2.9–9 Time = 60–80 minutes	Pseudo-first-order $E_a = 103$ kJ/mol			Milh et al. (2020)
9	Iohexol	[Iohexol] = 0.01 mM [PS] = 0.1–1 mM $T = 40°C–70°C$ pH 7. Time = 2–3 hours	$E_a = 122.2 \pm 2.8$ kJ/mol	85%	The presence of chloride inhibited iohexol degradation	Hu et al. (2020)
10	Thiamethoxam (TMX)	[TMX] = 0.01 mM [PS] = 50–1,000 mg/L $T = 20–70°C$, pH 3–10 Ultrasound power density (20–42 W/L)	$E_a = 108.7$ kJ/mol	95%	Heat-activated system results in greater degradation rates than ultrasound irradiation	Lebik-Elhadi et al. (2020)

TABLE 19.2
UV-Activated PS/PMS for Contaminant Degradation

S. No	Pollutant	Light Source	Degradation Efficiency	Observations	References
1	Orange III	KrClexcilamp (222 nm)	Dye Orange III (100% removal) and TOC content (72% removal)	Orange III oxidation efficiency, the tested oxidative systems can be arranged in the following order: $PS/UV/Fe^{2+} > PS/UV > PS/Fe^{2+} > UV > PS$	Resources et al. (2018)
2	Penicillin G (PG)	UVC lamp with intensity 950 μw/m²	94.28%		Norzaee et al. (2017)
3	Sulfolane	UVC lamps 254 nm	100%	Presence of 5 ppm O_3 in solution, provided by bubbling O_3 into the solution, increases the rate of sulfolane removal (up to three times)	Izadifard, Achari, and Langford (2017)
4	Propyl paraben	UV-LEDs: 10 W, 365 ± 5 nm	100%	UV-A radiation from LEDs or solar light can activate persulfate	Ioannidi, Frontistis, and Mantzavinos (2018)
5	Sulfamethoxazole (SMX)	Medium-pressure mercury UV lamp	Complete removal	$UV/PMS > UV/PS > UV/H_2O_2$	Ao and Liu (2017)
6	Sulfamethoxazole (SMX)	10 W low pressure mercury lamps (254 nm)	Complete removal		Yang et al. (2017)

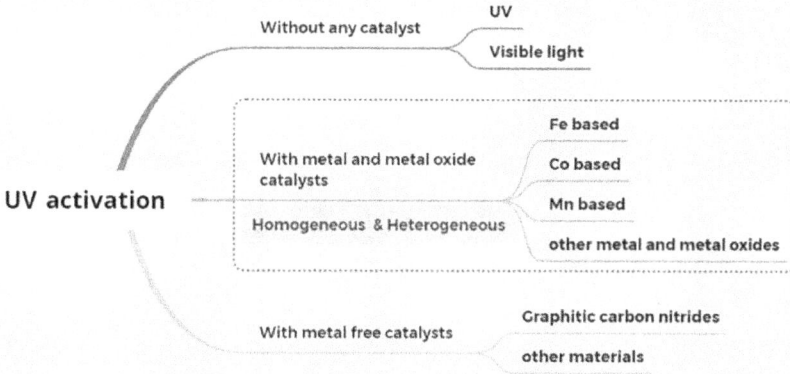

FIGURE 19.3   Different ways of UV activated PS/PMS.

Huang 2019; Wang et al. 2019; Xu et al. 2020; Y. T. Li et al. 2020; Malakootian, Ahmadian, and Khatami 2020). The activation of persulfates in the United States can be accomplished by two mechanisms: thermal activation and temporary cavitation. At the moment of the bubble's transient failure, the temperature and pressure inside it are exceedingly high (Chakma, Praneeth, and Moholkar 2017). The ultrasonic-aided AOP process is supposed to result in faster reaction times and better pollutant degradation rates. Temperature, pH, ultrasonic frequency and pressure, PS concentration, and the presence of inorganic ions were all investigated as influencing factors in removal efficiency (Hao et al. 2014; Gągol, Przyjazny, and Boczkaj 2018). The optimum frequency in US/SR-AOP is determined by the physical and chemical properties of the wastewater. Increased ultrasonic power strength causes increased sulfate radical formation, implying higher pollutant removal rates (Chen and Su 2012). The optimal pH conditions vary from acidic-alkaline conditions. Table 19.3 highlights the literature findings of optimal conditions reported by several studies. The presence of inorganic ions quenches sulfate and hydroxyl radicals. There have been experiments of the most frequently found inorganic ions such as Cl^-, HCO^{3-}/CO_3^{2-}, NO_3^-, SO_4^{2-}, and $H_2PO_4^-$ (Chakma, Praneeth, and Moholkar 2017). Overall, sono-activated PS oxidation is a promising potential technology for the treatment of refractory organic wastewater.

19.2.4   TRANSITION METAL CATALYSTS

The existing literature discusses the use of homogeneous and heterogeneous transition metal and metal oxide catalysts such as iron, cobalt, manganese, silver, copper, and nickel in triggering sulfate radicals due to their low-energy requirement

TABLE 19.3

Ultra Sonic Activation of PS/PMS

S. No	Pollutant	Source of Cavitation	Optimal Conditions	Degradation Efficiency	Observations	References
1	Dinitrotoluenes	800 W, 20 kHz	US power intensity = 126 W/cm^2 T = 45°C [PS] = 2 wt%	100		Chen and Su (2012)
2	Ammonium perfluorooctanoate (APFO)	300W, 20kHz	US power = 300 W [PS] = 10 mmol/L [APFO] = 11.6 µmol/L pH = 6.0	Defluorination ratio = 15%		Hao et al. (2014)
3	C.I. Direct Red 23	120 W, 60 kHz	US power intensity = 106 W/cm^2 [PS] = of 5×10^{-3}M[Fe$_0$] = 0.5 g/L pH = 6.0	95%	The dye depletion rate in the PS/Fe$_0$/US system was double that of the PS/Fe$_0$ system	Weng and Tsai (2016)
4	4-Nitroaniline	550 W, 60 Hz	US power = 60 W [PS] = 1.68 mM [NS] = 300 mg/L pH 5 Time = 90 minutes	93%	The US/PS system is combined with nanosilver (NS)	Malakootian, Ahmadian, and Khatami (2020)
5	Petroleum hydrocarbon		US power = 230 W [PS] = 2 mol/L pH 5 Water soil ratio = 2:3	56.41% and 82.23% in US/PS and US/Fe/PS system	Difficulty of oxidative degradation of organic compounds was alcohols > esters > cycloalkanes	Y. T. Li et al. (2020)

(Continued)

TABLE 19.3 (Continued)
Ultra Sonic Activation of PS/PMS

S. No	Pollutant	Source of Cavitation	Optimal Conditions	Degradation Efficiency	Observations	References
6	Bisphenol A (BPA) and Dimethyl Phthalate (DMP)	400 kHz Power density 0.11 W/L	[PMS] BPA = 1.2 mM [PMS] DMP = 5.0 mM	100%		Xu et al. (2020)
7	Organosulfur, nitro derivatives of benzene, BTEX, and phenol	1000 W, 25 kHz	Power density = 200 W/L Time = 60 minutes pH 10.5	100%		Gągol, Przyjazny, and Boczkaj (2018)
8	Carbamazepine (CBZ)	200 W, 40 kHz	US power = 200 W [PS] = 5.0 mmol/L pH 5.0 $T = 50°C$ Time = 2 hours	89.4%		Wang and Zhou (2016)
	Propranolol (PRO)	350 W, 20 kHz	US power = 250 W [PRO] = 40 µM [nZVI] = 0.15 g/L [PS] = 1 mM pH 3 Time = 30 minutes	94.2%	Inhibition on degradation $HCO_3^- > Cl^- > SO_4^{2-} > H_2PO_4^{2-} > NO_3^-$	Yu-qiong Gao

and ease of operation (Chen et al. 2019). PS/PMS activation with homogeneous metal catalysts is a standard practice because it is convenient and inexpensive. This method has its own advantages, such as using less energy, being easy to scale up, and possessing a high reactivity. Because of the low mass transfer rates, the homogeneous catalytic activation process will have higher reaction rates. Since iron and its oxides are less toxic, less costly, and less polluting than other transition metals, they have received the most attention (Xiao et al. 2020). PS/PMS activation by homogeneous metal ions and metal oxides is limited. Large amounts of metal ions are required to manage large amounts of wastewater collected, resulting in abundant metal ions in difficult-to-recover water supplies. The metal ions can be affected by the composition of the wastewater and the pH when they precipitate and hydrate in basic and acidic environments. As a result, the research focused on the development of heterogeneous metal and metal oxide complexes capable of effectively addressing these limitations.

Heterogeneous catalysts are easier to recover and more stable in the treatment process, can work over a broad pH spectrum, and can be reused for more treatment cycles, eliminating the need for additional treatment steps to recover metals from water (Xiao et al. 2020). In the literature, the use of dual metal oxides in PS/PMS activation to produce reactive oxygen species (ROS) due to synergetic coupling of the active sites is established. Chen et al. (2019) describe the use of S-doped Cu-Co bimetallic oxides (SCuCoO) for sulfate reduction generation in order to degrade chloramphenicol (CAP). Within 15 minutes of reaction time, the degradation was 98%. In the case of bimetallic catalysts, the crystalline structure and stoichiometric ratio of metals are critical during the catalytic activation phase Chen et al. (2018) generated three Co/Fe bimetallic oxides by adjusting the stoichiometric Fe/Co ratios using a simple hydrothermal and post-calcination procedure. All of the obtained products could easily decompose PMS to produce sulfate radicals, with the last one having the highest degradation rate against norfloxacin (NOR), a popular fluoroquinolone antibiotic used in animal husbandry.

As shown in Figure 19.4, different modifications in metal and metal oxides are reported in literature for SR activation. Bimetallic/trimetallic oxides have been synthesized and found superior in degradation of organic contaminants. L. Zhang et al. (2019) through a series of comparative degradation tests showed that the synergistic effect exhibited in trimetallic oxides Co_3MnFeO_6/PMS system was demonstrated to be indeed superior to that in bimetallic oxides/PMS system. The layered structure, greater stability, and easier tunability of layered double hydroxide (LDH) catalysts have got enhanced focus in sulfate radical activation.

Binary and ternary LDH have been employed for efficient oxidation of persulfates (Yang et al. 2020). The metal-based catalysts are proved to be versatile and multifunctional and have shown features such as high redox activity towards PS/PMS activation. Table 19.4 presents the literature review of several metallic catalysts employed for SR-AOP. Efforts are to be placed on improved synthesis procedure and support materials, and understanding of the interaction between metal particles and support materials is needed for higher catalytic efficiency.

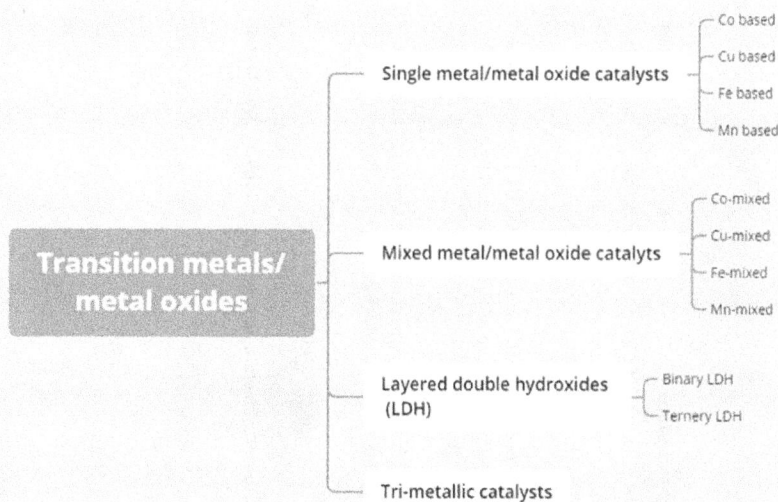

FIGURE 19.4 Various kinds of metal/metal oxides for PS/PMS activation.

The environmental impacts associated with higher metallic leaching need to be addressed.

19.2.5 CARBON MATERIALS

The metal and metal oxide catalysts are highly selective, consume lot of energy, and hence wastage of resources (Yu et al. 2020). Carbon-based materials have distinct advantages as PS/PMS activators because they are less expensive, more environmentally friendly (can activate sulfate radical without metal leaching concern), and more soluble (Wang and Wang 2018; Zhao et al. 2017). Oxygenated carbonyl, carboxyl, and hydroxyl groups can activate persulfate to produce $SO_4^{\bullet-}$. Carbonaceous-based materials such as carbon nanotubes (CNT), activated carbon, biochar, and graphene oxide have been widely used as catalysts for organic pollutant decomposition in water and wastewater treatment systems via the SR-AOP process. The functional groups on carbon surfaces influence the kinetics of the catalysts. Even though these carbon-based materials provide a promising alternative to wastewater treatment through sulfate radical production, there is still more work to be done in developing cost-effective methods for manufacturing new carbon materials in larger quantities, as well as understanding the activation mechanism. Each activation mechanism, such as heat, ultraviolet, metal, and ultrasonic, has advantages and drawbacks for the activation of PS or PMS, so mixing them would optimize their benefits. Hybrid activation approaches are gaining traction. Carbonaceous materials, as opposed to metal oxides, can prevent metal dissolution. It is important to investigate the relationship between the surface structure, elemental composition, and activation performance of carbonaceous-based materials.

TABLE 19.4
Catalytic Metal and Metal Oxides in SR-AOP

S. No	Pollutant	Catalyst	Optimal Conditions	Degrdation Efficiency	Observations	References
1	Carbofuran (CBF)	Fe (III) impregnated N-doped TiO2	[PMS] = 0–0.5 mM Catalyst dosage = 0.5 g/L pH 7.4 Time = 7 minutes	100	The efficiency of CBF degradation is still more than 70% after adding NO_3^-, SO_4^{2-}, HCO_3^-, and Cl^-	Abdelhaleem and Chu (2019)
2	Chloramphenicol (CAP)	Sulfur-doped copper-cobalt bimetallic oxides (S-CuYO)	[PMS] = 150 mg/L Catalyst dosage = 0.1 g/L Time = 15 min	98%	With the addition of 5 and 10 mmol/L Cl^-, the degradation rate of CAP decreased by 66% and 75%	Chen et al. (2019)
3	Norfloxacin (NOR)	magnetic Co/Fe bimetallic oxide	[PMS] = 500 μM [NOR] = 15 μM pH 8.5 Time = 3 minutes $T = 40°C$	100%	12 degradation intermediates/products (besides NOR) were detected	Chen et al. (2019)
4	Phenol	Co-ZSM-5	[PMS] = 2 g/L Catalyst dosage = 25 mg/L pH 3–3.5 Time = 5.8 hours	100%	Zero-order kinetics, activated energy = 69.7 kJ/mol	Shukla et al. (2010)
5	2,4-Dichlorophenol	Iron nanoparticles	[PMS] = 12.5 mM Catalyst dosage = 1 g/L [pollutant] = 30 mg/L pH 3 Time = 3 hours	37.8%		R. Li et al. (2015)

(Continued)

TABLE 19.4 (*Continued*)
Catalytic Metal and Metal Oxides in SR-AOP

S. No	Pollutant	Catalyst	Optimal Conditions	Degrdation Efficiency	Observations	References
6	Bisphenol A	$Fe_{0.8}Co_{2.2}O_4$	[PMS] = 0.2 g/L Catalyst dosage = 0.1 g/L [pollutant] = 20 mg/L pH 6 Time = 1 hour	>95		X. Li et al. (2016)
7	Phenol	Nano–Fe_0@CS	[PMS] = 0.5 g/L Catalyst dosage = 0.5 g/L [pollutant] = 20 ppm $T = 25°C$	100%	Catalytic activity of nano–Fe_0@CS is better than ZVI particles, iron ions, iron oxides, and cobalt oxide	Sun et al. (2012)
8	4-Chlorophenol	Fe_3O_4/MnO_2	[PMS] = 0.5g/L Catalyst dosage = 0.2 g/L [pollutant] = 50 mg/L pH 7.5 Time = 30 minutes	>95%		J. Liu et al. (2015)
9	Acid red G	Mn_3O_4	[PMS] = 1 g/L Catalyst dosage = 0.5 g/L [pollutant] = 50 mg/L pH 7 $T = 25°C$	100%	Mn_3O_4 nanoparticles had a remarkable stability and maintained high catalytic activity after five recycles	Tang, Zhang, and Guo (2015)
10	Rhodamine B	$CoFe_2O_4$/titanate nanotubes	[PMS] = 4g/L Catalyst dosage = 0.2 g/L [pollutant] = 100 mg/L pH 10 $T = 20°C$	100%	Co leaching was ~928 μg/L	Du et al. (2016)

(*Continued*)

TABLE 19.4 (Continued)
Catalytic Metal and Metal Oxides in SR-AOP

S. No	Pollutant	Catalyst	Optimal Conditions	Degrdation Efficiency	Observations	References
11	Phenol	$CoFe_2O_4/rGO$	[PMS] = 2 g/L Catalyst dosage = 67 mg/L [pollutant] = 20 mg/L Time = 40 minutes	>99%	Catalytic activity of $CoFe_2O_4/rGO > CoFe_2O_4 > rGO$	Yao et al. (2012)
12	Bisphenol A	CuFe2O4	[PMS] = 0.5g/L. Catalyst dosage = 0.4g/L [pollutant] = 50mg/L pH=6.72 Time= 60 min T = 20°C	>95%	Catalytic activity of $CuFe_2O_4 > Fe_2O_3 > CuO$ Cu leaching in $CuFe_2O_4$ was 1.27 (0.795%) and 0.7 mg/L, for first and second cycle, respectively	Xu, Ai, and Zhang (2016)
13	Atrazine	Co-doped mesoporous Fe3PO4	[PMS] = 0.2 mM Catalyst dosage = 0.1 g/L [pollutant] = 10 µm pH 7 Time = 30 minutes $T = 200°C$	100%	Co-leaching 52 µg/L Addition of Cl^-, NO_3^-, SO_4^{2-}, and PO_4^{3-} improved the degradation	Zhu et al. (2019)
14	Carbamazepine	Magnetic $Co_3MnFeO6$ nanoparticles	[PMS] = 0.4g/L Catalyst dosage = 0.2 g/L [pollutant] = 5 mg/L pH 7 Time = 30 minutes $T = 300°C$	99.2%		L. Zhang et al. (2019)

19.3 SR-AOPs IN INDUSTRIAL SALINE WASTEWATER TREATMENT: CURRENT STATUS AND CHALLENGES

Dumping untreated synthetic saline chemical waste streams will be hazardous to the ecosystem. Traditional membrane technology for salt removal can experience membrane blockage due to the presence of organic pollutants. As a result, novel methods for removing organic effluents from saline wastewaters are needed. In the area of contaminant remediation, persulfate-based AOPs have recently received a lot of publicity. The sulfate and hydroxyl radicals activated by the aforementioned mechanisms will oxidize organic pollutants through the electron transfer reaction and hydrogen abstraction. However, the impact of inorganic compounds on pollutant degradation remains unknown. Yuan et al. (2011) carried out the first analysis of its kind on the dual role of chloride (i.e., inhibitory and accelerating effect) on the advanced oxidation process of acid orange 7. The research discovered that even in the absence of a cobalt catalyst, there is the existence of undesirable aromatic chlorinated compounds at elevated chloride dosages. Yang et al. (2014) compared the effect of halides on the efficiency of organic contaminant degradation for UV/H_2O_2 and UV/$S_2O_8^{2+}$ on saline waters, and the results were compared. Contaminant degradation was higher in the prepared sample by UV/$S_2O_8^{2+}$ treatment. This was due to higher quantum efficiency of $S_2O_8^{2+}$ photolysis over H_2O_2 photolysis. Halides have little/no impact on the degradation of 3-cyclohexene-1-carboxylic acid. Lutze, Kerlin, and Schmidt (2015) investigated the sulfate radical oxidation mechanism with chlorine and bicarbonate ions presence. The formation of ClO_3^- was investigated, and the findings revealed that the inclusion of Cl^- and HCO_3^- reduced the oxidation potential. The experiments also demonstrate that the UV-assisted sulfate radical process has a higher pollutant degradation capability than the hydroxyl process at 254 nm UV radiation. The UV/persulfate method's ability to remove high amounts of phenol from saline wastewater was studied (Seid-Mohammadi et al. 2016). The findings showed that increasing the NaCl concentration marginally increased the removal rate. Since the oxidation in AOPs varies when different types of salt are present, there is ample scope to study the effect of different types of salt on AOP performance can be analyzed more closely.

The efficiency and, accordingly, the success of the advanced oxidation processes (AOPs) have generally been evaluated on the basis of degradation kinetics. The presence of chlorides and carbonates have negatively impacted the kinetics. The chloride content is a significant inhibitor of the degradation mechanism for industrial saline wastewater treatments by AOPs. The research by Fang et al. (2017) could be beneficial for future developments, as they eliminated chlorine involvement in the UV/PS degradation of monochlorophenols. Huang et al. (2017) studied the effect of chlorine on phthalic acid degradation by Co^{2+}/peroxymonosulfate (PMS). Higher amounts of chlorine have a detrimental effect, resulting in lower levels of AOX (adsorbable organic halogen) and mineralization as well as leads to the formation of chlorinated by-products. F. Yang et al. (2018) investigated the performance of methylene blue (MB) degradation in PMS, PMS/Cl$^-$, and PMS/base systems. They concluded that the formation of free usable chlorine species is serving as disinfectants and bleaching agents, hastening the deterioration of MB. Activation of PMS is dominated by chlorides under less alkaline situations. At higher base concentrations, reduced degradation rates are observed as the PMS becomes less stable.

L. Zhou et al. (2019) published a systematic study on the effects of halides (chloride (Cl^-), bromide (Br^-), bicarbonate (HCO_3^-), and natural organic matter (NOM)) on salbutamol (SAL) and terbutaline (TBL) degradation. The findings revealed that chlorine had no impact on the oxidation efficiencies, whereas the other ions had inhibitory effects. The detrimental impact of bromide was mostly attributed to $SO_4^{\bullet-}$ scavenging. The function of Cl^- in the $SO_4^{\bullet-}$ oxidation process must be thoroughly investigated. Cai et al. (2020) investigated the viability of a hybrid process of AOP combined with biological method for handling the reverse osmosis concentrate. The formation of adsorbable organic halogens (AOX) was investigated in two AOPs on a comparative basis by Xie et al. (2020). With total organic content (TOC) kept at constant, both AOPs showed a substantial rise in AOX concentration. Despite the latter's superior TOC removal performance, the UV/PMS process emitted more AOX than the UV/H_2O_2 process under all experimental conditions. As a result of the literature studies, it is important to infer that attempts to develop methods for preventing the harmful effects of chloride in the SR-AOP process should be undertaken as soon as possible before the large-scale application of this new advanced oxidation technology.

19.4 CONCLUDING REMARKS

SR-AOPs have been shown to be more effective than traditional AOPs, particularly in the treatment of saline industrial wastewater. The processing of singlet oxygen, as well as the presence of free available chlorine, will result in higher degradation potentials. In the SR-AOP process, increased alkaline conditions would restrict the formation of chlorinated byproducts. Aside from the complete removal of the pollutant under consideration, the process kinetics and parameters such as AOX content and toxicity must be validated to assess the feasibility of AOPs, especially in saline environments. Identification and quantification of byproducts throughout the oxidation phase necessitate extensive research.

REFERENCES

Alipour, Vali, Faride Moein, and Leila Rezaei. 2016. "Determining the Salt Tolerance Threshold for Biological Treatment of Salty Wastewater." *Health Scope* 6 (1): 4–8. doi:10.17795/jhealthscope-36425.

Aljuboury, Dheeaa Al Deen Atallah, Puganeshwary Palaniandy, Hamidi Bin Abdul Aziz, and Shaik Feroz. 2015. "Treatment of Petroleum Wastewater Using Combination of Solar Photo-Two Catalyst TiO_2 and Photo-Fenton Process." *Journal of Environmental Chemical Engineering* 3 (2). Elsevier B.V.: 1117–24. doi:10.1016/j.jece.2015.04.012.

Aljuboury, Dheeaa Al Deen Atallah, Puganeshwary Palaniandy, Hamidi Bin Abdul Aziz, and Shaik Feroz. 2016a. "Comparative Study of Advanced Oxidation Processes to Treat Petroleum Wastewater." *Hungarian Journal of Industry and Chemistry* 43 (2): 97–101. doi:10.1515/hjic-2015-0016.

Aljuboury, Dheeaa Al Deen Atallah, Puganeshwary Palaniandy, Hamidi Bin Abdul Aziz, and Shaik Feroz. 2016b. "Evaluation of the Solar Photo-Fenton Process to Treat the Petroleum Wastewater by Response Surface Methodology (RSM)." *Environmental Earth Sciences* 75 (4). Springer: 1–12. doi:10.1007/s12665-015-5192-y.

Al-Raad, Abbas A., Marlia M. Hanafiah, Ahmed Samir Naje, Mohammed A. Ajeel, Alfarooq O. Basheer, Thuraya Ali Aljayashi, and Mohd Ekhwan Toriman. 2019. "Treatment of Saline Water Using Electrocoagulation with Combined Electrical Connection of Electrodes." *Processes* 7 (5). doi:10.3390/pr7050242.

Asgari, Ghorban, Javad Feradmal, Ali Poormohammadi, Mehrnaz Sadrnourmohamadi, and Somaye Akbari. 2016. "Taguchi Optimization for the Removal of High Concentrations of Phenol from Saline Wastewater Using Electro-Fenton Process." *Desalination and Water Treatment* 57 (56): 27331–38. doi:10.1080/19443994.2016.1170635.

Borges, Fulvia Jung, Hélène Roux-De Balmann, and Roberto Guardani. 2009. "Modeling Electrodialysis and Photochemical Process for Their Integration in Saline Wastewater Treatment." *Computer Aided Chemical Engineering* 27 (C): 741–46. doi:10.1016/S1570-7946(09)70344-X.

Borges, Fulvia Jung, Hélène Roux-De Balmann, and Roberto Guardani. 2010. "Modeling Electrodialysis and a Photochemical Process for Their Integration in Saline Wastewater Treatment." *Brazilian Journal of Chemical Engineering* 27 (3): 473–82. doi:10.1590/S0104-66322010000300011.

Brüninghoff, Robert, Alyssa K. Van Duijne, Lucas Braakhuis, Pradip Saha, Adriaan W. Jeremiasse, Bastian Mei, and Guido Mul. 2019. "Comparative Analysis of Photocatalytic and Electrochemical Degradation of 4-Ethylphenol in Saline Conditions." *Environmental Science and Technology* 53 (15): 8725–35. doi:10.1021/acs.est.9b01244.

Q.Q. Cai, M.Y. Wu, R. Li, S.H. Deng, B.C.Y. Lee, S.L. Ong, J.Y. Hu. 2020. "Potential of combined advanced oxidation – Biological process for cost-effective organic matters removal in reverse osmosis concentrate produced from industrial wastewater reclamation: Screening of AOP pre-treatment technologies." *Chemical Engineering Journal* 389. doi:10.1016/j.cej.2019.123419.

Chakma, Sankar, Sai Praneeth, and Vijayanand S. Moholkar. 2017. "Mechanistic Investigations in Sono-Hybrid (Ultrasound/Fe^{2+}/UVC) Techniques of Persulfate Activation for Degradation of Azorubine." *Ultrasonics Sonochemistry* 38. Elsevier B.V.: 652–63. doi:10.1016/j.ultsonch.2016.08.015.

Chen, Cheng, Li Liu, Jing Guo, Lixiang Zhou, and Yeqing Lan. 2019. "Sulfur-Doped Copper-Cobalt Bimetallic Oxides with Abundant Cu(I): A Novel Peroxymonosulfate Activator for Chloramphenicol Degradation." *Chemical Engineering Journal* 361. Elsevier: 1304–16. doi:10.1016/j.cej.2018.12.156.

Chen, Liwei, Xu Zuo, Liang Zhou, Yang Huang, Shengjiong Yang, Tianming Cai, and Dahu Ding. 2018. "Efficient Heterogeneous Activation of Peroxymonosulfate by Facilely Prepared Co/Fe Bimetallic Oxides: Kinetics and Mechanism." *Chemical Engineering Journal* 345: 364–74. doi:10.1016/j.cej.2018.03.169.

Chen, Wen Shing, and Chi Pin Huang. 2019. "Mineralization of Dinitrotoluenes in Aqueous Solution by Sono-Activated Persulfate Enhanced with Electrolytes." *Ultrasonics Sonochemistry* 51. Elsevier B.V.: 129–37. doi:10.1016/j.ultsonch.2018.10.035.

Chen, Wen Shing, and Yi Chang Su. 2012. "Removal of Dinitrotoluenes in Wastewater by Sono-Activated Persulfate." *Ultrasonics Sonochemistry* 19 (4). Elsevier B.V.: 921–27. doi:10.1016/j.ultsonch.2011.12.012.

Darvishmotevalli, Mohammad, Ahmad Zarei, Maryam Moradnia, Mohammad Noorisepehr, and Hamed Mohammadi. 2019. "Optimization of Saline Wastewater Treatment Using Electrochemical Oxidation Process: Prediction by RSM Method." *MethodsX* 6. Elsevier B.V.: 1101–13. doi:10.1016/j.mex.2019.03.015.

Fang, Changling, Xiaoyi Lou, Ying Huang, Min Feng, Zhaohui Wang, and Jianshe Liu. 2017. "Monochlorophenols Degradation by UV/Persulfate Is Immune to the Presence of Chloride: Illusion or Reality?" *Chemical Engineering Journal* 323. Elsevier B.V.: 124–33. doi:10.1016/j.cej.2017.04.094.

Gągol, Michał, Andrzej Przyjazny, and Grzegorz Boczkaj. 2018. "Highly Effective Degradation of Selected Groups of Organic Compounds by Cavitation Based AOPs under Basic PH Conditions." *Ultrasonics Sonochemistry* 45: 257–66. doi:10.1016/j.ultsonch.2018.03.013.

Gao, Yu qiong, Nai yun Gao, Wei Wang, Shi fei Kang, Jian hong Xu, Hui ming Xiang, and Da qiang Yin. 2018. "Ultrasound-Assisted Heterogeneous Activation of Persulfate by Nano Zero-Valent Iron (NZVI) for the Propranolol Degradation in Water." *Ultrasonics Sonochemistry* 49. Elsevier: 33–40. doi:10.1016/j.ultsonch.2018.07.001.

Giannakis, Stefanos, Kun Yi Andrew Lin, and Farshid Ghanbari. 2021. "A Review of the Recent Advances on the Treatment of Industrial Wastewaters by Sulfate Radical-Based Advanced Oxidation Processes (SR-AOPs)." *Chemical Engineering Journal* 406. Elsevier B.V.: 127083. doi:10.1016/j.cej.2020.127083.

Hao, Feifei, Weilin Guo, Anqi Wang, Yanqiu Leng, and Helian Li. 2014. "Intensification of Sonochemical Degradation of Ammonium Perfluorooctanoate by Persulfate Oxidant." *Ultrasonics Sonochemistry* 21 (2). Elsevier B.V.: 554–58. doi:10.1016/j.ultsonch.2013.09.016.

Hu, Chen Yan, Yuan Zhang Hou, Yi Li Lin, Yan Guo Deng, Shuang Jing Hua, Yi Fan Du, Chiu Wen Chen, and Chung Hsin Wu. 2020. "Investigation of Iohexol Degradation Kinetics by Using Heat-Activated Persulfate." *Chemical Engineering Journal* 379. Elsevier: 122403. doi:10.1016/j.cej.2019.122403.

Hu, Peidong, and Mingce Long. 2016. "Cobalt-Catalyzed Sulfate Radical-Based Advanced Oxidation: A Review on Heterogeneous Catalysts and Applications." *Applied Catalysis B: Environmental* 181. Elsevier B.V.: 103–17. doi:10.1016/j.apcatb.2015.07.024.

Hu, Peidong, Hanrui Su, Zhenyu Chen, Chunyang Yu, Qilin Li, Baoxue Zhou, Pedro J.J. Alvarez, and Mingce Long. 2017. "Selective Degradation of Organic Pollutants Using an Efficient Metal-Free Catalyst Derived from Carbonized Polypyrrole via Peroxymonosulfate Activation." *Environmental Science and Technology* 51 (19): 11288–96. doi:10.1021/acs.est.7b03014.

Huang, Ying, Zhaohui Wang, Qingze Liu, Xiaoxiao Wang, Zhijun Yuan, and Jianshe Liu. 2017. "Effects of Chloride on PMS-Based Pollutant Degradation: A Substantial Discrepancy between Dyes and Their Common Decomposition Intermediate (Phthalic Acid)." *Chemosphere* 187: 338–46. doi:10.1016/j.chemosphere.2017.08.120.

Ioannidi, Alexandra, Zacharias Frontistis, and Dionissios Mantzavinos. 2018. "Destruction of Propyl Paraben by Persulfate Activated with UV-A Light Emitting Diodes." *Journal of Environmental Chemical Engineering* 6 (2). Elsevier: 2992–97. doi:10.1016/j.jece.2018.04.049.

Izadifard, Maryam, Gopal Achari, and Cooper H. Langford. 2017. "Degradation of Sulfolane Using Activated Persulfate with UV and UV-Ozone." *Water Research* 125. Elsevier Ltd: 325–31. doi:10.1016/j.watres.2017.07.042.

Kakavandi, Babak, and Mehdi Ahmadi. 2019. "Efficient Treatment of Saline Recalcitrant Petrochemical Wastewater Using Heterogeneous UV-Assisted Sono-Fenton Process." *Ultrasonics Sonochemistry* 56. Elsevier: 25–36. doi:10.1016/j.ultsonch.2019.03.005.

Laskar, Imranul I. 2014. "Chemical Science Review and Letters Treatment of Composite Reverse Osmosis (RO) Reject Water for Chemical Oxygen Demand (COD) Reduction." *Chemical Science Review and Letters* 2014 (10): 314–21.

Lebik-Elhadi, Hafida, Zacharias Frontistis, Hamid Ait-Amar, Farid Madjene, and Dionissios Mantzavinos. 2020. "Degradation of Pesticide Thiamethoxam by Heat – Activated and Ultrasound – Activated Persulfate: Effect of Key Operating Parameters and the Water Matrix." *Process Safety and Environmental Protection* 134. Institution of Chemical Engineers: 197–207. doi:10.1016/j.psep.2019.11.041.

Li, Yong Tao, Dan Li, Lian Jue Lai, and Yu Hang Li. 2020. "Remediation of Petroleum Hydrocarbon Contaminated Soil by Using Activated Persulfate with Ultrasound and Ultrasound/Fe." *Chemosphere* 238. Elsevier Ltd: 124657. doi:10.1016/j. chemosphere.2019.124657.

Li, Zhendong, Dongfang Liu, Wenli Huang, Xiaocheng Wei, and Weiwei Huang. 2020. "Biochar Supported CuO Composites Used as an Efficient Peroxymonosulfate Activator for Highly Saline Organic Wastewater Treatment." *Science of the Total Environment* 721. Elsevier B.V.: 137764. doi:10.1016/j.scitotenv.2020.137764.

Linaric, M., M. Markic, and L. Sipos. 2013. "High Salinity Wastewater Treatment." *Water Science and Technology* 68 (6): 1400–05. doi:10.2166/wst.2013.376.

Liu, Lin, Sen Lin, Wei Zhang, Usman Farooq, Genxiang Shen, and Shuangqing Hu. 2018. "Kinetic and Mechanistic Investigations of the Degradation of Sulfachloropyridazine in Heat-Activated Persulfate Oxidation Process." *Chemical Engineering Journal* 346: 515–24. doi:10.1016/j.cej.2018.04.068.

Lu, Jie, Linhua Fan, and Felicity A. Roddick. 2013. "Potential of BAC Combined with UVC/ H2O2 for Reducing Organic Matter from Highly Saline Reverse Osmosis Concentrate Produced from Municipal Wastewater Reclamation." *Chemosphere* 93 (4). Elsevier Ltd: 683–88. doi:10.1016/j.chemosphere.2013.06.008.

Lutze, Holger V., Nils Kerlin, and Torsten C. Schmidt. 2015. "Sulfate Radical-Based Water Treatment in Presence of Chloride: Formation of Chlorate, Inter-Conversion of Sulfate Radicals into Hydroxyl Radicals and Influence of Bicarbonate." *Water Research* 72. Elsevier Ltd: 349–60. doi:10.1016/j.watres.2014.10.006.

Malakootian, Mohammad, Mohammad Ahmadian, and Mehrdad Khatami. 2020. "Activation of Ultrasound Enhanced Persulfate Oxidation by Biogenic Nanosilvers for Degradation of 4-Nitroaniline." *Desalination and Water Treatment* 174: 240–47. doi:10.5004/dwt.2020.24886.

Milh, Hannah, Ben Schoenaers, Andre Stesmans, Deirdre Cabooter, and Raf Dewil. 2020. "Degradation of Sulfamethoxazole by Heat-Activated Persulfate Oxidation: Elucidation of the Degradation Mechanism and Influence of Process Parameters." *Chemical Engineering Journal* 379. Elsevier B.V.: 122234. doi:10.1016/j.cej.2019.122234.

Mirbolooki, Hanieh, Reza Amirnezhad, and Ali Reza Pendashteh. 2017. "Treatment of High Saline Textile Wastewater by Activated Sludge Microorganisms." *Journal of Applied Research and Technology* 15 (2): 167–72. doi:10.1016/j.jart.2017.01.012.

Myneni, Venkata Ratnam, Nagamalleswara Rao Kanidarapu, and Meena Vangalapati. 2020. "Methylene Blue Adsorption by Magnesium Oxide Nanoparticles Immobilized with Chitosan (CS-MgONP): Response Surface Methodology, Isotherm, Kinetics and Thermodynamic Studies." *Iranian Journal of Chemistry and Chemical Engineering (IJCCE)* 39 (6): 29–42. doi:10.30492/ijcce.2019.36342.

Norzaee, Samira, Bazrafshan Edris, Djahed Babak, Kord Ferdos, Razieh Khaksefidi Mostafapour. 2017. "UV Activation of Persulfate for Removal of Penicillin G Antibiotics in Aqueous Solution." *The Scientific World Journal*, 2017, Article ID 3519487, 6 pages. doi:10.1155/2017/3519487.

Norzaee, Samira, Mahmoud Taghavi, Babak Djahed, and Ferdos Kord Mostafapour. 2018. "Degradation of Penicillin G by Heat Activated Persulfate in Aqueous Solution." *Journal of Environmental Management* 215. Elsevier Ltd: 316–23. doi:10.1016/j. jenvman.2018.03.038.

Pan, Xiaoxue, Liqing Yan, Ruijuan Qu, and Zunyao Wang. 2018. "Degradation of the UV-Filter Benzophenone-3 in Aqueous Solution Using Persulfate Activated by Heat, Metal Ions and Light." *Chemosphere* 196. Elsevier Ltd: 95–104. doi:10.1016/j. chemosphere.2017.12.152.

Resources, Natural, IISD (International Institute for Sustainable Development), Founex Report, Birka Wicke, Richard Sikkema, Veronika Dornburg, André Faaij, 2018. "An ecological and bacterial study of the waters of the Tigris and Diyala rivers, south of Baghdad" *Director* 15 (40): 6–13.

Saravanakumar, Krishnan, Ramalingham Senthil Kumar, Donipathi Mogili Reddy Prasad, Balakrishna Sankari Naveen Prasad, Sathyamoorthy Manickam, and Vasu Gajendiran. 2020. "Batch and Column Arsenate Sorption Using Turbinaria Ornata Seaweed Derived Biochar: Experimental Studies and Mathematical Modeling." *ChemistrySelect* 5 (12): 3661–68. doi:10.1002/slct.202000548.

Senthil Kumar, R., D. M. Reddy Prasad, L. Govindarajan, K. Saravanakumar, and B. S. Naveen Prasad. 2020. "Synthesis of Green Marine Algal-Based Biochar for Remediation of Arsenic(V) from Contaminated Waters in Batch and Column Mode of Operation." *International Journal of Phytoremediation* 22 (3). Taylor & Francis: 279–86. doi:10.10 80/15226514.2019.1658710.

Stasinakis, A. S. 2008. "Use of Selected Advanced Oxidation Processes (AOPs) for Wastewater Treatment – A Mini Review." *Global Nest Journal* 10 (3): 376–85. doi:10.30955/gnj.000598.

Tan, Chaoqun, Naiyun Gao, Yang Deng, Wenlei Rong, Shengdong Zhou, and Naxin Lu. 2013. "Degradation of Antipyrine by Heat Activated Persulfate." *Separation and Purification Technology* 109. Elsevier B.V.: 122–28. doi:10.1016/j.seppur.2013.03.003.

Tan, Xu, Isaac Acquah, Hanzhe Liu, Weiguo Li, and Songwen Tan. 2019. "A Critical Review on Saline Wastewater Treatment by Membrane Bioreactor (MBR) from a Microbial Perspective." *Chemosphere* 220. Elsevier Ltd: 1150–62. doi:10.1016/j.chemosphere.2019.01.027.

Venkata Ratnam, M., Karthikeyan, C., Nagamalleswara Rao, K., and V. Meena. 2020. "Magnesium oxide nanoparticles for effective photocatalytic degradation of methyl red dye in aqueous solutions: Optimization studies using response surface methodology." *Materials Today: Proceedings* 26: 2308–2313. doi:10.1016/j.matpr.2020.02.498.

Wacławek, Stanisław, Holger V. Lutze, Klaudiusz Grübel, Vinod V. T. Padil, Miroslav Černík, and Dionysios D. Dionysiou. 2017. "Chemistry of Persulfates in Water and Wastewater Treatment: A Review." *Chemical Engineering Journal.* doi:10.1016/j.cej.2017.07.132.

Wang, Jianlong, and Shizong Wang. 2018. "Activation of Persulfate (PS) and Peroxymonosulfate (PMS) and Application for the Degradation of Emerging Contaminants." *Chemical Engineering Journal.* Elsevier B.V. doi:10.1016/j.cej.2017.11.059.

Wang, Qun, Ye Cao, Han Zeng, Youheng Liang, Jun Ma, and Xiaohui Lu. 2019. "Ultrasound-Enhanced Zero-Valent Copper Activation of Persulfate for the Degradation of Bisphenol AF." *Chemical Engineering Journal* 378. Elsevier B.V.: 122143. doi:10.1016/j.cej.2019.122143.

Wang, Qun, Xiaohui Lu, Ye Cao, Jun Ma, Jin Jiang, Xiaofeng Bai, and Tao Hu. 2017. "Degradation of Bisphenol S by Heat Activated Persulfate: Kinetics Study, Transformation Pathways and Influences of Co-Existing Chemicals." *Chemical Engineering Journal* 328: 236–45. doi:10.1016/j.cej.2017.07.041.

Xiao, Sa, Min Cheng, Hua Zhong, Zhifeng Liu, Yang Liu, Xin Yang, and Qinghua Liang. 2020. "Iron-Mediated Activation of Persulfate and Peroxymonosulfate in Both Homogeneous and Heterogeneous Ways: A Review." *Chemical Engineering Journal* 384. Elsevier B.V.: 123265. doi:10.1016/j.cej.2019.123265.

Xie, Yawei, Ranyun Xu, Rui Liu, Hongyuan Liu, Jinping Tian, and Lujun Chen. 2020. "Adsorbable Organic Halogens Formed during Treatment of Cl^--Containing Wastewater by Sulfate and Hydroxyl Radical-Based Advanced Oxidation Processes." *Chemical Engineering Journal* 389. Elsevier: 124457. doi:10.1016/j.cej.2020.124457.

Xu, Lijie, Xiaotian Wang, Yang Sun, Han Gong, Mingzhi Guo, Xiaomeng Zhang, Liang Meng, and Lu Gan. 2020. "Mechanistic Study on the Combination of Ultrasound and Peroxymonosulfate for the Decomposition of Endocrine Disrupting Compounds."

Ultrasonics Sonochemistry 60 (February 2019). Elsevier: 104749. doi:10.1016/j. ultsonch.2019.104749.

Yang, Fei, Ying Huang, Changling Fang, Ying Xue, Luoyan Ai, Jianshe Liu, and Zhaohui Wang. 2018. "Peroxymonosulfate/Base Process in Saline Wastewater Treatment: The Fight between Alkalinity and Chloride Ions." *Chemosphere* 199. Elsevier Ltd: 84–88. doi:10.1016/j.chemosphere.2018.02.023.

Yang, Lie, Jianming Xue, Liuyang He, Li Wu, Yongfei Ma, Huan Chen, Hong Li, Pai Peng, and Z. Zhang. 2019. "Review on Ultrasound Assisted Persulfate Degradation of Organic Contaminants in Wastewater: Influences, Mechanisms and Prospective." *Chemical Engineering Journal*. doi:10.1016/j.cej.2019.122146.

Yang, Qi, Yinghao Ma, Fei Chen, Fubing Yao, Jian Sun, Shana Wang, Kaixin Yi, Lihua Hou, Xiaoming Li, and Dongbo Wang. 2019. "Recent Advances in Photo-Activated Sulfate Radical-Advanced Oxidation Process (SR-AOP) for Refractory Organic Pollutants Removal in Water." *Chemical Engineering Journal*. doi:10.1016/j.cej.2019.122149.

Yang, Shanshan, Pingxiao Wu, Junqin Liu, Meiqing Chen, Zubair Ahmed, and Nengwu Zhu. 2018. "Efficient Removal of Bisphenol A by Superoxide Radical and Singlet Oxygen Generated from Peroxymonosulfate Activated with Fe_0-Montmorillonite." *Chemical Engineering Journal* 350. doi:10.1016/j.cej.2018.04.175.

Yang, Yi, Gourab Banerjee, Gary W. Brudvig, Jae Hong Kim, and Joseph J. Pignatello. 2018. "Oxidation of Organic Compounds in Water by Unactivated Peroxymonosulfate." *Environmental Science and Technology* 52 (10): 5911–19. doi:10.1021/acs. est.8b00735.

Yang, Yi, Xinglin Lu, Jin Jiang, Jun Ma, Guanqi Liu, Ying Cao, Weili Liu. 2017. "Degradation of Sulfamethoxazole by UV, UV/H_2O_2 and UV/Persulfate (PDS): Formation of Oxidation Products and Effect of Bicarbonate." *Water Research* 118. Elsevier Ltd: 196–207. doi:10.1016/j.watres.2017.03.054.

Yang, Yi, Joseph J. Pignatello, Jun Ma, and William A. Mitch. 2014. "Comparison of Halide Impacts on the Efficiency of Contaminant Degradation by Sulfate and Hydroxyl Radical-Based Advanced Oxidation Processes (AOPs)." *Environmental Science and Technology* 48 (4): 2344–51. doi:10.1021/es404118q.

Yang, Zhong Zhu, Chang Zhang, Guang Ming Zeng, Xiao Fei Tan, Hou Wang, Dan Lian Huang, Kai Hua Yang, Jing Jing Wei, Chi Ma, and Kai Nie. 2020. "Design and Engineering of Layered Double Hydroxide Based Catalysts for Water Depollution by Advanced Oxidation Processes: A Review." *Journal of Materials Chemistry A* 8 (8): 4141–73. doi:10.1039/c9ta13522g.

Yu, Jiangfang, Haopeng Feng, Lin Tang, Ya Pang, Guangming Zeng, Yue Lu, Haoran Dong. 2020. "Metal-Free Carbon Materials for Persulfate-Based Advanced Oxidation Process: Microstructure, Property and Tailoring." *Progress in Materials Science* 111. Elsevier: 100654. doi:10.1016/j.pmatsci.2020.100654.

Yuan, Ruixia, Sadiqua N. Ramjaun, Zhaohui Wang, and Jianshe Liu. 2011. "Effects of Chloride Ion on Degradation of Acid Orange 7 by Sulfate Radical-Based Advanced Oxidation Process: Implications for Formation of Chlorinated Aromatic Compounds." *Journal of Hazardous Materials* 196: 173–79. doi:10.1016/j.jhazmat.2011.09.007.

Zhang, Lei, Xiufei Zhao, Chenggang Niu, Ning Tang, Hai Guo, Xiaoju Wen, and Chao Liang. 2019. "Enhanced Activation of Peroxymonosulfate by Magnetic Co_3MnFeO_6 Nanoparticles for Removal of Carbamazepine: Efficiency, Synergetic Mechanism and Stability." *Chemical Engineering Journal* 362: 851–64. doi:10.1016/j.cej.2019.01.078.

Zhang, Tong, Yuanyuan Liu, Shan Zhong, and Lishan Zhang. 2020. "AOPs-Based Remediation of Petroleum Hydrocarbons-Contaminated Soils: Efficiency, Influencing Factors and Environmental Impacts." *Chemosphere* 246. Elsevier B.V. doi:10.1016/j. chemosphere.2019.125726.

Zhao, Qingxia, Qiming Mao, Yaoyu Zhou, Jianhong Wei, Xiaocheng Liu, Junying Yang, Lin Luo. 2017. "Metal-Free Carbon Materials-Catalyzed Sulfate Radical-Based Advanced

Oxidation Processes: A Review on Heterogeneous Catalysts and Applications." *Chemosphere*. doi:10.1016/j.chemosphere.2017.09.042.

Zhou, Lei, Chenzhi Yan, Mohamad Sleiman, Corinne Ferronato, Jean Marc Chovelon, Xingbao Wang, and Claire Richard. 2019. "Sulfate Radical Induced Degradation of B2-Adrenoceptor Agonists Salbutamol and Terbutaline: Implication of Halides, Bicarbonate, and Natural Organic Matter." *Chemical Engineering Journal* 368: 252–60. doi:10.1016/j.cej.2019.02.183.

Zhou, Zhou, Xitao Liu, Ke Sun, Chunye Lin, Jun Ma, Mengchang He, and Wei Ouyang. 2019. "Persulfate-Based Advanced Oxidation Processes (AOPs) for Organic-Contaminated Soil Remediation: A Review." *Chemical Engineering Journal* 372. Elsevier: 836–51. doi:10.1016/j.cej.2019.04.213.

Zhuo, Yakun, Mei Sheng, Xueke Liang, and Guomin Cao. 2017. "Treatment of High Salinity Wastewater Using CWPO Process for Reuse." *Journal of Advanced Oxidation Technologies* 20 (2). doi:10.1515/jaots-2017-0024.

Songlin, Wang and Ning, Zhou. 2016. "Removal of carbamazepine from aqueous solution using sono-activated persulfate process." *Ultrasonics Sonochemistry* 29: 156–162.

Xu, Y, Ai, J, Zhang, H. 2016. "The mechanism of degradation of bisphenol A using the magnetically separable CuFe2O4/peroxymonosulfate heterogeneous oxidation process." *Journal of Hazardous Materials* 309: 87–96.

20 The Fenton-Based Approaches Focusing Industrial Saline Wastewater Treatment

Shadab Usmani, Mo Washeem,
Bilal Rajput, Mohd Salim Mahtab,
Saif Ullah Khan, and Izharul Haq Farooqi
Aligarh Muslim University

CONTENTS

20.1   INTRODUCTION

Water is among the fundamental requirements for the sustenance of humans and all other living organisms. Our planet is covered with water for nearly 70%, almost 97% of which are salty [1]. Salinization is a significant environmental problem that has an impact on agriculture, land and water supply, and natural ecosystems. High concentrations of certain elements, such as arsenic, sodium, sulfate, boron, fluoride, selenium, and radioactivity, are also linked to increased groundwater salinity. Although salinity is a global problem, it is particularly severe in semi-arid and arid water-scarce zones where water's primary source is groundwater. Rising pressures on freshwater are causing the water table to deplete with salinity to increase [2]. Saline water consists of brackish water with a TDS concentration of 10,000 ppm and seawater with a TDS concentration of 50,000 ppm [3]. There are three main types of groundwater salinity, such as: (i) primary/natural salinity, (ii) secondary/dryland salinity, and (iii) irrigated/tertiary salinity. Primary salinity is induced by natural phenomena such as the deposition of salt (fluoride salts, halite, anhydrite, carbonates, gypsum, and sulfate salts) from decades of runoff or rock weathering. After raining, some of its water evaporates off the earth, water sources, and surfaces of plants, whereas some infiltrate into the groundwater beneath the soil, while even the rest fall into the oceans after joining streams and rivers. The soils accumulate small amounts of salt carried by rain and pass into groundwater with time . Secondary salinity is caused as the salt rises to the surface deposited by primary salinity cycles after an increase in groundwater levels. This is exacerbated by the removal of seasonal (long-lived) plants in drier regions, i.e., areas where the soil profile and groundwater begin to absorb salt over time. When vegetation is eliminated, the amount of water wastage from the landscape with plants is considerably reduced. Rather, additional water drains into the land, which increases the levels of groundwater [4]. When the amount of groundwater increases, the salt is also carried and dissolved into the existing unsaturated profile of the soil. Floors of the low-lying valley can eventually be saturated, and the flow of the river channel increases in volume and length. The groundwater that is discharging combines with the fresh surface water, resulting in inflows that range from marginal to brine. When these wet regions dry after the monsoon season, salt crystals are left behind and salt scalds ensue. Repeated water irrigation with salts persistent after evaporation and with the time causes irrigated/tertiary salinity. With runoff, these deposited salts leach back into the groundwater. As water is reused for a long time, explicitly or allowing it to filter through groundwater until it is drained away for reuse, tertiary salinity develops. When added, a portion of the water will evaporate and concentrate the salts in the residual water; repetitive cycles will lead to excessive concentrations of salt.

20.1.1 Occurrences and Effects of Salinity

The intrusion of saltwater into surface water is a major problem because it not only degrades our freshwater supplies but also harms the region's socioeconomic situation [5,6]. Osmosis is a process that allows water to enter plant roots and is driven by the salt concentration of the water in the soil and the plant water. When salt levels are excessively high in the groundwater, water can seep back into the ground from plant roots. This dehydrates the plant and produces poor yields or perhaps death [7]. High salt levels can change the flavour of drinking water. Chloride has an extremely low tolerance for taste. Constipation can be caused by high sodium and magnesium sulfate levels in drinking water, making it unfit for grazing animals. Consequences are more severe in drier climates, where evaporation rates are typically very high. Thus, proper disposal and treatment of saline water are very essential to minimize its severe effects on the environment and human health. Numerous techniques have been implemented for the treatment of saline water. Classic separation techniques, such as eutectic electrodialysis (ED), freeze crystallization (EFC), membrane distillation (MD), forward osmosis (FO), membrane crystallization (MCr), and advanced oxidation processes (AOPs) occupied a large portion of the analysis work. These innovative methods have a high water recovery rate from concentrated brines, i.e., 90%–98%. The literature also discusses the removal of chemical compounds from saline effluents [8,9]. Since saline water contains a variety of toxic and recalcitrant contaminants, traditional wastewater treatment schemes are restricted, necessitating the development of new technology that can effectively remove these pollutants. Several kinds of research on the treatment of saltwater environments indicated that advanced oxidation processes (AOPs) could be used to efficiently remove a wide range of refractory pollutants from the saline samples [10–13]. AOPs are systems of physical and chemical treatments that operate under the same chemical theory that generates extremely reactive species of radical substances to degrade a variety of organic compounds in water and wastewater [14–17]. The high oxidation capacity of hydroxyl radicals in AOPs degrades compounds and oxidizes them into simplified substrates, eventually resulting in CO_2 and water [12,18]. Electrochemical oxidation, ozonation, hydrogen peroxide oxidation, peroxidation, photocatalysis, ultraviolet irradiation, microwave enhanced AOP, wet air oxidation, and Fenton-based techniques are some of the most widely used AOPs. AOPs may be employed as a pre- and post-treatment system, along with supplement biological therapies by enhancing decomposition and compound removal effectiveness [10,19,20]. The Fenton method for the successful handling of saline compounds in salt and other residue waters is considered to be an important part of AOPs [12,21,22]. The Fenton process is thought to be a viable option for oxidizing intransigent compounds and converting them to biodegradable intermediates [23]. The fundamental mechanisms and implementations of various Fenton-based methods, with an emphasis on saline water treatment, are clarified in this chapter. The impact of contributing factors on the efficacy of processes is also highlighted. Furthermore, the scope of prospective research projects is recommended, as well as a discussion of very critical issues relevant to these approaches and procedures. This chapter covers a wide range of Fenton-based approaches, which would be useful to learners conducting researches on wastewater treatment, especially AOPs for saline water treatment alternatives.

20.2 FENTON PROCESS

H. J. H. Fenton during 1984 developed the Fenton process. To achieve highly reactive hydroxyl radicals ($OH^•$), the Fenton reaction requires ferrous ions (Fe^{2+}) as catalysts for active hydrogen peroxide (H_2O_2) decomposition in an acidic state [15]. $^•OH$ is capable of mineralizing refractory synthetic and an organic compound to CO_2 and H_2O by triggering the oxidative degradation reactions because of their greater

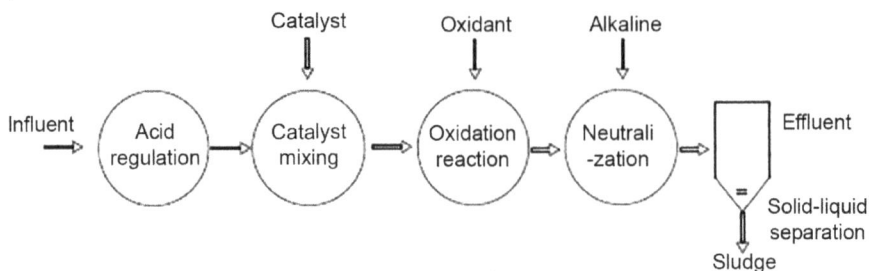

FIGURE 20.1 Systematic representation of the Fenton process. (See [27].)

potential to oxidize aqueously (+2.80 eV vs. NHE) [24,25]. The pH, reaction time, temperature, H_2O_2, and Fe^{2+} proportion (reagent ratio) and organic pollutant concentration, etc. are influencing factors [26] (Figure 20.1).

20.2.1 FENTON'S PROCESS MECHANISM

The Fenton process is based on a reaction between Fenton's reagents (Fe^{2+} and H_2O_2) that resulting in the production of highly reactive hydroxyl radicals (OH) as shown in Equations 20.1 and 20.2

$$Fe^{2+} + H_2O_2 \rightarrow Fe^{3+} + \ ^{\bullet}OH + OH^- \tag{20.1}$$

$$Fe^{3+} + H_2O_2 \rightarrow Fe^{2+} + HO_2^{\bullet} \tag{20.2}$$

In a Fenton-like reaction, Fe^{3+} (ferric ion) reacts with hydrogen peroxide to regenerate Fe^{2+} ions as well as create hydroperoxyl radical ($HO_2^{\bullet}$) Equation 20.2. $HO_2^{\bullet}$ is involved in the oxidation of a wide range of organic contaminants, and it is less reactive than OH. The Fenton-like reaction has a rate constant several orders lower than Equation 20.1, and an excess of Fe^{3+} induces iron sludge formation [28].

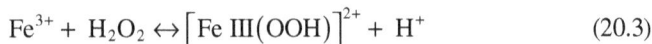

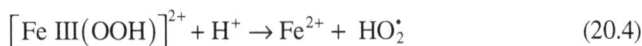

$$Fe^{3+} + H_2O_2 \leftrightarrow \left[Fe\ III(OOH) \right]^{2+} + H^+ \tag{20.3}$$

$$\left[Fe\ III(OOH) \right]^{2+} + H^+ \rightarrow Fe^{2+} + HO_2^{\bullet} \tag{20.4}$$

The production of the [Fe III (OOH)]$^{2+}$ complex, as seen in Equation 20.3, is linked to the reaction seen in Equation 20.2, which follows the equilibrium theory. They caused complex splits, which resulted in the formation of Fe^{2+} and the hydroperoxyl radical ($HO_2^{\bullet}$), as seen in Equation 20.4 [29]. The three types of pathways used by hydroxyl radicals to dissolve organic compounds are hydrogen abstraction, hydroxyl incorporation, and electron transfer [30]. The abstraction process of hydrogen entails hydrogen reactions of unsaturated organic compounds as shown in Equation 20.5. On the other hand, the direct hydroxyl radical reacts with carbon-carbon multiple bonds or the aromatic system of organic compounds in the hydroxyl addition mechanism as shown in Equation 20.6. Mechanisms for electron transmission are generally linked to electron inorganic ion transfer reactions Equation 20.7 [26].

$$^\bullet OH + RH \rightarrow R^\bullet + H_2O \qquad (20.5)$$

$$^\bullet OH + R \rightarrow R\,(OH) \qquad (20.6)$$

$$^\circ OH + Fe^{2+} \rightarrow Fe^{3+} + OH^- \qquad (20.7)$$

Investigations were carried out to determine how diverse chloride ion concentrations affect methyl t-butyl ether (MTBE)'s oxidation during the Fenton method. The MTBE as an aqueous solution with (5×10^4) M was used for oxidation. Initially, the concentration of hydrogen peroxide was kept at 7.5×10^3 M. NaCl was used to change the chloride content to 0.05, 0.1, 0.2, and 1 M. The thermostat and well-stirred batch reactor were used to carry out the experiments, and the temperature was kept at $25°C \pm 1°C$ in a batch reactor of volume 0.25 L. To stop photoreactions, all reactions were carried out in complete darkness. Reaction solutions were created by having the appropriate aliquot of the stock MTBE solution; introducing Fe^{2+}, perchloric acid was used to fix pH at 2.8. H_2O_2 inclusion kicked off the reaction. After different reaction times, samples were obtained and immediately dipped in $Na_2S_2O_3$ (0.01N) while experimenting. After 90 minutes, the depletion of MTBE came to a halt. During the lack of chloride ions, oxidation was quick and effective. The MTBE level has dropped to 10% of its starting level within 10 minutes, and it had reduced to 3% of its starting level within 90 minutes. In the presence of 0.05–0.2 M chloride ions, the effectiveness of MTBE degradation dropped to 85%–87% within 90 minutes of reaction. For 90 minutes of interaction with a 1 M level of chloride ions, the breakdown of MTBE was reduced to 68%. The rise in response time to 90 minutes might counteract the inhibiting impact. After this period, the reaction effectiveness in a solution containing 1 M Cl was remained very low. The inhibition of Fenton's reaction by chloride ions was discovered to be dependent on complexation and radical scavenging [31].

20.2.2 Factors Affecting the Fenton Process

The efficiency of the process is influenced by a variety of operating parameters, with Fenton's reagent ratio and pH being the most important determinants of the process's oxidation capacity.

20.2.2.1 pH

The pH value from 2.8 to 3.5 is thought to be ideal [32]. The method becomes inefficient and costly where there is too much acidic and alkaline environment. Very low pH levels cause the development of $[Fe\,(H_2O)\,6]^{2+}$ species, which delays the mechanism and leads to low hydroxyl radical yields [33]. Conversely, at pH levels greater than 3.5, ferric ions (Fe^{3+}) begin to precipitate as ferric hydroxide, which prevents the formation of OH radicals and the regeneration of Fe^{2+} [34].

20.2.2.2 H_2O_2 Concentration

H_2O_2 seems to be the most significant of the reagents used in the Fenton process as it decides the total mass of OH formed. Large doses of H_2O_2 result in greater pollution depletion, but only to a particular degree. H_2O_2 accumulation induces OH scrounging, resulting in less reactive HO_2^* which obstructs the system's efficiency as explained in the Equation below [35]

$$H_2O_2 + {}^\circ OH \rightarrow H_2O + HO_2^\circ \qquad (20.9)$$

A high dose of H_2O_2 reduces Fe^{2+} concentrations, causing the system to be influenced by the composition of relatively more Fe^{3+} ions. Fe^{3+} catalytic activity is thought to be several ways lower from Fe^{2+}; therefore, it decreases the generation of OH radicals [36].

20.2.2.3   Concentration of Fe^{2+}

Since Fe^{2+} serves as a catalyst in the decomposition of H_2O_2, greater Fe^{2+} levels lead to greater OH radical production. Greater Fe^{2+} levels cause greater oxidation of toxins, but only to a degree. Fe^{2+} accumulation induces OH scrounging, resulting in less reactive HO_2^* which obstructs the system's efficiency as explained in the following Equation [37].

$$Fe^{2+} + {}^\circ OH \rightarrow Fe^{3+} + OH^- \qquad (20.10)$$

Total dissolved solids (TDS) and higher electrical conductivity of effluent also increase with greater Fe^{2+} concentration value. It also results in an excessive quantity of sludge being produced at the end of the operation [16].

20.2.2.4   Reagents Ratio (Fe^{2+}/H_2O_2)

For the overall productivity of the Fenton operation, the concentration of Fe^{2+} and H_2O_2 is viewed as a critical factor, and the reagent ratio (i.e., Fe^{2+}/H_2O_2) is crucial in determining the optimal necessary amounts of OH and preventing its scavenging. The iron dose influences the process reaction rate, while the mineralizing degree of the contaminants depends primarily on the amount of H_2O_2. The large Fe^{2+}/H_2O_2 ratio speeds up the catalytic decay of H_2O_2, resulting in a significant number of OH radicals. The abundance of radicals, on the other hand, is subjected to scavenging effects and has an impact on mineralization capability. Furthermore, at the lesser Fe^{2+}/H_2O_2 proportion, the heavy accumulated amount of H_2O_2 induces OH radical scavenging and reduces operational efficiency [16]; therefore, it is essential to have an optimum ratio.

20.3   FENTON-BASED DIFFERENT APPROACHES

Several advanced approaches to the Fenton process are given below.

20.3.1   ELECTRO–FENTON PROCESS

To supplement the traditional Fenton procedure, electrons are used in this process. Fe^{2+} and H_2O_2 are produced with the use of the concept of cathodic reduction of Fe^{3+} and O_2 [38]. Electro-Fenton is divided into four forms based on reagents formation. The first form for the generation of H_2O_2 and Fe^{2+} includes the use of oxygen splitting cathode and sacrificial anode without any additional reagents [39]. The second from Fe^{2+} is produced by the sacrificial anode, and H_2O_2 is inserted separately. In

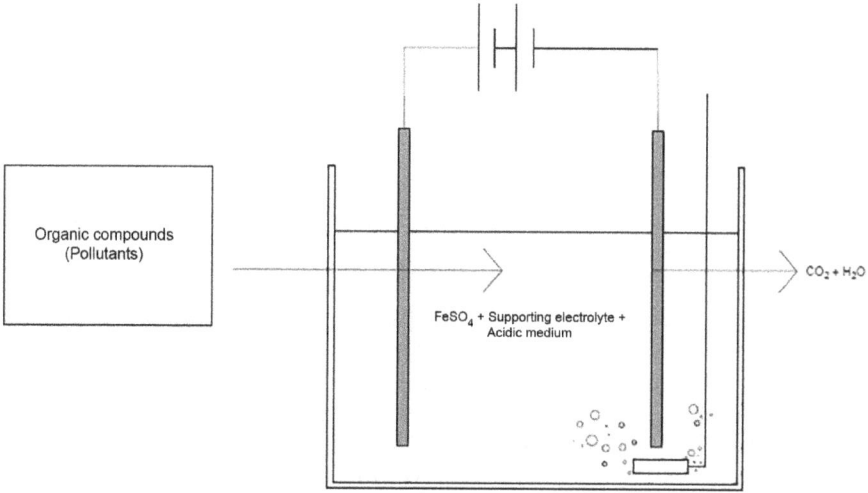

FIGURE 20.2   Schematic diagram of an Electro-Fenton reactor. (See [43].)

the third form, H_2O_2 is produced by oxygen sparging cathodes but Fe^{2+} is inserted separately [26]. In the fourth form, Fe^{3+} cathodic reduction produces Fe^{2+} [40]. The most widely used EFP for persistent electro-production of H_2O_2 is of form third. As seen in Equation below, a regular flow of oxygen gas at the cathode in an acidic medium induces two-electron reduction and the development of H_2O_2 [41]. At first, the cell is provided with small little ferrous salt to react with H_2O_2 and produces the Fe^{3+}, which, in turn, is continuing to regenerate Fe^{2+} cathodically as seen in Equation below [42] (Figure 20.2).

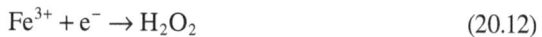

$$O_2 + 2H^+ + 2e^- \rightarrow H_2O_2 \tag{20.11}$$

$$Fe^{3+} + e^- \rightarrow H_2O_2 \tag{20.12}$$

In addition, electricity is being used as a renewable energy supplier so that secondary contaminants are not generated altogether [44]. The electronic production of H_2O_2 reduces the costs of shipping and storage by 80%. The efficient use of Fe^{3+} and constant regeneration of Fe^{2+} reduce sludge output while increasing OH generation [45]. This is an environmentally sustainable form of water and wastewater treatment as this E-Fenton technique doesn't use toxic reagents [46]. For the removal of phenol present in the saline wastewater, the electro-Fenton method (EFP) was evaluated for effectiveness. The influence of factors such as current density (4–16 mA/cm²), phenol level (250–2,000 mg/L), reaction duration (20–100 minutes), salinity (1%–4%), H_2O_2 level (75–300 mg/L), and pH (2–8) on the EFP was examined using an L25 orthogonal array design. The Taguchi procedure was used to assess the right conditions for phenol removal. Minitab 16 statistical software was used to design the initial phenol level, current density, pH, salinity, contact time, and H_2O_2 dose. The signal-to-noise (S/N) ratio was used to assess the experimental results. The highest of these S/N ratio values obtained was chosen as the optimum condition. The effect of salinity (NaCl) percentage on phenol elimination in the band of 0%–4% was investigated using EFP. The *S/N* elimination of phenol was 2.808 while the salinity was 2%. When the salinity level approached 2%, a lower S/N ratio was obtained. As a result, the optimum level of salinity was chosen as 2% [47].

20.3.2 Sono-Fenton Process

High-frequency ultrasound waves are sent in this step to dissociate water molecules to form hydrogen radicals (H°) and hydroxyl radicals (OH°) [48]. When high-frequency ultrasound waves (>15 kHz) are irradiated on liquid media, they induce adiabatic compression and rarefaction. The reduced pressure in the rarefaction region of the ultrasonic wave causes the formation of microbubbles [49]. During the compression process, these microbubbles expand and suddenly collapse. As a result, local high temperatures and pressures of up to 5,000 K and 1,000 atm are possible (Figure 20.3).

As result, these short-term conditions result in the production of H, OH°, and $HO_2^\bullet$ radicals as shown in Equation below [41]. As seen in Equations 20.13 and 20.14, the Sono-Fenton process (SFP) entails the sonolytic production of H_2O_2 and Fe^{2+} in the presence of Fe^{3+} ions [51].

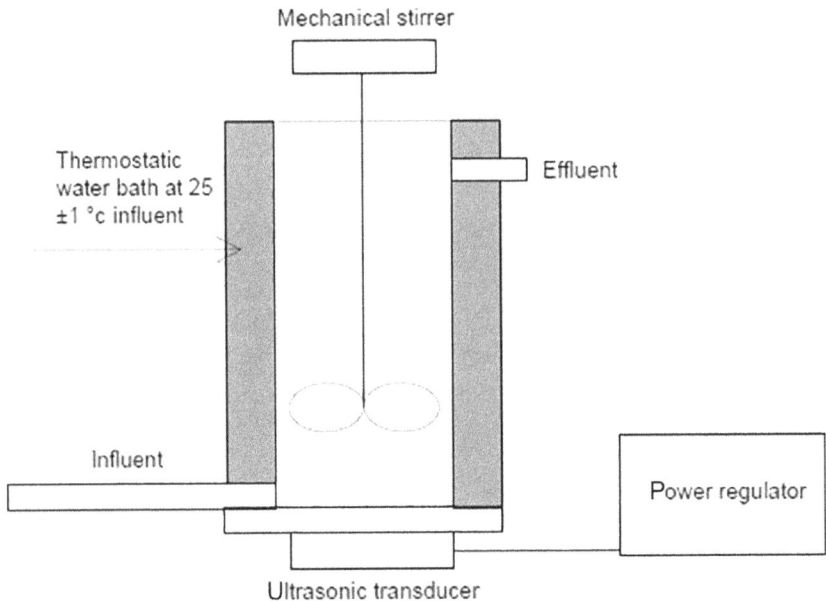

$$H_2O+ \rightarrow OH^\circ + H^\circ \qquad (20.13)$$

$$H^\circ + O_2 \rightarrow HO_2^\circ \qquad (20.14)$$

$$H^\circ + Fe^{3+} \rightarrow Fe^{2+} + H^+ \qquad (20.15)$$

FIGURE 20.3 Schematic diagram showing Sono-Fenton reactor ultrasonic transducer. (See [50].)

$$HO_2^\circ \leftrightarrow O_2^{\circ-} + H^+ \qquad (20.16)$$

$$Fe^{3+} + O_2^{\circ-} \rightarrow Fe^{2+} + O_2 \qquad (20.17)$$

$$HO_2^\circ + O_2^{\circ-} + H^+ \rightarrow H_2O_2 + O_2 \qquad (20.18)$$

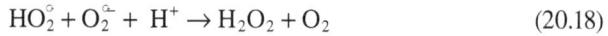

The synergistic effects of sonolysis with the Fenton process result in a greater generation of OH as compared to the standard Fenton process [49]. In the treatment of saline petrochemical effluent by the electro-Fenton (EF) technique, the influence of operating parameters on the integration of persulfate and ultrasound (US) was studied. The effects of electrode distance (ED), applied voltages pH, reaction time, and US irradiation on COD elimination were studied first for a mixture of US and EF (US/EF). The impact of persulfate (PUEF) addition on COD elimination efficiency was next examined. With an initial 850 mg/L COD level and reaction time of 3 hrs and 30 min, the US/EF procedure removed 94.1% of COD. The addition of persulfate reduced the reaction time greatly, in 2 hours, 91.7% of COD was eliminated. A greater pseudo-first-order constant of 0.01691/min reaction rate for the PUEF method relative to 0.0126 for the US/EF method further proved the effectiveness of the PUEF method against the US/EF method. Both processes were pH-dependent, with acidic pH conditions yielding the best results. Increasing the US irradiation, voltage, and response time also significantly enhanced COD elimination by increasing the rate of HO° radicals and H_2O_2 output [52].

20.3.3 PHOTO-FENTON PROCESS

In this approach, UV radiation is employed for boosting the production of OH and regenerate Fe^{2+} ions [25]. In this procedure, it is preferred to use UV or visible light radiation of less than 450 nm [53]. UV exposure causes photochemical regeneration of Fe^{2+}, which interacts with H_2O_2 to release OH and Fe^{3+} ions, as explained in the reaction below. The cycle is continued by Fe^{3+} regeneration, which results in higher OH efficiency, which improves Fenton's oxidation output as a result of successful pollutant degradation [54] (Figure 20.4).

As shown in Equation 20.20, the procedure also includes direct photolysis of H_2O_2 to produce OH°. Moreover, the existence of iron complexes in solution consumes a significant amount of light and influences H_2O_2 photolysis [55]. The pH is critical in the photo-Fenton process (PFP), which determines the proportion of different iron complexes. The Fe^{3+} ions are effectively transformed into the water complex of the most photo reactive ferric ion, the $[Fe(OH)]^{2+}$ species, at a pH of 3. As seen in Equation 20.21 [54,56], metal charge transfer excitation of $[Fe(OH)]^{2+}$ by UV radiation regenerates the Fe^{2+} and induces OH. Acidic conditions (pH 3) favour the oxidation of carbonates and bicarbonates into carbonic acid, which has reduced susceptibility to OH radicals in comparison [57].

$$Fe^{3+} + h\nu + H_2O \rightarrow Fe^{2+} + OH^\circ + H^+ \qquad (20.19)$$

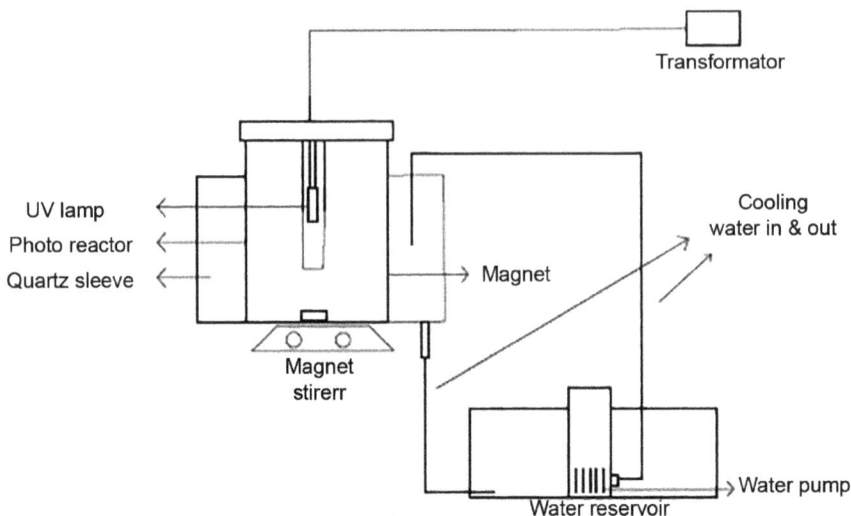

FIGURE 20.4 Systematic representation of the complete system for the process of photo-Fenton. (See [22].)

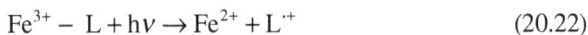

$$H_2O_2 + h\nu \rightarrow 2\,OH^c \qquad\qquad (20.20)$$

$$\left[Fe(OH)\right]^{2+} + h\nu \rightarrow Fe^{2+} + OH^c \qquad\qquad (20.21)$$

$$Fe^{3+} - L + h\nu \rightarrow Fe^{2+} + L^{+} \qquad\qquad (20.22)$$

For the treatment of saline wastewater contaminated with hydrocarbons, the photo-Fenton approach is being investigated. Aqueous saline solutions containing raw gasoline have been employed as the basis for oil-field-produced water. The dependency on the levels of iron ions (0.5–1 mM), hydrogen peroxide (100–200 mM), and sodium chloride (200–2,000 ppm) (0.5–1 mM) has been effectively analyzed. Absorption spectra and comprehensive organic carbon (TOC) measurements were used to trace the processes. In the start, the initial level of ferric ion (Fe^{2+}) had a significant influence on the decomposition processes. When the Fe^{2+} concentration was reduced from 1.0 to 0.5 mM, the percent decay at 1 hour decreased from 62% to 46% at the highest concentrations of sodium chloride (2,000 ppm) and hydrogen peroxide (200 mM); the sum of hydrogen peroxide, on the other hand, had no effect on photodegradation efficiency over the range tested (100–200 mM). The impact of salinity was much greater and harmed hydrocarbon photodegradation. At two different sodium chloride levels, experiments were conducted out with the least quantities of H_2O_2 and Fe^{2+}. Decomposition reached 78% after 1 hour of reaction at 200 ppm NaCl, and it reached 90% after the trial. After 1 hour, the deterioration at 2,000 ppm NaCl reached 50%, and by the conclusion of the experiment, it was 83% [58]. Another investigation was conducted in which pyrimethanil (PYR-20 mg/L) was treated with photo-Fenton in two water matrices, water containing 5g/L of NaCl (DWNaCl) and demineralized water (DW); the development of chlorinated transition

products was investigated. A compound parabolic collector (CPC) was used for conducting all these experiments with an H_2O_2 level of 150–350 mg/L and a Fe^{2+} level of
5 mg/L. Toxicity, high-performance liquid chromatography with diode-array detection (HPLC–DAD), dissolved organic carbon (DOC), and biodegradability studies
were all used to track the photocatalytic process. During DW, with 33 mM H_2O_2
absorbed, PYR was removed after 11.8 minutes of illumination and with 79 minutes
of illumination, the initial DOC was decreased by 50%. The equivalent decrease in
DOC in the DW-NaCl water matrix, on the other hand, needed 39 mM H_2O_2 and 110
minutes of illumination, with complete depletion of PYR occurring after 12 minutes.
LC–TOFMS is used to identify PYR transition products (TPs). In addition to the
non-chlorine TPs observed during DW degradation, photo-Fenton in a DW-NaCl
was discovered to also include specific chlorinated TPs. During photo-Fenton, all
TPs shaped were destroyed. Furthermore, the presence of chlorinated TPs has little
effect on the water's toxicity, and the TPs produced are more biodegradable than
PYR [59].

20.3.4 Heterogeneous Fenton Process

Solid Fenton catalysts are used in this method to generate Fe^{2+} ions efficiently. The
homogeneous Fenton method entails the external inclusion of Fe^{2+}, which is existing in the effluent of the treated wastewater and cannot be isolated. In addition, the
neutralization of Fe^{2+} in wastewater contributes to Fe^{3+} production, which ultimately
leads to increased production of sludge and requires an adequate disposal method.
The heterogeneous Fenton (HF) method is usually chosen to solve the issue of Fe^{2+}
inefficiency. The Fe^{2+} is produced from iron oxides and other sponsored catalyst
active sites in this method. The Fe^{2+} ions remain stable and immobilize over the
surface and pores of the catalyst. Fe^{2+} immobilization limits the formation of ferrous
oxide sludge production and extends the viability of the procedure across a broad
variety of pH. The reduction of catalyst loss was also advantageous [60]. The two different methods for producing highly reactive OH are combined in this method. The
direct heterogeneous reactions of Fenton over the catalyst surface are one method.
The surface-leached iron begins the Fenton homogeneous phase beyond the catalyst
in the other process [61]. The H_2O_2 interacts with the iron species on the surface of
solid Fenton catalysts to produce highly reactive OH in the direct heterogeneous
Fenton phase [41]. The heterogeneous Fenton mechanism also supports the decrease
from Fe^{3+} to Fe^{2+} via catalyst electron transfer processes [62] (Figure 20.5).

During a research study, an effective photo membrane (MPBR) with efficient
PSB (PSB) and PVDF fluidized bed was constructed and successfully applied to
handle real refractory wastewater with high salinity. This was done in conjunction
with the heterogeneous Fenton fluidized bed. Heterogeneous Fenton was developed as an effective pre-treatment to remove organic substances and to reduce
iron-sludge dissipation. The GO/PVDF membrane manufactured with the use of
GO nanosheet grafting was first used in MPBR for the production of salt-tolerated
PSB. COD concentration was 3,335 (mg/L) at the start. MPBR with GO/PVDF
membrane showed COD and NH_3–N removal rates of 95% and 98% respectively
and biomass production of 105 mg/L-d on average. The use of a heterogeneous

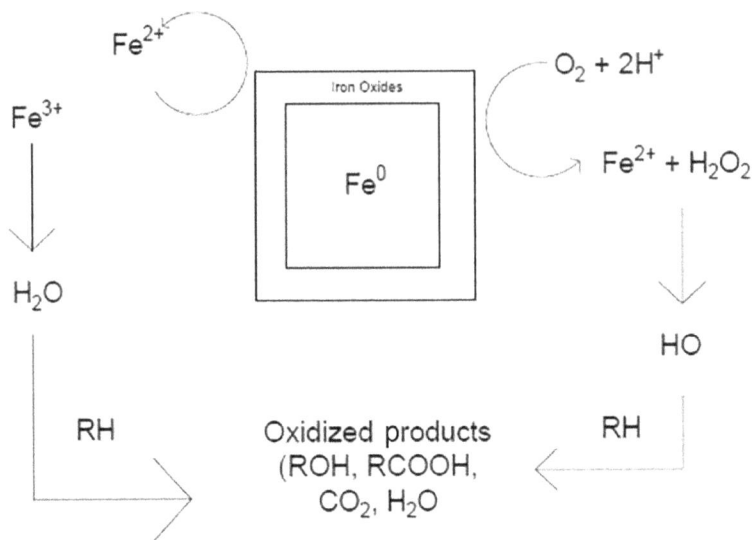

FIGURE 20.5 Simplistic diagram for showing the mechanism of heterogeneous Fenton process. (See [63].)

Fenton fluidized bed considerably improved biodegradability (from 0.21 to 0.43) and reduced iron sludge output. A GO/PVDF membrane with increased hydrophilicity, larger permeability (4.4 times), and appealing flux recovery rate (94%) was successfully employed for high-active PSB harvesting in the MPBR technique [64]. An experimental design is used to explore the influence of NaCl on the performance of the operation. Different H_2O_2 and Fe^{2+} starting concentrations are used to treat a solution containing 200 ppm 4-CP as a model chemical and various NaCl concentrations. It's noteworthy that the TOC elimination accomplished is higher when $[H_2O_2]_0$ is higher. On the other hand, the same TOC conversion is also achieved with different $[Fe^{2+}]_0$; however, when $[Fe^{2+}]_0$ increases, the rate of elimination accelerates. The chloride's inclusion slows the operation but it has no impact on the overall amount of TOC removed, according to the findings. Due to iron-chloride complexes formation, the anion appears to have a direct effect on the iron system but does not produce hydroxyl radical scavenging [65].

20.3.5 Solar Photo Fenton Process

This process uses solar energy to boost the concentrations of hydroxyl radicals and other photoactive complexes, allowing the Fenton process to run more efficiently [66]. As free solar renewable energy is available in this process, the contaminants are mineralized quickly. The solar-powered Fenton method is thought to be very effective, ensuring higher mineralization (up to 90%) in a time of about 30 minutes [67]. The method is very ecofriendly since it reduces the amount of electricity used in other applications. The use of light decreases the quantity of reagents and electricity used, allowing for faster reaction speeds [68]. Many lab-scale experiments have

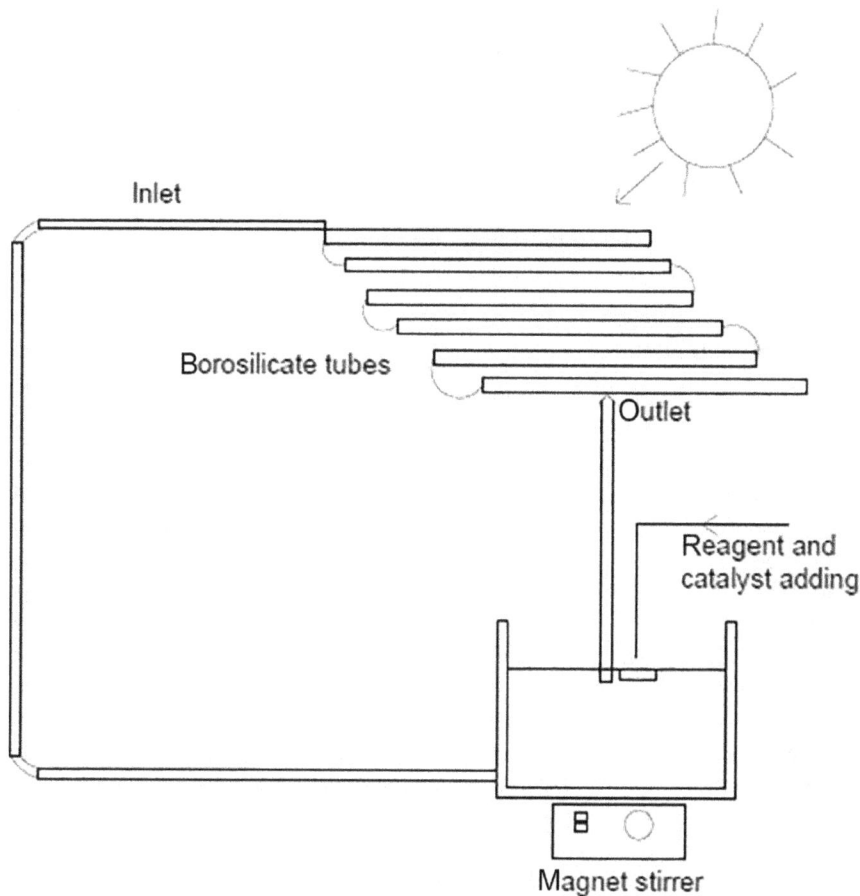

FIGURE 20.6 Schematic diagram of solar-photo Fenton process. (See [70].)

shown the amazing ability of sunlight energy as a photo-Fenton process alternative
to UV radiation [69] (Figure 20.6).

The photo-Fenton method was utilized to treat a saline aqueous solution
of phenol in a falling-film solar reactor, 2,4-dichlorophenol (2,4-DCP) and
2,4-dichlorophenoxyacetic acid (2,4-D). initial Fe^{2+} concentration (1–2.5 mM), initial
pH (5–7), and H_2O_2 insertion intensity (1.87–3.74 mmol/min) effects were all examined. The effectiveness of photodegradation was measured by the elimination of dissolved organic carbon (DOC), as determined by the species degradation of, 2,4-DCP,
2,4-D, and phenol. The findings of the variables evaluated in the photo-Fenton phase
with solar irradiation, namely pH, $[H_2O_2]$, and $[Fe^{2+}]$, were assessed using response
surface methodology. At an original pH of 5–7, sunlight/Fe^{2+}, sunlight/H_2O_2, and
photolysis techniques were used to degrade the 2,4-D/phenol/2,4-DCP combination
in the saline medium in early testing. These methods resulted in low consumption of
organic matter and a carbon reduction rate of roughly 10%–15% after 180 minutes
of reaction. There were also blank tests (Fenton reaction) carried out. The usage of

H_2O_2/Fe^{2+} resulted in a cumulative carbon elimination of roughly 33% after 180 minutes of reaction in the absence of solar irradiation. Only in the presence of sunlight/ H_2O_2/Fe^{2+} did the phenol/2,4-D/2,4-DCP mixture degrade faster and remove more organic carbon. The findings indicate that the variables' initial Fe^{2+} and H_2O_2 concentrations have the greatest impact on pollutant degradation as DOC elimination and pH, on the other hand, have no impact. Using the reaction method used here, about 85% DOC removal was achieved in 180 minutes under the most favourable laboratory conditions, and in roughly 30 minutes, the 2,4-D, phenol, and 2,4-DCP combination were completely removed [71].

20.3.6 Hybrid Fenton Process

The hybrid Fenton method (HFP) looked at how different Fenton's methods could be combined. To boost the Fenton process's degradation efficiency, it uses a mixture of electrolytic, photolytic, and sonolytic Fenton processes. The sono-photo-Fenton hybrid process entails the combination of sonolysis and UV light radiation system to improve the effectiveness of pollutant degradation. In the sono-Fenton method, water is sonolysed, resulting in the formation of OH and H° radicals, which then recombine to form H_2O_2, reducing the supply of OH radicals. But the use of UV light in the hybrid sono-Fenton method helps the formed H_2O_2 to decompose back into OH and thus increase the amounts of the radical. The combination of sonolysis and photolysis process also promote the regeneration of Fe^{2+} by reducing intermediate complexes produced by Fenton [72]. Combining photochemical and electrochemical Fenton processes is another significant hybrid technology. The hybrid photo-electron Fenton method yields more hydroxyl radicals, which improves the process's oxidation capacity. The use of UV light to photoreduction Fe^{3+} species in the EFP promotes the regeneration of Fe^{2+} ions and induces photo-decarboxylation of Fe (III)-carboxylate complexes. The photolytic hemolytic breakup of H_2O_2 facilitates higher OH output and improves the efficiency of the method [73] (Figure 20.7).

For the effective treatment of recalcitrant petrochemical wastewater (PCW), an effective hybrid method was used (PCW). The heterogeneous catalyst in this work was magnetic powdered activated carbon (MPAC), which was coupled with ultrasonic (US) and UV irradiations to produce H_2O_2 (marked as MPAC/US/UV/H_2O_2). Various controlling operating factors, such as solution pH, MPAC and H_2O_2 concentrations, quenchers, and US capacity, were used to determine the chemical oxygen demand (COD) removal ratio. The Zahn-Wellens test was used to measure the biodegradability of both raw and treated PCWs in the enabled sludge inhibition studies. In the system under investigation, MPAC revealed high reusability, stability, and catalytic activity. Original COD concentration in PCW was 680 ± 24 mg/L. Under optimum conditions, more than 87% of COD was eliminated in just 80 minutes, with a COD level of 82.9 mg/L remaining. MPAC's catalytic activity was greatly enhanced in the presence of US irradiation, H_2O_2, and UV light. The majority of the damage was caused by HO percent radicals, according to the findings of quenching studies. In contrast to H_2O_2, the removal of COD was more efficient when PMS and PS oxidants were present. PCW that has been treated is more biodegradable than PCW that has not been treated, according to the Zahn-Wellens test. Overall, combining

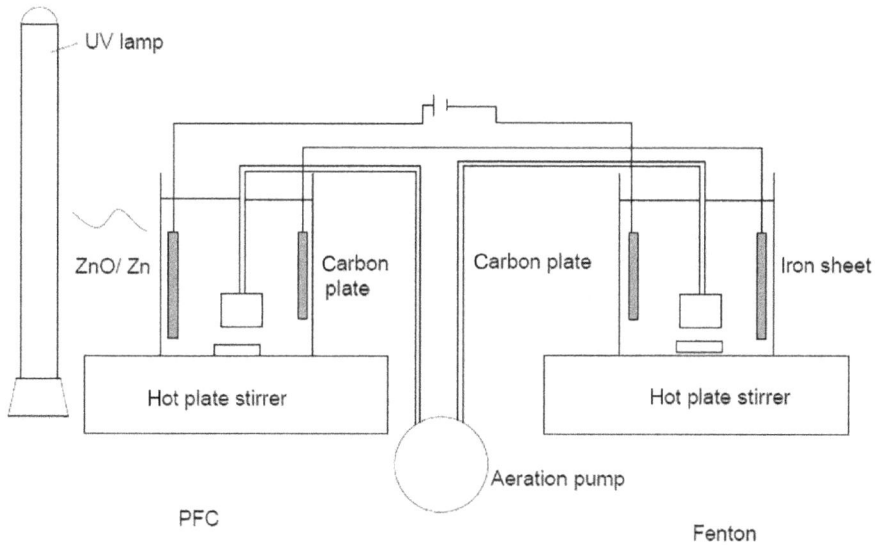

FIGURE 20.7 Schematic diagram of photocatalytic fuel cell (PFC) and Fenton hybrid process. (See [74].)

UV, US, and MPAC to activate H_2O_2 was an effective method for treating refractory industrial wastewaters efficiently and effectively [75].

Some previous studies for saline water treatment by different Fenton's approaches are arranged in Table 20.1.

20.4 COMBINATION OF PROCESSES

As conventional Fenton-process is combined with other technologies, such as electrolysis, UV radiation, and sonolysis, more hydroxyl radicals are produced, which lead to greater pollution degradation. The combined processes increased the rate at which the target pollutants degraded. Other benefits could include ferrous ion regeneration, low sludge output, and in situ H_2O_2 generation. Many of the researchers have found in their studies that combining two or more than two different Fenton's processes often led to an increase in the efficiency to remove the brine solution. The sample's biodegradability was also improved as a result of the procedure. The hybrid method combined sonolytic, electrolytic, and photolytic approaches to achieve a rapid removal efficiency than the individual sono, photo, and electro-Fenton processes. Different types of wastewater were treated using a combination of Fenton oxidation and biological processes. Wang et al. [76] studied surfactant wastewater treatment using Fenton oxidation and aerobic-bio-processing and identified Fenton Oxidation as an efficient wastewater treatment, with large amounts of surfactants and sulfate, which allowed a considerable reduction of COD. This treatment was also highly efficient in improving wastewater biodegradability. Also at Fenton reagents' small doses and limited UV exposure times, Nousheen [77] found that photo-Fenton treatment of anaerobically pre-treated commercial, domestic, and mixed wastewater was more

TABLE 20.1
Fenton Based Removal of Pollutants from Saline Wastewater

References	Sample Type	Sample Characteristics	Method Used	Remarks
[31]	Saline wastewaters containing methyl tert-butyl ether	MTBE 5×10^{-4} M dilute aqueous solution of pH 2.8, $T = 25°C$	Conventional Fenton	Chloride ions = 0.05–0.2 M, time = 90 minutes degradation 85%–87%; Chloride ions = 1 M, time = 90 minutes degradation = 68%
[58]	Saline wastewater contaminated with hydrocarbons	Organics concentration = 80–90 ppmC, pH 3.0, $T = 30°C$	Photo-Fenton	At NaCl = 200 ppm degradation = 90% At NaCl = 2,000 ppm degradation = 83%
[71]	Saline solution containing mixture of 2,4-dichlorophenol, phenol and 2,4-dichlorophenoxyacetic acid	2,4-D strength = (98%), 2,4-DCP strength = (98%) and Phenol strength = (>99.5%) NaCl = 60 g/L pH 5–6 $T = (24 – 31)°C$	Solar Photo-Fenton	2,4 dichloro-phenol, 2,4-dichlorophenoxyacetic acid and phenol removal = 100% time = 30 minutes dissolved organic carbon = 85%
[47]	Phenol from saline wastewater	Phenol = 250 mg/L, pH 3 and salinity 2%	Electro-Fenton process (Taguchi optimization)	Phenol removal = 91% Time = 20 minutes and Current density = 8 mA/cm^2
[52]	Recalcitrant organics in saline petrochemical wastewater	COD = 800 mg/L pH 7.4–8.8 and Cl = 13,890–19,600 (mg/L)	Persulfate and ultrasonic intensified Electro-Fenton process	COD removal efficiency = 94.1%. Time = 210 minutes Integration of persulfate irradiation with electro-Fenton (PUEF), COD effectively removed = 91.7%. Time = 120 minutes
[65]	4-chlorophenol (4-CP) aqueous solution	4-CP = 200 ppm [NaCl] 50,000 ppm pH 3 $T = 27°C$	Photo-Fenton	TOC removal = 60%

(Continued)

TABLE 20.1 (*Continued*)
Fenton Based Removal of Pollutants from Saline Wastewater

References	Sample Type	Sample Characteristics	Method Used	Remarks
[59]	Pyrimethanil under saline conditions	Pyrimethanil = 20 mg/L, NaCl = 5 g/L, pH 2.6–2.8	Photo-Fenton	Dissolved organic carbon reduction = 50%, time = 110 minutes, n H_2O_2 consumption of 39 mM
[75]	Saline recalcitrant petrochemical wastewater (PCW)	COD = 680 mg/L, pH 7.4, Conductivity = 58.7 µs/cm	Heterogeneous UV-assisted sono-Fenton process	COD removal efficiency = 87%, Time = 80 minutes
[64]	Refractory seafood-processing wastewater with high salinity	COD = 3,335 (mg/L), pH 5.0, NaCl = 0.5 g/L	Heterogeneous Fenton	COD removal efficiency = 95%, Time = 90 minutes

successful than the Fenton method alone in meeting limits of discharge. High salinity in wastewater, for example, may impede the decomposition of organic pollutants such as petroleum hydrocarbons and polycyclic aromatic hydrocarbons [78,79]. To achieve the desired degrading efficiency, appropriate acclimation of sludge and optimization of the process such as dilution of feedwater are required.

The chemical industry in Shanghai encountered excess phosphorus emissions, even though chemical and biological wastewater treatment technologies were used concurrently. After on-site treatment of industrial wastewater, the total phosphorus influenced by triethyl phosphate reached as high as 50–60 mg/L, much higher than the maximum contaminant limit (MCL) of 8 mg/L. There was an urgent need to investigate new strategies for reducing phosphorus to a safe level. A researcher [80] investigated the remediation of industrial effluents with high concentrations of organic phosphorus (Org.-P, dominated by triethyl phosphate (TEP)) using Fenton oxidation and a traditional (i.e., aerobic biodegradation) treatment approach keeping implementation potentials in parallel. Fenton significantly degraded Org.-P from 58 to 5 mg/L using the first-order kinetic model with a reaction rate constant of 0.07/minutes and the ideal reaction parameters of $H_2O_2 = 20\,mM$, $Fe^{2+} = 14\,mM$, taking reaction time as 120 minutes using the technique of continuous dosing ($N = 4$) at pH 3. Nonetheless, pre-treatment (e.g., desalination) before the Fenton reaction is advised since excessive salinity slowed TEP breakdown, probably due to the creation of Fe-Cl complexation and associated OH scavenging effect. Aerobic biodegradation was used to produce a 98% Org.-P mineralization rate. The outstanding performance was measured accordingly to the salinity of 4.6% (w/w), at which point the mineralization began to decrease significantly.

Microbial development might be hampered by excessive salinity. For practical wastewater biotreatment, an 8:1 dilution ratio was shown to be best. Coupling usage of biodegradation and coagulation contributed 405 mg/L COD and 6 mg/L Org.-P to a good wastewater quality, which complies with municipal sewer release standards [81]. Studies combining the ZVI and Fenton processes for nitroaromatic compound treatment have received a lot of interest in recent years. Para-chloronitrobenzene was effectively degraded by ZVI and Fenton oxidation using the sequential procedure reported by Le et al. (2011). Para-chloroaniline, a reductive product, was more vulnerable to Fenton agent oxidation than para-chloronitrobenzene [82]. Performed a study for the pre-treatment of 2,4-dinitroanisole (DNAN) generating wastewater by applying a combination of processes such as Fenton oxidation with zero-valent iron (ZVI) reduction process. The entire performance of the integrated process was analysed and operating parameters were adjusted. At an empty bed contact time (EBCT) of 8 hours for the ZVI procedure, it was noticed practically that nitroaromatic compounds were reduced 100%. The ideal pH was 3.0, H_2O_2 to Fe (II) molar ratio was 15, H_2O_2 dose $= 0.216$ mol/L, and hydraulic retention time (HRT) for the Fenton process was found to be 5 hours, respectively. A moving-bed biofilm reactor (MBBR) and an anaerobic baffled reactor (ABR) were used in a combined anaerobic-aerobic process that was fed with ZVI-Fenton outflow for 3 months. The results proved the ZVI-Fenton-ABR-MBBR system substantially effective in terms of rectifying the effluent's primary characteristics of interest, such as acute toxicity, COD concentration, aromatic compounds concentration, and colour. The combined ZVI-Fenton method appears to have promising potential for pre-treatment of wastewater comprising nitroaromatic chemicals, based on these findings [82]. The principal benefit of the combined procedures seems to be related to the initial transformation of nitroaromatic compounds in DNAN-generating wastewater into organic compounds that are more susceptible to oxidative breakdown, which is then encouraged for the Fenton reaction. The studies for treatability of bench-scale or pilot-scale aimed at actual wastewater were required to offer important information to aid develop and plan full-scale technology installation [83].

20.5 CONCLUDING REMARKS

Following a thorough review of the literature, we may conclude that Fenton-based methods are among the most appealing and successful AOPs for removing a wide range of recalcitrant compounds found in saline water. The CFP's widespread use is aided by widely accessible and non-toxic nature of Fenton's reagents such as H_2O_2 and Fe^{2+}. The method can be improved by combining it with some other techniques such as electrolysis, sonolysis, and photolysis, among others. The integrated method reduces sludge production, aids Fe^{2+} recovery, and increases the generation of hydroxyl radical, resulting in better optimal treatment efficiencies. However, almost every evolved methodology has its own set of constraints, and no one method is adequate to achieve the ultimate goal. As a result, we must choose the Fenton-based methods appropriately, either individually or combined, based on the required discharge standards, sample characteristics, and available resources. As a long-term approach to the crisis, the options must be cost-effective and environmentally sustainable. The CFP can extract about 75% of salts, whereas the hybrid Fenton processes may eliminate up to 95% of salts from salty water. Furthermore, the procedure improved brine biodegradability, allowing for additional biological therapies. The oxidation mechanism of the process is influenced by a variety of factors such as pH, temperature, H_2O_2 concentrations, Fe^{2+} dosage, optimum reagents ratio (H_2O_2/Fe^{2+}), and reaction period. In certain cases, pH and Fenton's reagents ratio are critical factors that decide the overall performance of the operation. The scrounging of hydroxyl radicals is hampered by an unnecessary dose of H_2O_2 and Fe^{2+}, which reduces the efficiency of the operation. As a result, for optimal treatment, it is important to thoroughly refine all operating parameters, which also eliminates the constraints correlated with the procedures. Additional research is needed to address the shortcomings of each Fenton-based methodology, as mentioned earlier, to render the method more user-friendly, reliable, cost-effective, and energy-efficient. As a long-term approach for strong wastewater treatment which can also be applied on a wide scale. Furthermore, integrated processes will be more suited for improved organics elimination, so more attention could be focused on them.

REFERENCES

1. Goosen MF, Mahmoudi H, Ghaffour N. Today's and future challenges in applications of renewable energy technologies for desalination. *Critical Reviews in Environmental Science and Technology*. 2014 May 3;44(9):929–999.

2. Khan SU, Noor A, Farooqi IH. GIS application for groundwater management and quality mapping in rural areas of District Agra, India. *International Journal of Water Resources and Arid Environments*. 2015;4(1):89–96.
3. Sharon H, Reddy KS. A review of solar energy-driven desalination technologies. *Renewable and Sustainable Energy Reviews*. 2015 Jan 1;41:1080–1118.
4. Van Weert F, Van der Gun J, Reckman J. *Global Overview of Saline Groundwater Occurrence and Genesis*. International Groundwater Resources Assessment Centre, Utrecht, 2009 Jul.
5. Bonsor HC, MacDonald AM, Ahmed KM, Burgess WG, Basharat M, Calow RC, Dixit A, Foster SS, Gopal K, Lapworth DJ, Moench M. Hydrogeological typologies of the Indo-Gangetic basin alluvial aquifer, South Asia. *Hydrogeology Journal*. 2017 Aug;25(5):1377–1406.
6. MacDonald AM, Bonsor HC, Ahmed KM, Burgess WG, Basharat M, Calow RC, Dixit A, Foster SS, Gopal K, Lapworth DJ, Lark RM. Groundwater quality and depletion in the Indo-Gangetic Basin mapped from in situ observations. *Nature Geoscience*. 2016 Oct;9(10):762–766.
7. https://www.qld.gov.au/environment/land/management/soil/salinity/impacts.
8. Lefebvre O, Moletta R. Treatment of organic pollution in industrial saline wastewater: A literature review. *Water Research*. 2006 Dec 1;40(20):3671–3682.
9. Xiang Q, Nomura Y, Fukahori S, Mizuno T, Tanaka H, Fujiwara T. Innovative treatment of organic contaminants in reverse osmosis concentrate from water reuse: A mini-review. *Current Pollution Reports*. 2019 Dec;5(4):294–307.
10. Deng Y. Advanced oxidation processes (AOPs) for reduction of organic pollutants in landfill leachate: A review. *International Journal of Environment and Waste Management*. 2009 Jan 1;4(3–4):366–384.
11. Husain Khan A, Abdul Aziz H, Khan NA, Ahmed S, Mehtab MS, Vambol S, Vambol V, Changani F, Islam S. Pharmaceuticals of emerging concern in hospital wastewater: Removal of Ibuprofen and Ofloxacin drugs using MBBR method. *International Journal of Environmental Analytical Chemistry*. 2020 Dec 19:1–5.
12. Mahtab MS, Islam DT, Farooqi IH. Optimization of the process variables for landfill leachate treatment using Fenton-based advanced oxidation technique. *Engineering Science and Technology, an International Journal*. 2021 Apr 1;24(2):428–435.
13. Sharma S, Ruparelia JP, Patel ML. *A General Review on Advanced Oxidation Processes for Wastewater Treatment*. Institute of Technology, Nirma University, Ahmedabad. 2011 Dec 8;481:8–10.
14. Ataei A, Mirsaeed MG, Choi JK, Lashkarboluki R. Application of ozone treatment in cooling water systems for energy and chemical conservation. *Advances in Environmental Research*. 2015;4(3):155–172.
15. da Costa FM, Daflon SD, Bila DM, da Fonseca FV, Campos JC. Evaluation of the biodegradability and toxicity of landfill leachates after pretreatment using advanced oxidative processes. *Waste Management*. 2018 Jun 1;76:606–613.
16. Gogate PR, Pandit AB. A review of imperative technologies for wastewater treatment I: Oxidation technologies at ambient conditions. *Advances in Environmental Research*. 2004;8(3–4):501–551.
17. Hussain M, Mahtab MS, Farooqi IH. The applications of ozone-based advanced oxidation processes for wastewater treatment: A review. *Advances in Environmental Research*. 2020;9(3):191–214.
18. Neyens E, Baeyens J. A review of classic Fenton's peroxidation as an advanced oxidation technique. *Journal of Hazardous Materials*. 2003;98(1–3):33–50. doi:10.1016/S0304-3894(02)00282-0.

19. Khan AH, Khan NA, Ahmed S, Dhingra A, Singh CP, Khan SU, Mohammadi AA, Changani F, Yousefi M, Alam S, Vambol S. Application of advanced oxidation processes followed by different treatment technologies for hospital wastewater treatment. *Journal of Cleaner Production.* 2020 Oct 1;269:122411.
20. Verma M, Haritash AK. Review of advanced oxidation processes (AOPs) for treatment of pharmaceutical wastewater. *Advances in Environmental Research.* 2020;9(1):1–17. doi:10.12989/aer.2020.9.1.001.
21. Sharma A, Verma M, Haritash AK. Degradation of toxic azo dye (AO7) using Fenton's process. *Advances in Environmental Research.* 2016;5(3):189–200. doi:10.12989/aer.2016.5.3.189.
22. Zazouli MA, Yousefi Z, Eslami A, Ardebilian MB. Municipal solid waste landfill leachate treatment by Fenton, photo-Fenton and Fenton-like processes: Effect of some variables. *Iranian Journal of Environmental Health Science and Engineering.* 2012;9(1):3. doi:10.1186/1735-2746-9-3.
23. Ismail S, Tawfik A. Treatment of hazardous landfill leachate using Fenton process followed by a combined (UASB/DHS) system. *Water Science and Technology.* 2016;73(7):1700–1708. doi:10.2166/wst.2015.655.
24. Buxton GV, Greenstock CL, Helman WP, Ross AB. Critical review of rate constants for reactions of hydrated electrons, hydrogen atoms and hydroxyl radicals ($\bullet$OH/$\bullet$O$^-$) in aqueous solution. *Journal of Physical and Chemical Reference Data.* 1988;17:513–886.
25. Kim SM, Geissen SU, Vogelpohl A. Landfill leachate treatment by a photoassisted Fenton reaction. *Water Science and Technology.* 1997 Jan 1;35(4):239–248.
26. Bello MM, Raman AAA, Asghar A. A review on approaches for addressing the limitations of Fenton oxidation for recalcitrant wastewater treatment. *Process Safety and Environmental Protection.* 2019;126:119–140. doi:10.1016/j.psep.2019.03.028.
27. Xu, M., Wu, C., & Zhou, Y. (2020). Advancements in the Fenton process for wastewater treatment. *Advanced Oxidation Processes – Applications, Trends, and Prospects,* Bustillo-Lecompte CF, 61. IntechOpen. doi:10.5772/intechopen.90256.
28. Deng Y, Zhao R. Advanced oxidation processes (AOPs) in wastewater treatment. *Current Pollution Reports.* 2015;1(3):167–176. doi:10.1007/s40726-015-0015-z.
29. De Laat J, Gallard H, Ancelin S, Legube B. Comparative study of the oxidation of atrazine and acetone by H_2O_2/UV, Fe (III)/UV, Fe (III)/H_2O_2/UV and Fe (II) or Fe (III)/H_2O_2. *Chemosphere.* 1999;39(15):2693–2706. doi:10.1016/S0045-6535(99)00204-0.
30. Huang CP, Dong C, Tang Z. Advanced chemical oxidation: Its present role and potential future in hazardous waste treatment. *Waste Management.* 1993;13(5–7):361–377. doi:10.1016/0956-053X(93)90070-D.
31. Siedlecka EM, Stepnowski P. Decomposition rates of methyl tert-butyl ether and its by-products by the Fenton system in saline wastewaters. *Separation and Purification Technology.* 2006 Dec 1;52(2):317–324.
32. Pouran SR, Aziz AA, Daud WMAW. Review on the main advances in photo-Fenton oxidation system for recalcitrant wastewaters. *Journal of Industrial and Engineering Chemistry.* 2015;21:53–69. doi:10.1016/j.jiec.2014.05.005.
33. Xu XR, Li XY, Li, XZ, Li HB. Degradation of melatonin by UV, UV/H_2O_2, Fe^{2+}/H_2O_2, and UV/Fe^{2+}/H_2O_2 processes. *Separation and Purification Technology.* 2009;68(2):261–266. doi:10.1016/j.seppur.2009.05.013.
34. Szpyrkowicz L, Juzzolino C, Kaul SN. A comparative study on oxidation of disperse dyes by electrochemical process, ozone, hypochlorite and Fenton reagent. *Water Research.* 2001;35(9):2129–2136. doi:10.1016/S0043-1354(00)00487-5.
35. Ahmadi M, Ramavandi B, Sahebi S. Efficient degradation of a biorecalcitrant pollutant from wastewater using a fluidized catalyst-bed reactor. *Chemical Engineering Communications.* 2015;202(8):1118–1129. doi:10.1080/00986445.2014.907567.

36. Biglarijoo N, Mirbagheri SA, Ehteshami M, Ghaznavi SM. Optimization of Fenton process using response surface methodology and analytic hierarchy process for landfill leachate treatment. *Process Safety and Environmental Protection*. 2016;104:150–160. doi:10.1016/j.psep.2016.08.019.

37. Muangthai I, Ratanatamsakul C, Lu MC. Removal of 2, 4-dichlorophenol by fluidized-bed Fenton process. *Sustainable Environment Research*. 2010;20(5):325.

38. He H, Zhou Z. Electro-Fenton process for water and wastewater treatment. *Critical Reviews in Environmental Science and Technology*. 2017;47(21):2100–2131. doi:10.1016/j.jes.2015.12.003.

39. Ting WP, Lu MC, Huang YH. The reactor design and comparison of Fenton, electro-Fenton and photoelectro-Fenton processes for mineralization of benzene sulfonic acid (BSA). *Journal of Hazardous Materials*. 2008;156(1–3):421–427. doi:10.1016/j.jhazmat.2007.12.031.

40. Zhang H, Zhang D, Zhou J. Removal of COD from landfill leachate by electro-Fenton method. *Journal of Hazardous Materials*. 2006;135(1–3):106–111. doi:10.1016/j.jhazmat.2005.11.025.

41. Pliego G, Zazo JA, Garcia-Muñoz P, Munoz M, Casas JA, Rodriguez JJ. Trends in the intensification of the Fenton process for wastewater treatment: An overview. *Critical Reviews in Environmental Science and Technology*. 2015;45(24), 2611–2692. doi:10.1080/10643389.2015.1025646.

42. Brillas E, Sirés I, Oturan MA. Electro-Fenton process and related electrochemical technologies based on Fenton's reaction chemistry. *Chemical Reviews*. 2009;109(12), 6570–6631. doi:10.1021/cr900136g.

43. Nidheesh PV, Gandhimathi R. Trends in electro-Fenton process for water and wastewater treatment: An overview. *Desalination*. 2012;299:1–15, ISSN 0011-9164. doi:10.1016/j.desal.2012.05.011.

44. Cheng-Chun J, Jia-fa Z. Progress and prospect in electro-Fenton process for wastewater treatment. *Journal of Zhejiang University Science A*. 2007;8(7):1118–1125.

45. Huang CP, Chu CS. Indirect electrochemical oxidation of chlorophenols in dilute aqueous solutions. *Journal of Environmental Engineering*. 2012;138(3), 375–385. doi:10.1061/(ASCE)EE.1943–7870.0000518.

46. Ghoneim MM, El-Desoky HS, Zidan NM. Electro-Fenton oxidation of sunset yellow FCF azo-dye in aqueous solutions. *Desalination*. 2011;274:22–30.

47. Asgari G, Feradmal J, Poormohammadi A, Sadrnourmohamadi M, Akbari S. Taguchi optimization for the removal of high concentrations of phenol from saline wastewater using electro-Fenton process. *Desalination and Water Treatment*. 2016 Dec 1;57(56):27331–27338.

48. Eren Z. Ultrasound as a basic and auxiliary process for dye remediation: A review. *Journal of Environmental Management*. 2012;104:127–141. doi:10.1016/j.jenvman.2012.03.028.

49. Gogate PR. Treatment of wastewater streams containing phenolic compounds using hybrid techniques based on cavitation: A review of the current status and the way forward. *Ultrasonics Sonochemistry*. 2008;15(1):1–15. doi:10.1016/j.ultsonch.2007.04.007.

50. Ji GD, Zhang XR, Guo F. Investigation of carbazole degradation by the sono-Fenton process. *Water Science and Technology*. 2013;68(3):608–613.

51. Gligorovski S, Strekowski R, Barbati S, Vione D. Environmental implications of hydroxyl radicals (•OH). *Chemical Reviews*, 2015;115(24):13051–13092. doi:10.1021/cr500310b.

52. Ahmadi M, Haghighifard NJ, Soltani RD, Tobeishi M, Jorfi S. Treatment of a saline petrochemical wastewater containing recalcitrant organics using Electro-Fenton process: Persulfate and ultrasonic intensification. *Desalination and Water Treatment*. 2019 Nov 1;169:241–250.

53. Zepp RG, Faust BC, Hoigne J. Hydroxyl radical formation in aqueous reactions (pH 3–8) of iron (II) with hydrogen peroxide: The photo-Fenton reaction. *Environmental Science and Technology*. 1992;26(2):313–319. doi:10.1021/es00026a011.

54. Faust BC, Hoigné J. Photolysis of Fe (III)-hydroxy complexes as sources of OH radicals in clouds, fog and rain. *Atmospheric Environment. Part A. General Topics*. 1990;24(1):79–89. doi:10.1016/0960-1686(90)90443-Q.

55. Safarzadeh-Amiri A, Bolton JR, Cater SR. Ferrioxalate-mediated photodegradation of organic pollutants in contaminated water. *Water Research*. 1997;31(4):787–798. doi:10.1016/S0043-1354(96)00373-9.

56. Avetta P, Pensato A, Minella M, Malandrino M, Maurino V, Minero C, Vione D. Activation of persulfate by irradiated magnetite: Implications for the degradation of phenol under heterogeneous photo-Fenton-like conditions. *Environ. Sci. Technol.*, 2015;49(2):1043–1050. doi:10.1021/es503741d.

57. Legrini O, Oliveros E, Braun AM. Photochemical processes for water treatment. *Chemical Reviews*. 1993;93(2):671–698. doi:10.1021/cr00018a003.

58. Moraes JE, Quina FH, Nascimento CA, Silva DN, Chiavone-Filho O. Treatment of saline wastewater contaminated with hydrocarbons by the photo-Fenton process. *Environmental Science & Technology*. 2004 Feb 15;38(4):1183–1187.

59. Sirtori C, Zapata A, Malato S, Agüera A. Formation of chlorinated by-products during photo-Fenton degradation of pyrimethanil under saline conditions. Influence on toxicity and biodegradability. *Journal of Hazardous Materials*. 2012 May 30;217:217–223.

60. Garrido-Ramírez EG, Theng BKG, Mora ML. Clays and oxide minerals as catalysts and nanocatalysts in Fenton-like reactions – A review. *Applied Clay Science*. 2010;47(3–4):182–192. doi:10.1016/j.clay.2009.11.044.

61. He J, Yang X, Men B, Wang D. Interfacial mechanisms of heterogeneous Fenton reactions catalyzed by iron-based materials: A review. *Journal of Environmental Sciences*. 2016;39:97–109. doi:10.1016/j.jes.2015.12.003.

62. Zhang MH, Dong H, Zhao L, Wang DX, Meng D. A review on Fenton process for organic wastewater treatment based on optimization perspective. *Science of the Total Environment*. 2019;670:110–121. doi:10.1016/j.scitotenv.2019.03.180.

63. Litter MI, Slodowicz M. An overview on heterogeneous Fenton and photoFenton reactions using zerovalent iron materials. *Journal of Advanced Oxidation Technologies*. 2017;20(1):20160164. doi:10.1515/jaots-2016-0164.

64. Li C, Li X, Qin L, Wu W, Meng Q, Shen C, Zhang G. Membrane photo-bioreactor coupled with heterogeneous Fenton fluidized bed for high salinity wastewater treatment: Pollutant removal, photosynthetic bacteria harvest, and membrane anti-fouling analysis. *Science of the Total Environment*. 2019 Dec 15;696:133953.

65. Bacardit J, Stötzner J, Chamarro E, Esplugas S. Effect of salinity on the photo-Fenton process. *Industrial & Engineering Chemistry Research*. 2007 Nov 7;46(23):7615–7619.

66. Fernandes L, Lucas MS, Maldonado MI, Oller I, Sampaio A. Treatment of pulp mill wastewater by Cryptococcus podzolicus and solar photo-Fenton: A case study. *Chemical Engineering Journal*, 2014;245:158–165. doi:10.1016/j.cej.2014.02.043.

67. Kuo WS, Wu LN. Fenton degradation of 4-chlorophenol contaminated water promoted by solar irradiation. *Solar Energy*, 2010;84(1):59–65. doi:10.1016/j.solener.2009.10.006.

68. Serra A, Domènech X, Brillas E, Peral J. Life cycle assessment of solar photo-Fenton and solar photoelectro-Fenton processes used for the degradation of aqueous α-methylphenylglycine. *Journal of Environmental Monitoring*. 2011;13(1):167–174. doi:10.1039/C0EM00552E.

69. Wang P, Zeng G, Peng Y, Liu F, Zhang C, Huang B, Lai M. 2,4,6-Trichlorophenolpromoted catalytic wet oxidation of humic substances and stabilized landfill leachate. *Chemical Engineering Journal*. 2014;247:216–222. doi:10.1016/j.cej.2014.03.014.

70. Alalm MG, Tawfik A, Ookawara S. Degradation of four pharmaceuticals by solar photo-Fenton process: Kinetics and costs estimation. *Journal of Environmental Chemical Engineering.* 2015;3(1):46–51, ISSN 2213-3437. doi:10.1016/j.jece.2014.12.009.

71. Luna AJ, Nascimento CA, Foletto EL, Moraes JE, Chiavone-Filho O. Photo-Fenton degradation of phenol, 2, 4-dichlorophenoxyacetic acid and 2, 4-dichlorophenol mixture in saline solution using a falling-film solar reactor. *Environmental Technology.* 2014 Feb 1;35(3):364–371.

72. Wu C, Liu X, Wei D, Fan J, Wang L. Photosonochemical degradation of phenol in water. *Water Research.* 2001;35(16):3927–3933. doi:10.1016/S0043-1354(01)00133-6.

73. Boye B, Dieng MM, Brillas E. Anodic oxidation, electro-Fenton and photoelectro-Fenton treatments of 2, 4, 5-trichlorophenoxyacetic acid. *Journal of Electroanalytical Chemistry.* 2003;557:135–146. doi:10.1016/S0022-0728(03)00366-8.

74. Nordin N, Ho LN, Ong SA, Ibrahim AH, Wong YS, Lee SL, Oon, YS, Oon YL. Hybrid system of photocatalytic fuel cell and Fenton process for electricity generation and degradation of Reactive Black 5. *Separation and Purification Technology*, 2017;177:135–141.

75. Kakavandi B, Ahmadi M. Efficient treatment of saline recalcitrant petrochemical wastewater using heterogeneous UV-assisted sono-Fenton process. *Ultrasonics Sonochemistry.* 2019 Sep 1;56:25–36.

76. Wang XJ, Song Y, Mai JS. Combined Fenton oxidation and aerobic biological processes for treating surfactant wastewater containing abundant sulfate. *Journal of Hazardous Materials.* 2008;160:344–348.

77. Nousheen R, Batool A, Rehman MSU, Ghufran MA, Hayat MT, Mahmood T. Fenton-biological coupled biochemical oxidation of mixed wastewater for color and COD reduction. *Journal of the Taiwan Institute of Chemical Engineers.* 2014;45:1661–1665.

78. Adelaja O, Keshavarz T, Kyazze G. The effect of salinity, redox mediators, and temperature on anaerobic biodegradation of petroleum hydrocarbons in microbial fuel cells. *Journal of Hazardous Materials.* 2015;283:211–217.

79. Hadibarata T, Khudhair AB, Kristanti RA, Kamyab H. Biodegradation of pyrene by Candida sp S1 under high salinity conditions. *Bioprocess and Biosystems Engineering.* 2017;40:1411–1418.

80. Yang L, Sheng M, Zhao H, Qian M, Chen X, Zhuo Y, Cao G. Treatment of triethyl phosphate wastewater by Fenton oxidation and aerobic biodegradation. *Science of the Total Environment.* 2019;678:821–829.

81. Le C, Liang J, Wu J, Li P, Wang X, Zhu N, Wu P, Yang B. Effective degradation of para-chloronitrobenzene through a sequential treatment using zero-valent iron reduction and Fenton oxidation. *Water Science and Technology.* 2011;64:2126–2131.

82. Shen J, Ou C, Zhou Z, Chen J, Fang K, Sun X, Li J, Zhou L, Wang L. Pretreatment of 2, 4-dinitroanisole (DNAN) producing wastewater using a combined zero-valent iron (ZVI) reduction and Fenton oxidation process. *Journal of Hazardous Materials.* 2013;260:993–1000.

83. Somensi CA, Simionatto EL, Bertoli Jr. SLAW, Radetski CM. Use of ozone in a pilot-scale plant for textile wastewater pre-treatment: Physico-chemical efficiency, degradation by-products identification and environmental toxicity of treated wastewater. *Journal of Hazardous Materials.* 2010;175:235–240.

Index

For Product Safety Concerns and Information please contact our EU
representative GPSR@taylorandfrancis.com
Taylor & Francis Verlag GmbH, Kaufingerstraße 24, 80331 München, Germany

www.ingramcontent.com/pod-product-compliance
Lightning Source LLC
Chambersburg PA
CBHW060746220326
41598CB00022B/2342

* 9 7 8 1 0 3 2 0 2 8 3 6 1 *